Modeling and Testing of Reinforced Concrete or Composite Structures Using Advanced New Materials

Modeling and Testing of Reinforced Concrete or Composite Structures Using Advanced New Materials

Editor

Ayman El-Zohairy

MDPI

Basel • Beijing • Wuhan • Barcelona • Belgrade • Novi Sad • Cluj • Manchester

Editor
Ayman El-Zohairy
Department of Engineering
and Technology
Texas A&M University-Commerce
Commerce
United States

Editorial Office
MDPI
St. Alban-Anlage 66
4052 Basel, Switzerland

This is a reprint of articles from the Special Issue published online in the open access journal *Materials* (ISSN 1996-1944) (available at: www.mdpi.com/journal/materials/special_issues/modeling_test_reinf_concr_compos_struct_adv_mater).

For citation purposes, cite each article independently as indicated on the article page online and as indicated below:

Lastname, A.A.; Lastname, B.B. Article Title. *Journal Name* **Year**, *Volume Number*, Page Range.

ISBN 978-3-0365-9940-3 (Hbk)
ISBN 978-3-0365-9939-7 (PDF)
doi.org/10.3390/books978-3-0365-9939-7

Contents

About the Editor ix

Preface xi

Teghreed H. Ibrahim, Abbas A. Allawi and Ayman El-Zohairy
Impact Behavior of Composite Reinforced Concrete Beams with Pultruded I-GFRP Beam
Reprinted from: *Materials* **2022**, *15*, 441, doi:10.3390/ma15020441 1

Dimos Triantis, Ilias Stavrakas, Andronikos Loukidis, Ermioni D. Pasiou and Stavros K. Kourkoulis
Spatio-Temporal Distribution of the Sources of Acoustic Events in Notched Fiber-Reinforced Concrete Beams under Three-Point Bending
Reprinted from: *Materials* **2023**, *16*, 5118, doi:10.3390/ma16145118 25

Carlos D. A. Loureiro, Hugo M. R. D. Silva, Joel R. M. Oliveira, Nuno L. S. Costa and Carlos A. O. Palha
The Effect of Microwave Radiation on the Self-Healing Performance of Asphalt Mixtures with Steel Slag Aggregates and Steel Fibers
Reprinted from: *Materials* **2023**, *16*, 3712, doi:10.3390/ma16103712 44

Sameer Neve, Jiang Du, Rojyar Barhemat, Weina Meng, Yi Bao and Dibyendu Sarkar
Valorization of Vetiver Root Biochar in Eco-Friendly Reinforced Concrete: Mechanical, Economic, and Environmental Performance
Reprinted from: *Materials* **2023**, *16*, 2522, doi:10.3390/ma16062522 65

Mazaher Salamat-Talab, Alireza Akhavan-Safar, Ali Zeinolabedin-Beygi, Ricardo J. C. Carbas and Lucas F. M. da Silva
Effect of Through-the-Thickness Delamination Position on the R-Curve Behavior of Plain-Woven ENF Specimens
Reprinted from: *Materials* **2023**, *16*, 1811, doi:10.3390/ma16051811 85

Adeel Faisal, Safeer Abbas, Syed Minhaj Saleem Kazmi and Muhammad Junaid Munir
Development of Concrete Mixture for Spun-Cast Full-Scale Precast Concrete Pipes Incorporating Bundled Steel and Polypropylene Fibers
Reprinted from: *Materials* **2023**, *16*, 512, doi:10.3390/ma16020512 102

Zuzanna Zofia Woźniak, Adrian Chajec and Łukasz Sadowski
Effect of the Partial Replacement of Cement with Waste Granite Powder on the Properties of Fresh and Hardened Mortars for Masonry Applications
Reprinted from: *Materials* **2022**, *15*, 9066, doi:10.3390/ma15249066 125

Balamurali Kanagaraj, Tattukolla Kiran, Jayakumar Gunasekaran, Anand Nammalvar, Prince Arulraj and Beulah Gnana Ananthi Gurupatham et al.
Performance of Sustainable Insulated Wall Panels with Geopolymer Concrete
Reprinted from: *Materials* **2022**, *15*, 8801, doi:10.3390/ma15248801 138

Juan J. Santana, Natalia Rodríguez-Brito, Concepción Blanco-Peñalver, Vicente F. Mena and Ricardo M. Souto
Durability of Reinforced Concrete with Additions of Natural Pozzolans of Volcanic Origin
Reprinted from: *Materials* **2022**, *15*, 8352, doi:10.3390/ma15238352 155

Ilona Szewczak, Malgorzata Snela and Patryk Rozylo
Influence of Adhesive Layer Thickness on the Effectiveness of Reinforcing Thin-Walled Steel Beams with CFRP Tapes—A Pilot Study
Reprinted from: *Materials* **2022**, *15*, 8365, doi:10.3390/ma15238365 173

Ibadur Rahman, Priyanka Singh, Nirendra Dev, Mohammed Arif, Faiz Noor Khan Yusufi and Ameer Azam et al.
Improvements in the Engineering Properties of Cementitious Composites Using Nano-Sized Cement and Nano-Sized Additives
Reprinted from: *Materials* **2022**, *15*, 8066, doi:10.3390/ma15228066 184

Kaffayatullah Khan, Mudassir Iqbal, Muhammad Raheel, Muhammad Nasir Amin, Anas Abdulalim Alabdullah and Abdullah M. Abu-Arab et al.
Prediction of Axial Capacity of Concrete Filled Steel Tubes Using Gene Expression Programming
Reprinted from: *Materials* **2022**, *15*, 6969, doi:10.3390/ma15196969 206

Gebrail Bekdaş, Celal Cakiroglu, Sanghun Kim and Zong Woo Geem
Optimization and Predictive Modeling of Reinforced Concrete Circular Columns
Reprinted from: *Materials* **2022**, *15*, 6624, doi:10.3390/ma15196624 225

Gebrail Bekdaş, Celal Cakiroglu, Sanghun Kim and Zong Woo Geem
Optimal Dimensioning of Retaining Walls Using Explainable Ensemble Learning Algorithms
Reprinted from: *Materials* **2022**, *15*, 4993, doi:10.3390/ma15144993 243

Wutong Yan, Liangjiang Chen, Bing Han, Huibing Xie and Yue Sun
Numerical Model for Flexural Analysis of Precast Segmental Concrete Beam with Internal Unbonded CFRP Tendons
Reprinted from: *Materials* **2022**, *15*, 4105, doi:10.3390/ma15124105 258

Odey Alshboul, Ghassan Almasabha, Ali Shehadeh, Omar Al Hattamleh and Ali Saeed Almuflih
Optimization of the Structural Performance of Buried Reinforced Concrete Pipelines in Cohesionless Soils
Reprinted from: *Materials* **2022**, *15*, 4051, doi:10.3390/ma15124051 279

Arivoli Masilamani, Malathy Ramalingam, Parthiban Kathirvel, Gunasekaran Murali and Nikolai Ivanovich Vatin
Mechanical, Physico-Chemical and Morphological Characterization of Energy Optimised Furnace (EOF) Steel Slag as Coarse Aggregate in Concrete
Reprinted from: *Materials* **2022**, *15*, 3079, doi:10.3390/ma15093079 294

Huda Al Qader, Ahmed M. Jasim, Hani Salim, Yangchuan Xing and David Stalla
Enhanced Mechanical and Durability Properties of Cement Mortar by Using Alumina Nanocoating on Carbon Nanofibers
Reprinted from: *Materials* **2022**, *15*, 2768, doi:10.3390/ma15082768 314

Celal Cakiroglu, Kamrul Islam, Gebrail Bekdaş, Sanghun Kim and Zong Woo Geem
Interpretable Machine Learning Algorithms to Predict the Axial Capacity of FRP-Reinforced Concrete Columns
Reprinted from: *Materials* **2022**, *15*, 2742, doi:10.3390/ma15082742 333

Pietro Croce, Paolo Formichi and Filippo Landi
Experimental Assessment and Numerical Modeling of the Bond–Slip Correlation for Steel Rebars in r.c. Members
Reprinted from: *Materials* **2022**, *15*, 951, doi:10.3390/ma15030951 353

Sylvia E. Kelechi, Musa Adamu, Abubakar Mohammed, Yasser E. Ibrahim and Ifeyinwa I. Obianyo
Durability Performance of Self-Compacting Concrete Containing Crumb Rubber, Fly Ash and Calcium Carbide Waste
Reprinted from: *Materials* **2022**, *15*, 488, doi:10.3390/ma15020488 . . . . **375**

Bing Shangguan, Qingtian Su, Joan R. Casas, Hang Su, Shengyun Wang and Rongxin Zhao
Modeling and Testing of a Composite Steel–Concrete Joint for Hybrid Girder Bridges
Reprinted from: *Materials* **2023**, *16*, 3265, doi:10.3390/ma16083265 . . . . **392**

About the Editor

Ayman El-Zohairy

Prof. Ayman El-Zohairy is an Assistant Professor at Texas A&M University-Commerce, USA. His research focuses on material testing, steel-concrete composite structures, experimental testing, and the numerical analysis of structural elements and performance and strengthening of steel-concrete composite bridges.

Preface

It is possible to improve the inherent brittleness and crack control ability of normal concrete by incorporating new materials. Fiber, rubber, and cementitious materials are recognized as high-performance construction materials because of their high toughness levels under compressive and tensile loads, as well as their cost effectiveness. Therefore, incorporating these new materials into reinforced concrete or composite structures will affect the structural performance of these structures.

Ayman El-Zohairy
Editor

materials

MDPI

Article

Impact Behavior of Composite Reinforced Concrete Beams with Pultruded I-GFRP Beam

Teghreed H. Ibrahim [1], Abbas A. Allawi [1] and Ayman El-Zohairy [2,*]

[1] Department of Civil Engineering, University of Baghdad, Baghdad 17001, Iraq; tagreed.hassan@coeng.uobaghdad.edu.iq (T.H.I.); A.Allawi@uobaghdad.edu.iq (A.A.A.)
[2] Department of Engineering and Technology, Texas A&M University-Commerce, Commerce, TX 75429, USA
* Correspondence: Ayman.Elzohairy@tamuc.edu; Tel.: +1-903-468-8683

Abstract: The present study experimentally and numerically investigated the impact behavior of composite reinforced concrete (RC) beams with the pultruded I-GFRP and I-steel beams. Eight specimens of two groups were cast in different configurations. The first group consisted of four specimens and was tested under static load to provide reference results for the second group. The four specimens in the second group were tested first under impact loading and then static loading to determine the residual static strengths of the impacted specimens. The test variables considered the type of encased I-section (steel and GFRP), presence of shear connectors, and drop height during impact tests. A mass of 42.5 kg was dropped on the top surface at the mid-span of the tested beams from five different heights: 250, 500, 1000, 1500, and 1900 mm. Moreover, nonlinear Finite Element (FE) models were developed and validated using the experimental data. Static loading was defined as a displacement-controlled loading and the impact loading was modeled as dynamic explicit analysis with different drop velocities. The validated models were used to conduct a parametric study to investigate the effect of the concrete compressive strength on the performance of the composite beams under static and impact loadings. For the composite specimen with steel I-sction, the maximum impact force was 190% greater than the reference specimen NR-I at a drop height of 1900 mm, whereas the maximum impact forces for the specimens composite specimens with GFRP I-sction without and with shear connectors were 19% and 77%, respectively, more significant than the reference beam at the same drop height. The high stiffness for the steel I-beams relative to the GFRP I-beam was the reason for this difference in behavior. The concrete compressive strength was more effective in improving the impact behavior of the composite specimens relative to those without GFRP I-beams.

Keywords: pultruded GFRP I-beam; composite beam; impact; static; experiments; deflections; finite element analysis

check for updates

Citation: Ibrahim, T.H.; Allawi, A.A.; El-Zohairy, A. Impact Behavior of Composite Reinforced Concrete Beams with Pultruded I-GFRP Beam. *Materials* **2022**, *15*, 441. https://doi.org/10.3390/ma15020441

Academic Editor: Giovanni Garcea

Received: 17 December 2021
Accepted: 5 January 2022
Published: 7 January 2022

1. Introduction

Hybrid construction using combined materials is essential for achieving performance targets such as durability, sustainability, seismic resistance, and speedy construction. Because composite materials are simple, buildable, and cost-competitive, they give a consistent performance and are the best choice. Pultruded Glass Fiber Reinforced polymer (GFRP) I-beams are a new type of Fiber Reinforced Polymer (FRP) composite that are developed using pultrusion. These new beams are recommended for bridge construction due to their superior corrosion resistance and higher compressive and flexural strength [1,2]. The behavior of concrete beams under the effect of impact loading is different than under the effect of static loading [3]. The materials are subjected to a higher strain rate when loaded dynamically than statically. There is little work in the literature on the impact and load rate sensitivity of pultruded composite materials. Understanding the behavior of such materials is becoming increasingly important as the use of these unconventional materials for infrastructure applications is developed.

Previous studies explored different types of composite beams under the effect of impact loading. Tabiei et al. [4] investigated the influence of loading velocity on the GFRP beam section of a square box without concrete to determine the critical design parameters. Static and impact tests at different impact velocities were performed to determine the loading rate sensitivity of the pultruded box beams. The ultimate load decreased as the impact velocity increased. The impact behavior of GFRP pultruded rectangular hollow profile filled with concrete was investigated experimentally by Li et al. [5]. The hollow box profile provided tensile strength and protection for the internal concrete block from chemical attacks and the filled concrete provided the system with structural stability. Soleimani and Roudsari [6] studied the impact behavior of reinforced concrete (RC) beams with and without externally bounded sprayed and fabric GFRP under impact and quasi-static loading. A mass of 591 kg was dropped with various heights and loading rates. The stiffness of the tested beams decreased with increasing the drop height. Moreover, applying sprayed GFRP (with and without mechanical stiffeners) and fabric GFRP on the surface of RC beams increased the stiffness. The concrete strength affected the bending characteristics under impact loading [7]. It increased the maximum impact load by 59% and decreased the damping ratio by 47% when the compressive strength increased from 20 MPa to 50 MPa.

Nonlinear Finite Element (FE) models were developed and calibrated to analyze the impact behavior of composite RC beams with pultruded GFRP materials [5–8]. Analytical models were developed using Abaqus to analyze structural members strengthened with GFRP sheets and subjected to different dynamic loading conditions (quasi-static, impact) [8]. The developed models could be an effective tool to predict the performance of retrofitted beams under dynamic loading conditions. Furthermore, it showed that FRP retrofitting RC beams subjected to repetitive impact loads could effectively improve their dynamic performance and slow damage progress. Abaqus used the load-time curves of the experimental study to check parameters such as deflection, strain, and stresses as a function of time for analysis. A hollow solution was used to solve the equation set to identify the unknown variable (this method was included in the Abaqus program) [7].

Composite RC beams with pultruded GFRP beams are widely used for bridge construction due to their superior corrosion resistance and higher compressive and flexural strength [1,2]. However, the research on the impact performance of these types of beams is still very limited. Moreover, comparisons between the impact flexural properties of encased steel and GFRP I-sections as well as the effect of using shear connectors on the composite interaction need in-depth investigation. Therefore, the present study experimentally and numerically investigated the impact behavior of these beams. Eight specimens of two groups were cast in different configurations. The first group consisted of four specimens and was tested under static load to provide reference results for the second group. The four specimens in the second group were tested first under impact loading and then static loading to determine the residual static strengths of the impacted specimens. The test variables considered the type of encased I-section (steel and GFRP), presence of shear connectors, and drop height during impact tests. A mass of 42.5 kg was dropped on the top surface at the mid-span of the tested beams from five different heights: 250, 500, 1000, 1500, and 1900 mm. Moreover, nonlinear Finite Element (FE) models were developed and validated using the experimental data. Static loading was defined as a displacement-controlled loading and the impact loading was modeled as dynamic explicit analysis with different drop velocities. The validated models were used to conduct a parametric study to investigate the effect of the concrete compressive strength on the performance of the composite beams under static and impact loadings.

2. Experimental Program

Eight specimens of two groups were cast in different configurations, as listed in Table 1. The first group consisted of four specimens, which were tested under static load to provide reference results for the second group. The four specimens in the second group were tested first under impact loads and then static loads to determine the residual static strengths of the impacted specimens.

Table 1. Details of the tested specimens.

Group	Specimens	Cross-Section (mm)	Specimen Weight (kg/m)	Type of Test	Type of Encased Beam	Connection *
I	NR	200 × 300	503.89	Static	N/A	NSC
	CG	200 × 300	509.83	Static	GFRP I-beam	NSC
	CGC	200 × 300	510.51	Static	GFRP I-beam	SC
	CS	200 × 300	529.8	Static	Steel I-beam	NSC
II	NR-I	200 × 300	503.89	Impact and Static	N/A	NSC
	CG-I	200 × 300	509.83	Impact and Static	GFRP I-beam	NSC
	CGC-I	200 × 300	510.51	Impact and Static	GFRP I-beam	SC
	CS-I	200 × 300	529.8	Impact and Static	Steel I-beam	NSC

* NSC: No shear connectors, SC: with shear connectors.

2.1. Details of Specimens

The overall length of the specimens was 3000 mm with a clear span of 2750 mm, as shown in Figure 1a. All specimens had the same cross-sectional dimensions with a width of 200 mm and a total thickness of 300 mm, as illustrated in Figure 1b. Conventional reinforcements were used for the reference specimens NR and NR-I. The flexural reinforcements were two rebars with a diameter of 16 mm in tension and two rebars with a diameter of 10 mm in compression. The transverse was stirrups with a diameter of 10 mm diameter at a spacing of 125 mm to prevent premature shear failure. The flexural and shear reinforcement were the same for the other specimens. Four specimens CG, CG-I, CGC, and CGC-I were reinforced with pultruded GFRP I-beams positioned at the centroid of the cross-section as shown in Figure 1b. Dimensions of the pultruded GFRP I-sections are illustrated in Figure 1c. Shear connectors were provided at the top flange of the GFRP I-section in specimens CGC and CGC-I to increase the composite interaction between the GFRP I-beam and concrete. The diameter of these connectors was 12 mm with a height of 70 mm and a spacing of 375 mm. The shear connectors were stiffened with washers and nuts after being inserted through drilled holes on both sides of the top flange of the GFRP I-beam. The remaining two specimens CS and CS-I were reinforced with steel I-beams positioned at the center of the cross-section as shown in Figure 1b.

2.2. Material Properties

Concrete specimens were prepared and tested to determine the compressive strength, splitting tensile strength, modulus of rupture, and modulus of elasticity of concrete. Three standard cubes of dimensions 150 mm × 150 mm × 150 mm and three standard cylinders of dimensions 150 mm × 300 mm were tested by a universal compression machine to determine the compressive strength of concrete. The splitting tensile strength of concrete was determined according to ASTM C496-96 [9]. The modulus of elasticity of concrete was calculated according to ASTM C469 [10] by using standard concrete cylinders 150 mm × 300 mm. The mechanical properties of concrete are listed in Table 2.

The mechanical properties of steel bars and steel plates, such as yield tensile strength and ultimate tensile strength, were evaluated according to ASTM A370-19 [11]. Three steel bars with a diameter of ∅ 10 mm and ∅ 16 mm with 0.5 m length and two steel plates with flat coupons were tested. The results are listed in Table 3.

The mechanical properties of the GFRP I-section, such as compressive and tensile strength and elastic modulus, were examined. The compression tests were conducted following ASTM D695 [12]. Fifteen specimens with a coupon dimension of 10 mm × 12.7 mm × 38.1 mm were tested. Ten coupons were cut from the longitudinal direction of the flange and the web of the I-section and five coupons were cut from the web in the transverse direction. Tension tests were conducted following ISO 527 [13] on ten specimens with coupon dimensions of 10 mm × 25 mm × 250 mm. All coupons were cut from the longitudinal direction, five from the flange and five from the web. The average mechanical properties of the tested coupons are listed in Table 3. Moreover, the longitudinal and transverse modulus of elasticity of the GFRP beam were 27.1 GPa and 6.8 GPa, respectively, as provided by the manufacturer (DURA composites, Clacton On Sea, UK).

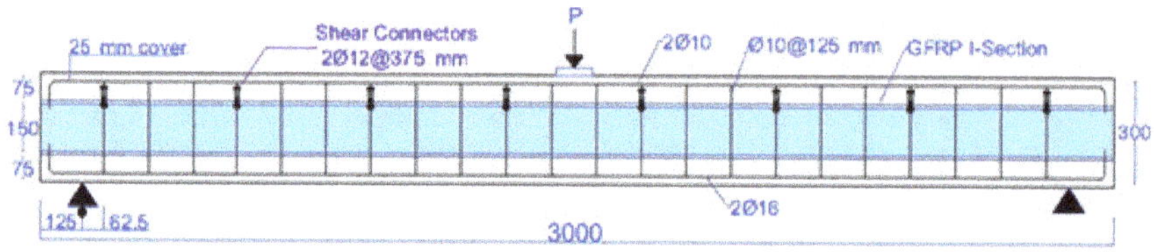

(a)

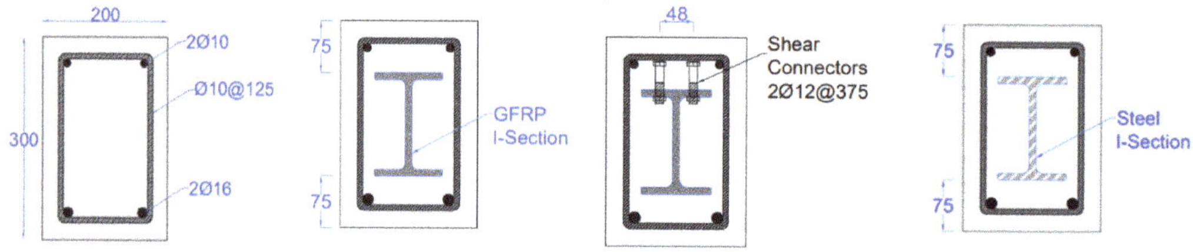

Specimens NR and NR-I | Specimens CG and CG-I | Specimens CGC and CGC-I | Specimens CS and CCS-I

(b)

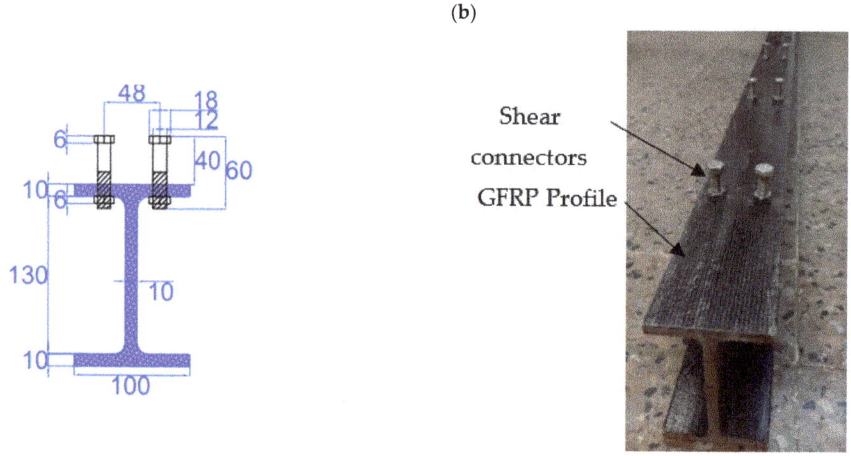

(c)

Figure 1. Details of the tested specimens: (**a**) Elevation of the specimens; (**b**) Cross -sections of the tested specimens; (**c**) Geometrical dimensions of the GFRP profile and installment of the shear connectors (Unit: mm).

Table 2. Mechanical properties of concrete.

Slump Test (mm)	Compressive Strength * (MPa)	Compressive Strength ** (MPa)	Splitting Tensile Strength (MPa)	Modulus of Rupture (MPa)	Modulus of Elasticity (MPa)
100	23.43	29.29	2.46	2.41	20,754.64

* Concrete cylinders, ** Concrete cubes.

Table 3. Mechanical properties of steel reinforcement and GFRP.

Material	Diameter (mm)	Thickness (mm)	Yield Strength (MPa)	Transverse Compressive Strength (MPa)	Longitudinal Compressive Strength (MPa)	Longitudinal Tensile Strength (MPa)	Longitudinal Modulus of Elasticity (GPa)	Transverse Modulus of Elasticity (GPa)
Steel	16	-	520.73	-	-	687.07	200	-
	10	-	407.7	-	-	465.63	200	-
	-	10	375.9	-	-	479.63	200	-
GFRP	-	10	-	118.3	326.14	347.5	27.1	6.8

2.3. *Instrumentation and Test Setup*

2.3.1. Static Test

Beam specimens of group one were tested as simply-supported beams under a monotonic concentrated load at the mid-span up to failure using a 1000 kN capacity hydraulic universal testing machine, as shown in Figure 2. The applied load was monitored using a load cell with a 1000 kN capacity. Moreover, a linear variable differential transformer (LVDT) was used to measure the vertical deflection at the mid-span of the tested specimen.

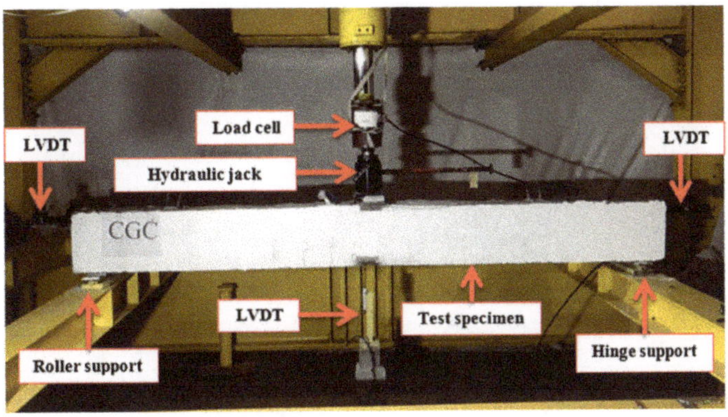

Figure 2. Test setup for static tests.

2.3.2. Impact Test

Figure 3 illustrates the loading frame of impact tests performed with a drop mass impactor. The mass of 42.5 kg was dropped on the top surface at the mid-span of the tested beams from five different heights: 250, 500, 1000, 1500, and 1900 mm. A hoist and chain system was used to raise the hammer over the specimen during testing. The dropped mass was manually released according to the specified height and steel guide rails were used to vertically drop the mass.

The hammer has the potential energy $m_h \times a_h \times H$ (mass of the hammer × acceleration of the hammer under gravity × height with respect to the specimen's top surface). When the hummer was raised up and released, the potential energy was converted to kinetic energy as the hammer fell with an acceleration a_h. Due to the frictional forces of the machine, the downward acceleration of the hammer was less than the acceleration of gravity (9.81 m/s^2). Just before the hammer struck the beam, the velocity and kinetic energy were calculated by Equations (1) and (2), respectively.

$$v = \sqrt{2a_h H} \quad (1)$$

$$Ek = \frac{1}{2}m_h \cdot v^2 = m_h \cdot a_h \cdot H \quad (2)$$

When the hammer hit the beam, the hammer's momentum was rapidly transferred to the beam. As a response, the hammer's momentum was reduced. This resulted in a reduction in the hammer's kinetic energy and an increase in the beam's energy. This sudden energy transfer between the hammer and the beam led to a rapid build-up of stress in the beam. The momentum (M) is obtained from Equation (3). The impact characteristics of the impact tests are listed in Table 4.

$$M = m \cdot v \tag{3}$$

Table 4. Characteristics of the impact test.

Specimens	Striker Mass (kg)	Drop Height (m)	Tested Acceleration of Striker m/s^2	Impact Velocity (m/s)	Kinetic Energy (Joule)	Momentum (kg·m/s)
NR-I	42	0.25	7.63	1.95	80.09	82.02
	42	0.5	6.31	2.51	132.53	105.51
	42	1.0	6.72	3.67	282.15	153.95
	42	1.5	6.55	4.43	412.66	186.18
	42	1.9	4.89	4.31	389.93	180.98
CG-I	42	0.25	6.56	1.81	68.91	76.08
	42	0.5	6.96	2.64	146.25	110.84
	42	1.0	5.45	3.30	228.84	138.64
	42	1.5	5.32	3.99	335.11	167.78
	42	1.9	5.86	4.72	467.28	198.12
CGC-I	42	0.25	8.38	2.05	87.96	85.96
	42	0.5	7.01	2.65	147.11	111.16
	42	1.0	7.51	3.87	315.32	162.75
	42	1.5	5.86	4.19	369.28	176.12
	42	1.9	8.26	5.60	659.38	235.35
CS-I	42	0.25	6.14	1.75	64.50	73.61
	42	0.5	7.57	2.75	158.94	115.55
	42	1.0	8.12	4.03	340.99	169.24
	42	1.5	6.68	4.48	420.94	188.04
	42	1.9	6.22	4.86	496.47	204.21

The specimens were tested at a clear span of 2750 mm. The setup of the two supports allowed for rotation while preventing twisting of the specimens. The vertical movement of supports was restrained using two steel yokes (see Figure 3b,c). The impact loads were recorded while striking the specimen using two load cells. A dynamic load cell with a capacity of (300 kN) was rigidly connected to the drop mass system between the steel shaft and the impactor (Figure 3a). The second load cell with a capacity of 1000 kN was placed under one support to measure the reaction under support (see Figure 3c). The mid-span deflection was measured for each stepped time using a laser velocity sensor from Keyence Company LK-081 (Itasca, IL, USA) with the controller LK-2101 (Itasca, Illinois, U.S.A). The measuring range of the laser velocity sensor is $\pm$ 15mm, as provided by the manufacturer. In order to measure the impactor velocity, an accelerometer was attached to the upper surface of the impactor. All output results were collected by a data recorder of type DATAQ DI-710 (Dataq Instruments Inc., Akron, OH, USA). After conducting the impact tests, the residual load-carrying capacity of the beam specimens was determined using static loading tests until failure.

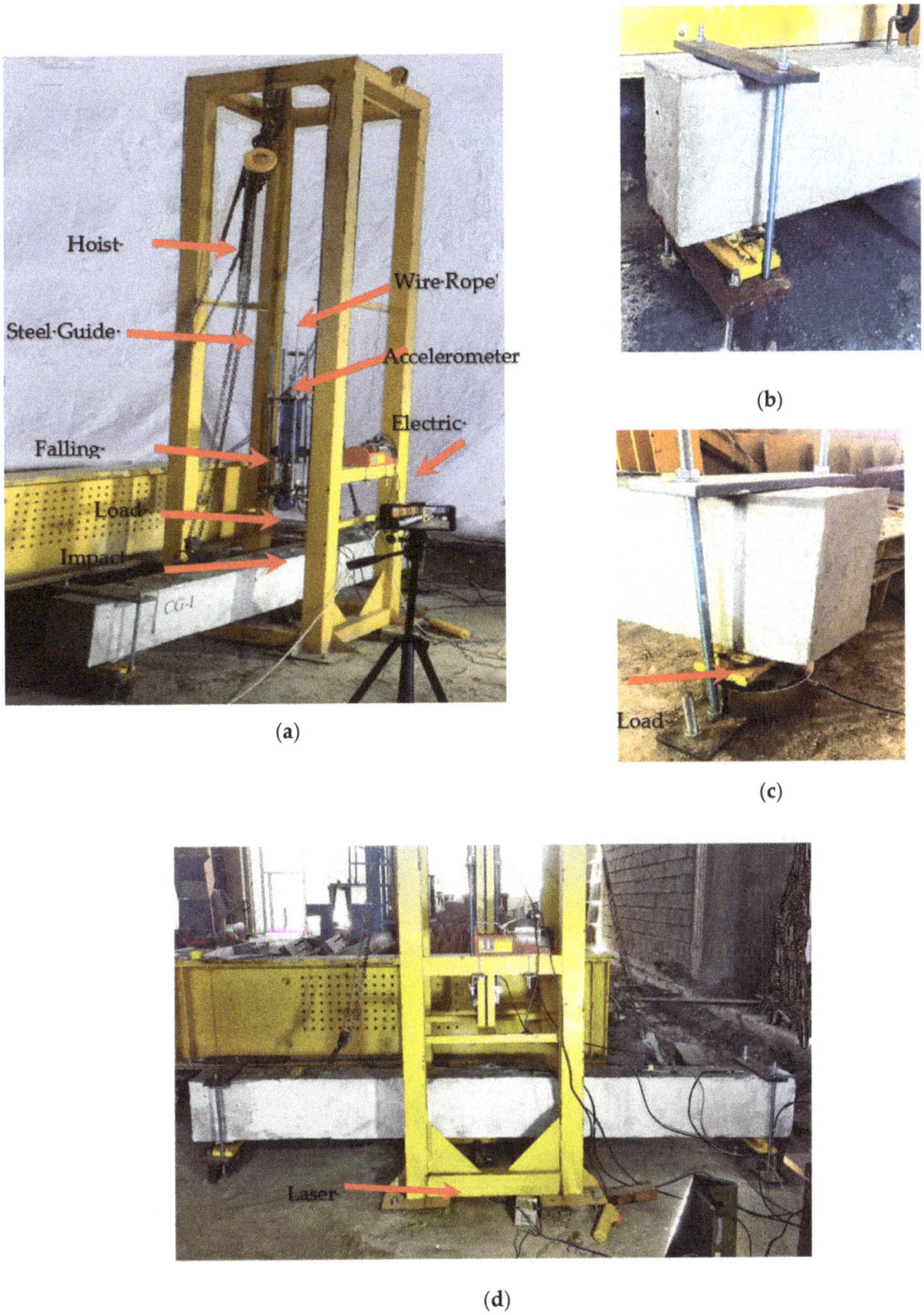

Figure 3. Setup of the impact test: (**a**) Steel frame of the impact test; (**b**) Left support with steel yokes; (**c**) Right support with load cell; (**d**) Front view of the test.

3. Experimental Results

3.1. Static Test Results

3.1.1. Crack Patterns and Modes of Failure

The first crack was flexural cracking and formed at the mid-spans of the tested specimens. The cracking loads of the composite specimens CG, CGC, and CS were 4%, 5%, and 53% higher than that one for the reference specimen NR, respectively. Then, more cracks were created and propagated from the loading point along the specimen's longitudinal axis. Shear cracks appeared at the final stage of loading. The failure mechanism of the reference specimen NR was yielding in the tension steel rebars followed by crushing in concrete in the compression zone as shown in Figure 4a. For composite specimens CG and CGC and after reaching the ultimate loads, concrete in the compression zones was crushed, buckling of the compression steel rebars and cover spalling occurred, initial inter-laminar failure and rupture of the GFRP profile was accompanied by a loud noise as described in Figure 4b,c. For composite specimen, CS with steel I-beam, crushing of concrete in the compression zone started at the load of 140 kN. After that, a plastic region was created in the steel I-beam, where the deflection increased rapidly while the applied load ranged from 160 to 170 kN (see Figure 4d).

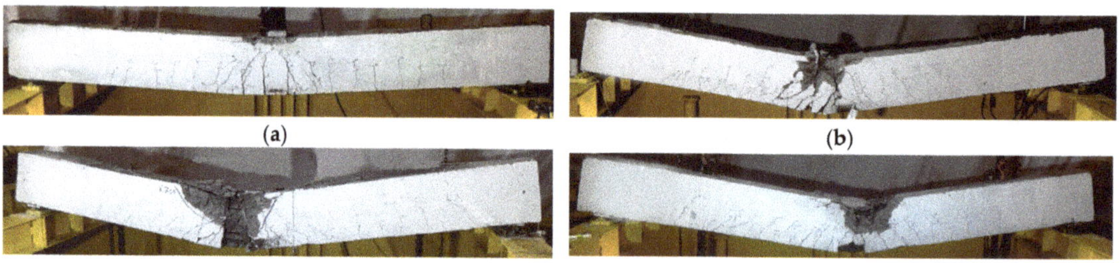
(a) (b) (c) (d)

Figure 4. Modes of failure of the tested specimens under static test: (**a**) Specimen NR; (**b**) Specimen CG; (**c**) Specimen CGC; (**d**) Specimen CS.

3.1.2. Load-Deflection Behavior

Figure 5 illustrates the load-deformation relationships of the tested specimens. The specimens showed linear relationships between the applied loads and deflections until the yielding of the steel reinforcement. For composite specimens with GFRP I-beam and steel I-beam, the yielding loads increased by 38%, 51%, and 104% for specimens CG, CGC, and CS relative to the reference specimen NR, respectively. Moreover, the stiffness was enhanced for these specimens. Table 5 lists a summary of the experimental results. The ultimate capacities of the composite specimens CG and CGC were 134.62 kN and 147.23 kN, respectively. The embedded GFRP I-beam in these specimens increased the load-bearing capacity by 51% and 65%, respectively. Increasing the composite interaction between the GFRP I-beam and concrete by means of shear connectors improved the ultimate capacity by 9.5% relative to the composite specimen without shear connectors. For specimen CS with steel I-beam, the steel reinforcement in tension yielded at a load of 162.5 kN, which is 87.5% higher than that one for the reference specimen NR. Subsequently, the load increased to the ultimate value of 178.3 kN, which is 100% higher than that for the reference specimen NR. Then the applied load decreased by 10% due to the yielding of the steel I-beam at the mid-span.

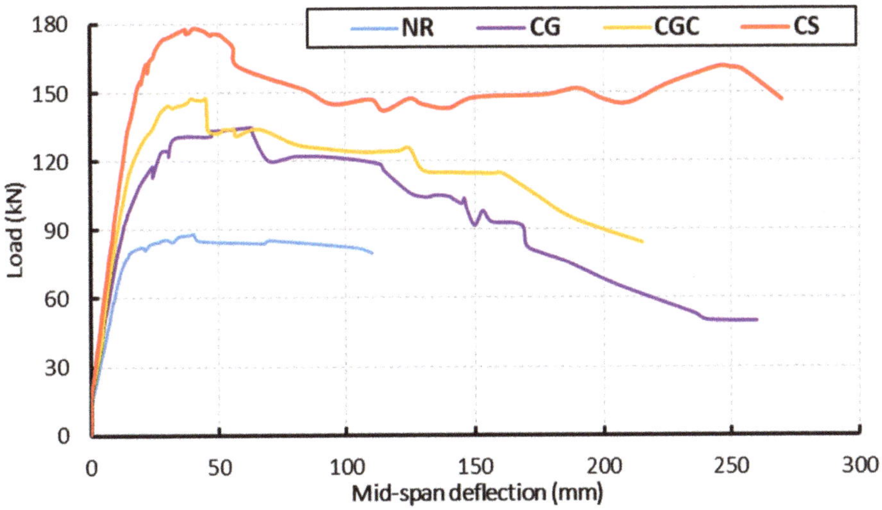

Figure 5. Load-deflection relationships for the tested specimens under static loads.

Table 5. Summary of the experimental results under static tests.

Specimen	Crack Load P_{cr} (kN)	Yield Load P_y (kN)	Ultimate Load P_u (kN)	Increase in Ultimate Load (%)	Mode of Failure
NR	19.82	79.5	89.17	-	Yield to steel bars and concrete crushing.
CG	20.62	110.4	134.62	51	Delamination and transverse shear failure in the web of the GFRP I-section.
CGC	20.89	120.2	147.23	65	Delamination and transverse shear failure in the web of the GFRP I-section.
CS	30.03	162.5	178.3	100	Yielding of steel I-section, crushing of compressive concrete.

3.2. *Impact Test Results*

3.2.1. Crack Patterns and Modes of Failure

Cracks were initially formed at the point of impact on the top surfaces of the tested specimens followed by visible flexural cracks at the beam soffit as shown in Figure 6. For the NR-I specimen, cracks initially appeared when the mass hit the beam from a height of 1.0 m followed by flexural cracks, which spread along the beam. However, the shear cracks appeared when the mass hit the beam from a height of 1.5 m. Moreover, spalling of the concrete occurred at the impact zone and cracks on the proximal face of concrete caused penetration of 8 mm with an 82.5 mm boring diameter (see Figure 6a). For CG-I and CGC-I specimens, cracks occurred at the middle area of the specimen when the mass hit the beam from a height of 1.5 m. However, the flexural cracks propagated down the beam and appeared when the height was 1.9 m. These specimens experienced penetrations of 3 mm and 4.5 mm and boring diameters of 57.5 mm and 67.5 mm, respectively (see Figure 6b,c). For specimen CS-I, cracks began to appear after striking from a height of 1.5 m. Moreover, cracks appeared at the locality of the strike and spread to the top of the specimen with a total penetration of 17 mm and a 93.2 mm boring diameter (see Figure 6d). The crack patterns on the side faces of the specimens are illustrated in Figure 7.

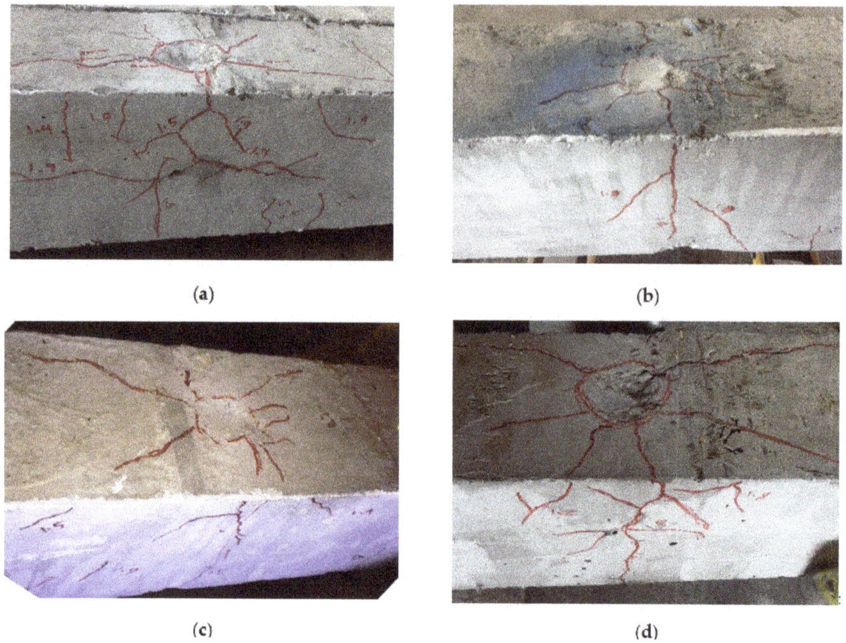

Figure 6. Local damage at the impact zone of the impact tested specimens: (**a**) NR-I; (**b**) CG-I; (**c**) CGC-I; (**d**) CS-I.

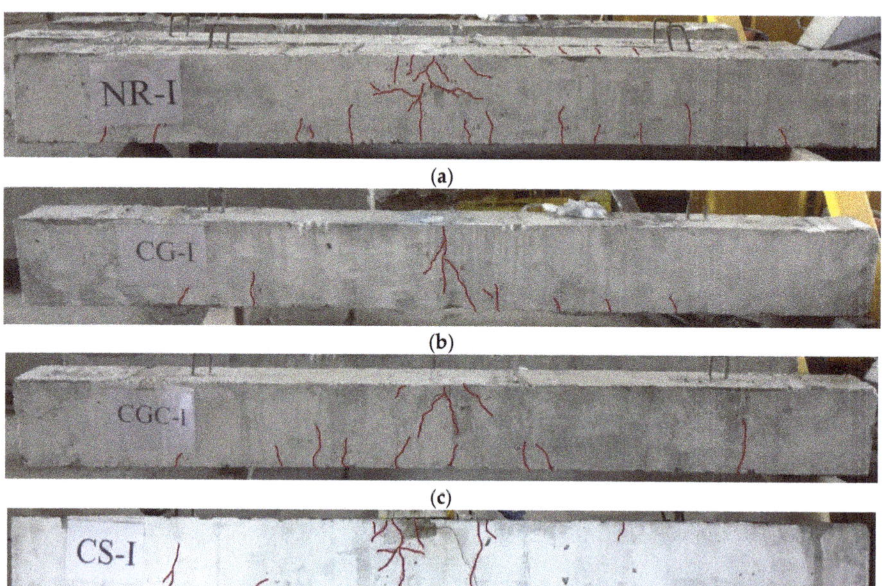

Figure 7. Crack patterns on the side face of the impacted specimens: (**a**) NR-I; (**b**) CG-I; (**c**) CGC-I; (**d**) CS-I.

3.2.2. Impact Response

The impact load and deflection time histories for each specimen were recorded for each height. Figure 8 illustrates the specimen NR-I's impact force and mid-span deflection time histories as an example of the impact response. The first pulse's peak grew in proportion to the drop height, whereas the initial pulse-like waveform had a duration of about 25 ms regardless of the drop height. The length of the blunt waveform grew as the drop height increased, but the peaks of the blunt waveforms were approximated independently of the drop height. The maximum mid-span deflection increased as the drop height increased. Comparisons of the maximum impact force and maximum mid-span deflection of specimens NR-I, CG-I, CGC-I, and CS-I for different heights are listed in Table 6 and shown in Figures 9 and 10, respectively.

Figure 9 shows that the maximum impact force increased as the drop height increased. For the composite specimen CS-I, the maximum impact force was 130% greater than the reference specimen NR-I at a drop height of 1900 mm whereas the maximum impact forces for specimens CG-I and CGC-I were 19% and 77%, respectively, more significant than the reference beam at the same drop height. The high stiffness for the steel I-beams relative to the GFRP I-beams was the reason for this difference in behavior, where the impact force is proportional to the structural stiffness. As seen in Figure 10, the composite specimens CG-I and CGC-I showed reductions in the corresponding mid-span deflections relative to the reference specimen NR-I. For the 1900 mm drop height, reductions of 10.7% and 14.1% were obtained, respectively, whereas the composite specimen CS-I showed a 45% increase in the corresponding mid-span deflection due to the high impact force caused by the section's high stiffness.

Table 6. Impact test results.

Specimen	Height of Drop (mm)	Impact Force (kN)	% Change	Corresponding Mid-Span Deflection (mm)	% Change
NR-I	250	28.52	-	1.20	-
	500	30.21	-	1.80	-
	1000	43.02	-	2.00	-
	1500	66.53	-	2.60	-
	1900	95.82	-	2.70	-
CG-I	250	46.79	+64%	1.09	−9.1%
	500	62.83	+30%	1.59	−11.6%
	1000	86.58	+100%	1.95	−2.5%
	1500	95.32	+43%	2.39	−8.1%
	1900	113.98	+19%	2.41	−10.7%
CGC-I	250	33.25	+17%	1.01	−15.8
	500	42.70	+41%	1.20	−33.3%
	1000	68.57	+59%	1.95	−2.5%
	1500	107.56	+62%	2.13	−18.1%
	1900	169.79	+77%	2.32	−14.1
CS-I	250	63.72	+123%	1.23	2.5%
	500	111.96	+270%	2.1	16.1%
	1000	142.81	+232%	2.34	17.2%
	1500	195.61	+194%	3.04	16.9%
	1900	221.24	+130%	3.92	45.1%

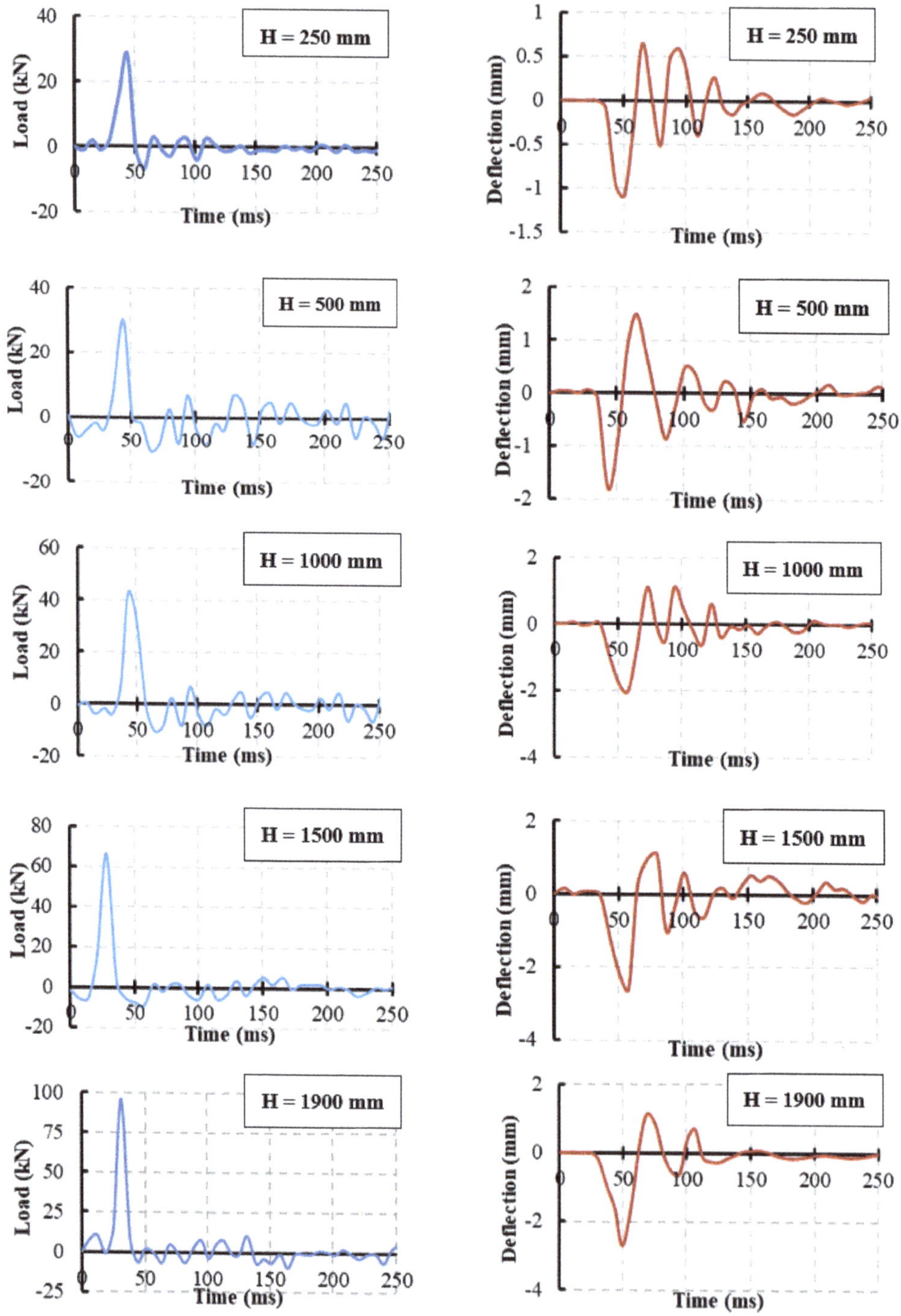

Figure 8. The specimen NR-I's impact force and mid-span deflection time histories.

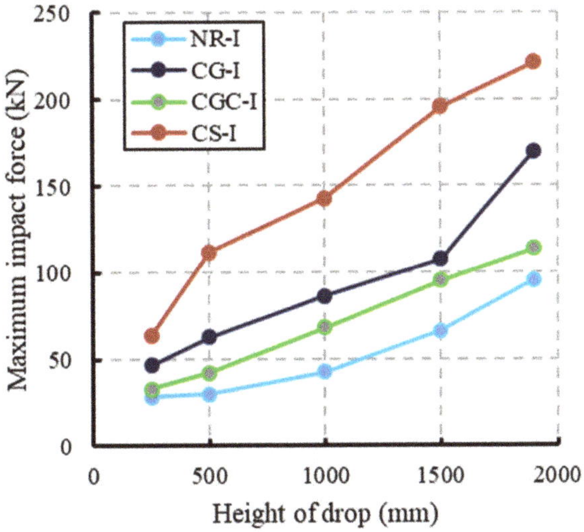

Figure 9. Comparisons of the maximum impact force for different heights of drop.

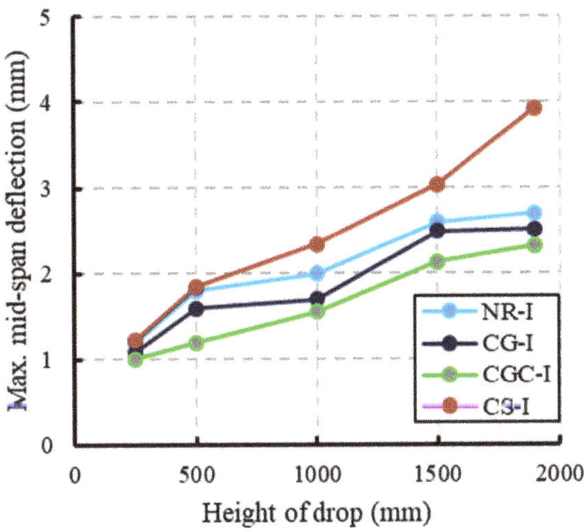

Figure 10. Comparisons of the maximum mid-span deflection for different heights of drop.

3.2.3. Residual Static Test after the Impact

The residual static strengths were explored by conducting static tests until failure after the completion of the impact tests. The ultimate capacities are compared in Table 7 between the specimens without and with previous impact loads. More reduction in the reference specimen's capacity occurred and the specimen with the GFRP I-beam without shear connectors showed the least reduction. Providing shear connectors to specimen CGC-I led to higher stiffness and significant damage relative to the specimen without shear connectors CG, which caused more reduction in the residual beam strength (see Table 7). Figure 11 shows the crack patterns of the impacted specimens after the residual static tests. Additional cracks were formed and propagated vertically and diagonally, which increased in width and depth until failure. All specimens failed in flexural mode. The specimen CS-I exhibited lateral buckling due to the high level of the applied load, which decreased the steel I-beam strength as illustrated in Figure 12.

Table 7. Ultimate capacities of the specimens without and with previous impact loads.

Specimens	Without Impact Loads (kN)	With Impact Loads (kN)	Residual Static Strength (%) *
NR/NR-I	89.17	75.2	84
CG/CG-I	134.62	133.1	99
CGC/CGC-I	147.23	137.5	93
CS/CS-I	178.3	163.3	91

* Percentage of the static strength of specimens without impact loading.

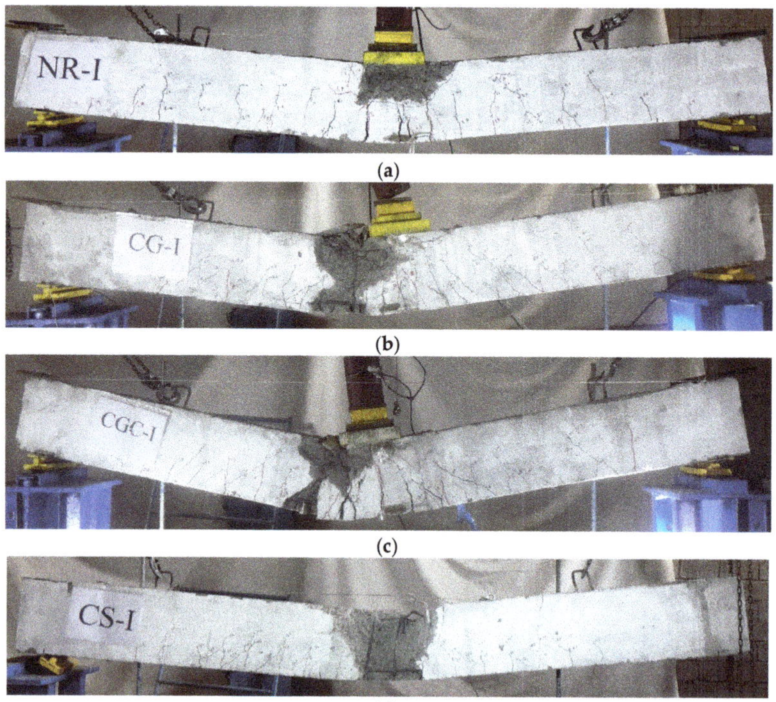

(a)

(b)

(c)

(d)

Figure 11. Crack patterns for impact specimens after the residual static tests: (**a**) Specimen NR-I; (**b**) Specimen CG-I; (**c**) Specimen CGC-I; (**d**) Specimen CS-I.

Figure 12. Lateral buckling failure of specimen CS-I (top view).

4. Numerical Modeling

4.1. Finite Element Model

Finite element analysis was performed using the Abaqus software [14]. The developed model considered two groups of loading conditions. The first group was the static loading defined in terms of general static procedures. The loading was applied as displacement-controlled loadings at the center of the beam whereas the second group was the repeated impact loading determined as dynamic explicit analysis. The loading was applied using the impactor at the mid-spans of the beams as initial velocities. The proposed models were validated using the experimental test results.

The concrete, shear connectors, and steel I-profile were modeled using three-dimensional eight-node elements with reduced integration (C3D8R). The longitudinal and transverse steel reinforcement were modeled using three-dimensional truss elements (T3D2) with a linear kinematic hardening model. The pultruded GFRP I-profile was simulated using an 8-node doubly curved thick shell element with reduced integration (S8R). The impactor in the impact loading case was modeled by a discrete rigid shell that rotated 360° to create the cylindrical shape.

4.2. Material Models

The concrete damage plasticity (CDP) model was employed to model the behavior of the tested specimens to consider the concrete cracking and crushing. The response of the CDP model under uniaxial tensile loading was characterized by a linear–elastic stress-strain relationship up to the value of the failure stress. The stress-strain curves were used to describe the compressive and tensile behaviors, as shown in Figure 13 [15]. Table 8 lists the different damage parameters for cracking and crushing. These parameters included the dilation angle (φ), eccentricity (ε), compressive strength to uniaxial pressure ratio biaxial (f_{bo}/f_{co}), coefficient (K), and viscosity parameters (μ). These parameters were established from previous analyses and were implemented in this study.

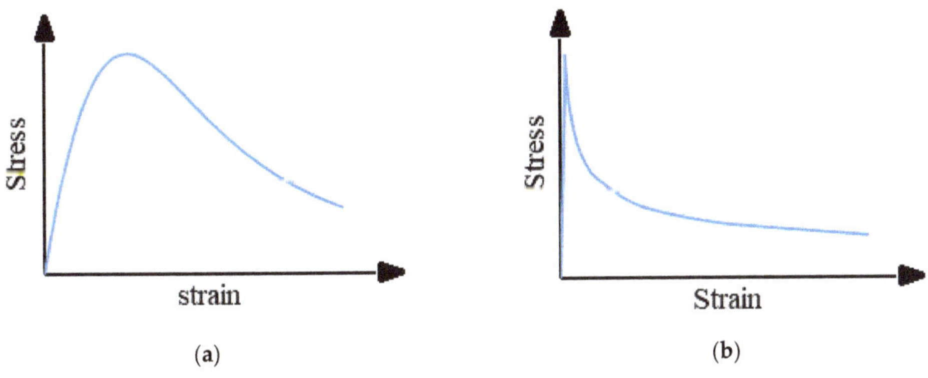

Figure 13. Stress-strain relationships for concrete [16]. (**a**) Compressive stress-strain curve; (**b**) Tensile stress-strain curve.

Table 8. Input parameters for the CDP model.

Parameter	φ	Eccentricity ε	f_{bo}/f_{co}	K	μ
Value	31°	0.1	1.16	0.667	0.001

The steel I-beam, longitudinal steel rebars, and stirrups were modeled as a non-linear relationship adopted by [15], as shown in Figure 14. The linear isotropic part is defined by the modulus of elasticity of the reinforcement and Poisson's ratio. The plastic part is defined by the yield stress f_y, ultimate stress f_u, and plastic strains, as illustrated in Figure 14. Using a combination of damage initiation criteria, the progressive damage model was employed to characterize the behavior of the pultruded profiles. Hashin's criteria [17] were used to determine when the pultruded profile begins to degrade. After satisfying the damage initiation criteria, the material stiffness of the pultruded profile degrades according to the damage evolution law. Table 9 lists the values of engineering constants for elastic properties of GFRP material, the strength properties to represent the damage intuition criteria, and the parameters for the progressive damage model presented by damage evolution (fracture energy) and damage stabilization indicated by viscosity coefficients. The values in this table were determined by experimental tests and from the available works of literature [16,18] based on the type of fibers and matrix as well as trial and error.

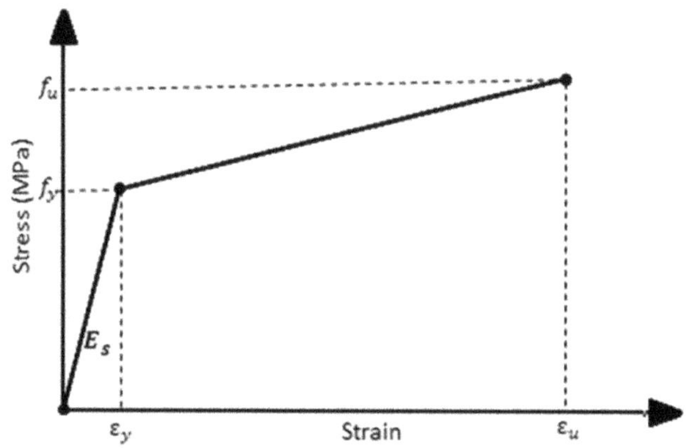

Figure 14. Stress-strain relationship of steel I-beam, rebars, and stirrups.

Table 9. GFRP materials properties and progressive damage parameters input in ABAQUS.

	Definition	Value
Engineering Elastic Constants	Longitudinal Modulus of Elasticity (E_z)E	27.1 GPa **
	Transverse Modulus of Elasticity ($E_x = E_y$)	6.8 GPa
	Transverse Shear Modulus of Elasticity (G_{xy})	17.5 GPa
	In-Plane Shear Modulus of Elasticity ($G_{zx} = G_{zy}$)	2.7 GPa
	Major Poisson' Ratio ($\upsilon_{zx} = \upsilon_{zy}$)	0.23
	Minor Poisson' Ratio (υ_{xy})	0.1
Strength Values	Longitudinal Tensile Strength	347.5 MPa
	Longitudinal Compressive Strength	326.14 MPa
	Transverse Tensile Strength	50 MPa
	Transverse Compressive Strength	118.3 MPa
	Transverse Shear Strength	8.04 MPa
	In-Plane Shear Strength	104.23 MPa
Damage Evolution	Longitudinal Tensile Fracture Energy	18.3
	Longitudinal Compressive Fracture Energy	5.8
	Transverse Tensile Fracture Energy	5
	Transverse Compressive Fracture Energy	5.5
Damage Stabilization	Viscosity Coefficient in the Longitudinal Tensile Direction	0.001
	Viscosity Coefficient in the Longitudinal Compressive Direction	0.001
	Viscosity Coefficient in the Transverse Tensile Direction	0.001
	Viscosity Coefficient in the Transverse Compressive Direction	0.001

4.3. Static Analysis

The numerical model employed the general static processes by using the displacement-controlled loading. The suggested model was validated using experimental data from the tested specimens NR, CG, and CGC. The assembled FE model for the composite specimen CGC is shown in Figure 15. The embedded technique with a complete bond was employed to simulate the interaction between the steel rebars and concrete as well as the GFRP profile and concrete.

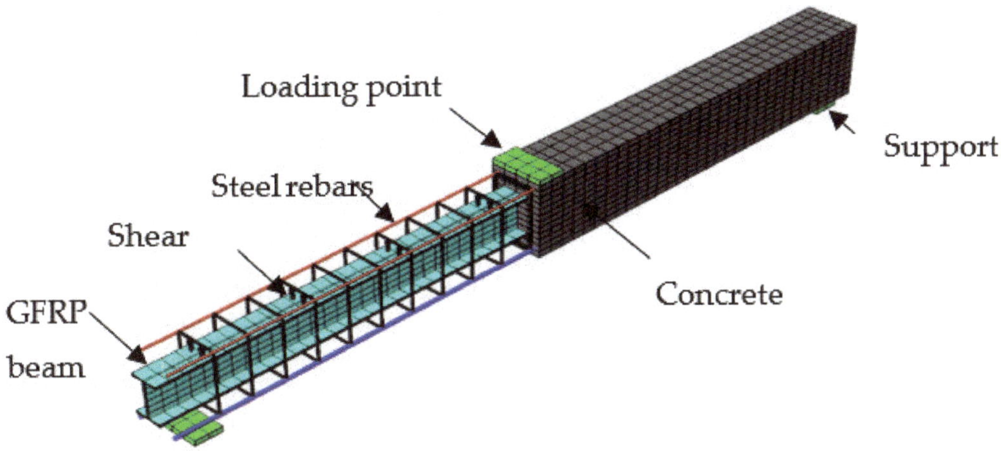

Figure 15. FE mesh of the analyzed specimen CGC for static analysis.

Model Verifications

Figure 16 presents comparisons between the experimental and FE results in terms of the load-deflection relationships. The analyzed beams showed initially linear elastic behavior with a higher stiffness than the experimental results. These beams showed ±6% or less high capacities relative to the experimental results. The perfect bond assumed between the concrete and steel rebars as well as the GFRP profile was the main reason for this difference in response. Moreover, the boundary conditions, as well as the material constitutive models, were considered ideal in the numerical analyses relative to those in the experiments. In general, there was good agreement between the experimental and FE results under static loading.

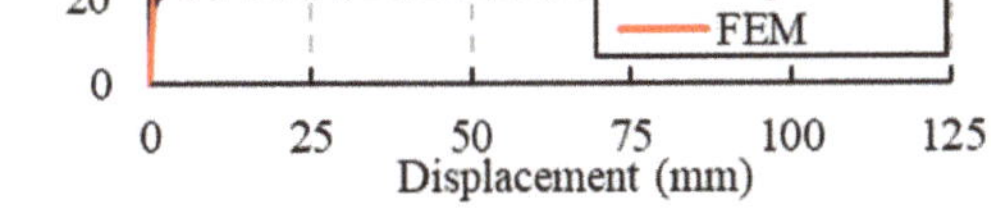

(a)

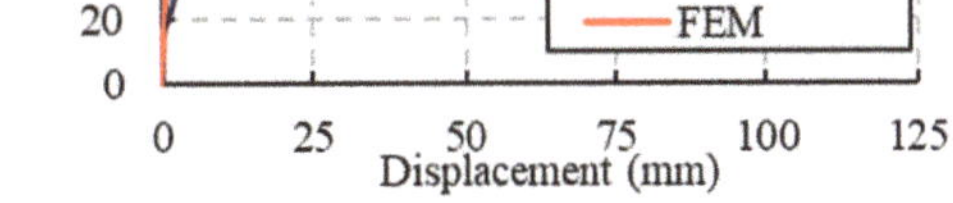

(b)

Figure 16. *Cont.*

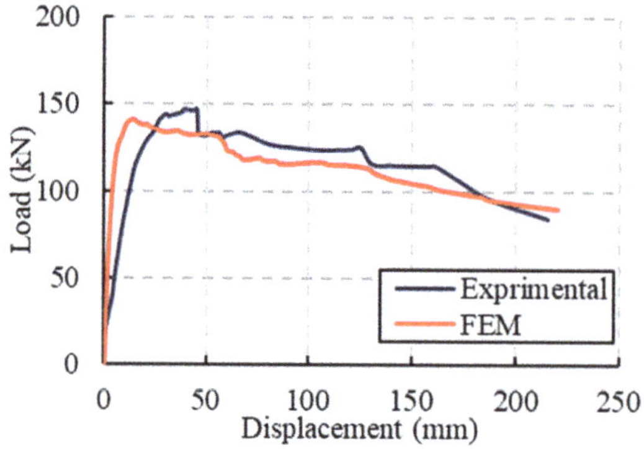

(**c**)

Figure 16. Validations of the FE results for static loading: (**a**) Specimen NR; (**b**) Specimen CG; (**c**) Specimen CGC.

4.4. Impact Analysis

The numerical model employed the dynamic explicit step to apply the impact loading. The suggested model was validated using the experimental data of specimens NR-I, CG-I, and CGC-I. FE models were developed to represent five series impacts by five identical impactors that hit the model at different drop velocities. The contact between the hammer and top surface of the beam was employed as surface-to-surface contact, which used a friction coefficient of 0.35 with tangential behavior. Figure 17 shows the assembled FE model for the composite specimen CGC. The time period used was 100 ms for each hit. During the repeated blows, the mid-span deflection and impact force were recorded for each hit.

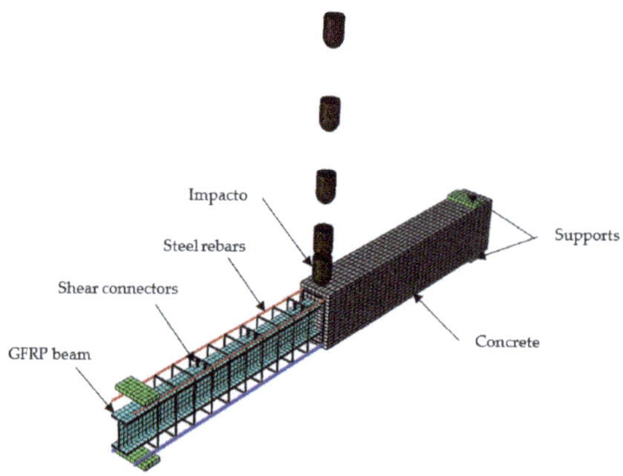

Figure 17. FE mesh of the composite specimen CGC-I for impact analysis.

Model Verifications

Comparisons between the FE and experimental results of the impact force and mid-span deflection time histories for specimen NR-I, as an example, are shown in Figure 18. The numerical simulations and experiment results were validated using five impact hits. Good agreement between the experimental and FE results was obtained. The difference between the maximum impact loads in the FE model and the experimental result was 10%

of all specimens. Moreover, this difference was about 6% of the mid-span deflections for all specimens.

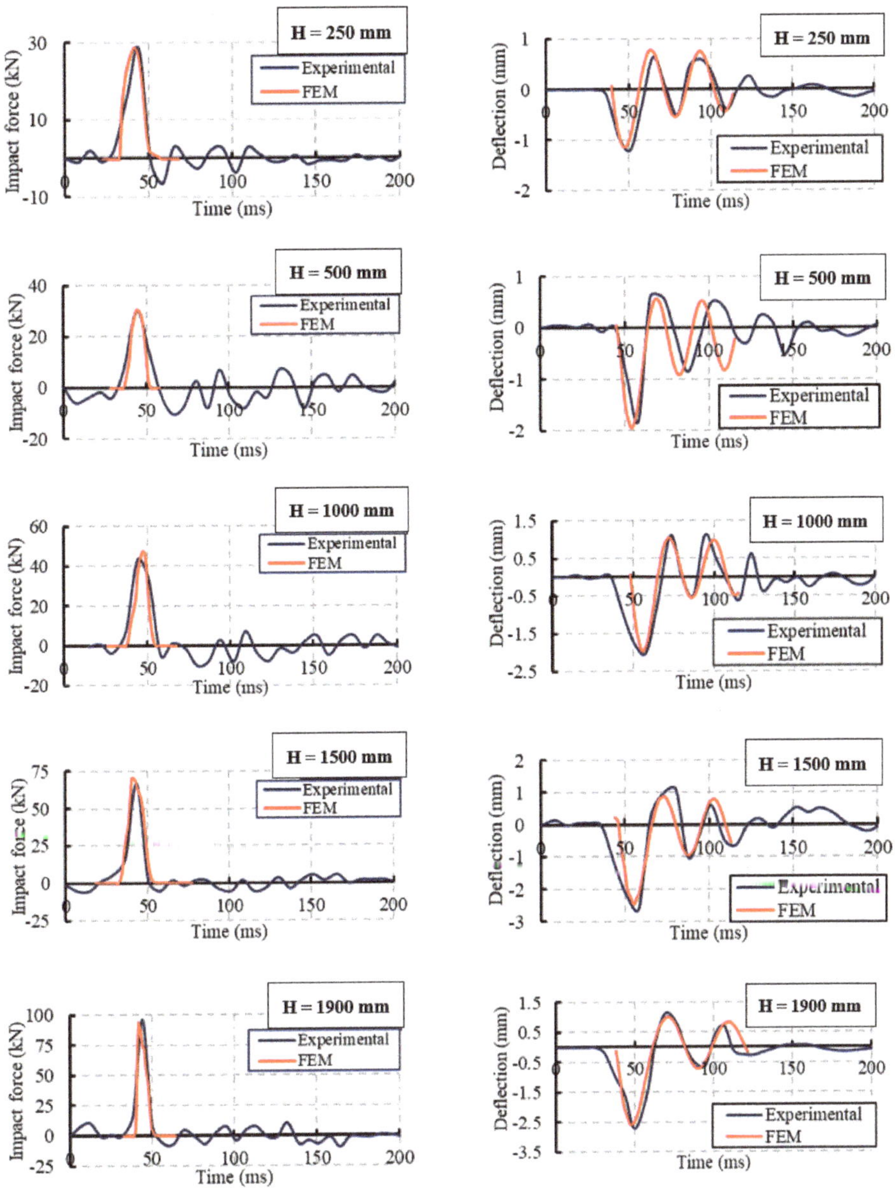

Figure 18. Validations of the experimental and FE results of the impact force and mid-span deflection time histories for specimen NR-I.

5. Parametric Study

A parametric study was carried out to investigate the influence of the concrete compressive strength on the ultimate strength of the composite specimens with the GFRP I-beams under the effect of static and impact loadings.

5.1. Static Analysis

The influence of different concrete compressive strengths of 25 MPa, 35 MPa, and 45 MPa on the static flexural behavior of the analyzed specimens was investigated. Table 10 and Figure 19 illustrate the concrete compressive strength effect on the ultimate load capacity and the corresponding deflection. As the compressive strength increased, the ultimate load capacity increased and the corresponding deflection decreased. The increasing and decreasing percentages were calculated based on the 25 MPa as the reference compressive strength. For the composite specimens, the ultimate load capacity was improved by 23% and 40% for compressive strengths 35 MPa and 45 MPa, respectively. However, these improvements were 13% and 22% for the reference specimen NR, respectively. On the other hand, the corresponding deflections were reduced by 60% for the specimen CG. However, these reductions were 23% and 48% for the concrete strength 35 MPa and 45 MPa, respectively, for the specimen CGC. These results confirmed the effectiveness of concrete strength to improve the stiffness of the composite specimens is less pronounced when the shear connectors were used. However, the same effect on the ultimate load capacity was obtained. For specimen NR, the corresponding mid-span deflections were reduced only by 4.4% and 10.4% for the concrete strengths 35 MPa and 45 MPa, respectively. The concrete compressive strength was more effective in improving the behavior of the composite specimens relative to those without GFRP I-beams.

Table 10. Summary of the concrete compressive strength effects.

Specimens	Compressive Strength (MPa)	Max. Load (kN)	Increase in Max. Load (%)	Deflection at Max. Load (mm)	Decrease in Deflection (%)
NR	25	94.43	-	22.63	-
	35	107.00	13.31	21.64	4.4
	45	115.64	22.46	20.27	10.4
CG	25	127.5	-	18	-
	35	156.71	23	7.28	59.6
	45	180.1	41.8	6.84	62
CGC	25	141.19	-	14.127	-
	35	173.88	23.15	10.83	23.3
	45	196.73	39.33	7.28	48.47

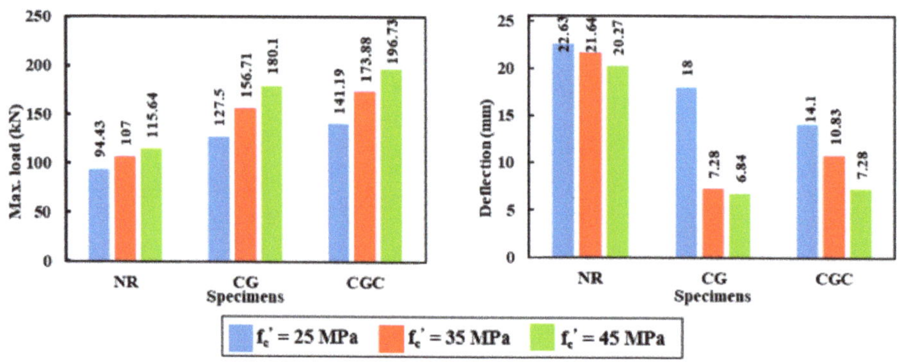

Figure 19. The effect of concrete compressive strength on the ultimate load capacity and corresponding deflection for the analyzed specimens.

5.2. Impact Analysis

The concrete compressive strengths 25 MPa, 35 MPa, and 45 MPa were examined to determine their influences on the load capacity and corresponding mid-span deflection under the effect of impact loads. Figures 20–22 show these influences for the five impact hits. Relative reductions were noticed in the deflections at the maximum impact load for the first hit (H = 250 mm), whereas increases were observed for the other hits. The first hit had a more negligible effect on the behavior of the analyzed model. Therefore, the deflection in the first hit had a behavior similar to that of deflection in the static loading.

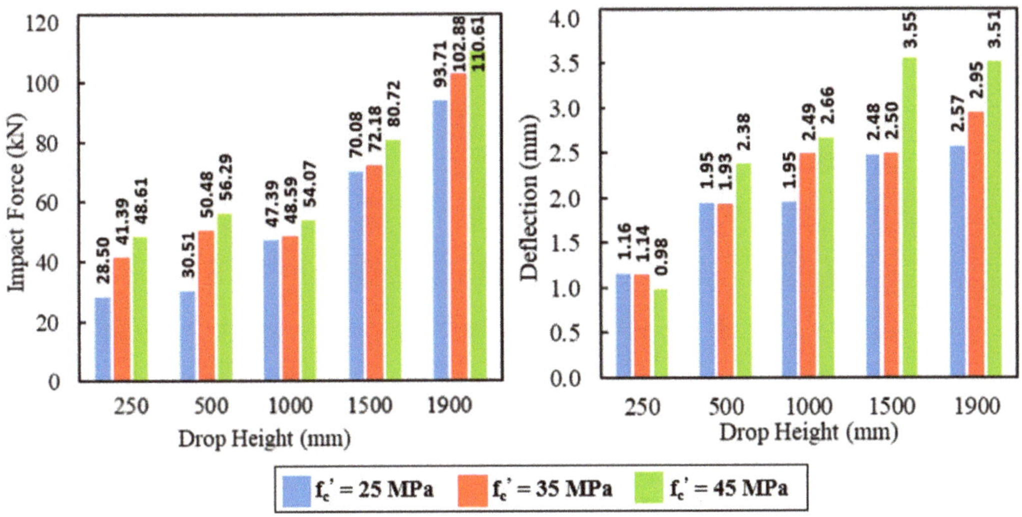

Figure 20. The effect of the concrete compressive strength on the impact forces and mid-span deflections for the reference specimen NR-I.

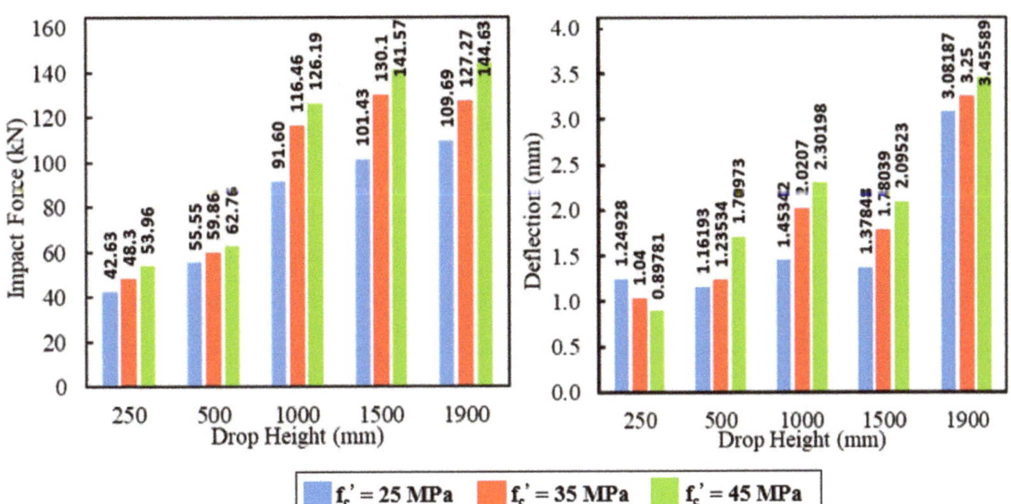

Figure 21. The effect of concrete compressive strength on the impact forces and mid-span deflections for the composite specimen CG-I.

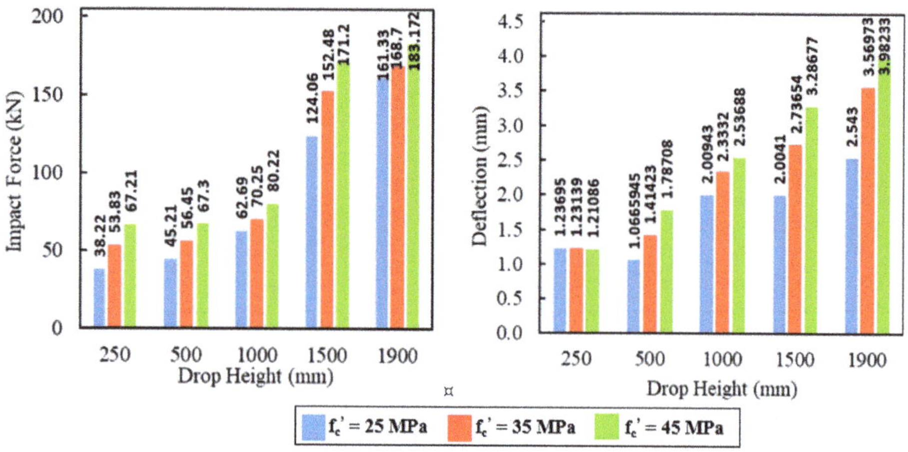

Figure 22. The effect of concrete compressive strength on the impact forces and mid-span deflections for the composite specimen CGC-I.

Variation percentages in the impact load and mid-span deflection were calculated based on the results of the compressive strength 25 MPa as listed in Tables 11 and 12, respectively. The average increase in the maximum impact load was between 18% and 25% for all specimens when the compressive strength was 35 MPa. However, it was between 30% and 40% when the compressive strength was 45 MPa. Likewise, for deflection, the average difference in the maximum deflection at the maximum impact load was between 9% and 25% for all specimens when the compressive strength was 35 MPa. This difference was ranged between 30% and 43% when the compressive strength was 45 MPa. The impact characteristics of the RC composite specimens with GFRP I-beam were influenced effectively by increasing the compressive strength of concrete.

Table 11. Summary of the concrete compressive strength effects on the impact force.

Specimens	Compressive Strength (MPa)	Increase in Max. Impact Load (%) for Drop Height (mm)					Average Increasing Ratio (%)
		250	500	1000	1500	1900	
NR-I	35	45.25	65.44	2.53	3	9.78	25.2
	45	70.56	84.48	14.1	15.18	18.03	40.47
CG-I	35	13.31	7.76	27.14	28.27	16.02	18.5
	45	26.59	12.98	37.76	39.58	31.85	29.75
CGC-I	35	40.83	24.87	12.06	22.91	4.57	21.05
	45	75.83	48.87	27.96	38	13.54	40.84

Table 12. Summary of the concrete compressive strength effects on the deflection.

Specimens	Compressive Strength (MPa)	The Difference in Max. Deflection (%) for Drop Height (mm)					Average Difference Ratio (%)
		250	500	1000	1500	1900	
NR-I	35	−1.4	−0.9	27.27	0.89	14.59	9.03
	45	−14.9	22.18	36.48	43.5	36.56	30.72
CG-I	35	−16.7	6.32	39.03	29.15	5.46	19.34
	45	−28.1	47.15	58.38	52	12.14	39.56
CGC-I	35	−0.4	32.6	16.11	36.55	40.37	25.21
	45	−2.1	67.55	26.25	64	56.9	43.3

6. Conclusions

The present study experimentally and numerically investigated the impact behavior of composite RC beams with the pultruded I-GFRP and I-steel beams. Eight specimens of two groups were cast in different configurations. The first group consisted of four specimens and was tested under static load to provide reference results for the second group. The four specimens in the second group were tested first under impact loading and then static loading to determine the residual static strengths of the impacted specimens. Moreover, nonlinear FE models were developed and validated using the experimental data. The validated models were used to conduct a parametric study to investigate the effect of the concrete compressive strength on the performance of the composite beams under static and impact loadings. The main conclusions are summarized as follows:

1. The embedded GFRP I-beam in the RC specimens with and without shear connectors increased the load-bearing capacity by 51% and 65%, respectively, relative to the reference specimen NR. However, this improvement was 100% in the case of using steel I-beam.
2. Increasing the composite interaction between the GFRP I-beam and concrete by means of shear connectors improved the ultimate capacity by 9.5% relative to the composite specimen without shear connectors.
3. For the composite specimen CS-I, the maximum impact force was 190% greater than the reference specimen NR-I at a drop height of 1900 mm whereas the maximum impact forces for specimens CG-I and CGC-I were 19% and 77%, respectively, more significant than the reference beam at the same drop height. The high stiffness for the steel I-beams relative to the GFRP I-beams was the reason for this difference in behavior, where the impact force is proportional to the structural stiffness.
4. The composite specimens CG-I and CGC-I showed reductions in the maximum mid-span deflections relative to the reference specimen NR-I. For the 1900 mm drop height, reductions of 9.4% and 21% were obtained, respectively, whereas the composite specimen CS-I showed a 17% increase in the maximum mid-span deflection due to the high impact force caused by the section's high stiffness.
5. As the compressive strength increased, the static ultimate load capacity increased and the corresponding deflection decreased. For the composite specimens with GFRP I-beam, the ultimate load capacity was improved by 23% and 40% for compressive strengths 35 MPa and 45 MPa, respectively. However, these improvements were 13% and 22% for the reference specimen NR, respectively. On the other hand, the corresponding deflections were reduced by 60% for the specimen CG. However, these reductions were 23% and 48% for the concrete strength 35 MPa and 45 MPa, respectively, for the specimen CGC. These results confirmed the effectiveness of concrete strength to improve the stiffness of the composite specimens was less pronounced when the shear connectors were used.
6. The impact characteristics of the RC composite specimens with GFRP I-beam were influenced effectively by increasing the compressive strength of concrete. The average increase in the maximum impact load was between 18% and 25% for all specimens when the compressive strength was 35 MPa. However, it was between 30% and 40% when the compressive strength was 45 MPa. Therefore, the concrete compressive strength was more effective in improving the impact behavior of the composite specimens relative to those without GFRP I-beams.

It is worth noting that these impact tests were performed to initially unloaded specimens while in reality, the composite RC beams would be loaded and then an impact load would be applied. Therefore, pre-loading before administering impact loads will be the next research focus.

Author Contributions: Investigation, T.H.I. and A.A.A.; resources, T.H.I.; software, T.H.I.; supervision, A.A.A. and A.E.-Z.; validation, T.H.I.; visualization, A.A.A. and A.E.-Z.; analysis, A.E.-Z.; writing—original draft, A.A.A.; writing—review and editing, A.E.-Z. All authors have read and agreed to the published version of the manuscript.

Funding: This research received no external funding.

Institutional Review Board Statement: Not applicable.

Informed Consent Statement: Not applicable.

Data Availability Statement: The data presented in this study are available on request from the corresponding author.

Acknowledgments: The support provided by the Department of Civil Engineering at the University of Baghdad is gratefully acknowledged. In addition, the authors are grateful to thank the Consulting Engineering Bureau at the University of Baghdad (CEB-UB) for assistance in fulfilling the material test.

Conflicts of Interest: The authors declare no conflict of interest.

References

1. Neagoe, C.A.; Gil, L.; Pérez, M.A. Experimental study of GFRP-concrete hybrid beams with low degree of shear connection. *Constr. Build. Mater.* **2015**, *101*, 141–151. [CrossRef]
2. Yuan, J.S.; Hadi, M.N.S. Flexural behavior of composite beam reinforced with GFRP I-beam. In Proceedings of the 6th Asia-Pacific Conference on FRP in Structures APFIS 2017, Singapore, 19–21 July 2017; pp. 19–21.
3. Yu, Y.J.; Kim, C.-H.; Cho, J.-Y. Investigation of Behavior of RC Beams Subjected to Impact Loading Considering Combination of Mass and Impact Velocity. *Procedia Eng.* **2017**, *210*, 353–359. [CrossRef]
4. Tabiei, A.; Svanson, A.; Hargrave, M. Impact behavior of pultruded composite box-beams. *Compos. Struct.* **1996**, *36*, 155–160. [CrossRef]
5. Li, Z.; Khennane, A.; Hazell, P.J.; Remennikov, A.M. Performance of a hybrid GFRP-concrete beam subject to low-velocity impacts. *Compos. Struct.* **2018**, *206*, 425–438. [CrossRef]
6. Soleimani, S.M.; Roudsari, S.S. Analytical study of reinforced concrete beams tested under quasi-static and impact loadings. *Appl. Sci.* **2019**, *9*, 2838. [CrossRef]
7. Allawi, A.A.; Ali, S.I. Flexural behavior of composite GFRP pultruded I-section beams under static and impact loading. *Civ. Eng. J.* **2020**, *6*, 2143–2158. [CrossRef]
8. Roudsari, S.; Hamoush, S.; Soleimani, S.; Abu-Lebdeh, T.; Haghighifar, M. Analytical Study of Reinforced Concrete Beams Strengthened by FRP Bars Subjected to Impact Loading Conditions. *Am. J. Eng. Appl. Sci.* **2018**, *11*, 407–425. [CrossRef]
9. *ASTM C496/C496M-17*; Standard Test Method for Splitting Tensile Strength of Cylindrical Concrete Specimens. ASTM International: West Conshohocken, PA, USA, 2017.
10. *ASTM C469/C469M-14e1*; Standard Test Method for Static Modulus of Elasticity and Poisson's Ratio of Concrete in Compression. ASTM International: West Conshohocken, PA, USA, 2014.
11. *ASTM-A370-19*; Standard Test Methods and Definitions for Mechanical Testing of Steel Products. ASTM International: West Conshohocken, PA, USA, 2019.
12. *ASTM-D695-15*; Standard Test Method for Compressive Properties of Rigid Plastics. ASTM International: West Conshohocken, PA, USA, 2015.
13. *International Standard. ISO 527-4*; Determination of Tensile Properties of Plastics Part 4. No. 1109. ISO: Geneva, Switzerland, 1997; p. 14.
14. Abaqus. *Abaqus 6.2: Computer Software for Finite Element Analysis*; Dassault Systems Simulia: Johnston, RI, USA, 2020.
15. *EN 1992-1-1*; Eurocode 2: Design of Concrete Structures—Part 1-1: General Rules and Rules for Buildings. British Standards Institution: London, UK, 2011.
16. Barbero, E.J.; Cosso, F.A.; Roman, R.; Weadon, T.L. Determination of material parameters for ABAQUS progressive damage analysis of E-glass epoxy laminates. *Compos. Part B Eng.* **2013**, *46*, 211–220. [CrossRef]
17. Hashin, Z. Failure criteria for unidirectional fiber composites. *J. Appl. Mech. Trans. ASME* **1980**, *47*, 329–334. [CrossRef]
18. Barbero, E.J. *Finite Element Analysis of Composite Materials Using ABAQUS*; CRC Press Taylor and Francis Group: Boca Raton, FL, USA, 2013.

materials

MDPI

Article

Spatio-Temporal Distribution of the Sources of Acoustic Events in Notched Fiber-Reinforced Concrete Beams under Three-Point Bending

Dimos Triantis [1,*], Ilias Stavrakas [1], Andronikos Loukidis [1], Ermioni D. Pasiou [2] and Stavros K. Kourkoulis [2]

1 Electronic Devices and Materials Laboratory, Department of Electrical & Electronics Engineering, University of West Attica, 250 Thivon Avenue, 122 44 Athens, Greece; ilias@uniwa.gr (I.S.); a.loukidis@uniwa.gr (A.L.)
2 Laboratory for Testing and Materials, Department of Mechanics, School of Applied Mathematical and Physical Sciences, National Technical University of Athens, Zografou Campus, 157 73 Athens, Greece; epasiou@central.ntua.gr (E.D.P.); stakkour@central.ntua.gr (S.K.K.)
* Correspondence: triantis@uniwa.gr; Tel.: +30-2105381316

Abstract: The acoustic activity, generated in notched, beam-shaped concrete specimens, loaded under three-point bending, is studied in terms of the position of the sources of acoustic events, and the frequency of their generation. Both plain specimens (without any internal reinforcement) and specimens reinforced with various types of short fibers were tested. The target of the study is to investigate the existence of indices that could be considered as pre-failure indicators of the upcoming fracture. In addition, an attempt is undertaken to classify the damage mechanisms activated to tensile or shear nature. Considering comparatively the spatio-temporal evolution of the position of the acoustic sources and the respective temporal evolution of the frequency of generation of acoustic events, it was concluded that for relatively low load levels the acoustic sources are rather randomly distributed all over the volume of the specimens. As the load increases toward its maximum value, the acoustic sources tend to accumulate in the immediate vicinity of the crown of the notch and the average distance between them approaches a minimum value. When this minimum value is attained, the load is maximized and the generation frequency of the acoustic events increases rapidly. The simultaneous fulfillment of these three conditions is observed a few seconds before the onset of propagation of the catastrophic macrocrack for all classes of specimens tested, providing a kind of warning signal about the upcoming fracture. Moreover, the classification of the damage mechanisms to tensile and shear ones revealed a crucial difference between the plain and the reinforced specimens after the maximization of the load applied. Indeed, while for the plain specimens, the prevailing damage mechanism is tensile microcracking, for the reinforced specimens a balance between tensile and shear damage mechanisms is observed after the load applied has attained its peak and starts decreasing.

Keywords: acoustic emissions; acoustic sources; plain concrete; fiber-reinforced concrete; three-point bending

Citation: Triantis, D.; Stavrakas, I.; Loukidis, A.; Pasiou, E.D.; Kourkoulis, S.K. Spatio-Temporal Distribution of the Sources of Acoustic Events in Notched Fiber-Reinforced Concrete Beams under Three-Point Bending. *Materials* **2023**, *16*, 5118. https://doi.org/10.3390/ma16145118

Academic Editor: Sergei Tarasov

Received: 17 June 2023
Revised: 12 July 2023
Accepted: 18 July 2023
Published: 20 July 2023

1. Introduction

As it is experimentally observed, elastic waves are generated at the interior of materials while they are subjected to increasing mechanical loading. These elastic waves are detected as acoustic signals by means of proper sensors and are known as Acoustic Emissions (AEs). The AEs are related to the onset of formation, growth, and coalescence of microcracks. It is beyond any doubt that proper analysis of AEs provides interesting information concerning the level of damage accumulation within the loaded material. In addition, it provides useful indices (known as pre-failure indicators), related to the entrance of the specimen (or the structure) into its critical loading stage, namely, the stage of impending catastrophic fracture.

Various characteristics and statistical parameters of the acoustic activity (for example, the frequency of generation of acoustic events, the energy release rate, the b-value, etc.), have been proven to be quite valuable and flexible tools, for the investigation of the evolution of microcracking processes within the loaded material (especially at the stage at which macroscopic fracture is approaching [1–7]), providing, in addition, reliable pre-failure indicators [8–13]. Moreover, it has been proven that the combined consideration of the Rise Angle (RA) and the Average Frequency (AF) of an acoustic signal permits the classification of the damage mechanisms activated to tensile or shear nature [14,15]. Intensive research activity is nowadays in progress in the direction of clarifying the interconnection between the AEs and the respective electric/electromagnetic activity (in the form of electric currents [16–18], or of electromagnetic emissions [19–21], or of variations of the electric resistance [22–24]), which is detected while specific classes of building materials are subjected to various mechanical loading schemes.

Along the same lines, the simultaneous use of more than one acoustic sensor (depending on the geometry of the loaded specimen) during the experimental procedure, permits the determination of the spatial coordinates (x, y, z) of the source of an acoustic event. In other words, it makes it possible to localize the points where the damage mechanisms are activated, at any time instant from the onset of loading up to the final macroscopic fracture. The spatial evolution of the position of the sources of the acoustic events has been used as a tool for the localization of the areas of intense microcracking activity by many researchers worldwide [1,25–28], as well as an index for the quantification of the degree of spatial concentration or dispersion (or, in other words, of the spatial "density") of the sources of microcracking [29,30].

In general, acoustic events are detected during the whole loading procedure, already from relatively low load levels (compared to the loads causing macroscopic catastrophic fracture). In most cases, the acoustic events detected at very early loading steps are not directly related to the final fracture. In fact, there is a randomness in the spatial distribution of these early acoustic events, strongly related to the initial mechanical properties of the material, mainly its (an)isotropy and (in)homogeneity, pre-existing discontinuities, and the respective local stress fields around them. Clearly, the successful prediction of a macroscopic fracture depends on some knowledge of the density of these pre-existing "anomalies" of the structure of the material [31].

The spatial distribution of the acoustic events and other characteristics of the signals that are collected during the loading of a specimen until its fracture have been studied thoroughly and attempts are made to classify the results with respect to the remaining life of the specimen. The theory of "three stages" proposed by Mogi [26] is the one most widely adopted, and can be briefly described as follows: During the first stage, the acoustic activity is weak, and critical acoustic events are not detected or they are very rare. In the second one, the rate of generation of acoustic events appears relatively increased; however, their sources are still randomly distributed all over the mass of the loaded specimen. Finally, in the third stage, the sources of the acoustic events exhibit strongly increased concentration in the close vicinity of the points from which macroscopic cracks will start propagating. During this last stage, the frequency of generation of acoustic events is strongly intensified, and assuming that it is described in terms of the "time-to-failure" parameter, it is governed by a specific power law [32]. Moreover, the average Euclidean distance between the acoustic sources is proven to significantly decrease, while the level of the load applied tends to attain the critical level that causes macroscopic fracture [30].

In this direction, an attempt to quantify the spatio-temporal evolution of the position of the sources of acoustic events is undertaken in this study. To achieve this target, a novel approach is adopted, based on the average position of the sources of the acoustic events, which belong to a sliding window of n successive events. In parallel, for each "window" (group of acoustic events), the average frequency of generation of events is calculated in an attempt to enlighten relative correlations, which could be valuable for the detection of pre-failure indices. For the implementation of the above goals, advantage is taken of the data of an experimental protocol with beam-shaped notched specimens made of concrete, fiber reinforced or plain (i.e., without any kind of reinforcing elements), which were subjected to bending. The aforementioned analysis of the experimental data indicated that a series of conditions are simultaneously fulfilled a few seconds before the onset of propagation of the fatal macrocrack, suggesting that a proper combination of them could serve as an interesting pre-failure signal. The innovative aspect of the present study is the simultaneous and comparative study of the spatial and the temporal evolution of the position of the sources of acoustic events. To achieve this target, a novel experimentally determined parameter is introduced, representing the average Euclidean distance between the sources of the acoustic signals, i.e., the sources of damage mechanisms. The temporal evolution of this parameter is proven to provide valuable insight into the process of damage accumulation within the mass of the loaded specimen. Equally important innovative aspect of the study is the spatial localization of the average position of the acoustic sources with respect to the crown of the notch. It is the simultaneous analysis of these two parameters that permits detection of pre-failure indicators, namely, signals which are assumed to timely warn about the upcoming macroscopic fracture.

2. The Specimens and the Experimental Protocol

The protocol from which the experimental data were obtained has been analytically described in a previous paper by the authors' team [12], and thus it is only synoptically outlined here. It included three-point bending experiments with specimens made of concrete. The specimens were beam-shaped of length equal to 700 mm and their cross-section was square (150 mm × 150 mm). All specimens were mechanically notched at their central section, using a rotating cutting disc. The depth of the notch was equal to 25 mm and its breadth (width) was equal to about 5 mm. The main role of the notch is to create an area of severe stress concentration, and thus to predefine the area where fracture will appear. Moreover, the existence of the notch enhances the activation of damage mechanisms from relatively early loading stages due to the locally intense stress field. Clearly, the novel parameters introduced in this study (i.e., the average Euclidean distance between the sources of the acoustic signals, and the average position of the acoustic sources with respect to the crown of the notch) could be equally well employed for the analysis of data obtained from protocols with intact beams.

Four different compositions of the concrete mixture were considered as shown in Table 1. The protocol was initially designed to include sixteen experiments (four specimens per composition); however, only fourteen were implemented successfully.

Table 1. The four types of specimens tested.

Test	Sample/Code	Fiber's Density	Characteristics of the Fibers
A	Plain concrete/ ref-b02	---	---
B	Concrete reinforced with short plastic fibers/ force60-4-b02	4 kg/m^3	Polyolefin Length ≈ 60 mm, Diameter ≈ 0.84 mm Tensile strength ≈ 465 MPa
C	Concrete reinforced with short plastic fibers/ pp940-50-6-b02	4 kg/m^3	Polypropylene Length ≈ 50 mm, Diameter ≈ 0.75 mm Tensile strength ≈ 600–650 MPa
D	Concrete reinforced with short metallic fibers/ metal-25-B01	25 kg/m^3	Steel Length ≈ 50 mm, Diameter ≈ 1.30 mm Tensile strength ≈ 690 MPa

The four specimens that will be described analytically in the next sections are typical ones for each composition. The response of the remaining specimens is more or less similar to that of the typical specimens chosen per composition.

It is noted at this point that the data of this protocol have been analyzed in previous works according to approaches based on quite different physical principles. It is mentioned, for example, that in ref. [8] the data were analyzed based on concepts of the "Natural Time Domain" (an approach which ignores the conventional time), while in ref. [12] the analysis is based on concepts of "Non-Extensive Statistical Mechanics" (founded on a class of entropies which violate the additivity property, namely, the property constituting the foundation of the familiar Boltzmann–Gibbs Statistical Mechanics—these entropies are commonly denoted as non-additive entropies).

The specimens were supported with the aid of two metallic cylinders (Figure 1) and the load was applied at the central section (i.e., the notched one) by means of a third identical metallic cylinder. An Instron-Satec servo-hydraulic frame was used to apply the load, adopting displacement-control conditions at a rate of 0.08 mm/min. The acoustic activity was detected using eight acoustic sensors (R15a), the spatial arrangement of which is shown in Figure 1a. A characteristic photo of a specimen before loading is shown in Figure 1b.

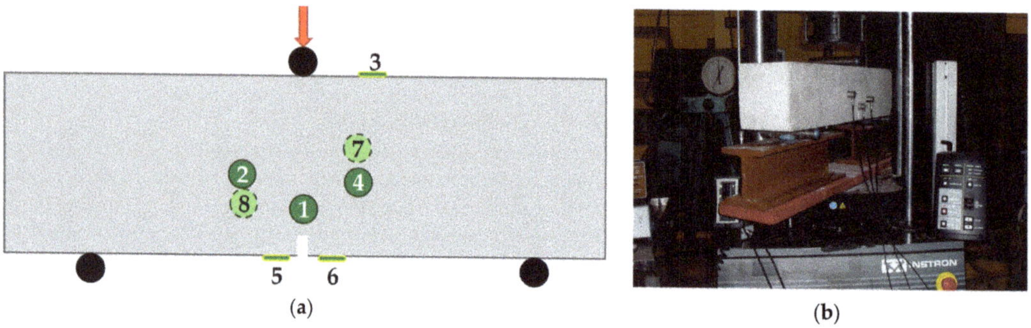

Figure 1. (**a**) The spatial arrangement of the eight acoustic sensors; dashed lines and light green color indicate sensors mounted at the rear face of the specimens; (**b**) a typical specimen before loading.

It is mentioned at this point that, although eight acoustic sensors were used (rendering possible the determination of all three spatial coordinates of each acoustic source), the following analysis is implemented in a two-dimensional sense (in the present study in the (x, y) plane). In other words, the results are projected on the vertical plane of symmetry including the longitudinal axis of the beams. This is a first approach based on the assumption that the role of the third dimension (in the present study of the z-direction) is of minor importance, compared to that of the remaining two dimensions. Although this assumption sounds arbitrary, one should recall that, according to the elementary Bernoulli–Euler bending theory, the stress field developed is, in fact, two dimensional. Clearly, the validity of this assumption is to be further validated by means of additional protocols.

The mechanical response of four typical specimens (one from each class of Table 1) is shown synoptically in Figure 2, in which the load applied is plotted versus the displacement induced. Excluding the specimen made of plain concrete (test A), which was fractured abruptly and the load was zeroed suddenly, the remaining three types of specimens failed according to a different manner: After attaining its peak value the load was reduced to a level equal to about 30% to 60% of the respective peak value, and then it was kept almost constant. In other words, the macroscopic crack that emanated from the crown of the notch did not propagate instantly toward the upper face of the beams, but rather it started propagating smoothly and slowly, due to the action of the fibers that kept together the two fragments, assigning a kind of significant "ductility" to the response of the beams.

It is to be mentioned here, that from the Strength of Materials point of view, there is no statistically significant difference between the values of the maximum load, L_{max}, recorded for the three classes of fiber-reinforced specimens (specimens B, C, D). It is, perhaps, surprising that the role of the characteristics of the reinforcing fibers is rather marginal, at least concerning the strength of the product. However, this is by no means a definite conclusion. It is valid exclusively for the specimens of the present protocol and the authors did not attempt to explain it, since the mechanical response of the specimens was well beyond the scope of this study. In general, concerning the characteristics of the fibers, it is known that they play a crucial role, especially regarding the mechanical response and strength of the final product. However, the purpose of the present study was to analyze the acoustic activity in terms of alternative parameters (for example, the average distance between the acoustic sources and the average position of them), rather than to analyze the mechanical characteristics of the four classes of specimens. In this direction, it was not considered crucial to choose reinforcing fibers with specific characteristics.

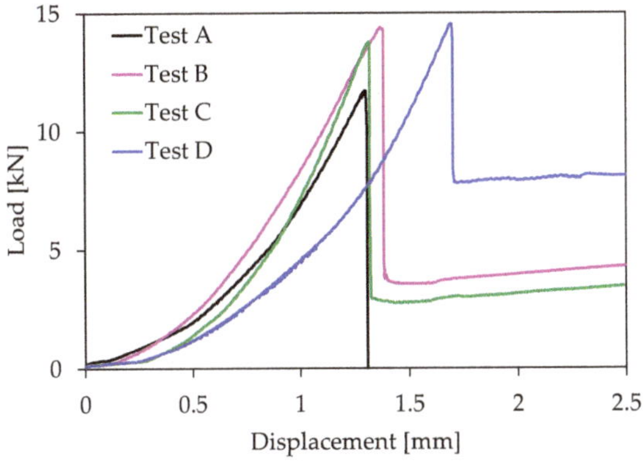

Figure 2. The mechanical response of typical specimens: The load applied versus the displacement.

3. The Methodology Adopted to Study the Spatio-Temporal Evolution of the Average Position of the Acoustic Sources

Taking advantage of the data simultaneously recorded by the eight acoustic sensors, and considering a reference system with its origin at the center of symmetry of the beam (see Figure 3a), the coordinates (x_i, y_i) of the sources of each acoustic event were determined, starting from the onset of the loading procedure up to the instant of macroscopic fracture, t_f. It is clarified here that for the reinforced specimens (for which fracture is not instantaneous but rather the crack propagates slowly), a conventional fracture instant is defined, as the instant at which the load drops from its peak value to the value at which it is kept constant, as it is seen in Figure 3b, for a typical specimen of the fourth class.

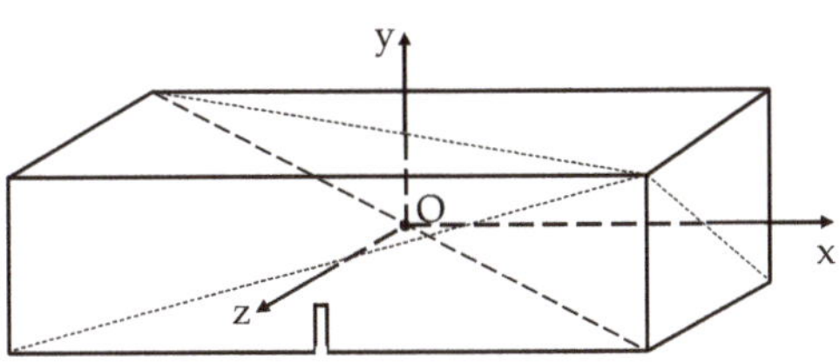

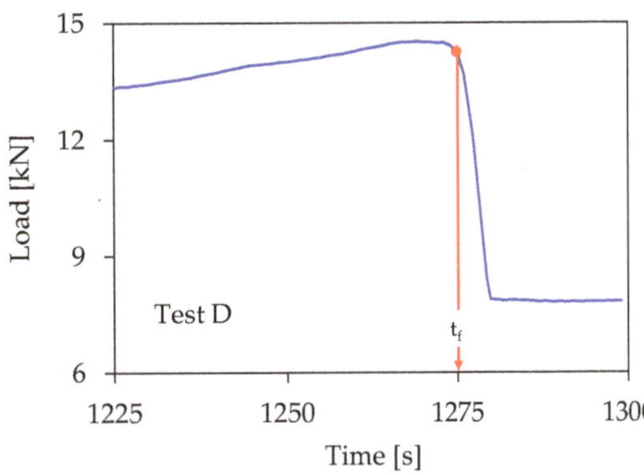

(a) (b)

Figure 3. (**a**) The reference system adopted; (**b**) defining the conventional fracture instant for the reinforced specimens.

The number of acoustic events, N, recorded for each one of the tests that will be analyzed in this study is shown in Table 2, along with the respective maximum load attained, L_{max}. The N acoustic events for each experiment are divided into k groups, each one of which includes n events, where (for the needs of the present study) it was set n = 10 (the only exception is the last two or three groups, which may include n = 9 or n = 11 events, depending on the value of the N/k ratio, in the case that it is not an integer).

Table 2. The number of acoustic events recorded and the maximum load attained for each specimen.

Specimen	Test	N	L_{max} [kN]
3pb-ref-b02	A	71	11.70
force60-4- b02	B	89	14.37
pp940-50-6-b02	C	87	13.75
metal-25-B01	D	88	14.53

In order to study the spatio-temporal evolution of the average position of the acoustic sources in the vertical plane Oxy, the coordinates (x_i, y_i) of all acoustic sources, detected during the specific experiment, are retrieved from the software of the AE set up and the Euclidean distance between any two successive acoustic sources is determined as:

$$D_{i-1} = \sqrt{(x_i - x_{i-1})^2 + (y_i - y_{i-1})^2}, \quad i = 2, 3, \ldots \tag{1}$$

For each one of the k groups defined above, the average distance $\overline{D_j}$ between the acoustic sources (of the specific group) is determined as the mean value of the (n − 1) distances between the n successive sources of the group:

$$\overline{D_j} = \frac{D_1 + D_2 + \ldots + D_{n-1}}{n-1}, \quad j = 1, 2, \ldots, k \tag{2}$$

Finally, the average value of the coordinates of the acoustic sources for each one of the k groups is calculated as:

$$\overline{x_j} = \frac{x_1 + x_2 + \ldots + x_n}{n}, \quad \overline{y_j} = \frac{y_1 + y_2 + \ldots + y_n}{n}, \quad j = 1, 2, \ldots, k \tag{3}$$

considering that each AE event (and thus each acoustic source) comprises many AE hits in the form of clusters. Therefore, using average values for x and y appears to be more appropriate in the direction of highlighting the evolution of the position of the AE sources inside the material and the way they "travel" around the notch's crown. In addition, the average frequency $\overline{f_j}$ of generation of acoustic events is calculated for each one of the *k* groups, as:

$$\overline{f_j} = \frac{n}{\Delta t}, \quad j = 1, 2, \ldots, k \tag{4}$$

where Δt is the time interval between the first and last acoustic event of the specific group.

As a final step, each one of the $\overline{x_j}$, $\overline{y_j}$, $\overline{f_j}$, and $\overline{D_j}$ (j = 1, 2, . . . , k) values is paired to an average time instant τ_j, which is the average value of the time instants, t_i, at which each one of the n events of the specific group was recorded.

$$\tau_j = \frac{t_1 + t_2 + \ldots + t_n}{n}, \quad j = 1, 2, \ldots, k \tag{5}$$

4. Results and Discussion

In this section, the acoustic activity generated in four typical specimens (one for each class) is studied according to a common pattern. As a first step, the temporal evolution of the frequency of generation of acoustic events is considered. Then, the temporal evolution of the average Euclidean distance between the acoustic source is analyzed. Both quantities are plotted versus the time-to-failure parameter, in juxtaposition to the respective evolution of the load applied. Finally, the spatial evolution of the acoustic sources (with respect to the crown of the notch) is plotted and discussed.

4.1. The Unreinforced Concrete (3pb-ref-b02)

For the specific experiment, the N = 71 acoustic events recorded during the loading procedure, were divided into k = 7 groups (with 10 events for the first six of them and 11 events for the last one). The temporal evolution of the load applied is plotted in Figure 4a, in juxtaposition to the frequency of generation of acoustic events. Each group is colored with a different color, and the specific color code is kept constant for all figures describing the specific test.

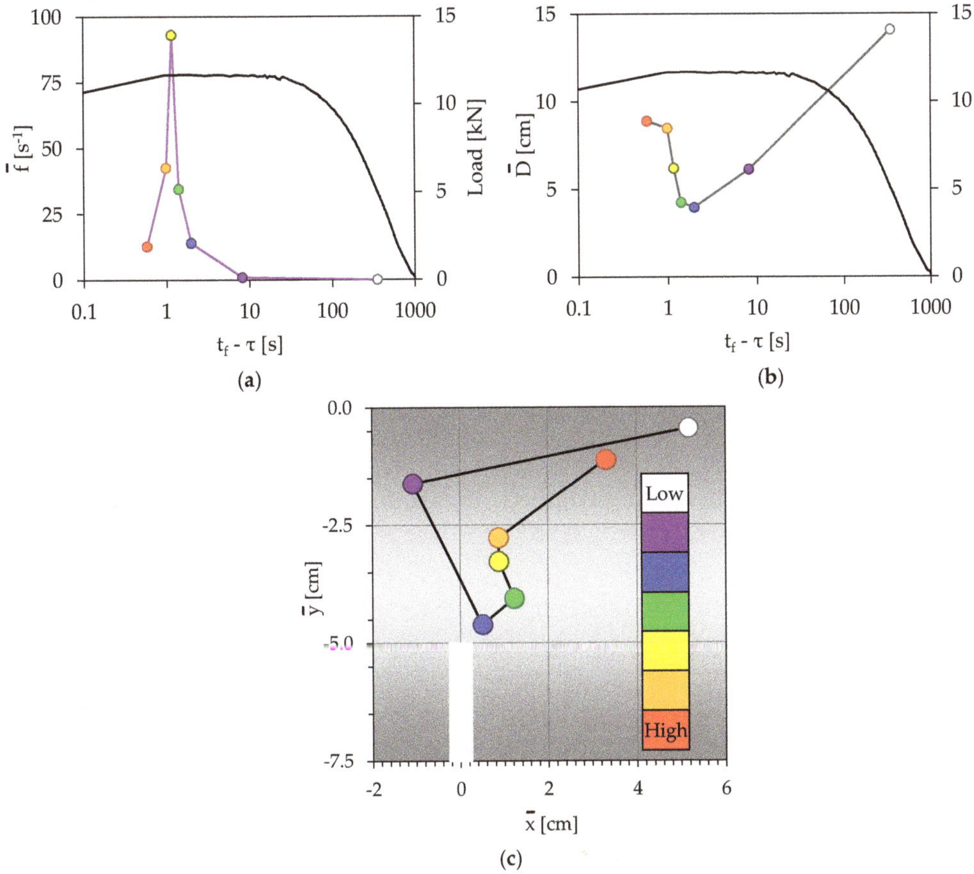

Figure 4. The temporal evolution of (**a**) the frequency of generation of acoustic events, and (**b**) the average Euclidean distance between the acoustic sources, in juxtaposition to the respective evolution of the load applied. (**c**) The spatial evolution of the acoustic sources (specimen without reinforcing fibers).

Three time intervals can be distinguished in Figure 4a: During the first one ($t_f - \tau > 2$ s), which covers the whole period of gradual increase in the load applied (until it attains its maximum value), the specimen appears to be completely "silent" concerning the acoustic activity, with $\bar{f}$ equal to almost zero (for $t_f - \tau > 9$ s), or the acoustic activity is very weak

($2\ s < t_f - \tau < 9\ s$) with $\bar{f} < 12\ s^{-1}$. The first three groups of events (colored white, magenta, and blue) belong to this interval, in which the load is an increasing time function.

The second time interval ($1\ s < (t_f - \tau) < 2\ s$) is characterized by a very steep increase in the frequency of generation of acoustic events, from $\bar{f} = 12\ s^{-1}$ to $\bar{f} = 90\ s^{-1}$, while the load applied is kept almost constant and equal to the peak value attained ($L_{max,A} = 11.7$ kN). Two groups of acoustic events (green and yellow) belong to this second interval.

Finally, during the third interval ($0.4\ s < (t_f - \tau) < 1\ s$), both the frequency of generation of acoustic events and the load applied exhibit a decreasing trend until the specimens are macroscopically fractured. The last two groups of acoustic events (orange and red) belong to this third interval.

Concerning the temporal evolution of the average distance between the acoustic sources, which is plotted in Figure 4b, it is clearly seen that at the early loading stages its value is relatively high (it attains a maximum equal to about 14 cm), indicating that any damage mechanism activated (microcracking, debonding, fiber cracking etc.) produces results which are not as yet accumulated around specific material points of the specimen. Therefore, macroscopic failure is not impending. With the increasing load level, the average distance between the acoustic sources decreases toward a minimum value equal to about 4 cm, attained at about 2 s before fracture. Therefore, it is indicated that at this time instant the results of the damage mechanisms have been accumulated in the immediate vicinity of the crown of the notch (as it can be definitely concluded, also, from Figure 4c), rendering coalescence of microcracks significantly easier, and thus macroscopic cracking is now impending. After this critical time instant, the average distance between the acoustic sources starts increasing again, almost until the fracture of the specimen.

The spatial evolution of the average position of the acoustic sources is plotted in Figure 4c. It is clear from this figure that during the first time interval ($t_f - \tau > 2$ s), the acoustic sources are randomly distributed and their average position is relatively far from the crown of the notch. As the frequency of generation of acoustic events starts increasing very rapidly (second time interval, $1\ s < (t_f - \tau) < 2\ s$), the acoustic sources appear as being attracted by the crown of the notch. Their average distance from the crown of the notch tends to be a minimum, attained at the third group of acoustic events. Then, the distance of the acoustic sources from the crown of the notch starts increasing again, maximized at the last group of events, i.e., just before the fragmentation of the specimen.

Comparative consideration of Figure 4a–c reveals that the time instant at which the average position of the acoustic sources (considered in the vertical axial plane of symmetry) is closer to the crown of the notch corresponds to:

(i) Maximization of the load applied;
(ii) Minimization of the distance between the acoustic sources;
(iii) Onset of rapid increase in the frequency of generation of acoustic events.

All these three conditions are simultaneously satisfied (for the specific tests) during the third group of events, which is "located" about 2 s before the macroscopic fracture. Alternatively, it can be said that the simultaneous fulfillment of these three conditions designates the onset of propagation of the fatal crack that will split the specimen into two fragments. During the last interval (in fact, during the last second before fragmentation), the acoustic activity seems to become significantly weaker in terms of the evolution of the average frequency of generation of acoustic events.

The above-described response can be attributed to several co-existing reasons. Indeed, at this last time interval, the failure plane is already "formed" (during the previous loading stages and due to the geometry of the specimens and the loading scheme adopted) and it is characterized by a local stress field with quite intense stress concentrations. Therefore, it is reasonable to assume that the AE activity at remote locations of the specimen (with respect to the fracture plane) is significantly reduced since the energy is mainly consumed around the tip of the macrocrack that has already started propagating. Moreover, the damages within and close to the fracture plane are very severe, and thus the elastic energy released due to any event is very high, reducing the energy locally available for an additional event to be generated (increasing simultaneously the time interval required for this generation). This behavior could explain the experimentally observed high energy and long duration of the acoustic events recorded during the last loading stages. Similar behavior to that described here has been observed, as well as in recent works of the team of authors [12,23], in which completely different methods of analysis of the acoustic activity were adopted.

4.2. Concrete Reinforced with Short Polyolefin Fibers (Force-60-4-b02)

During the loading procedure of this experiment, N = 89 acoustic events were recorded. They were divided into k = 9 groups with n = 10 events for each one excluding the 9th group which contained only nine events. The temporal evolution of the load applied is plotted in Figure 5a versus the time-to-failure parameter ($t_f - \tau$), in juxtaposition to the frequency of generation of acoustic events. As previously mentioned, each group is marked with a different color, which is kept constant for all figures devoted to this test.

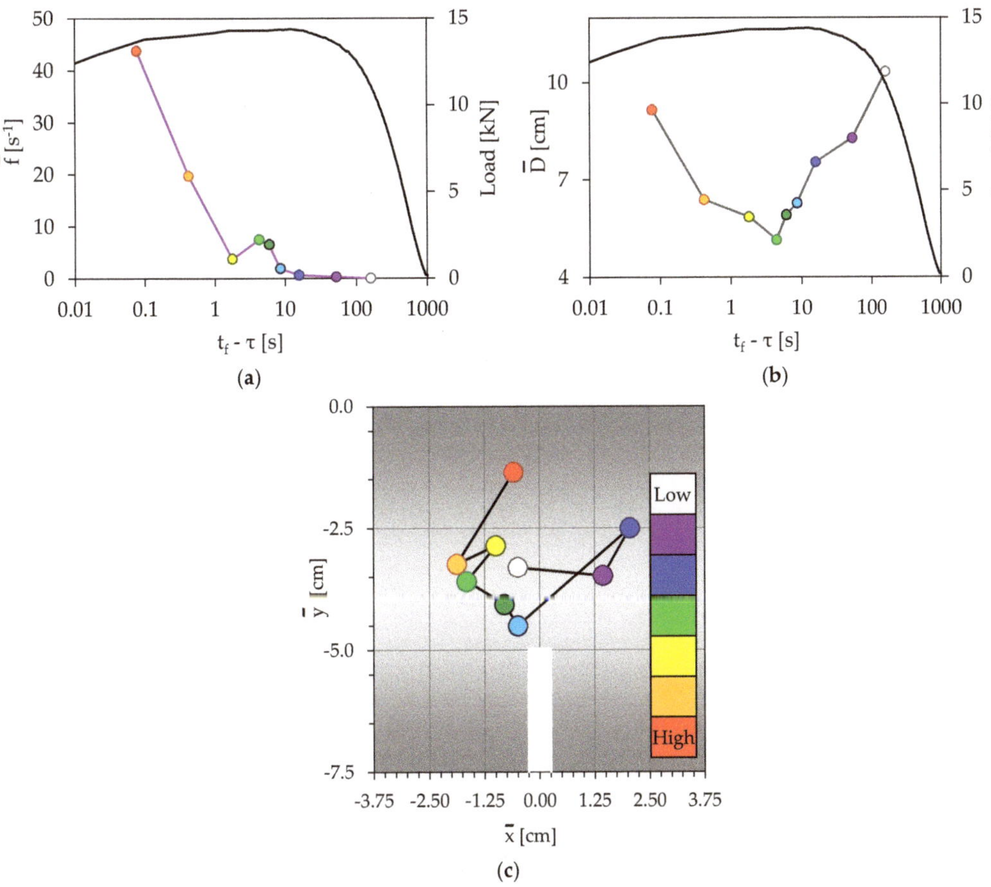

Figure 5. (**a**) The temporal evolution of (**a**) the frequency of generation of acoustic events, and (**b**) the average Euclidean distance between the acoustic sources, in juxtaposition to the respective evolution of the load applied. (**c**) The spatial evolution of the acoustic sources (specimen with polyolefin reinforcing fibers).

Consistently with the previous test, the acoustic activity is negligible, almost for the whole period during which the load applied is increasing, i.e., for the interval $t_f - \tau > 8$ s. The first three groups (white, magenta, and blue) of events belong to this interval. Then, while the load has already attained its maximum value ($L_{max,B}$ = 14.37 kN), the acoustic activity starts increasing and the frequency of generation of acoustic events attains values approaching 8 s^{-1}. During this interval (which includes three groups of acoustic events, namely, the light blue, the green, and the light green ones), the duration of which is equal to 4 s (4 s < $t_f - \tau$ < 8 s), the load applied has exceeded its peak value and it has started decreasing, although imperceptibly, from 100% to 99.6% of $L_{max,B}$.

For both these intervals, i.e., for $t_f - \tau > 4$ s, the temporal evolution of the average distance between the acoustic sources, which is plotted in Figure 5b, exhibits a steadily decreasing trend, i.e., the acoustic sources are approaching each other: the average distance between them decreases from a relatively high value, equal to about 10.5 cm, to a global minimum value equal to about 5 cm.

During the last 4 s of the loading procedure, i.e., for $t_f - \tau < 4$ s, the distance between the acoustic sources starts increasing again, while the acoustic activity keeps intensifying after a relatively small drop. The specific behavior is explained by taking into account that although the macrocrack has started propagating at about $t_f - \tau \approx 4$ s, the presence of reinforcing fibers does not permit abrupt fragmentation of the specimens. Therefore, the crack propagates very slowly and additional events are now generated due to two additional reasons: Fracture of fibers which keep the two parts of the specimen together and shear effects along the fibers–mortar interfaces.

The spatial evolution of the average position of the acoustic sources is plotted in Figure 5c. It is seen that for the three first groups of acoustic events (corresponding to negligible or weak acoustic activity, i.e., for $t_f - \tau > 8$ s), their sources are rather randomly distributed. At the instant that the acoustic activity is intensified (fourth group of acoustic events, or $t_f - \tau \approx 4$ s) the average distance of acoustic sources from the crown of the notch is minimized, indicating clearly that the macrocrack starts propagating. During the last interval of the loading procedure ($t_f - \tau < 4$ s), the distance of the acoustic sources from the crown of the notch starts increasing again; however, it remains relatively close to the plane of symmetry, since the acoustic events are now generated due to failure (fracture or relative slip) of fibers within the plane of symmetry (i.e., of fibers that keep the two fragments together while the crack propagates slowly toward the upper surface of the specimen).

Considering, again, comparatively Figure 5a–c, similar conclusions are drawn with those obtained from the analysis of the unreinforced specimen: The time instant at which the average position of the acoustic sources (considered in the vertical axial plane of symmetry) is closer to the crown of the notch corresponds to maximization of the load applied, a tendency of minimization of the distance between the acoustic sources and the onset of increase in the frequency of generation of acoustic events. In this test, all these three conditions are satisfied at an instant with $t_f - \tau \approx 4$ s. The only difference between this test and that with the unreinforced specimen is that the acoustic activity keeps increasing even during the very last loading stage, i.e., after the macrocrack has started propagating; however, the mechanisms responsible for this activity are related to the failure of the fibers in the transverse vertical plane of symmetry, namely, the fibers that prevent abrupt fracture of the beam.

4.3. *Concrete Reinforced with Short Polypropylene Fibers (pp940 50 6 b02)*

In this experiment, typical for the class of specimens reinforced with short polypropylene fibers, N = 87 acoustic events were recorded. They were divided, again, into k = 9 groups with n = 10 events for the first six of them and n = 9 events for the last three groups of acoustic events. The same as in previous quantities (i.e., the temporal evolution of the load applied, that of the frequency of generation of acoustic events, and that of the average Euclidean distance between the acoustic sources as well as the spatial evolution of the acoustic sources) are shown in Figure 6. It is clear that the response of this specimen is absolutely compatible, if not identical (with minor differences of quantitative nature), with that of the previous section (namely, the specimen with reinforcing fibers made of polyolefin): The acoustic activity is negligible or very weak almost until the load attains a value closely approaching the maximum value attained ($L_{max,C}$ = 13.75 kN), i.e., a value equal to about 99.5% of $L_{max,C}$ ($t_f - \tau > 8$ s). The average distance between the acoustic sources is steadily decreasing toward a global minimum value, which is finally attained at $t_f - \tau \approx 6$ s, the instant at which the acoustic activity attains its local maximum (light blue group). Then, exactly as in the previous specimen, both the acoustic activity and the distance between the acoustic sources keep increasing, for the same previous reasons (failure due to fracture or slip of the fibers in the transverse vertical plane of symmetry, i.e., of the fibers that prevent abrupt fracture, despite the fact that the macrocrack is already propagating although very slowly).

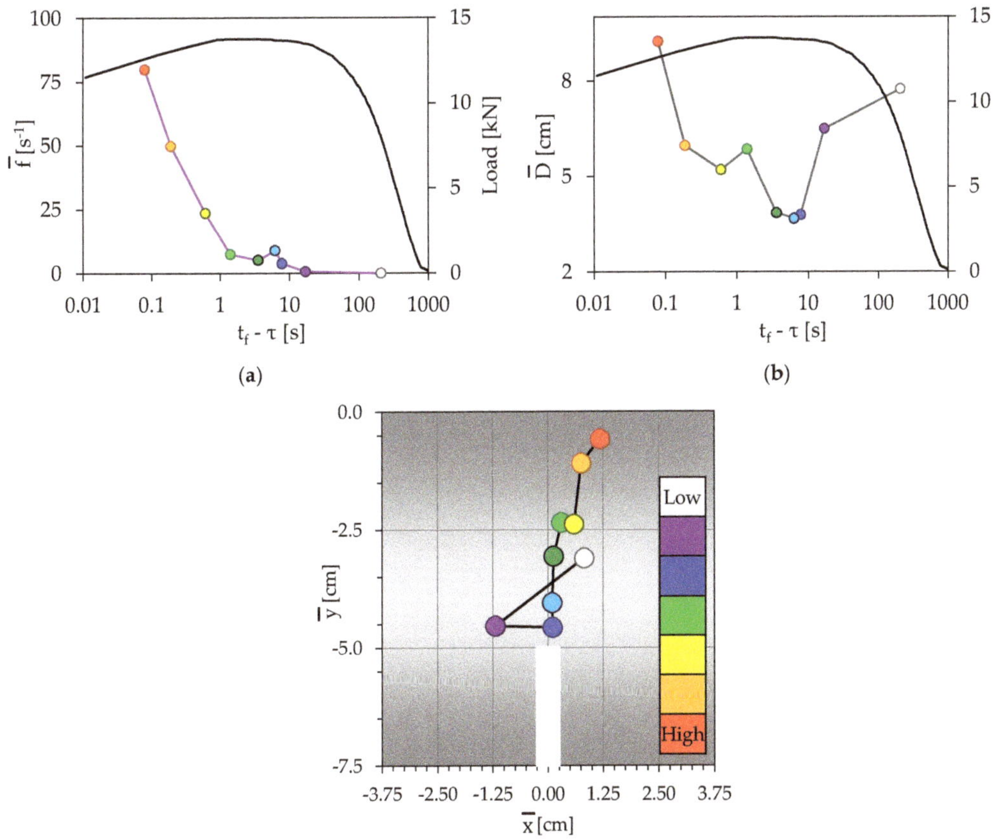

(c)

Figure 6. (**a**) The temporal evolution of (**a**) the frequency of generation of acoustic events, and (**b**) the average Euclidean distance between the acoustic sources, in juxtaposition to the respective evolution of the load applied. (**c**) The spatial evolution of the acoustic sources (specimen with polypropylene reinforcing fibers).

Moreover, the spatial distribution of the acoustic sources follows the same pattern: The sources are initially randomly distributed and gradually they are attracted by the crown of the notch, attaining their smallest distance at the onset of the intensification of the acoustic activity (blue group of events), i.e., at about $t_f - \tau \approx 6$ s. As mentioned previously, from this instant on, the acoustic sources are moving away from the crown of the notch, remaining within the transverse plane of symmetry (i.e., the plane of fracture), following, in fact, the front of the slowly propagating (due to the action of the reinforcing fibers) macrocrack.

All three conditions (mentioned for the previous two specimens), which are fulfilled while the distance of the acoustic sources from the crown of the notch tends to be minimized (thus triggering the onset of propagation of the fatal macrocrack), i.e., maximization of the load, minimization of the distance between the acoustic sources and onset of rapid increase in the frequency of generation of acoustic events, are simultaneously satisfied at the instant $t_f - \tau \approx 6$ s.

4.4. *Concrete Reinforced with Short Steel Fibers (Metal-25-B01)*

During the last experiment of the protocol discussed in this study, N = 88 acoustic events were recorded. As mentioned previously, they were divided into k = 9 groups with n = 10 events for the first seven of them and n = 9 events for the last two groups.

The temporal evolution of the load applied is plotted in Figure 7a, in juxtaposition to the respective evolution of the frequency of generation of acoustic events. The evolution of the average Euclidean distance between the acoustic sources is shown in Figure 7b. Finally, the spatial evolution of the acoustic sources is depicted in Figure 7c. Although the response of the specific specimen exhibits considerable similarities with the previously discussed specimens reinforced with plastic fibers (made of polyolefin or polypropylene ones), these are some of the differences that should be highlighted:

- The first is that one detects now acoustic activity from relatively earlier loading steps, although it is again very weak for the $t_f - \tau > 8$ s interval, during which the load applied attains 99.5% of its maximum value ($L_{max,D}$ = 14.53 kN). The average distance between the acoustic sources decreases from a value equal to about 7.7 cm to a global minimum value equal to about 3.3 cm.
- The second difference in the response of the specific specimen (compared to the ones reinforced with plastic fibers) is that the acoustic activity does not increase monotonously up to the final fracture of the specimen. At about $t_f - \tau \approx 2$ s, a global maximum is attained and then the activity starts decreasing, although not monotonically, until the fracture instant. During both these periods (i.e., the period of rapid increase in the acoustic activity and the period of its rapid decrease), namely, for the interval $t_f - \tau < 8$ s, the average distance between the acoustic sources exhibits an increasing trend from the global minimum value of 3.3 cm toward a maximum value approaching (for the last group of events) the value of 7.7 cm, which was attained during the very early loading steps (first group of events).

Concerning the spatial distribution of the acoustic sources, it follows exactly the same as the previous pattern: For the first group of events ($t_f - \tau < 8$ s), the sources are relatively far away from the crown of the notch. Then, during the next two groups of events (magenta and blue), the acoustic sources are densely packed very close to the crown of the notch. This trend is terminated at the instant $t_f - \tau \approx 6$ s, corresponding to a local maximum of the acoustic activity. From this instant on, the acoustic sources are moving away from the crown of the notch; however, they remain within (or very close) to the transverse plane of symmetry (in other words, within the fracture plane), following again the front of the slowly (due to the presence of the reinforcing fibers) propagating macrocrack.

The three conditions, that were fulfilled for all previous specimens (reinforced or not) in order for the distance of the acoustic sources from the crown of the notch to be minimized (namely, the maximization of the externally applied load, the minimization of the distance between the acoustic sources, and the onset of the increase in the frequency of generation of acoustic events), are here simultaneously satisfied in a narrow time interval, with 6 s$< t_f - \tau < 8$ s. Assuming that the instant, at which the sources of the acoustic events approach very closely the crown of the notch, is the triggering of the onset of propagation of the final catastrophic macrocrack, it can be safely stated that in this experiment the macroscopic fracture started at an instant between 6 and 8 s before the conventional fracture instant (as it was previously defined in Section 3).

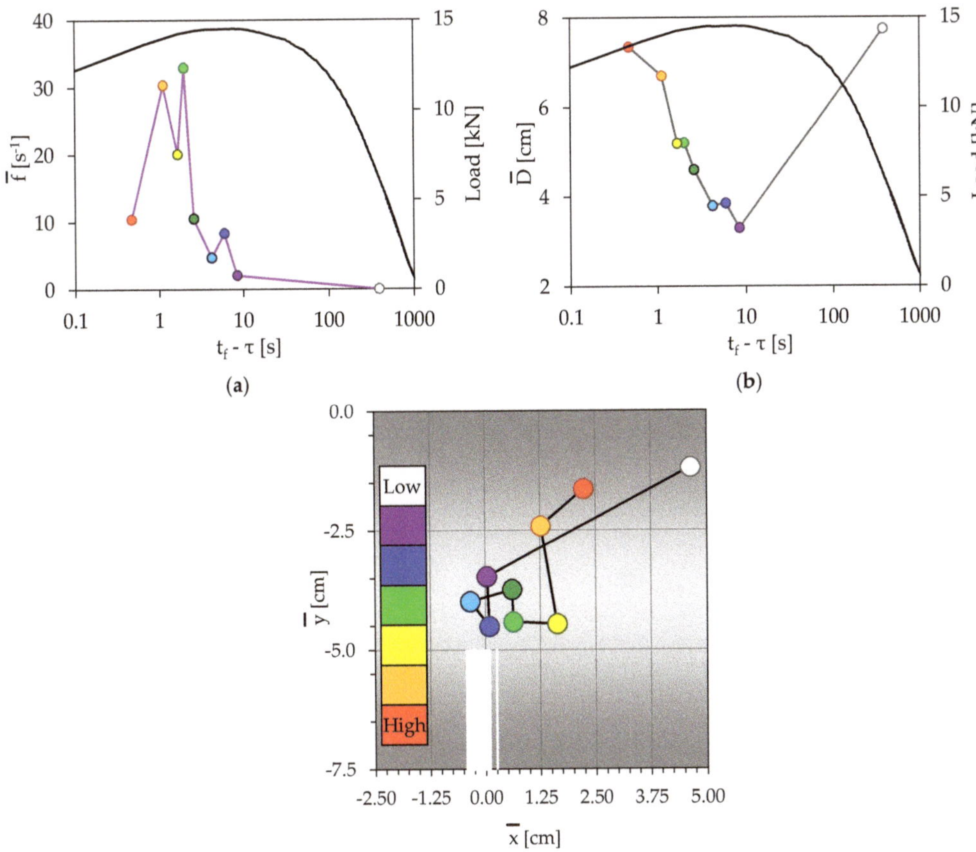

Figure 7. (**a**) The temporal evolution of (**a**) the frequency of generation of acoustic events, and (**b**) the average Euclidean distance between the acoustic sources, in juxtaposition to the respective evolution of the load applied. (**c**) The spatial evolution of the acoustic sources (specimen with steel reinforcing fibers).

5. Crack Classification Using the Relationship between the Average Frequency and RA Value

In this section, an attempt is undertaken to classify the microcracking activity as tensile or shear, in order to validate the previous assumption about the nature of the damage mechanisms activated before and after the applied load attains its maximum value. This classification is based on the individual hit waves, as recorded by the AE sensors, and projects a relation between the Average Frequency, AF, and the RA parameter (namely, the Rise Time (RT) per Amplitude) according to RILEM [33]. The Average Frequency (AF) is defined as the number of counts divided by the total duration of an acoustic hit and it is measured in kHz, while the RA value is defined as the Rise time divided by the peak Amplitude of an AE hit and is measured in ms/V.

According to this approach, mode-II damage mechanisms (frictional or equivalently shear mechanisms) correspond to acoustic signals with high RA and intermediate AF values. On the contrary, mode-I mechanisms (tensile microcracking) correspond to acoustic signals with low RA values combined with high AF ones. It is to be emphasized at this point, that in the original contributions, in which the AF to RA relation was introduced as a characteristic that distinguishes mode-I from mode-II acoustic events [14,15], a unique numerical value for the slope of the "dividing line" was not suggested. It is merely a choice of each researcher and, usually, it is assumed to be the diagonal line of the AF-RA plots, assuming that a ratio of the respective scales is equal to 10.

In the present study, two groups of acoustic hit signals are considered: The first one (denoted as Group I and represented here with grey rectangles) includes signals which are generated until the time instant at which the load attains its peak value, L_{max}, while the second group (denoted as Group II, and represented here with red rectangles) includes the signals generated while the load imposed starts decreasing until the fracture (actual or conventional) of the specimen. The results of this classification for the four experiments described in the previous section are shown in Figure 8, while in Table 3 one can see the percentage of acoustic signals that belongs to each one of the two categories (i.e., shear or tensile) for both A and B groups.

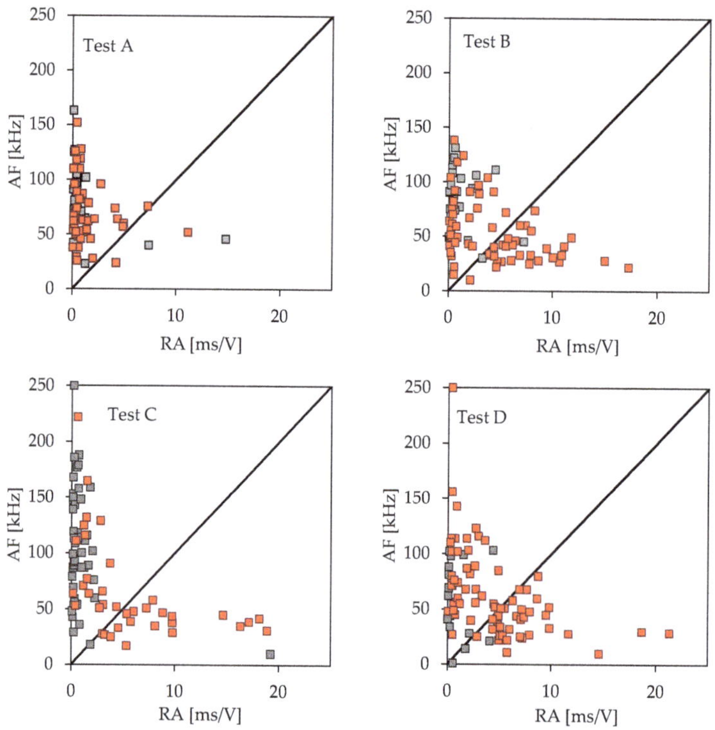

Figure 8. Classification of the acoustic signals detected as tensile (above the diagonal, the slope of which is equal to 10) and as shear (below the diagonal) for the four tests analyzed in Section 4 (grey rectangles correspond to Group I while red ones to Group II).

Table 3. The number of acoustic events recorded prior and after the maximization of the load applied for each test.

Test ↓	**Group**→ **Type of Signal**→	*Group I (up to L_{max})* *Tensile*	 *Shear*	*Group II (after L_{max})* *Tensile*	 *Shear*
A		92%	8%	96%	4%
B		92%	8%	52%	48%
C		94%	6%	49%	51%
D		89%	11%	51%	49%

It is clearly concluded from Figure 8 and Table 3, that for the unreinforced specimen, the vast majority of acoustic signals are classified as tensile, both before the load attains its peak value (about 92% of the total signals), and while the load starts decreasing (about 96% of the total signals). On the contrary, for all three types of reinforced specimens, this is true only before the maximization of the load applied: Indeed, during this time interval, the signals detected are classified as tensile (with a percentage ranging from 89% to 94%). However, after the load has been maximized and its value has started decreasing, the situation changes dramatically: About half of the signals generated in this interval are classified as shear ones, thus verifying the conclusions previously drawn in Section 4, that while the load decreases an additional damage mechanism (in addition to that of tensile microcracking) is now activated, namely, that of debonding of the reinforcing fibers from the concrete mass. Moreover, an almost perfect balance between the two damage mechanisms (i.e., tensile microcracking and debonding) is observed.

In order to gain a deeper insight into the gradual transition from the purely tensile damage mechanism to the combined action of the microcracking and debonding, an index m is introduced, defined as the ratio of the Average Frequency (AF) over the RA value of the "primary acoustic hit" of each event. In order to clarify the term "primary acoustic hit", one should recall that any acoustic event is defined with the aid of a number of acoustic hits, recorded by a series of acoustic sensors. The first one of these hits, i.e., the one that arrived first at one of the sensors is characterized as the "primary acoustic hit".

Considering the k groups of acoustic events (as it is analytically described in Section 4), one can determine the average value $\overline{m}$ of the index m for each one of the k groups of each experiment. The temporal evolution of the average value $\overline{m}$ of the index m is plotted in Figure 9 versus the $(t_f - \tau)$ parameter.

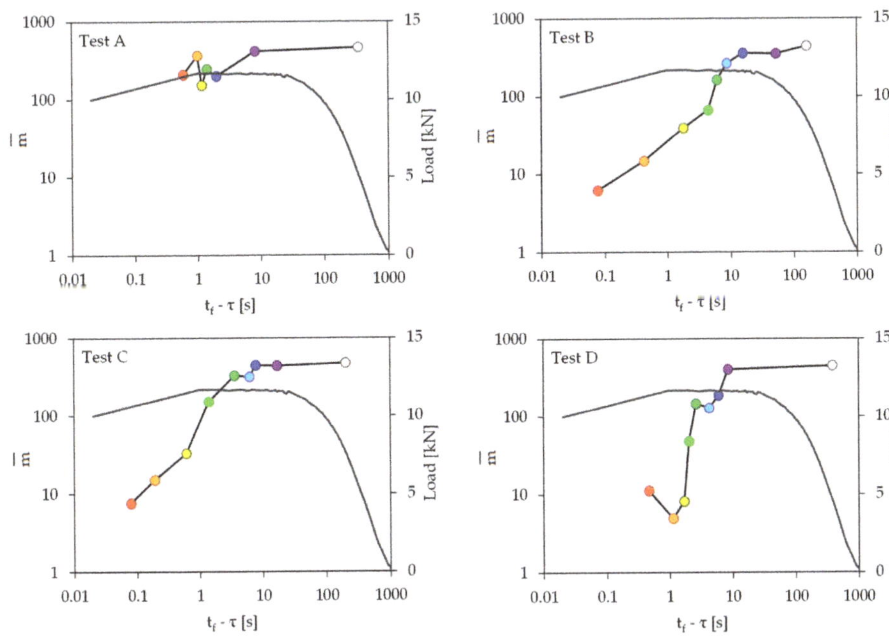

Figure 9. The temporal evolution of the index $\overline{m}$ in juxtaposition to the respective evolution of the load applied versus the $(t_f - \tau)$ parameter for all four experiments considered.

It is once again definitely concluded from Figure 9 that for the unreinforced specimen, the dominant damage mechanism activated throughout the whole loading procedure (i.e., from the onset of loading until the catastrophic fracture of the specimen) is the mode-I mechanism (namely, tensile microcracking). Indeed, for all k = 7 groups of acoustic events of this experiment, the average value of m attains extremely high values ranging in the $150 < \overline{m} < 450$ interval. The maximum value $\overline{m} \approx 450$ corresponds to the first group of events, while the minimum value $\overline{m} \approx 150$ corresponds to the fifth group of events (recall that this fifth group of acoustic events corresponds to the maximization of the intensity of the acoustic activity).

On the other hand, for all three classes of reinforced specimens, the average value of the index m exhibits a constantly decreasing trend after the load has been maximized. Indeed, starting from an extremely high value (in the range of $400 < \overline{m} < 500$ for all three experiments) it approaches values even below $\overline{m} = 10$, for the very last groups of acoustic events. In other words, after the load has attained its peak value, the mode-I damage mechanism recedes gradually in favor of the mode-II one. The consistency of the temporal evolution of the m-index for all three classes of reinforced specimens (independently of the nature of the reinforcing fibers) is striking.

6. Concluding Remarks

Notched, beam-shaped specimens, made of plain or fiber-reinforced concrete were loaded under three-point bending. The evolution of the acoustic activity generated during the loading procedure was considered both in terms of time and space. More specifically, the position of the acoustic sources and the average distance between them were comparatively studied in juxtaposition to the temporal evolution of the average frequency of generation of acoustic events.

It was concluded that independently of the nature of the specimens tested (plain or fiber reinforced), some features of the acoustic activity are common. Indeed, at the time instant at which the load applied tends to be maximized three conditions are, almost, simultaneously and systematically fulfilled:

i. The average distance between the acoustic sources attains a global minimum value and then it starts increasing again;
ii. The acoustic sources are densely packed in the immediate vicinity of the crown of the notch and then they start gradually moving away from the specific area;
iii. The frequency of generation of acoustic events starts increasing quite abruptly.

As it is expected, the acoustic activity is very weak at the very early loading stages (see, for example, Figures 4a, 5a, 6a and 7a). Then, it is gradually intensified and the onset of its severe intensification is detected at the instant at which the average position of the acoustic sources from the crown of the notch is minimized (while simultaneously the distance between the acoustic sources is, also, minimized). After this instant, at least one macrocrack has started propagating producing "strong" acoustic signals, and thus "high" acoustic activity. In this context, Figures 4c, 5c, 6c, and 7c can be considered as representing the location of the tip of the propagating macrocrack, explaining why the characterization "high" for the acoustic activity is associated with locations which are moving gradually away from the crown of the notch.

The above conclusions are systematically observed for all specimens tested, and the differences between them are quantitative rather than qualitative. This is vividly seen in Figure 10, in which the average distance between the acoustic sources is plotted comparatively for all specimens analyzed in previous sections. It is clear that, despite the different compositions of the specimens, the Euclidean distance between the acoustic sources attains a global minimum, in a narrow time interval (ranging from about 2 s to 9 s before fracture). Moreover, the magnitude of this global minimum ranges within very narrow limits, i.e., in the range from 3.9 cm to 5.1 cm.

Of equal importance is the observation concerning the nature of the damage mechanisms activated, especially while the load applied approaches its maximum value.

- For the plain specimens, the damage mechanism prevailing is that of mode-I, i.e., tensile macrocracking, both before the load attains its peak value, and while the load enters its decreasing stage.

- For the reinforced specimens (with plastic or metallic fibers), the mode-I damage mechanism prevails only while the load is increasing. During the interval at which the load attains its maximum value and is kept almost constant (and during the time interval that it starts decreasing), acoustic events due to the mode-II damage mechanism are generated and gradually the two mechanisms reach an almost perfect balance between them. Clearly, the mode-II damage mechanism is attributed to the gradual debonding of the reinforcing fibers (that keep together the two parts of the beam on either side of the slowly propagating macrocrack) from the concrete matrix, while the mode-I events are attributed to the fracture of these fibers and not to microcracking within the body of the concrete matrix.

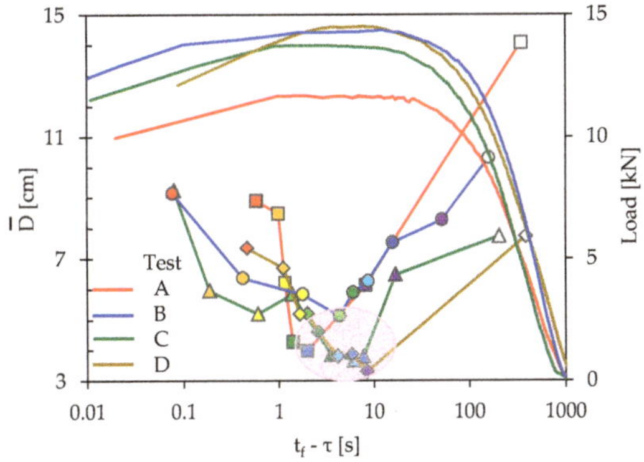

Figure 10. The temporal evolution of the average distance between the acoustic sources in terms of the time to-failure parameter, for all specimens analyzed, in juxtaposition to the respective evolution of the load applied.

In conclusion, it was clearly highlighted that a series of conditions were systematically fulfilled simultaneously a few seconds (ranging from about 2 s to about 10 s) before the propagation of the fatal macrocrack. At the same time instant, the average distance of the position of the sources of acoustic hits from the crown of the notch attains its minimum value (ranging from 2.5 mm to 4.0 mm). These observations were systematically noticed for all the specimens tested (plain or reinforced), suggesting that each one of the conditions fulfilled could be considered as a reliable pre-failure index, or, at least, that this idea deserves to be further (and more thoroughly) studied for wider classes of materials and different loading protocols, including in situ Structural Health Monitoring, of large-scale structures. It should be, however, kept in mind that moving from "sterilized" laboratory conditions to in situ Structural Health Monitoring is a challenging step. Various types of "noises" may "infect" the acoustic signals recorded. The specific issue is among the most critical ones of in situ Structural Health Monitoring, and although huge effort is made worldwide to objectively "clean" the signals recorded, the issue is not as yet closed.

Author Contributions: Conceptualization, D.T.; methodology, D.T., S.K.K. and A.L.; experimental, E.D.P., S.K.K., A.L. and I.S.; writing—original draft preparation, D.T.; writing—review and editing, S.K.K., A.L., E.D.P. and I.S.; visualization, S.K.K.; supervision, D.T. and S.K.K.; project administration, D.T. All authors have read and agreed to the published version of the manuscript.

Funding: This research received no external funding.

Institutional Review Board Statement: Not applicable.

Informed Consent Statement: Not applicable.

Data Availability Statement: Not applicable.

Conflicts of Interest: The authors declare no conflict of interest.

References

1. Lockner, D. The role of acoustic emission in the study of rock fracture. *Int. J. Rock Mech. Min. Sci. Geomech. Abstr.* **1993**, *30*, 883–899. [CrossRef]
2. Ohtsu, M. The history and development of acoustic emission in concrete engineering. *Mag. Concr. Res.* **1996**, *48*, 321–330. [CrossRef]
3. Colombo, I.S.; Main, I.G.; Forde, M.C. Assessing damage of reinforced concrete beam using "*b*-value" analysis of acoustic emission signals. *J. Mater. Civ. Eng.* **2003**, *15*, 280–286. [CrossRef]
4. Cai, M.; Morioka, H.; Kaiser, P.K.; Tasaka, Y.; Kurose, H.; Minami, M.; Maejima, T. Back-analysis of rock mass strength parameters using AE monitoring data. *Int. J. Rock Mech. Min. Sci.* **2007**, *44*, 538–549. [CrossRef]
5. Grosse, C.; Ohtsu, M. *Acoustic Emission Testing*; Grosse, C.U., Ohtsu, M., Eds.; Springer: Berlin/Heidelberg, Germany, 2008.
6. Rao, M.V.M.S.; Lakshmi, K.J.P.; Rao, G.M.N.; Vijayakumar, K.; Udayakumar, S. Precursory microcracking and brittle failure of Latur basalt and migmatite gneiss under compressive loading. *Curr. Sci.* **2011**, *101*, 1053–1059.
7. Sagar, R.V.; Prasad, B.K.R. A review of recent developments in parametric based acoustic emission techniques applied to concrete structures. *Nondestruct. Test. Eval.* **2012**, *27*, 47–68. [CrossRef]
8. Loukidis, A.; Pasiou, E.D.; Sarlis, N.V.; Triantis, D. Fracture analysis of typical construction materials in natural time. *Phys. A* **2020**, *547*, 123831. [CrossRef]
9. Lei, X.; Satoh, T. Indicators of critical point behavior prior to rock failure inferred from pre-failure damage. *Tectonophysics* **2007**, *431*, 97–111. [CrossRef]
10. Niu, Y.; Zhou, X.-P.; Zhou, L.-S. Fracture damage prediction in fissured red sandstone under uniaxial compression: Acoustic emission b-value analysis. *Fatigue Fract. Eng. Mater. Struct.* **2020**, *43*, 175–190. [CrossRef]
11. Friedrich, L.F.; Cezar, É.S.; Colpo, A.B.; Tanzi, B.N.R.; Sobczyk, M.; Lacidogna, G.; Niccolini, G.; Kosteski, L.E.; Iturrioz, I. Long-Range Correlations and Natural Time Series Analyses from Acoustic Emission Signals. *Appl. Sci.* **2022**, *12*, 1980. [CrossRef]
12. Kourkoulis, S.K.; Loukidis, A.; Pasiou, E.D.; Stavrakas, I.; Triantis, D. Response of fiber reinforced concrete while entering into the critical stage: An attempt to detect pre-failure indicators in terms of Non-Extensive Statistical Mechanics. *Theor. Appl. Fract. Mech.* **2022**, *123*, 103690. [CrossRef]
13. Chmel, A.; Shcherbakov, I. A comparative acoustic emission study of compression and impact fracture in granite. *Int. J. Rock Mech. Min. Sci.* **2013**, *64*, 56–59. [CrossRef]
14. Ohno, K.; Ohtsu, M. Crack classification in concrete based on acoustic emission. *Constr. Build. Mater.* **2010**, *24*, 2339–2346. [CrossRef]
15. Aggelis, D.G. Classification of cracking mode in concrete by acoustic emission parameters. *Mech. Res. Commun.* **2011**, *38*, 153–157. [CrossRef]
16. Li, D.; Wang, E.; Ju, Y.; Wang, D. Laboratory investigations of a new method using pressure stimulated currents to monitor concentrated stress variations in coal. *Nat. Resour. Res.* **2021**, *30*, 707–724. [CrossRef]
17. Triantis, D.; Pasiou, E.D.; Stavrakas, I.; Kourkoulis, S.K. Hidden affinities between electric and acoustic activities in brittle materials at near-fracture load levels. *Rock Mech. Rock Eng.* **2022**, *55*, 1325–1342. [CrossRef]
18. Mao, W.; Wu, L.; Xu, Y.; Yao, R.; Lu, J.; Sun, L.; Qi, Y. Pressure-stimulated rock current as loading diorite to failure: Particular variation and holistic mechanisms. *J. Geophys. Res. Solid Earth* **2022**, *127*, e2022JB024931. [CrossRef]
19. Carpinteri, A.; Lacidogna, G.; Manuello, A.; Niccolini, G.; Schiavi, A.; Agosto, A. Mechanical and electromagnetic emissions related to stress-induced cracks. *Exp. Tech.* **2012**, *36*, 53–64. [CrossRef]
20. Wei, M.; Song, D.; He, X.; Li, Z.; Qiu, L.; Lou, Q. Effect of rock properties on electromagnetic radiation characteristics generated by rock fracture during uniaxial compression. *Rock Mech. Rock Eng.* **2020**, *53*, 5223–5238. [CrossRef]
21. Qiu, L.; Zhu, Y.; Song, D.; He, X.; Wang, W.; Liu, Y.; Xiao, Y.; Wei, M.; Yin, S.; Liu, Q. Study on the nonlinear characteristics of emr and ae during coal splitting tests. *Minerals* **2022**, *12*, 108. [CrossRef]
22. Niccolini, G.; Potirakis, S.M.; Lacidogna, G.; Borla, O. Criticality hidden in acoustic emissions and in changing electrical resistance during fracture of rocks and cement-based materials. *Materials* **2020**, *13*, 5608. [CrossRef] [PubMed]
23. Loukidis, A.; Stavrakas, I.; Triantis, D. Electrical methods for sensing damage in cement mortar beams combined with acoustic emissions. *Materials* **2022**, *15*, 4682. [CrossRef] [PubMed]
24. Song, M.; Hu, Q.; Liu, H.; Li, Q.; Zhang, Y.; Hu, Z.; Liu, J.; Deng, Y.; Zheng, X.; Wang, M. Characterization and correlation of rock fracture-induced electrical resistance and acoustic emission. *Rock Mech. Rock Eng.* **2023**, 1–21. [CrossRef]
25. Chen, D.; Li, N.; Wang, E.-y. Temporal and spatial evolution of acoustic emission and waveform characteristics of specimens with different lithology. *J. Geophys. Eng.* **2018**, *15*, 1878–1888.
26. Mogi, K. *Experimental Rock Mechanics*; CRC Press: Boca Raton, FL, USA, 2006; Volume 3.
27. Hampton, J.; Gutierrez, M.; Matzar, L.; Hu, D.; Frash, L. Acoustic emission characterization of microcracking in laboratory-scale hydraulic fracturing tests. *J. Rock Mech. Geotech. Eng.* **2018**, *10*, 805–817. [CrossRef]
28. Wei, H.; Hu, B.; Wang, F.; Zheng, J.; Jin, J.; Liu, C. Temporal-spatial evolution characteristics of acoustic emission in asphalt concrete cracking process under low temperature. *Constr. Build. Mater.* **2020**, *248*, 118632. [CrossRef]
29. Gao, A.; Qi, C.; Shan, R.; Wang, C. Experimental study on the spatial–temporal failure characteristics of red sandstone with a cemented structural surface under compression. *ACS Omega* **2022**, *7*, 20250–20258. [CrossRef]

30. Triantis, D.; Stavrakas, I.; Loukidis, A.; Pasiou, E.D.; Kourkoulis, S.K. Exploring the acoustic activity in brittle materials in terms of the position of the acoustic sources and the power of the acoustic signals—Part I: Founding the approach. *Forces Mech.* **2022**, *7*, 100088. [CrossRef]
31. Lei, X.-L.; Kusunose, K.; Nishizawa, O.; Cho, A.; Satoh, T. On the spatio-temporal distribution of acoustic emissions in two granitic rocks under triaxial compression: The role of pre-existing cracks. *Geophys. Res. Lett.* **2000**, *27*, 1997–2000. [CrossRef]
32. Triantis, D.; Kourkoulis, S.K. An alternative approach for representing the data provided by the acoustic emission technique. *Rock Mech. Rock Eng.* **2018**, *51*, 2433–2438. [CrossRef]
33. RILEM Technical Committee. Recommendation of RILEM TC 212-ACD: Acoustic emission and related NDE techniques for crack detection and damage evaluation in concrete. *Mater. Des.* **2010**, *43*, 1187–1189.

materials

Article

The Effect of Microwave Radiation on the Self-Healing Performance of Asphalt Mixtures with Steel Slag Aggregates and Steel Fibers

Carlos D. A. Loureiro [1], Hugo M. R. D. Silva [1,*], Joel R. M. Oliveira [1], Nuno L. S. Costa [1] and Carlos A. O. Palha [2]

[1] ISISE—Institute for Sustainability and Innovation in Structural Engineering, Department of Civil Engineering, University of Minho, 4800-058 Guimaraes, Portugal; id9629@alunos.uminho.pt (C.D.A.L.); joliveira@civil.uminho.pt (J.R.M.O.); a79917@alunos.uminho.pt (N.L.S.C.)

[2] Department of Civil Engineering, University of Minho, 4800-058 Guimaraes, Portugal; cpalha@civil.uminho.pt

* Correspondence: hugo@civil.uminho.pt

Abstract: Self-healing in asphalt mixtures is a property that can be enhanced by external heating, which causes a thermal expansion that increases the flow of bitumen with reduced viscosity through the cracks. Therefore, this study aims to evaluate the effects of microwave heating on the self-healing performance of three asphalt mixtures: (1) conventional, (2) with steel wool fibers (SWF), and (3) with steel slag aggregates (SSA) and SWF. After evaluating the microwave heating capacity of the three asphalt mixtures with a thermographic camera, their self-healing performance was determined with fracture or fatigue tests and microwave heating recovery cycles. The results demonstrated that the mixtures with SSA and SWF promoted higher heating temperatures and presented the best self-healing capacity during the semicircular bending test and heating cycles, with significant strength recovery after a total fracture. In contrast, the mixtures without SSA presented inferior fracture results. Both the conventional mixture and that containing SSA and SWF presented high healing indexes after the four-point bending fatigue test and heating cycles, with a fatigue life recovery of around 150% after applying two healing cycles. Therefore, the conclusion is that SSA greatly influences the self-healing performance of asphalt mixtures after microwave radiation heating.

Keywords: self-healing; steel slag aggregates; steel wool fibers; asphalt mixtures; microwave radiation heating; self-healing assessment

Citation: Loureiro, C.D.A.; Silva, H.M.R.D.; Oliveira, J.R.M.; Costa, N.L.S.; Palha, C.A.O. The Effect of Microwave Radiation on the Self-Healing Performance of Asphalt Mixtures with Steel Slag Aggregates and Steel Fibers. *Materials* **2023**, *16*, 3712. https://doi.org/10.3390/ma16103712

Academic Editor: Ayman El-Zohairy

Received: 31 March 2023
Revised: 5 May 2023
Accepted: 8 May 2023
Published: 13 May 2023

1. Introduction

Industrial and technological development, combined with demands for a better quality of life, has increased the consumption of material resources, making it impossible for the planet to continue on this path. Thus, society has become aware that resources are finite and that reducing their consumption and use is imperative.

The construction industry is one of Europe's largest and most active economic sectors. However, it is also responsible for a significant environmental impact due to its high energy consumption, how it develops, uses, distributes, and disposes of virgin resources, and produces excessive greenhouse gas emissions [1,2].

More specifically, the road infrastructure sector damages the environment and the planet's sustainability, from transporting goods and people to constructing pavements. Thus, changing this sector's current design, management, and monitoring paradigm is essential. This sector must develop policies to achieve sustainable development objectives effectively, creating innovative processes aimed at reducing CO_2 emissions, reducing the use of virgin materials, and reusing waste in construction while ensuring positive effects from an economic, technical, and environmental point of view [3,4].

Based on this line of action, this paper aimed to study whether the incorporation of industrial by-products, steel slag aggregates, and steel fibers, steel wool fibers, promotes the enhancement of the self-healing of asphalt mixtures when these are subjected to heating by microwave radiation. This study is expected to contribute to a more sustainable paving area, promoting the increase of pavement lifetime and minimizing the consumption of virgin materials in pavement rehabilitation/conservation works.

Given the lack of previous studies on the self-healing performance of asphalt mixtures incorporating high amounts of by-products, this study will undoubtedly contribute to increasing the scientific community's knowledge of this research topic.

2. Literature Review

Sustainability and circular economy principles are increasingly present in the decision making of governmental agencies due to society's growing concern with the planet's environment, which will bring profound transformations to the road pavement industry. The environmental and economic costs associated with constructing new road infrastructures or the conservation/rehabilitation of existing pavements are considered excessive and inadequate. Consequently, there is an urgent need to develop innovative paving solutions to change the current consumption paradigm. From this perspective, this article aims to assess solutions that simultaneously incorporate industrial by-products (steel slag) in the production of asphalt mixtures and promote an increase in the pavement's service life through self-healing.

Steel slag aggregates (SSA) originate from the crude steel production process's by-product, and these could be used in several industrial sectors: road infrastructures, cement, concrete, and additives [5]. Correia et al. [6] reported that the steel industry produces about 11 to 15% black slag and 2 to 3% white slag (for this slag, its use in construction is discarded). According to the World Steel Association [7], in 2020, crude steel production was about 1900 million tons, so it was possible to extract about 200–285 million tons of steel slag. Therefore, given the current environmental concerns, using steel slag to produce aggregates for pavement construction is an excellent contribution to the planet's sustainability.

In a review paper [8], the authors concluded that several properties of steel slag aggregates (e.g., roughness, shape, angularity, hardness, polishing resistance, and wear resistance) make them suitable for use in asphalt mixtures. In brief, the SSAs have a low presence of fines, a high sand equivalent, a good shape with high angularity, rough texture, a high density (3.2–3.8 g/cm^3), and high porosity and water absorption [9]. In another study, the authors revealed that the SSAs are a by-product with great potential to be used as aggregates in asphalt mixtures due to their excellent wearing and polishing resistance and good affinity with bitumen. Moreover, based on the results obtained, the authors concluded that the mixtures showed excellent performance evaluated through water sensitivity and rutting resistance tests, surpassing the results obtained for conventional asphalt mixtures with identical compositions produced with natural aggregates [10].

Even though asphalt pavements have many advantages (cost efficiency, noise reduction, comfortable driving experience), asphalt concrete undergoes functional and structural degradation after a few years of service, mainly due to traffic and weather conditions. As the asphalt binder ages, it becomes more brittle, leading initially to micro-cracks and then to macro-cracks, which lead to pavement performance failure, both from a structural and functional point of view [11]. So, one of the solutions to increase the durability of infrastructures is to use high-quality materials with self-healing capabilities.

In recent years, self-healing has expanded from chemistry to engineering. This phenomenon has multiple and unique characteristics that make it a promising approach to increasing the durability of asphalt pavements by regenerating micro-cracks and recovering the initial performance without reconstructing the pavement [12].

This behavior is only possible since bitumen is a viscoelastic fluid with a relatively long surface wetting and diffusion capability, exhibiting the behavior of a self-healing material [13]. Therefore, bitumen can close micro-cracks and restore the asphalt's mixture strength and resistance. However, the mechanism that causes self-healing is not yet fully understood. There is still neither a single definition nor a standardized method for predicting the self-healing capacity of asphalt materials. In these materials, some internal factors influence the self-healing performance: molecular composition, structure, asphalt movement and diffusion, asphalt viscoelasticity, and thixotropy. Externally, factors such as humidity, temperature, induction conditions, and induction duration are also influential in the performance of this phenomenon [14]. Thus, to better understand this behavior, this article aims to study the influence of microwave radiation heating to enhance self-healing in sustainable asphalt mixtures.

Electromagnetic induction and microwave radiation are the processes most commonly used to promote heating asphalt mixtures to enhance their self-healing effectiveness. The induction heating method combines electromagnetic induction and Joule heating effects. Faraday's law states that an electromotive force is induced in a conductive material when an alternating magnetic field is present [15]. Therefore, electrically conductive asphalt materials can be developed based on that principle, but the key challenge lies in effectively improving their conductivity without reducing their mechanical performance on the pavement. There are different additives to incorporate into the mix design of asphalt mixtures. However, introducing metallic materials (e.g., steel fiber, magnetite, carbonyl iron powder, and steel slag) is the best option. As a result, heating by electromagnetic induction is impossible without the presence of a conductive material [15].

Meanwhile, the working principle of microwave radiation, called dielectric heating, refers to heating a dielectric medium by the electric field component of high-frequency radiation [15]. Interestingly, microwave radiation heating does not rely on conductive materials and works directly on the dielectric material. Thus, this technique can heat conventional asphalt mixtures without any conductive materials, although using conductive additives can accelerate heating [15].

Previous studies have focused on incorporating conductible materials into asphalt mixtures to speed up the heating of mixtures subjected to electromagnetic induction or microwave radiation. The heating efficiency depends on many factors, including the conductivity and other characteristics (e.g., type, diameter, length, and percentage) of the additives/fibers added to the mixture, their spatial distribution, and the type of aggregates and fillers used in the mixtures. Theoretically, it is feasible to repair cracks in asphalt concrete by adding conducting fibers, improving its conductivity, and heating the mixture with induction energy to increase its self-healing rate [16]. García et al. [17] defended that the amount of heat generated in induction heating depends mainly on the radius of the fibers rather than on their volume in the mixture. According to Norambuena-Contreras and Garcia [18], crack healing of asphalt mixtures by microwave heating can be achieved by adding metallic fibers to the asphalt mixture composition. These fibers can increase the heating rates of asphalt mixtures since they can absorb and conduct more thermal energy than the other components of the mixture (i.e., aggregates and bitumen). The authors concluded that a higher fiber content contributes to fiber clusters and mix oxidation.

Incorporating SSA in asphalt mixtures, besides reducing the consumption of natural aggregates, also contributes to enhancing the heating of asphalt mixtures through induction heating systems due to their chemical composition. Steel slag is a desirable low-cost absorber to trigger microwave heating with excellent magnetic properties [19]. In another study, the authors concluded that replacing 30% natural aggregates with steel slag aggregates provides better healing results and helps the mixture improve the load–displacement relationship with higher ductile behavior [20]. Wan et al. [21] studied how the composition and size of steel slag aggregates affect the asphalt mixtures' induction heating efficiency. This paper investigated the steel slag's heating efficiency based on its components, size, and iron element content. The authors concluded that the steel slags could be heated by induction and that iron and Fe_3O_4 were the active components. Besides, steel slag aggregates with higher iron content generally had higher heating efficiency. Li et al. [22] studied the feasibility of replacing limestone filler with steel slag filler to enhance the heat release and self-healing properties of asphalt mastics. The authors concluded that the steel slag filler-based asphalt mastics showed better healing rates than the limestone filler ones because the steel slag filler could convert the energy of microwave irradiation into more thermal energy.

Evaluating self-healing effectiveness is one of the most critical tasks in this type of research. Several specific tools, methods, and tests have been applied to quantify the success of self-healing in different asphalt materials (i.e., bitumen, mastics, and asphalt mixtures), which depend on the type of material studied. In particular, the semicircular bending (SCB) fracture test and the four-point bending fatigue test are the principal tests used for measuring the self-healing efficiency of asphalt mixtures [18,23,24].

In this type of scientific study, there are naturally some limitations. The main limitation is its applicability in a practical context, as microwave radiation heating is distributed more or less uniformly over the whole surface area of the specimen. In contrast, the heating of actual road pavements can only be done on the pavement's surface course. Thus, comparative studies should be made with other heating methods, e.g., electromagnetic induction, to see which is the most efficient in a laboratory and practical context. Furthermore, in the future, an environmental and economic study of the life cycle of an asphalt mixture with this heating method should be carried out to validate its feasibility.

3. Materials and Methods

3.1. Materials

The materials used in this study were a 35/50 pen-grade bitumen and those presented in Figure 1, which include steel slag aggregates (SSA), natural granitic aggregates (NGA), limestone filler (LF), and steel wool fibers (SWF).

(a) (b) (c) (d)

Figure 1. Different materials used in this study: (**a**) steel slag aggregates, (**b**) natural granite aggregates, (**c**) limestone filler, and (**d**) steel wool fibers.

3.2. Aggregates and Bitumen Characterization

This study used natural granite and steel slag aggregates formerly characterized at the University of Minho, as presented in a previous study [10]. In that study, several tests indicated the aggregates' properties, including the particle density and water absorption, the wear resistance through the micro-Deval coefficient, the resistance of coarse aggregates to polishing, and the binder–aggregate affinity. The authors concluded that SSA has excellent potential for incorporation into asphalt mixtures due to its excellent wear and polishing resistance and good bitumen affinity.

The particle size distribution of the different aggregate fractions for the SSA (0/10 and 10/14) and the natural aggregate fractions (0/4, 4/6, and 6/14) was determined using the EN 933-1 standard. Those results were essential to designing asphalt mixtures that comply with the grading envelope of the Portuguese specifications [25].

The bitumen used in this study was characterized by the penetration test at 25 °C, according to EN 1426, and the softening point test using the ring and ball method, according to EN 1427.

3.3. Definition of the Percentage of Steel Wool Fibers

The percentage and geometry of the fibers selected in this work to incorporate in the asphalt mixtures for self-healing assessment were based on previous studies.

Liu et al.'s [26] study used a fiber percentage of 6% of the bitumen volume with positive results since the asphalt mixtures showed higher strength values than those without fibers. In the same paper, the authors mentioned the excellent resistance to particle loss and acceptable induction speed of the mixtures with fibers. However, the fiber length was only 4.2 mm, which was not ideal for a faster temperature rise in asphalt mixtures.

Tabaković et al. [27] used three different fiber percentages for each asphalt mixture (5, 10, and 15% of the bitumen volume) with a fiber length of 10 mm. The asphalt mixture with 5% fibers had several advantages, showing the best self-healing results. Moreover, the good results were visible to the naked eye since the crack closed entirely after healing, and the asphalt mixture did not present additional problems, such as, for example, the formation of crusts or oxidation, observed in mixtures with a higher percentage of fibers.

Gallego et al. [28] produced several asphalt mixtures with different percentages of fibers between 0.2% and 1.8% of the mixture's total mass, using fibers with a variable length (5 mm or 10 mm) and diameter (0.04–0.06 mm and 0.10–0.12 mm) in each mixture. That paper demonstrated that the energy efficiency of the 10 mm-long fibers is higher than that of the 5 mm-long fibers, and the fibers with 0.04–0.06 mm thicknesses had the best healing performance. Finally, the authors confirmed that only 0.2% of the fibers over the mixture's total mass was sufficient to obtain the expected performance for 10 mm-long fibers.

Therefore, based on the mentioned articles, this work selected the following conditions to produce mixtures B and C with steel wood fibers: a percentage of fibers equal to 7% of bitumen mass or 0.35% of the asphalt mixture total mass, with lengths of nearly 10 mm and a diameter between 0.04 and 0.06 mm. The fibers were cut from steel wool into pieces with the chosen lengths and then separated manually to obtain the necessary amount of fiber.

3.4. Mix Design, Production, and Characterization of Asphalt Mixtures

The following asphalt mixtures (AC 14 surf 35/50) were selected for this study to evaluate the composition influence on their mechanical and self-healing performance:

- Mixture A (MA)—a conventional mixture with natural aggregates.
- Mixture B (MB)—a new mixture with natural aggregates and steel wool fibers.
- Mixture C (MC)—a new mixture with steel slag aggregates, natural aggregates, and steel wool fibers.

The asphalt mixtures design was based on the various aggregate fractions' density and particle size distribution, adjusting the final grading curves within the AC 14 surf grading envelope stated in the Portuguese standards [25]. Mixture C incorporated 50% steel slag aggregates, while mixtures A and B were produced with natural aggregates.

In this study, a 5.0% bitumen content was defined for all mixtures to reduce the number of independent variables under analysis. The bitumen content was defined based on the optimum content determined in a previous study through the Marshall mix design method [10]. Steel wool fibers (SWF) were added to the bitumen and aggregate blend for mixtures B and C. The mass of steel wool fibers added was the same for both mixtures, i.e., 7% of the bitumen mass of each mixture. The mixtures' production occurred at a mixing temperature of 165 °C, specified in EN 12697-35 standard, after heating the aggregates at 170 °C and the bitumen at 165 °C. The materials of each mixture were mixed roughly for two minutes until the aggregates were fully covered by bitumen. In mixtures B and C, the steel wool fibers were gradually placed into the mixer bowl during the total mixing time to distribute the fibers better.

Two pieces of equipment were used for asphalt mixture compaction. First, three rectangular slabs of each mixture were compacted at 150 °C with the roller compactor equipment, according to EN 12697-33. This compaction uses a segmented roller with alternated operated rotation, simulating a street roller's on-site action. The slabs were later cut into twelve beams ($380 \times 51 \times 64$ mm^3). Then, nine cylindrical specimens of each mixture (diameter of 101 mm and a height of nearly 63 mm) were compacted at around 150 °C, applying 75 blows per side using a Marshall impact hammer, according to EN 12697-30.

The cylindrical specimens were initially used to study their microwave induction heating and, after being cut diametrically, to study the fracture toughness in a semicircular bending (SCB) test. The beams were tested for induction heating and were essential to determining the fatigue resistance and stiffness modulus in a four-point bending test.

After compaction, the samples' bulk density (BD) was determined according to method B of EN 12697-6, and the mixture's maximum density (MD) was assessed according to method A of EN 12697-5. Both values were used to calculate the air voids content of each asphalt mixture, which must be controlled within the limits specified in the Portuguese specifications [25]. According to those specifications, AC 14 surface mixtures should have an air voids content between 3% and 5%. Furthermore, it is stated that the air voids content must be calculated using EN 12697-8 and EN 12698-6 standards.

3.5. *Preliminary Microwave Induction Heating Tests of Asphalt Mixtures*

Three cylindrical specimens of each mixture were used in a preliminary study to evaluate the thermal induction behavior of the different asphalt mixtures as a function of the microwave power and heating time. Several scientific studies analyzed different procedures for heating asphalt mixtures with microwave radiation. The ideal self-healing induction time varies depending on the equipment, specimen geometry, asphalt mixture type, and induction power [26–28].

Therefore, a preliminary analysis was performed in this work to assess the temperature evolution of the specimens when induced at different microwave power levels of 540 W, 720 W, and 900 W, with temperature measurements with a thermographic camera every 40 s for a total heating time of 160 s. Figure 2 shows the main steps for measuring the surface temperature of each test specimen. Before being tested, the specimens were placed in a thermal chamber at 20 °C to homogenize the initial test conditions. In addition, the results of the thermographic camera were previously calibrated in the laboratory.

(a) (b) (c)

Figure 2. Procedure for determining the surface temperature of specimens: (**a**) thermographic camera, (**b**) asphalt specimen in the microwave oven, and (**c**) thermal image of the specimen.

The microwave radiation distribution was uniform over the surface of the cylindrical specimen since a rotating plate was used during the microwave heating. On the other hand, the microwave rotating plate could not be used when heating the fatigue test beams (Section 3.6.3) due to their dimensions. Thus, radiation may have been unevenly distributed over the whole surface of the beams.

3.6. *Self-Healing Assessment and Mechanical Characterization of the Asphalt Mixtures*

Mechanical characterization is essential to analyzing the performance of any asphalt mixture, but it gained additional importance in this work. Some properties of the studied asphalt mixtures (e.g., fracture strength and fatigue life) were assessed at different damage stages to calculate the healing ratio derived from a microwave thermal induction process. Thus, the self-healing performance of the studied asphalt mixtures was assessed by a cyclic set of mechanical tests (i.e., semicircular bending, fatigue, and stiffness modulus tests) followed by microwave radiation heating periods applied to heal the damaged specimens, as detailed in the following sections.

3.6.1. Semicircular Bending Test

The asphalt mixtures were initially characterized through a three-point bending test performed on semicircular specimens, according to EN 12697-44 standard. This test evaluated the fracture toughness of the three asphalt mixtures, measuring the maximum fracture load and the corresponding displacement for the twelve specimens tested.

The test was performed in a Matest Unitronic universal loading structure in a bending test configuration with two support points spaced out 8 cm apart and a load application point at the specimen's midpoint, as presented in Figure 3. The load application speed was 0.5 mm/min, and the test temperature was −20 °C. This low temperature ensures a brittle fracture of the specimens, creating well-defined cracks that are easier to control during the healing phase. Before the fracture test, a 1 cm-deep notch was made at the semicircular specimens' half span to regulate the crack's start and progression.

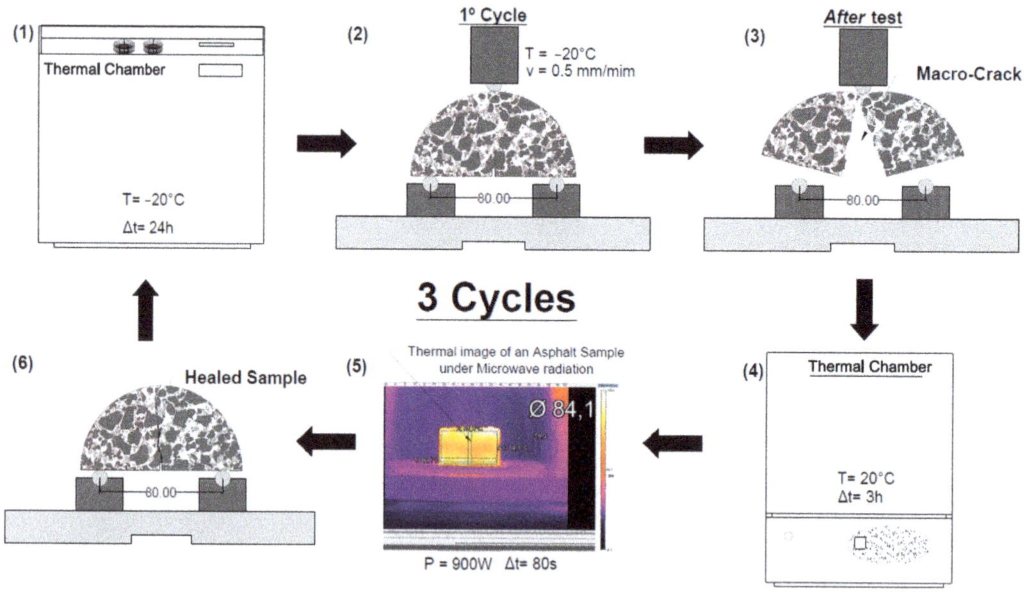

Figure 3. Cyclic setup of semicircular tests and microwave heating processes applied to assess the healing ratio.

The procedure adopted to calculate the healing capacity of asphalt mixtures through cycles of semicircular bending tests (fracture stage) and microwave radiation heating (healing stage) is summarized in the following steps and Figure 3.

1. The semicircular specimens are conditioned in a thermal chamber at −20 °C for 24 h.
2. The semicircular bending (SCB) test is performed at a speed of 0.5 mm/min.
3. The damaged specimens are placed in a ventilated thermal chamber at 20 °C for three hours to evaporate some moisture in the crack.
4. Microwave radiation heating is applied for 80 s using a microwave oven at a power of 900 W during the healing phase.

5. After induction, the specimens rest at room temperature for 2 h before repeating procedures 1 to 5 for three consecutive cycles.

The main objective of the fracture test was to determine the strength recovery (healing) capacity of the mixtures induced by heating after the complete rupture of the specimens. The average strength recovery was computed as the mixture healing ratio (HR), according to Equation (1). This parameter calculates the proportion between the peak load reached by the intact specimen before the healing cycles ($F_{initial}$) and the peak load achieved by the damaged sample after each induction heating cycle of the healing process (F_{final}).

$$HR = \frac{F_{final}}{F_{initial}} \quad (1)$$

Finally, after concluding the three healing cycles in this test configuration, the healing ratios of the three asphalt mixtures resulting from each healing stage were calculated to analyze the influence of each asphalt mixture composition on its healing capacity.

3.6.2. Stiffness Modulus Test

The stiffness modulus is a fundamental characteristic of asphalt mixtures used as input in road pavement design and is essential to assessing the structural behavior of infrastructures with asphalt layers. This property depends on several factors, such as (1) temperature, (2) loading frequency, and (3) material characteristics.

The analysis of this property is complex from the point of view of the structural performance of a pavement. For example, increasing the stiffness increases the structure's bearing capacity by reducing the resulting strain levels in the pavement layers. Nevertheless, asphalt mixtures with increased stiffness can also reduce the flexibility of the pavement, reducing fatigue cracking resistance.

Three beams of each mixture were tested in a four-point sinusoidal bending test configuration with two support points at the ends and two load application points in the middle span of the beam to evaluate the stiffness modulus of the studied asphalt mixtures. The test was performed under controlled strain at 20 °C according to the EN 13108-20 standard. In addition, a low peak-to-peak strain level (1×10^{-4}) was set to assess the stiffness modulus in the linear region of the asphalt mixtures' viscoelastic behavior. In this test, the amplitude of the sinusoidal strain wave was plus-minus 50 microstrains. A frequency sweep test (0.1, 0.2, 0.5, 1, 2, 5, 8, and 10 Hz) was carried out according to EN 12697-26 standard procedures to determine the stiffness modulus of asphalt mixtures.

Those three beams were tested twice again after being exposed to two healing cycles to determine the change in the stiffness modulus caused by induction heating with a microwave (210 s at a power level of 900 W).

3.6.3. Fatigue Cracking Test

The fatigue cracking test evaluates the behavior of asphalt mixtures concerning fatigue cracking caused by repetitive loads of heavy traffic under specific weather conditions. This test is performed following the EN 12697-24 standard, applying a sinusoidal cyclic load at a frequency of 10 Hz in a four-point bending configuration (similar to the stiffness modulus test) at 20 °C, as suggested by the EN 13108-1 standard. Unlike the stiffness modulus test, the fatigue test is destructive since it applies high strain levels (4×10^{-4}) that cause progressive fatigue cracking damage to the nine asphalt mixture beams tested.

The fatigue cracking test aims to determine the fatigue life of the mixtures, i.e., the number of cycles required for the asphalt mixture to reach fatigue failure. The number of load cycles at which the stiffness modulus reduces to half its initial value ($N_{0.5}$) is usually the fatigue failure criterion used to define the fatigue life of the beams for each strain level when fatigue tests are performed under strain control. However, induced healing of road pavements should be used as a maintenance technique and is not expected to be only applied after total fatigue cracking failure. Thus, other approaches were explored in the literature to define a less aggressive fatigue-stopping criterion, allowing an improved healing performance of the asphalt mixtures during their thermal induction in the microwave.

During the literature review, several papers advocated that the beams should not be tested until total fatigue cracking failure to obtain a better healing ratio. Garcia et al. [29] showed that the optimum moment to induce self-healing of asphalt mixtures occurs at 35% of the number of cycles of total fatigue failure ($0.35 \times N_{0.5}$). Liu et al. [30] presented positive self-healing results using a more practical approach to define the fatigue failure criterion before applying healing cycles. The authors set a new fatigue stop criterion as the number of load cycles causing a reduction of 35% in the initial stiffness modulus ($N_{0.35}$).

In this work, the fatigue tests considered the same fatigue failure criterion ($N_{0.35}$) before applying each healing stage, stopping the fatigue test when the initial stiffness modulus was reduced by 35%. During the healing stage, the beams were put in the microwave oven (Figure 4) inside a wooden mold to prevent permanent deformation at high temperatures. The asphalt mixtures were then heated for 210 s at a power level of 900 W since these were the optimum conditions observed for beams. Finally, the beam's temperature distribution was measured using a thermographic camera.

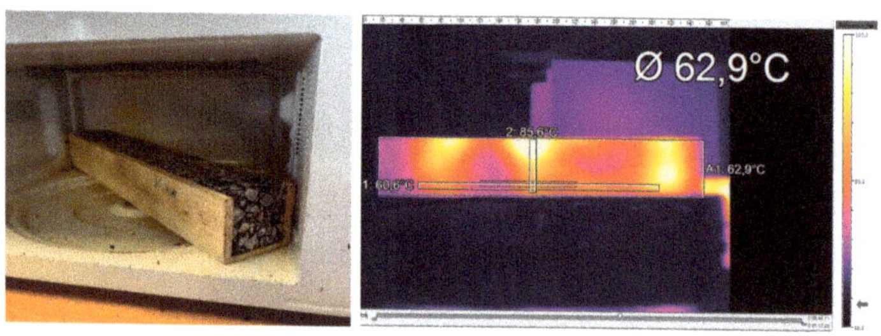

(a) (b)

Figure 4. Procedure for determining the surface temperature of specimens: (**a**) asphalt beam inside a wooden mold in the microwave oven and (**b**) thermal image of the beam.

After the first healing stage, the beams were tested again to evaluate asphalt mixtures' fatigue life, using the same fatigue failure criterion ($N_{0.35}$), before applying a second healing cycle. Finally, a third fatigue test was carried out on the healed beams to evaluate the fatigue life for two fatigue failure criteria: (1) $N_{0.35}$ to calculate the healing index after fatigue and (2) $N_{0.5}$ to assess the total fatigue failure of the mixtures. The cyclic setup of the fatigue tests and microwave heating stages is shown in Figure 5.

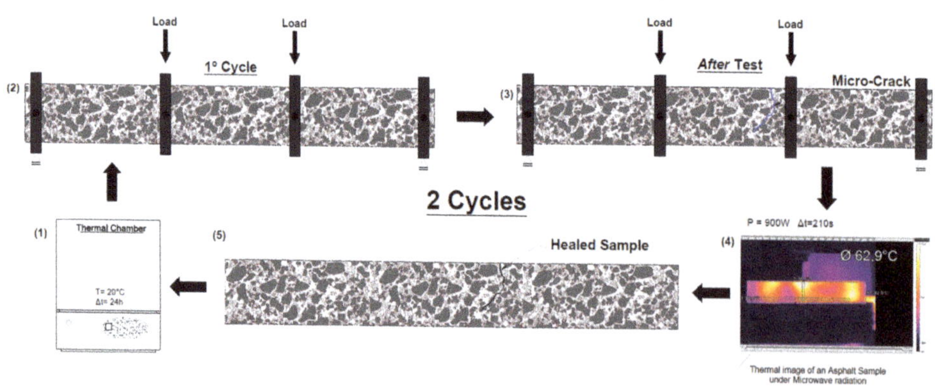

Figure 5. Cyclic setup of fatigue tests and microwave heating applied to assess the healing index.

This cyclic setup evaluated the pavement's service life gain after healing by comparing the fatigue life (number of cycles corresponding to $N_{0.35}$) of beams subjected to fatigue tests before and after microwave radiation heating. The regeneration or healing index (HI) was calculated to determine the gain in a pavement's service life [31].

After carrying out the consecutive fatigue tests, the HI value was computed as the ratio of the differential fatigue life after healing ($N_f - N_x$) to the fatigue life in the first test (N_x), as shown in Equation (2), for a given fatigue failure criterion (35% reduction of the initial stiffness). N_f is the fatigue life of the mixture after healing (no. of cycles), and N_x is the mixture fatigue life in the first fatigue test (no. of cycles) before healing.

$$\mathrm{HI} = \frac{N_f - N_x}{N_x} \tag{2}$$

4. Results and Discussion

4.1. Aggregates and Bitumen Characterization

The particle size distributions of the NGA and SSA fractions were obtained before designing the studied asphalt mixtures. The grading curves showed that the SSA material has fewer fines than the NGA, which increased the use of limestone filler in mixture C. The density values determined by Moura et al. [10] for these materials (2.65 Mg/m^3 for NGA and LF; 3.25 Mg/m^3 for SSA) were also used to design the asphalt mixtures.

The 35/50 pen-grade bitumen used in this work is the most widely applied in Portugal. The bitumen presented a penetration value of 44.9 tenths of a millimeter and a softening point temperature of 51.1 °C, and both values are within limits established by the EN 12591 standard for this type of bitumen.

4.2. Mix Design, Production, and Characterization of Asphalt Mixtures

Table 1 presents the compositions of the three asphalt mixtures (A, B, and C) studied in this work by showing the percentage volume of each aggregate fraction used. The corresponding mass values used to produce the mixtures were determined based on the aggregates' densities. The composition was defined based on the particle size distribution of the different aggregates to obtain a final grading curve of each asphalt mixture fitting the grading envelope of an AC 14 surf defined in the Portuguese standards [25].

Table 1. Volume percentages of the aggregates' fractions used in the different asphalt mixtures.

Fraction	NGA 0/4	NGA 4/6	NGA 6/14	SSA 0/10	SSA 10/14	LF
Mixture A	40%	14%	42%	0%	0%	4%
Mixture B	40%	14%	42%	0%	0%	4%
Mixture C	27%	0%	17%	36%	14%	6%

The asphalt binder content of mixtures A and B was 5.0%, using an equivalent value in mixture C, adjusted to 4.5%, to consider the higher density of the SSA (3.29 Mg/m^3) in comparison to the NGA (2.64 Mg/m^3). The percentage of steel wool fibers added to mixtures B and C was 7% of bitumen weight. The fibers were included as an additive and were not considered when adjusting the aggregate grading curve of those mixtures.

The volumetric characteristics of the asphalt mixtures were obtained by assessing their maximum density (MD) and bulk density (BD) of compacted specimens. Then, each specimen's air voids content (Vv) was computed from MD and BD values. Table 2 presents the mean volumetric characteristics for each asphalt mixture.

Table 2. Volumetric characterization of asphalt mixtures.

Property	Maximum Density (Mg/m^3)	Bulk Density (Mg/m^3)	Air Voids Content (%)
Mixture A	2.484	2.415	2.8
Mixture B	2.481	2.410	2.9
Mixture C	2.830	2.753	2.7

The three mixtures have similar air voids contents, slightly below the limit defined in the specifications. The high density and low air voids content of mixture C should be emphasized from these results. Good workability was observed for the mixture with 50% SSA during the laboratory production and compaction phases, which solved the high air voids content issue mentioned in Moura et al.'s [10] study for a similar asphalt mixture.

4.3. Preliminary Microwave Induction Heating Tests of Asphalt Mixtures

In the present work, the influence of the asphalt mixture composition was initially studied by assessing its thermal induction behavior when heated by microwave radiation. The temperature evolution of cylindrical specimens was evaluated every 40 s, for a total time of 160 s, with different microwave power levels (540 W, 720 W, and 900 W) to find differences in the heating rate between the studied mixtures. The tests were performed according to the procedures and techniques described previously. Figure 6 shows the mean surface temperature of three specimens of each asphalt mixture (A, B, and C) as a function of the heating time for the different microwave power levels used.

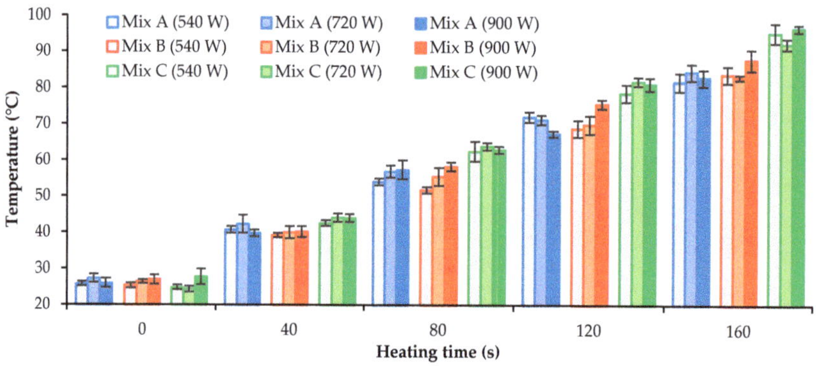

Figure 6. Evolution of surface temperature of the different asphalt mixtures with the heating time for three microwave power levels.

These results show that mixture B generally presented slightly higher temperatures than mixture A for identical induction times. Mixture C showed higher temperatures for all induction periods and power levels, mainly for longer induction times (80 s to 160 s). Thus, the combined use of SSA and SWF increases the heating rate in the microwave induction process. The temperatures measured with the thermographic camera had a very low deviation, assuring high confidence in the obtained results. The positive influence of incorporating SSA and SWF (mixture C) becomes evident in the thermal images of Figure 7, obtained for the same microwave heating conditions (160 s and 900 W).

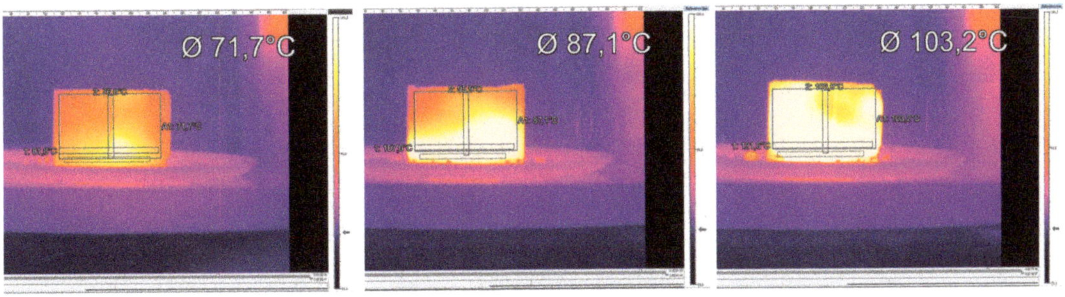

(**a**) (**b**) (**c**)

Figure 7. Thermal images of the different asphalt mixtures for the same radiation conditions (160 s and 900 W) in the microwave: (**a**) mixture A, (**b**) mixture B, and (**c**) mixture C.

The influence of microwave power levels on the heating results was not as substantial as expected, probably due to the equipment used. However, temperatures evolved more rapidly and uniformly for the 900 W power, particularly in mixtures B and C. Thus, that power level was selected for this study's remaining microwave heating tests. Long induction times (i.e., higher than 120 s) should be avoided when heating cylindrical specimens to reduce bitumen aging at high induction temperatures (near 100 °C) since bitumen becomes stiff and brittle, decreasing the fatigue and fracture resistance of the mixtures.

4.4. Self-Healing Assessment and Mechanical Characterization of the Asphalt Mixtures

This section mainly discusses the self-healing performance of the studied asphalt mixtures resulting from their fracture recovery or fatigue life increase after successive cycles of mechanical tests (damaging stage) and microwave heating (healing stage).

4.4.1. Cyclic Setup for Semicircular Bending Tests and Microwave Heating

In this study, a three-point semicircular bending test (Figure 8a) was performed to evaluate the fracture toughness of the different asphalt mixtures and their ability to recover (Figure 8c) their mechanical performance after total failure (Figure 8b). The macro-cracks observed in fracture tests may have inhibited the self-healing action compared to micro-cracks occurring in fatigue tests, but this work assumed that challenge.

(**a**) (**b**) (**c**)

Figure 8. Asphalt mixture specimen: (**a**) before the SCB test; (**b**) damaged after the fracture test; (**c**) after microwave healing.

The experimental procedure used thirty-six semicircular specimens (twelve of each mixture) that underwent several testing cycles, with four SCB fracture test repetitions, before and after three healing cycles of induction heating via microwave radiation (during 80 s at a power level of 900 W). An elastic band wrapped the specimens to facilitate the healing of the fractured faces of the semicircular samples during the induction period.

Table 3 presents the mean results of the four SCB fracture test repetitions performed over the three healing cycles for the studied asphalt mixtures.

Table 3. Data and mean results of the four SCB tests performed to evaluate the healing ratio.

Variable		**Mixture A**		**Mixture B**		**Mixture C**	
		Mean	**SD**	**Mean**	**SD**	**Mean**	**SD**
1st SCB test (intact, before healing)	F (kN) δ (mm)	9.8 1.0	0.6 0.3	10.2 1.2	1.2 0.2	9.7 1.3	2.1 0.2
2nd SCB test (after 1st healing cycle)	F (kN) δ (mm)	0.4 0.2	0.2 0.1	0.6 0.3	0.4 0.2	2.8 0.8	0.9 0.3
3rd SCB test (after 2nd healing cycle)	F (kN) δ (mm)	0.8 0.3	0.4 0.1	0.7 0.3	0.3 0.1	3.6 1.1	1.1 0.1
4th SCB test (after 3rd healing cycle)	F (kN) δ (mm)	0.9 0.3	0.5 0.1	0.8 0.3	0.6 0.2	3.1 1.0	0.6 0.4

After collecting and organizing the data (load and displacement) from the semicircular bending fracture tests, the most relevant results for each specimen, the maximum load and the peak displacement, were analyzed. The load vs. displacement graphs of the four SCB test repetitions were plotted for all the tested specimens. However, Figure 9 presents examples of only one characteristic sample of each asphalt mixture studied during this stage since showing all the results would be inviable.

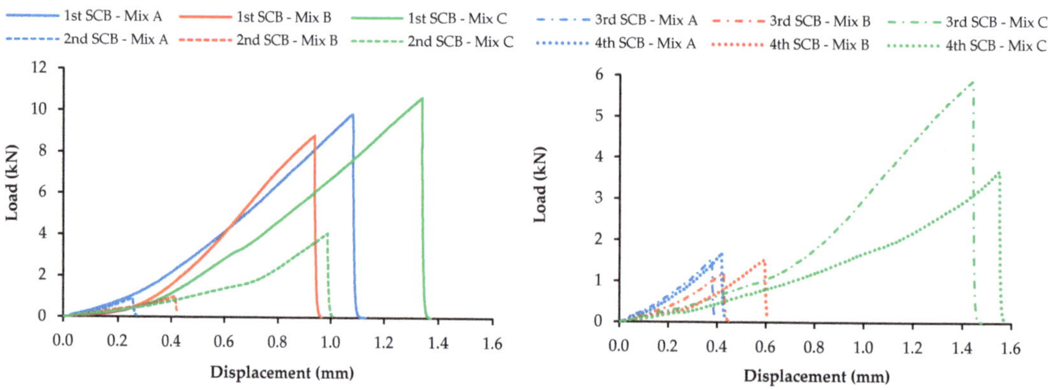

(a) (b)

Figure 9. Examples of load vs. displacement evolution of the different asphalt mixtures during the four fracture test repetitions: (**a**) 1st and 2nd SCB tests; (**b**) 3rd and 4th SCB tests.

The healing ratio (HR) value was calculated as the ratio of the peak load of the specimens after each healing cycle to the corresponding peak load of the intact asphalt mixture in the first SCB test. The HR values of the three asphalt mixtures after each healing process are presented in Figure 10. When analyzing the results, it should be noted that the semicircular specimens presented brittle behavior in the SCB fracture test, as expected, due to the low temperature (−20 °C) used.

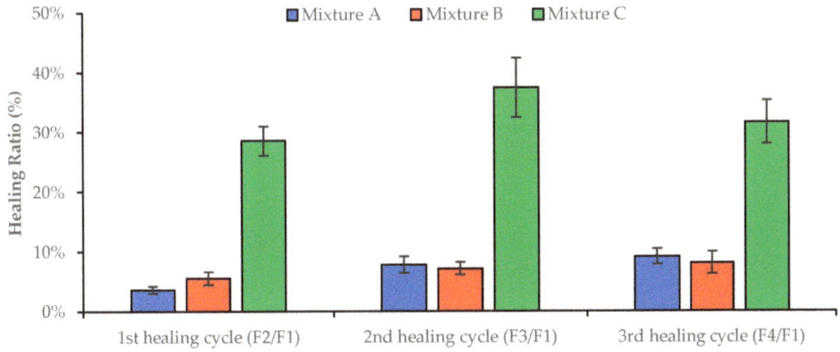

Figure 10. Healing ratio values of the semicircular test specimens after the three healing cycles.

Based on the data obtained, it is noticeable that the fracture damage caused by the SCB test recovered more significantly after healing in mixture C. This mixture recovered about 40% of its initial fracture toughness for increased peak displacements at rupture in all the SCB test repetitions performed after healing. Thus, multiple healing stages may be applied for road pavement maintenance. Nevertheless, the other asphalt mixtures (A and B) showed lower healing ratios (around 10%) after total damage.

Due to the destructive nature of SCB tests, the results of this test presented a higher deviation than other tests carried out in this work. A null hypothesis one-way ANOVA (alpha level of 0.05) compared the healing ratio of the three asphalt mixtures considering the three healing cycles, showing a statistically significant difference in the mean HR between at least two asphalt mixtures ($F(2, 33) = 33.077$ to 76.442, and $p = 0.000$). The Tukey's HSD test for multiple comparisons found that the HR was significantly different between mixtures A and C ($p = 0.000$, 95% C.I. = [16.02%, 33.33%] after the third healing cycle) and mixtures B and C ($p = 0.000$, 95% C.I. = [16.35%, 33.66%] after the third healing cycle). On the other hand, the mean healing ratios of mixtures A and B are similar ($p = 0.995$).

Finally, Figure 11 shows the mean temperatures of the three asphalt mixtures achieved during microwave heating (900 W power and 80 s heating time) of the semicircular specimens for each healing cycle between the SCB tests.

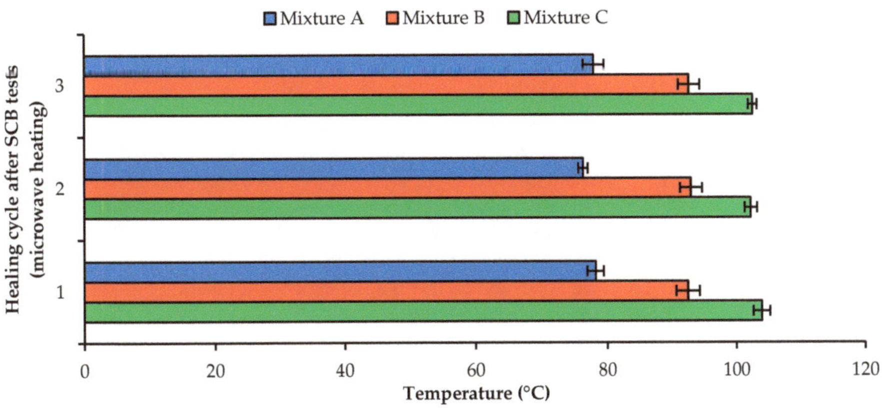

Figure 11. Mean temperatures of the mixtures induced by microwave heating in each healing cycle.

The results obtained at this stage align with those previously presented in the preliminary microwave heating tests. Once again, mixture C (combining SSA and SWF) showed a faster temperature growth for the same heating conditions, which may justify the best self-healing performance of this mixture after microwave heating. Moreover, the conductive materials (steel slag and steel fibers) used in mixtures B and C increased the microwave heating efficiency compared to conventional asphalt mixture A. The range of temperatures measured for different mixtures after 80 s of microwave radiation is very high (30 °C), demonstrating the importance of using conductive materials to improve heating. A similar one-way ANOVA assessment compared the microwave heating efficiency when heating the three asphalt mixtures considering the three healing cycles, clearly showing a statistically significant difference in the mean temperatures achieved by all asphalt mixtures ($F(2, 33) = 78.233$ to 123.876, and $p = 0.000$). Further research should be developed to define induction times according to the expected healing temperature because this factor may influence self-healing performance.

4.4.2. Cyclic Setup for Stiffness Modulus Tests and Microwave Heating

The stiffness modulus of the studied asphalt mixtures was measured based on the procedures previously presented. Figure 12 shows the evolution of the stiffness modulus and phase angle of mixtures A, B, and C with the frequency at 20 °C. Tests were made before healing (test 1) and after two healing cycles (tests 2 and 3) in the microwave.

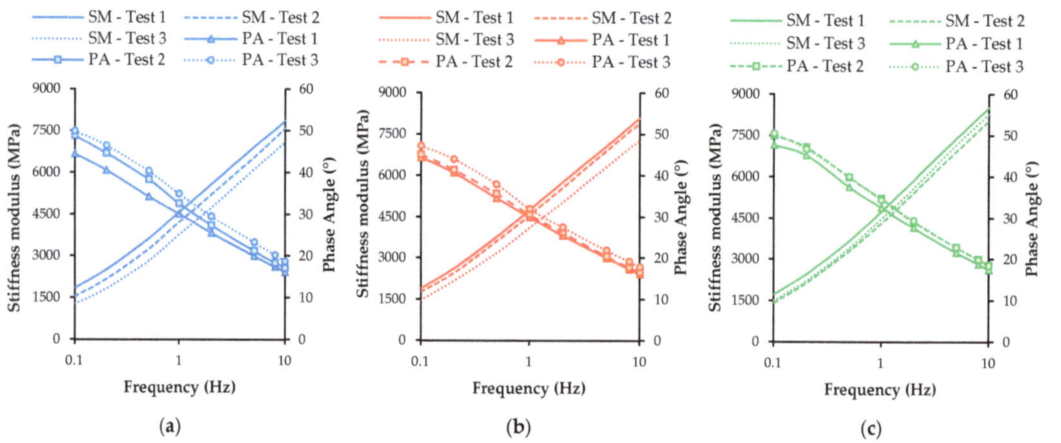

Figure 12. Evolution of the stiffness modulus (SM) and phase angle (PA) of the studied mixtures before healing (test 1) and after one (test 2) and two (test 3) healing cycles: (**a**) mixture A; (**b**) mixture B; (**c**) mixture C.

The results of these tests show the variation of the asphalt mixtures' stiffness modulus and phase angle due to the microwave heating process. The results demonstrated that the microwave heating applied to the asphalt mixtures during the healing stage did not significantly change the fundamental properties of the studied asphalt mixtures. Nevertheless, a slight reduction in the stiffness modulus can be observed for all asphalt mixtures, indicating that some structural damage from testing the asphalt mixtures was slightly more influential than the healing effect and bitumen aging (stiffening) during microwave heating.

Figure 13 presents the variation of the stiffness modulus (i.e., the mean value of individual results for three beams of each asphalt mixture) at standard test conditions of 8 Hz and 20 °C before and after two healing cycles. A null hypothesis one-way ANOVA (alpha level of 0.05) compared the stiffness modulus of the three asphalt mixtures considering the different healing cycles, revealing that there is no statistically significant difference in the mean stiffness modulus between the three asphalt mixtures ($F(2, 6) = 0.808$ to 3.177, and $p = 0.115$ to 0.489). The healing process's effect on the mixtures' mean stiffness modulus was only significantly different for mixture B between tests 1 and 3 ($p = 0.008$), presenting p values lower than the alfa level of 0.05 in all other situations.

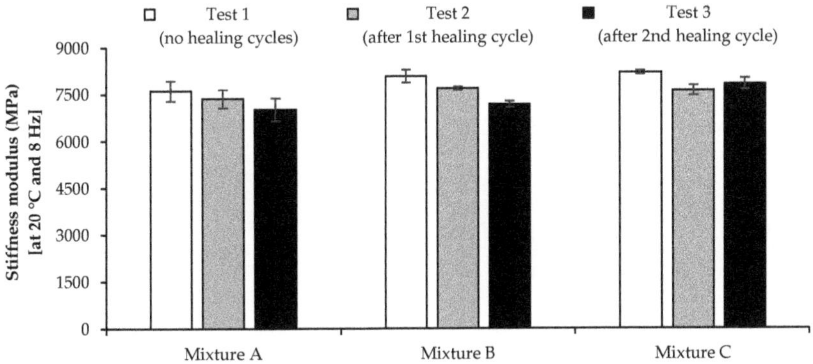

Figure 13. Stiffness modulus of the mixtures obtained in the test conditions mentioned in the standard (frequency of 8 Hz and temperature of 20 °C).

These results confirmed the low influence of microwave heating on the stiffness modulus of asphalt mixtures. However, it is possible to observe that mixture C slightly increased the stiffness modulus during the second healing cycle, probably due to the higher temperatures induced in mixture C by microwave heating, which may have improved the mixture healing and the binder–aggregate or binder–SSA bond while increasing bitumen aging.

4.4.3. Cyclic Setup for Fatigue Tests and Microwave Heating

A four-point sinusoidal bending test was performed on the studied asphalt mixtures to evaluate their fatigue cracking resistance before and after two healing cycles. The failure criterion in the fatigue test was set for a stiffness reduction of 35% from its initial value to optimize strength recovery during the healing stages. The corresponding number of loading cycles was recorded as the beam's fatigue life (N0.35). Then, after a healing stage with microwave heating (900 W for 210 s), the beams were tested again twice to evaluate the extended fatigue life ($N_{0.35}$) using the same failure criterion. The ratio between the extended fatigue life after healing (in the two fatigue test repetitions) and the intact beams' original fatigue life was defined as the mixtures' fatigue life gain (healing index).

For this cyclic setup of fatigue tests and microwave heating, nine intact beams of each asphalt mixture were tested for fatigue (1st fatigue test), stopping the test after a stiffness reduction of 35% ($N_{0.35}$). Then, the first stage of microwave heating was applied (1st healing cycle). The healed beams were tested again for fatigue (2nd fatigue test) using the same failure criterion ($N_{0.35}$) just before the second microwave healing cycle. Finally, the beams were tested again for fatigue (3rd fatigue test) until total fatigue failure of the mixtures for a stiffness reduction of 50% ($N_{0.5}$), but the fatigue life $N_{0.35}$ was also registered.

Table 4 presents the results from the fatigue tests performed in this study. The results show a significant stiffness recovery of the fatigue cracking damage after each fatigue test and microwave healing stage for all mixtures. Thus, the fatigue failure criterion and thermal radiation process used to induce healing were adequately selected.

Table 4. Average fatigue resistance of the studied mixtures before and after two healing cycles.

Mixture	Before Healing		After One Healing Cycle		After Two Healing Cycles		
	$E_{initial}$ (MPa)	$N_{0.35}$ (cycles)	$E_{initial}$ (MPa)	$N_{0.35}$ (cycles)	$E_{initial}$ (MPa)	$N_{0.35}$ (cycles)	$N_{0.5}$ (cycles)
A	7568	35,414	7757	28,423	7748	24,521	35,838
B	7370	23,971	7414	10,038	7392	5218	7439
C	7267	19,125	7294	15,563	7681	12,647	19,808

Moreover, the initial stiffness modulus ($E_{initial}$) in the third fatigue test (after two healing cycles) was higher than before healing for all mixtures, probably due to bitumen aging.

Figure 14 exemplifies the fatigue life gain of each asphalt mixture (A, B, and C) after two thermal healing cycles performed at a fatigue failure criterion of 35% reduction of the stiffness modulus. Figure 14d compares the behavior of the three asphalt mixtures.

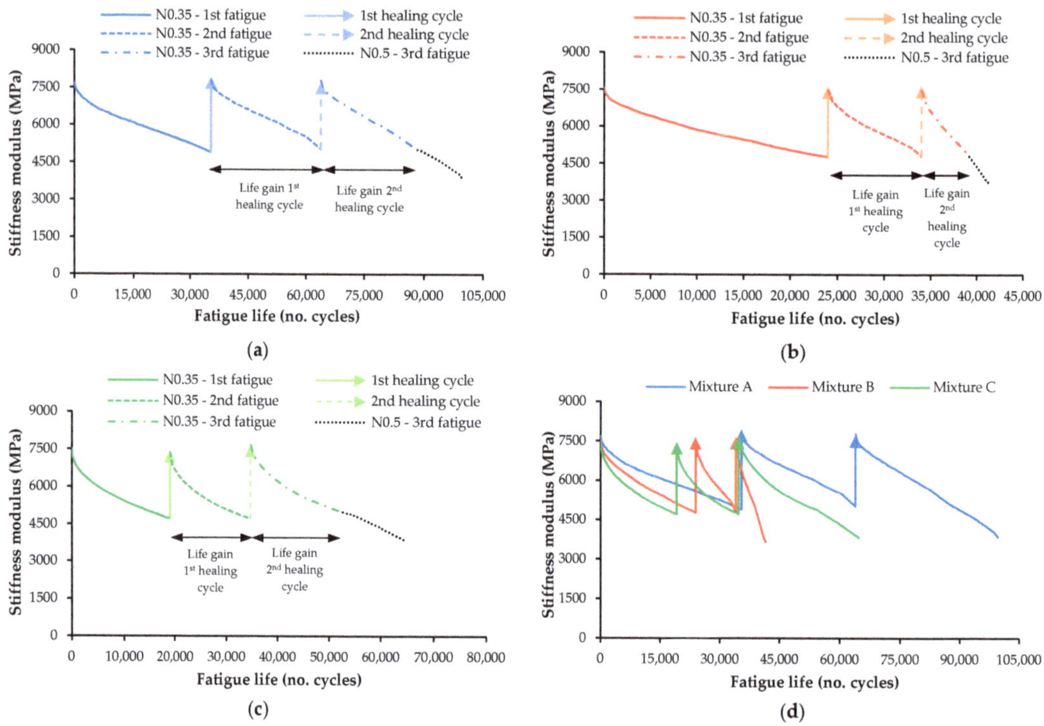

Figure 14. Evolution of the stiffness modulus during three fatigue tests performed before (intact beam) and after two healing cycles in the microwave: (**a**) mixture A; (**b**) mixture B; (**c**) mixture C; (**d**) comparison of all mixtures.

After applying the two microwave radiation healing cycles, the results show a noticeable fatigue life gain of asphalt mixtures, particularly for mixtures A and C. Therefore, a similar heating process may be applied for preventive maintenance of road pavements. Such an approach may improve the self-healing of asphalt mixtures and significantly extend their fatigue life and the durability of the pavement. Mixtures A and C increased their fatigue lives by far more than 100%, demonstrating the expected advantages of this solution. Therefore, by using these innovative solutions in new asphalt pavements, the environmental and economic costs could be reduced compared with the current expenses of traditional conservation works. Nevertheless, during the first fatigue cycle (before healing), the conventional mixture performed better than the alternative mixtures with SWF and SSA, demonstrating the importance of carefully designing conductive asphalt mixtures to increase their mechanical performance.

Based on the data acquired from the fatigue tests (fatigue life, mean number of cycles for fatigue failure at 35% reduction of the initial stiffness, or $N_{0.35}$) for the intact beams before healing (N_x) and the healed asphalt beams after each healing cycle (N_f), the healing index (HI) of the studied asphalt mixtures was computed with Equation (2). This study considered this HI value the most representative since it uses the same fatigue failure criterion ($0.35 \times E_{initial}$) before and after healing. Moreover, the data of the three fatigue test repetitions were obtained from the same beam for each asphalt mixture. Therefore, this index can verify the fatigue life gain (number of cycles) resulting from the self-healing properties of each asphalt mixture and enhanced by microwave heating radiation.

Figure 15 shows the HI values of the three asphalt mixtures after isolating the extended fatigue life gain of the first and second healing stages or adding both cycles.

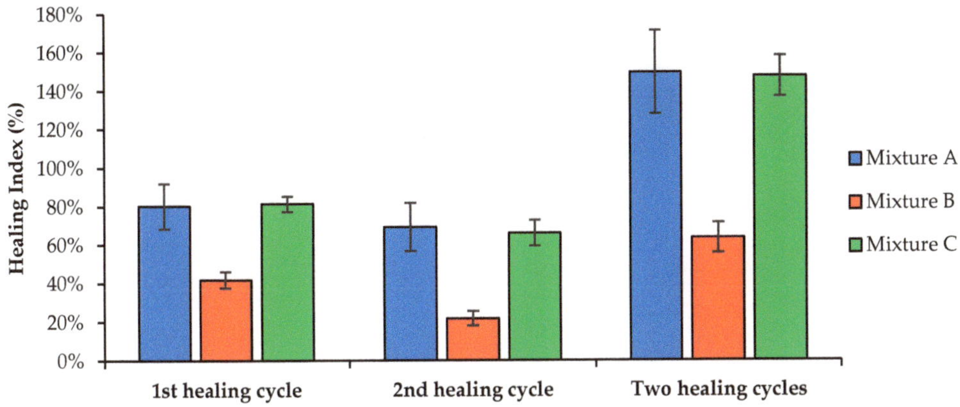

Figure 15. Healing index (HI) of each asphalt mixture in the fatigue test after two healing cycles.

The fatigue life gain is perceptible for all studied mixtures from the data presented. Mixture A showed excellent results, with an HI or fatigue life gain of 149% due to the self-healing phenomena enhanced by the microwave radiation heating system. Thus, microwave heating also boosts the regeneration capacity of conventional asphalt mixtures. Mixture C (incorporating SSA and SWF) also showed excellent self-healing results since the fatigue life increased by 147% after two microwave healing cycles. Mixture B showed the worst HI value in this test but raised the fatigue life by around 65%. The mean errors obtained for the HI results confirmed the reliability of the conclusions obtained from this test, even though Mixture A presented a slightly higher dispersion of results.

A null hypothesis one-way ANOVA (alpha level of 0.05) was performed to compare the type of asphalt mixtures' effect on the healing index after the different healing cycles, revealing that there was a statistically significant difference in the mean HI between at least two asphalt mixtures ($F(2, 6) = 5.350$, and $p = 0.010$ to 0.018) after healing. The Tukey's HSD test for multiple comparisons found that the healing index was significantly different between mixtures A and B ($p = 0.015$ to 0.030, 95% C.I. = [21.37%, 149.29%] after two healing cycles) and mixtures B and C ($p = 0.017$ to 0.028, 95% C.I. = [19.37%, 147.29%] after two healing cycles). There was no statistically significant difference between the mean healing indexes of mixtures A and C ($p = 0.960$ to 0.995).

The HI values decreased from the first to the second healing cycle in all mixtures, demonstrating some loss of efficiency when several healing repetitions are applied to recover fatigue cracking damage. Thus, further studies should be done to limit the number of healing cycles so that microwave radiation does not result in asphalt mixture performance problems. Figure 16 confirms that some beams experienced small deformations and crumbling in the load application zones of the fatigue test mechanism with the increasing number of fatigue tests and healing processes performed in this study.

(**a**) (**b**) (**c**) (**d**)

Figure 16. Sporadic degradation of beams after some cycles of fatigue tests and microwave heating. (**a**) Mixture A, (**b**) Mixture B, (**c**) Mixture B, (**d**) Mixture C.

5. Conclusions

This study mainly aimed to complement the current knowledge about asphalt mixtures self-healing using a microwave radiation heating method. New compositions for asphalt mixtures were studied using an industrial by-product, steel slag aggregates, and steel wool fibers.

The study's results revealed that the combined use of these conductive materials in asphalt mixtures could enhance their heating induced by microwave radiation and improve the self-healing capacity of asphalt mixtures for future maintenance of road pavements. The following points summarize the main conclusions of this study:

- The healing capacity of the asphalt mixtures can be assessed through cycles of mechanical damage tests (i.e., fracture or fatigue) and microwave radiation heating (healing stage), as the specimens always recovered some strength after damage.
- The fracture tests on semicircular specimens of asphalt mixtures exhibited some strength recovery after healing induced by microwave heating. However, only mixture C (with SSA and SWF) showed sufficiently high healing ratios, around 40%, while the other two mixtures showed lower healing ratios of 10%.
- The cyclic setup of fatigue tests and microwave heating confirmed a significant fatigue life gain after healing (nearly 150% for mixtures A and C and 65% for mixture B).
- The number of microwave heating cycles applied to recover fatigue cracking damage should be limited due to some loss of efficiency after several healing repetitions.
- Combining steel slag aggregates and wool fibers in new asphalt mixtures can enhance their microwave radiation heating and self-healing performance.

The results of this work suggest that a similar heating process should be replicated and applied for preventive maintenance of road pavements to improve the self-healing of asphalt mixtures and extend their fatigue life and durability of the pavement. Incorporating new materials (e.g., SSA and SWF) in the asphalt mixtures may enhance self-healing performance, with environmental and economic advantages. Nevertheless, additional work should be carried out on this topic to further improve the healing performance of asphalt mixtures, particularly for in situ applications.

Author Contributions: Conceptualization, C.D.A.L., C.A.O.P., H.M.R.D.S. and J.R.M.O.; methodology, C.D.A.L., C.A.O.P., H.M.R.D.S. and J.R.M.O.; validation, H.M.R.D.S. and J.R.M.O.; formal analysis, C.D.A.L., H.M.R.D.S. and J.R.M.O.; investigation, C.D.A.L., C.A.O.P., N.L.S.C., H.M.R.D.S. and J.R.M.O.; writing—original draft preparation, C.D.A.L., C.A.O.P., N.L.S.C., H.M.R.D.S. and J.R.M.O.; writing—review and editing, C.D.A.L., H.M.R.D.S. and J.R.M.O.; supervision, H.M.R.D.S. and J.R.M.O. All authors have read and agreed to the published version of the manuscript.

Funding: This research was funded by the PORTUGAL 2020 Partnership Agreement through the Competitiveness and Internationalization Operational Program (POCI) and the European Regional Development Fund (ERDF), under the R&D Project "RENEw—Construction waste for a circular economy," with reference POCI-01-0247-FEDER-033834, and by Fundação para a Ciência e a Tecnologia through the Ph.D. grant number 2021.06428.BD. This work was also partly financed by FCT/MCTES through national funds (PIDDAC) under the R&D Unit Institute for Sustainability and Innovation in Structural Engineering (ISISE), under reference UIDB/04029/2020, and under the Associate Laboratory Advanced Production and Intelligent Systems ARISE under reference LA/P/0112/2020.

Institutional Review Board Statement: Not applicable.

Informed Consent Statement: Not applicable.

Data Availability Statement: Data sharing does not apply to this article.

Acknowledgments: The authors would like to acknowledge the University of Minho staff involved in this work, who contributed to achieving the objectives of this study.

Conflicts of Interest: The authors declare no conflict of interest. The funders had no role in the study's design; in the collection, analyses, or interpretation of data; in the writing of the manuscript; or in the decision to publish the results.

Abbreviations

The following abbreviations are used in this manuscript:

BD	Bulk density
$E_{initial}$	Initial stiffness modulus of the asphalt mixture in the fatigue test
HI	Healing index
HR	Healing ratio
LF	Limestone filler
MD	Maximum density
NGA	Natural granitic aggregates
PA	Phase angle
SCB	Semicircular bending
SD	Standard deviation
SM	Stiffness modulus
SSA	Steel slag aggregates
SWF	Steel wool fibers
Vv	Air voids content

References

1. Eurostat. *Sustainable Development in the European Union—Monitoring Report on Progress towards the SDGs in an EU Context*; Eurostat: Luxembourg, 2020.
2. UNEP. *Global Material Flows and Resource Productivity—Assessment Report for the UNEP International Resource Panel*; United Nations Environment Programme: Paris, France, 2016.
3. Araújo, J.P.C.; Oliveira, J.R.M.; Silva, H.M.R.D. The importance of the use phase on the LCA of environmentally friendly solutions for asphalt road pavements. *Transp. Res. Part D Transp. Environ.* **2014**, *32*, 97–110. [CrossRef]
4. Mantalovas, K.; Di Mino, G.; Jimenez Del Barco Carrion, A.; Keijzer, E.; Kalman, B.; Parry, T.; Lo Presti, D. European National Road Authorities and Circular Economy: An Insight into Their Approaches. *Sustainability* **2020**, *12*, 7160. [CrossRef]
5. Eurofer. European Steel in Figures. Available online: www.eurofer.eu (accessed on 8 February 2023).
6. Correia, A.G.; Ferreira, S.; Araújo, N.; Castro, F.; Trigo, L.; Roque, A.J.; Fortunato, E.; Santayana, F.P. *Estudo de Viabilidade de Aplicação de Agregado Siderúrgico Inerte para Construção (ASIC) da Siderúrgia Nacional do Seixal em Camadasde Base, Sub-base, Leito de Pavimento e Aterro*; University of Minho: Guimarães, Portugal, 2006.
7. World Steel Association. *World Steel in Figures*; World Steel Association: Brussels, Belgium, 2021.
8. Loureiro, C.D.A.; Moura, C.F.N.; Rodrigues, M.; Martinho, F.C.G.; Silva, H.M.R.D.; Oliveira, J.R.M. Steel Slag and Recycled Concrete Aggregates: Replacing Quarries to Supply Sustainable Materials for the Asphalt Paving Industry. *Sustainability* **2022**, *14*, 5022. [CrossRef]
9. Skaf, M.; Manso, J.M.; Aragón, Á.; Fuente-Alonso, J.A.; Ortega-López, V. EAF slag in asphalt mixes: A brief review of its possible re-use. *Resour. Conserv. Recycl.* **2017**, *120*, 176–185. [CrossRef]

10. Moura, C.; Nascimento, L.; Loureiro, C.; Rodrigues, M.; Oliveira, J.; Silva, H. Viability of Using High Amounts of Steel Slag Aggregates to Improve the Circularity and Performance of Asphalt Mixtures. *Appl. Sci.* **2022**, *12*, 490. [CrossRef]
11. Li, J.; Liu, J.; Zhang, W.; Liu, G.; Dai, L. Investigation of thermal asphalt mastic and mixture to repair potholes. *Constr. Build. Mater.* **2019**, *201*, 286–294. [CrossRef]
12. Tabakovic, A.; Schlangen, E. Self-Healing Technology for Asphalt Pavements. In *Self-Healing Materials: Advances in Polymer Science*; Hager, M., van der Zwaag, S., Schubert, U., Eds.; Springer: Berlin/Heidelberg, Germany, 2015; Volume 273, pp. 285–306.
13. Varma, R.; Balieu, R.; Kringos, N. A state-of-the-art review on self-healing in asphalt materials: Mechanical testing and analysis approaches. *Constr. Build. Mater.* **2021**, *310*, 125197. [CrossRef]
14. Liang, B.; Lan, F.; Shi, K.; Qian, G.; Liu, Z.; Zheng, J. Review on the self-healing of asphalt materials: Mechanism, affecting factors, assessments and improvements. *Constr. Build. Mater.* **2021**, *266*, 120453. [CrossRef]
15. Chen, F.; Balieu, R. A state-of-the-art review of intrinsic and enhanced electrical properties of asphalt materials: Theories, analyses and applications. *Mater. Des.* **2020**, *195*, 109067. [CrossRef]
16. García, A.; Schlangen, E.; Ven, M.; Liu, Q. Electrical conductivity of asphalt mortar containing conductive fibers and fillers. *Constr. Build. Mater.* **2009**, *23*, 3175–3181. [CrossRef]
17. García, A.; Schlangen, E.; Ven, M.; Liu, Q. A simple model to define induction heating in asphalt mastic. *Constr. Build. Mater.* **2012**, *31*, 38–46. [CrossRef]
18. Norambuena-Contreras, J.; Garcia, A. Self-healing of asphalt mixture by microwave and induction heating. *Mater. Des.* **2016**, *106*, 404–414. [CrossRef]
19. Yang, C.; Wu, S.; Xie, J.; Amirkhanian, S.; Liu, Q.; Zhang, J.; Xiao, Y.; Zhao, Z.; Xu, H.; Li, N.; et al. Enhanced induction heating and self-healing performance of recycled asphalt mixtures by incorporating steel slag. *J. Clean. Prod.* **2022**, *366*, 132999. [CrossRef]
20. Phan, T.M.; Park, D.-W.; Le, T.H.M. Crack healing performance of hot mix asphalt containing steel slag by microwaves heating. *Constr. Build. Mater.* **2018**, *180*, 503–511. [CrossRef]
21. Wan, J.; Wu, S.; Xiao, Y.; Chen, Z.; Zhang, D. Study on the effective composition of steel slag for asphalt mixture induction heating purpose. *Constr. Build. Mater.* **2018**, *178*, 542–550. [CrossRef]
22. Li, C.; Wu, S.; Chen, Z.; Tao, G.; Xiao, Y. Enhanced heat release and self-healing properties of steel slag filler based asphalt materials under microwave irradiation. *Constr. Build. Mater.* **2018**, *193*, 32–41. [CrossRef]
23. Jiang, J.; Ni, F.; Wu, F.; Sadek, H.; Lv, Q. Evaluation of the healing potential of asphalt mixtures based on a modified semicircular bending test. *Constr. Build. Mater.* **2019**, *196*, 284–294. [CrossRef]
24. Xiang, H.; Gao, J.; Li, N.; Zheng, G.; Chen, J.; Zhuang, S.; Wang, C. Comparison of different astigmatic optical lenses by direct measurement of surface shape. *Opt. Eng.* **2020**, *59*, 024113. [CrossRef]
25. Estradas de Portugal. *Caderno de Encargos Tipo Obra. 14.03—Pavimentação, Características dos Materiais*; Estradas de Portugal: Almada, Portugal, 2014.
26. Liu, Q.; Chen, C.; Li, B.; Sun, Y.; Li, H. Heating Characteristics and Induced Healing Efficiencies of Asphalt Mixture via Induction and Microwave Heating. *Materials* **2018**, *11*, 913. [CrossRef]
27. Tabaković, A.; O'Prey, D.; McKenna, D.; Woodward, D. Microwave self-healing technology as airfield porous asphalt friction course repair and maintenance system. *Case Stud. Constr. Mater.* **2019**, *10*, e00233. [CrossRef]
28. Gallego, J.; del Val, M.A.; Contreras, V.; Páez, A. Heating asphalt mixtures with microwaves to promote self-healing. *Constr. Build. Mater.* **2013**, *42*, 1–4. [CrossRef]
29. Garcia, A.; Salih, S.; Gómez-Meijide, B. Optimum moment to heal cracks in asphalt roads by means electromagnetic induction. *Constr. Build. Mater.* **2020**, *238*, 117627. [CrossRef]
30. Liu, Q.; Schlangen, E.; van de Ven, M.; van Bochove, G.; van Montfort, J. Evaluation of the induction healing effect of porous asphalt concrete through four point bending fatigue test. *Constr. Build. Mater.* **2012**, *29*, 403–409. [CrossRef]
31. Onifade, I.; Birgisson, B. Damage and fracture characterization of asphalt concrete mixtures using the equivalent micro-crack stress approach. *Constr. Build. Mater.* **2017**, *148*, 521–530. [CrossRef]

materials

MDPI

Article

Valorization of Vetiver Root Biochar in Eco-Friendly Reinforced Concrete: Mechanical, Economic, and Environmental Performance

Sameer Neve, Jiang Du, Rojyar Barhemat, Weina Meng, Yi Bao * and Dibyendu Sarkar *

Department of Civil, Environmental and Ocean Engineering, Stevens Institute of Technology, Hoboken, NJ 07030, USA
* Correspondence: yi.bao@stevens.edu (Y.B.); dsarkar@stevens.edu (D.S.)

Abstract: Biochar has shown great promise in producing low-cost low-carbon concrete for civil infrastructure applications. However, there is limited research comparing the use of pristine and contaminated biochar in concrete. This paper presents comprehensive laboratory experiments and three-dimensional nonlinear finite element analysis on the mechanical, economical, and environmental performance of reinforced concrete beams made using concrete blended with biochar generated from vetiver grass roots after the roots were used in an oil extraction process. Both pristine biochar and biochar that were used to treat wastewater through adsorbing heavy metals (100 mg/L of Pb, Cu, Cd, and Zn) were investigated. The biochar was used to replace up to 6% Portland cement in concrete. Laboratory experiments were conducted to characterize the workability, mechanical properties, shrinkage, and leaching potential of the concrete blended with biochar. The results showed that using biochar could increase the compressive strengths and reduce the shrinkage of concrete without causing a leaching problem. The results from finite element analysis of the reinforced concrete beams showed that the use of biochar was able to increase the flexural performance of the beams as well as their economic and environmental performance. This research will promote the development and structural applications of low-cost low-carbon concrete.

Keywords: biochar; circular economy; finite element analysis; green concrete; leaching of heavy metals; reinforced concrete

check for updates

Citation: Neve, S.; Du, J.; Barhemat, R.; Meng, W.; Bao, Y.; Sarkar, D. Valorization of Vetiver Root Biochar in Eco-Friendly Reinforced Concrete: Mechanical, Economic, and Environmental Performance. *Materials* **2023**, *16*, 2522. https://doi.org/10.3390/ma16062522

Academic Editor: Ayman El-Zohairy

Received: 17 February 2023
Revised: 13 March 2023
Accepted: 19 March 2023
Published: 22 March 2023

1. Introduction

Concrete is the most widely used construction material worldwide because of its low cost, ease of use, and local availability [1]. Concrete is produced with carbon-intensive ingredients such as Portland cement whose manufacturing process involves combustion at high temperatures and releases large amounts of CO_2. The manufacturing of cement accounts for about 8% of global emissions caused by humans [2]. Substituting cement with low-carbon low-cost materials is promising to drastically lower the carbon footprint and cost of concrete for low-carbon cost-effective concrete structures.

Previous research showed that it is promising to valorize various types of waste to produce such materials to amend with concrete for structural applications. The widely used types of solid waste include industrial by-products (e.g., fly ash [3] and slag [4]), construction and demolition waste (e.g., waste concrete [5] and waste brick [6]), as well as municipal waste (e.g., waste glass [7] and waste plastic [8]). The various types of waste were utilized to replace Portland cement partially or fully. Extensive research has shown that the appropriate use of solid waste can improve the workability, mechanical properties, and durability of concrete while reducing the carbon footprint and material cost [9,10]. For example, Spiesz et al. used fly ash and slag into concrete blended with waste glass to mitigate the alkali-silica reaction, which is often induced by the waste glass, while sustaining the workability and mechanical properties of concrete [11]. Torres et al.

discovered that using fine foundry waste improved the compressive, tensile, and flexural strengths of concrete [12]. Ling and Nor valorized waste tire to increase the skid resistance of concrete paving blocks [13]. Modarres et al. found that utilizing coal waste ash and coal waste powder in concrete pavement increased the compressive strength and toughness of concrete [14].

Recently, biochar made using lignocellulosic waste has attracted increasing interest because of its potential to improve concrete properties. First, using biochar can enhance the mechanical properties of concrete. For example, using biochar to partially replace Portland cement increased the compressive and flexural strengths of cement paste and mortar [15,16]. The enhancement of the mechanical strengths is attributed to the filler effect of fine biochar particles that filled pores and thus refined the microstructures [17]. The filler effect not only densifies the matrix but also enhances the interfacial transition zones when aggregates are utilized [18]. Second, biochar with macro-pores (diameter > 0.1 mm) imparts an internal curing effect. Pre-soaked biochar particles that absorbed water released into the matrix during the hardening process [18]. The released water mitigated the shrinkage of the matrix [19], which in turn helped minimize the occurrence of cracking [20]. Third, biochar is usually highly porous, as volatiles escape during pyrolysis, and, therefore, the use of biochar increases the thermal spalling resistance of concrete as the pores alleviate vapor pressure and the damages caused by the internal pressure at high temperatures [17]. In addition to the benefits of concrete properties, biochar gains dimensional strength and resistance to biological degradation through pyrolysis by hindering inflation because of the removal of cellulosic elements, particularly hemicellulose, which can absorb moisture and are vulnerable to biodegradation [17].

Previous studies by Tan et al. [21] indicate that the addition of low amounts of biochar (1–3% by mass) increased the compressive strength, and the addition of high amounts of biochar (>5% by mass) decreased the compressive strength. Whereas the increase in biochar dosage did not significantly impact the flexural strength. The addition of biochar to green roofs also worked efficiently as a water retention medium, leading to a reduction in roof temperature [22]. Biochar derived from date palm leaves and seeds as up to 1% cement replacement led to an increase in compressive strength by about 7% [23]. Recent research has shown that it is possible to achieve negative-carbon concrete with biochar which was used to sequestrate carbon dioxide [24,25].

The above studies on using biochar in concrete production demonstrate the feasibility of using biochar to improve the mechanical properties and durability of concrete. Since biochar is often considered waste, the use of biochar in producing concrete will likely reduce the carbon footprint and cost of concrete, simultaneously. Meanwhile, it is promising to use biochar to treat wastewater with heavy metals [26,27], because the high specific surface area of biochar promotes the adsorption and immobilization of heavy metals. It is rational to imagine whether the biochar used to adsorb heavy metals can be used to produce concrete for structural applications. Despite the potential benefits, several challenges have been identified from utilizing biochar to absorb heavy metals and produce concrete: (1) the presence of heavy metals may compromise the mechanical properties and durability of concrete since heavy metals often hinder the hydration reactions of cement [28]. (2) The constructional performance of concrete blended with biochar is still unclear because the porous microstructure of biochar may absorb the mixing water and affect the workability of concrete adversely. (3) The heavy metal leaching potential of concrete blended with biochar remains unclear, thus leading to concerns about the environmental performance of concrete for structural applications.

This research is motivated by the hypothesis that the combination of biochar and concrete can achieve desired constructional, mechanical, economic, and environmental performance through the appropriate design of the concrete mixture and reinforced concrete for structural applications. To test this hypothesis, this research has four main objectives: (1) to evaluate the efficiency of the vetiver root biochar in removing heavy metals in an aqueous media; (2) to evaluate the workability, compressive strengths, shrinkage, and leaching potential of concrete blended with biochar; (3) to understand the underlying mechanisms of the effect of biochar on concrete properties; and (4) to evaluate the mechanical, economic, and environmental performance of reinforced concrete made using the developed concrete with biochar.

These research objectives reflect the intellectual merits of this research, which generates new knowledge about the employment of biochar on the constructional, mechanical, economic, and environmental performance of reinforced concrete. Regarding the broader impacts, the knowledge generated from this research will promote the valorization of various types of waste in producing low-carbon cost-effective concrete for structural applications, which will, on one hand, reduce the carbon footprint and cost of the construction industry, and on the other hand, minimize the landfill of waste and the environmental impact such as leaching of heavy metals from waste.

To achieve the above research objectives, this research conducted comprehensive laboratory experiments on biochar and biochar-modified concrete to investigate the effects of both pristine and heavy metal-contaminated biochar on the fresh and hardened properties as well as the leaching potential of concrete and performed finite element analysis to understand the effects of biochar on the mechanical, and environmental performance of reinforced concrete.

The remainder of the paper is organized as follows: Section 2 elaborates on the methods, including the investigated materials, laboratory experiments, and finite element analysis. Section 3 presents the results from the experiments and finite element analysis. Section 4 discusses the results, and Section 5 summarizes the conclusions.

2. Materials and Methods

2.1. Overview

This section provides an overview of the methods, as shown in Figure 1. This research conducted laboratory experiments to evaluate the effect of biochar on the workability, compressive strengths, shrinkage, and leaching potential of concrete. Both pristine (PBC) and heavy metal-contaminated biochar (CBC) were prepared and utilized to produce concrete. The heat of cement hydration was measured to understand the underlying mechanisms of the effect of biochar on the compressive strengths of concrete. The leaching potential of concrete was evaluated through the Synthetic Precipitation Leaching Procedure (SPLP), standard Toxicity Characteristic Leaching Procedure (TCLP), and progressive TCLP. Chemical analysis was performed through inductively coupled plasma optical emission spectroscopy (ICP-OES). Based on the mechanical properties of concrete, finite element analysis was performed using a three-dimensional (3D) finite element model to investigate the effects of biochar on the mechanical properties and environmental impacts of reinforced concrete beams for structural applications. The details of the adopted methods are elaborated on in the following subsections.

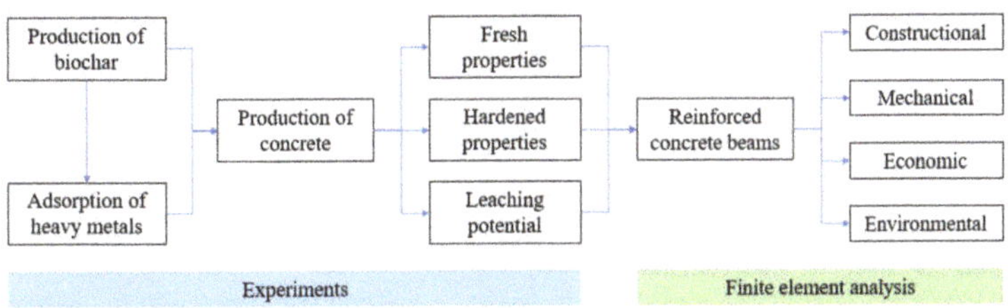

Figure 1. Illustration of the methods adopted in this research.

2.2. Preparation, Characterization, and Treatment of Biochar

2.2.1. Preparation

The biochar used in this research was generated from the roots of vetiver, which is a type of lignocellulosic, perennial, warm-season (or C4) grass with high efficiency at carbon sequestration for its dense root system with its roots having a high commercial value in the aromatherapy and perfumery industry [29]. After the oil is extracted, the roots are considered waste and thus disposed of. In a controlled greenhouse environment, vetiver grass was grown using slips acquired from Mosquito Hawk Farms LLC in Anahuac, TX, USA. After the roots were cleaned meticulously and air-dried for 3 days under ambient conditions, the roots were manually ground and subjected to hydro-distillation to extract the essential oil. The ground vetiver roots recovered from the oil extraction procedure were rinsed three times with deionized water to remove leftover contaminants. Using the optimized pyrolysis settings developed in a previous study [30], the roots were pyrolyzed at 500 °C for 60 min to produce biochar. A furnace (model: Lindberg 51862HR) and retort were employed to generate biochar from these feedstocks in N_2 gas at a flow rate of 0.1 mL/min.

2.2.2. Characterization of Physical and Chemical Properties

This research characterized the bulk density, yield, surface area, pH, electric conductivity, and cation exchange capacity as well as the physical and chemical properties of the biochar. The total elemental analysis for inorganics was performed following a standard method recommended by the Environmental Protection Agency (EPA) of the USA with a few minor modifications [31]. An elemental analyzer was used to determine the carbon, hydrogen, nitrogen, and oxygen elements of biochar (model: Costech 4010). Surface morphology analysis and elemental semi-quantitative analysis were performed via scanning electron microscope-energy dispersion spectroscopy (SEM-EDS) with an 80 mm^2 silicon drift detector (Oxford Instruments, Concord, MA, USA).

The surface chemical properties were determined via Fourier transform infrared spectroscopy using an FTIR spectrometer (model: PerkinElmer Spectrum 100, Waltham, MA, USA) with spectra collected between 400 cm^{-1} and 4000 cm^{-1}. Sample pellets were made for all samples using fused-KBr at a 100:1 KBr:Biochar ratio for transmission studies. An average of 100 scans at a resolution of 4 cm^{-1} were performed for each sample. The spectra were examined using commercial software, called OMNIC (v9.2.106 from Thermo Fisher Scientific Inc., Waltham, MA, USA).

2.2.3. Heavy Metal Loading

The biochar was used to conduct batch adsorption tests for four heavy metals, which are Cadmium (Cd), Copper (Cu), Lead (Pb), and Zinc (Zn). These heavy metals are representative of stormwater runoff from different sources and wastewater. The solutions of Cu^{2+}, Cd^{2+}, Pb^{2+}, and Zn^{2+} were prepared using $Cu(NO_3)_2{\cdot}3H_2O$ (Fisher Scientific, Branchburg, NJ, USA), $Cd(NO_3)_2{\cdot}4H_2O$ (Sigma-Aldrich, St. Louis, MO, USA), $Pb(NO_3)_2$ (Fisher Scientific, New Jersey, US), and $Zn(NO_3)_2{\cdot}6H_2O$ (Fisher Scientific, New Jersey, US) salts, which were dissolved in deionized water. A solution of 0.1 M sodium nitrate ($NaNO_3$) was added to the solutions to provide ionic strength. The temperature was controlled at 24 ± 2 °C, and 100 mg/L of each metal concentration was adsorbed onto 10 g biochar (1% *w*/*v*) in a 1 L solution for 24 h. Several batches were prepared to obtain the desired quantity of contaminated biochar. The initial and final concentrations of the solution were recorded. The removal percentage of heavy metals was calculated by Equation (1) [32]:

$$\text{Removal percentage (\%)} = (C_0 - C_e)/(C_0 \times 100) \quad (1)$$

where C_0 and C_e are the initial and the final concentrations of the solution, respectively.

2.3. *Design and Preparation of Concrete*

The mixture design of concrete blended with biochar is listed in Table 1. Portland cement was used as the binder, which was partially replaced by biochar. The water-to-binder ratio was set at 0.5. River sand was utilized as fine aggregate. A high-range water reducer (HRWR) was utilized to improve the flowability. Various cement replacement percentages were considered. This research is a part of our research on ultra-high-performance concrete, which does not contain coarse aggregate. Thus, although the studied mixtures do not use coarse aggregate, they are still called concrete.

Table 1. Mixture design of concrete with biochar (kg/m^3).

Designation	Cement	PBC	CBC	Sand	HRWR	Water
Control	450	0	0	1350	1.0	225
PBC-2	441	9	0	1350	1.0	225
PBC-4	432	18	0	1350	1.0	225
PBC-6	423	27	0	1350	1.0	225
CBC-2	441	0	9	1350	1.0	225
CBC-4	432	0	18	1350	1.0	225
CBC-6	423	0	27	1350	1.0	225

Note: PBC-2 represents concrete with 2% cement replaced by pristine biochar, by mass. CBC-6 represents concrete with 6% cement replaced by contaminated biochar, by mass.

The particle size gradation of the cement and the sand used in the cement mixture is shown in Figure 2. The maximum grain size of the sand was 4.75 mm in diameter.

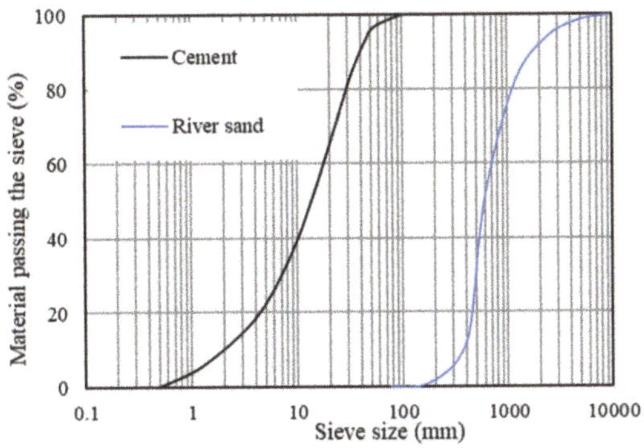

Figure 2. Particle size distribution of raw materials.

A Hobart HL200 mixer was used to prepare concrete in three steps: (1) the dry ingredients (i.e., cement, biochar, and sand) were introduced to the mixer and mixed at 107 revolutions per minute (RPM) for 2 min. (2) The HRWR was dissolved in 90% mixing water and gradually added to the mixer within 1 min. (3) The mixture was mixed at 107 RPM for 3 min.

The concrete was poured into standard molds in triplicates, including cubic molds that measure 50 mm × 50 mm × 50 mm and prism molds that measure 280 mm × 25 mm × 25 mm in dimensions. The cube specimens were used to test the compressive strengths, and the prism specimens were used to test the shrinkage. During the casting of concrete, the molds were placed on a vibrating table to ensure consolidation of concrete. Immediately after the casting, the specimens were wrapped with polythene sheets and then stored in a controlled environment at 25 °C and 70% relative humidity for 24 h. After that, the specimens were cured in lime-saturated water at 25 ± 2 °C until the date of testing. All the results characterizing the concrete performance were the mean values of triplicates. In addition, after the mixing process, all concrete mixtures were examined by hand, and no agglomeration or segregation was found.

2.4. Characterization of Concrete

2.4.1. Workability

The workability of the concrete was evaluated according to ASTM C143 [33]. Specifically, a cone with a base diameter of 200 mm, a top diameter of 100 mm, and a height of 300 mm was used. Concrete was cast and consolidated into a cone which was then removed to let the concrete slump. The slump of fresh concrete was measured and used to evaluate the workability of concrete. Three tests were duplicated, and their results were averaged.

2.4.2. Compressive Strength

The compressive strengths of concrete at 7 days and 28 days were evaluated using the cube specimens according to ASTM C109 [34]. The loading rate was kept constant at 1.8 kN/min. Three sample replicates were prepared for each test, and the average results were obtained.

2.4.3. Heat of Hydration

The evolution of the heat of cement hydration was evaluated using an isothermal calorimeter (model: Calmetrix I-Cal 4000 HPC), which was programmed to maintain the samples at 25 °C. About 60 g of fresh concrete was sealed in a plastic vial and placed into the calorimeter. The measurement of the heat of hydration started 2 min after the end of mixing and continued for 36 h. The results were normalized by the mass of the cement.

2.4.4. Autogenous Shrinkage

The autogenous shrinkage of concrete was evaluated in accordance with ASTM C1698 [35]. The final setting time was set as time zero. The test was conducted daily for the first week and weekly until 28 days. The prism specimens measuring 25 mm × 25 mm × 280 mm were sealed with water-proof alumina tape to prevent moisture loss. Three sample replicates were prepared for each test. The average results for each test were reported.

2.4.5. Leaching Potential

The Synthetic Precipitation Leaching Procedure (SPLP) method recommended by US EPA as a batch leaching test was used to evaluate the risk of heavy metals [36]. The pH of the reagent water used in the SPLP was adjusted to 4.25 ± 0.05 using a solution of sulfuric and nitric acid at a ratio of 60:40, by volume. Then, 200 mL of the extraction fluid and 10 g of concrete were put into 500 mL high-density polyethylene (HDPE) bottles. The samples were filtered using a vacuum filtration system and a 0.6–0.8 μm glass fiber filter (Whatman GF/F) after being shaken for 18 h. After filtering, the leachate was analyzed in terms of the pH value, RCRA8 metals (As, Ba, Cd, Cr, Pb, Hg, Ag, and Se), along with the metals that were adsorbed into the biochar (Zn and Cu) by an ICP-OES.

The leachability of the trace elements from concrete to determine if it is hazardous was assessed according to the Toxicity Characteristic Leaching Procedure (TCLP) recommended by EPA [37]. A piece of concrete pellet weighing about 10 gm was obtained from the middle of the concrete block at 1 day, 14 days, and 28 days. Each sample of crushed concrete was treated for 18 h with an acetic acid solution with a pH of 2.88 ± 0.05 and a liquid-to-solid ratio of 20:1.

Progressive TCLP tests were performed to investigate the behaviors of metal pollutants that leached from the concrete [38]. There were five successive steps throughout the test. Each stage followed the same process as the typical TCLP test as mentioned above. Following each extraction, the residuals were put back into the extraction bottles so that a new batch of leachate was used. The concrete samples were transferred without first cleaning them with deionized water. The leachate was then analyzed following the analytical procedure of the SPLP protocol.

Certified reference solutions were checked beside the samples. Both internal and external standards were examined in every ICP-OES analysis. To calibrate the system, standard solutions with $R^2 > 0.995$ were attained before each study. Three replicated measurements were prepared for each sample. ICP Expert Software (v7.1.0.6821) was used to operate and process the ICP-OES data (Agilent Technologies, Santa Clara, CA, USA). All the containers were cleaned with laboratory-grade detergent, steeped in 10% nitric acid solution for 12 h, soaked in deionized water for 6 h, and rinsed with deionized water.

The Tukey–Kramer honestly significant difference (HSD) test was performed to assess significant differences ($p < 0.05$) using the Origin software (v2022b) [39].

2.5. Finite Element Analysis

The flexural behaviors of reinforced concrete beams made using the concrete with biochar were investigated via 3D finite element analysis using software called ABAQUS 2017. The methods used to establish the finite element model were developed in previous research [40,41]. The cross-section of the beams was 200 mm × 300 mm, and the length of the beams was 3000 mm. Each beam was reinforced by five longitudinal steel bars and 21 stirrups. Three longitudinal bars were placed at the bottom, which was 22 mm (Φ22) in diameter and subject to tension. Two longitudinal bars were placed at the top, which was 16 mm (Φ16) in diameter and subject to compression. The diameter of the stirrups was 6 mm (Φ6). The spacing of the stirrups was 150 mm in the middle and 75 mm at the two ends of the beam. The concrete cover thickness was 20 mm. For the hybrid beam production on the job site, the authors suggest doing as follows to minimize the negative effect on the successive casting layers: (1) all designed mixtures should be mixed at the same time; (2) the workability of all mixtures should be controlled at the same level; (3) the casting interval between different concrete layers should be shortened as much as possible. The dimensions of the beams are consistent with those in the references [42,43]. Four-point loads were applied to investigate flexural behaviors. The beam design and finite element model are shown in Figure 3. The concrete and steel bars were modeled using solid elements (C3D8R) and truss elements (T3D2), respectively. The mesh size of the concrete and steel bars was determined through a mesh sensitivity analysis. The global mesh sizes of the concrete beam and steel bars were 5 mm and 10 mm, respectively.

The boundary conditions of the beams were defined as follows: (1) in the symmetrical plane at the mid-span section, the displacement along the length direction is zero. (2) In the symmetrical plane at the mid-width section, the displacement along the width direction is zero. (3) Along the lines over the rollers, the displacement along the vertical direction is zero. The vertical loads were applied to the beam under the displacement control mode. The study utilized the embedded region constraint in the software ABAQUS 2017 to model the interaction between concrete and steel reinforcement by using the keyword "embed", meaning that debonding was not considered. This method has been widely employed in finite element analysis on the behavior of reinforced concrete structures [44–47]. The model with the embedded region constraint does not need the input of the material properties of concrete and steel reinforcement.

The density and Poisson's ratio of concrete were 2600 kg/m^3 and 0.2, respectively. The density and Poisson's ratio of steel bars were 7800 kg/m^3 and 0.3, respectively. The yielding strengths of the steel bars were related to their diameters. The yielding strengths of the steel bars with diameters 6 mm, 16 mm, and 22 mm were 420 MPa, 445 MPa, and 430 MPa, respectively. The elastic modulus of steel bars was 205 GPa. These material properties are consistent with previous research on experiments and finite element analysis of the flexural behaviors of reinforced concrete [42]. The compressive strength of concrete is not listed because the value was obtained from experiments, as elaborated in Section 2.2. The elastic modulus of concrete was calculated according to ACI 318 [48].

The concrete damage plasticity model was applied to consider the damage in concrete during the loading process. More detailed information about the concrete damage plasticity model is available in references [49,50] and is replicated in this paper. The constitutive relationship of concrete is associated with compressive strength, as shown in Figure 4a. The parameters of the adopted concrete damage plasticity model are listed in Table 2. The elastic-perfectly-plastic constitutive model was employed in this study, as it is a common and simplified model to describe steel behavior [51,52], as shown in Figure 4b.

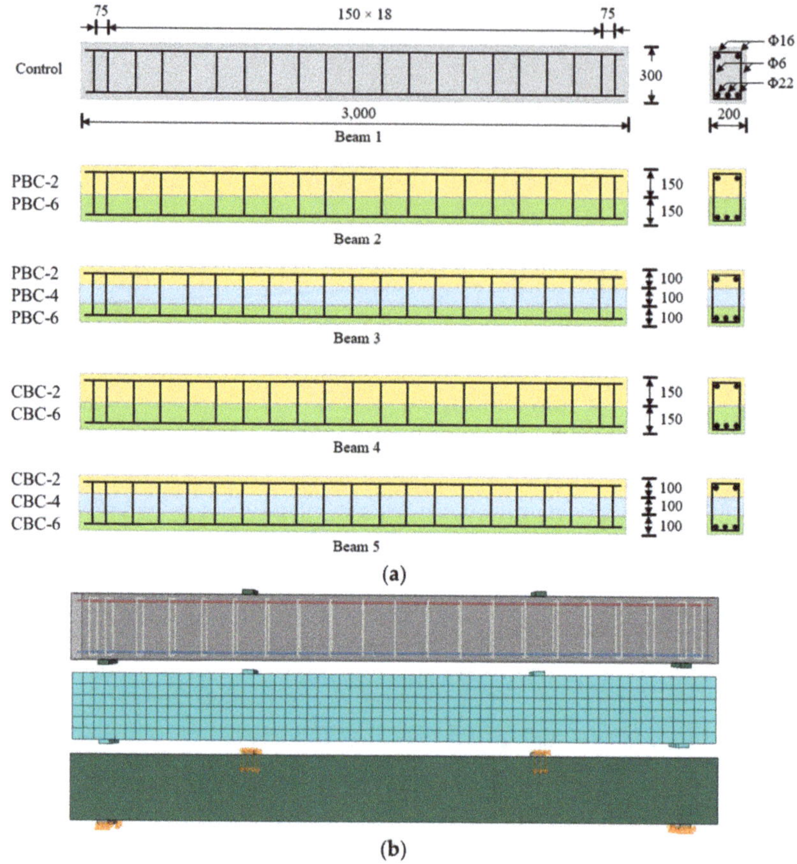

Figure 3. Depiction of the beams: (**a**) design and dimensions (unit: mm); (**b**) finite element model with the meshing, loading, and boundary conditions.

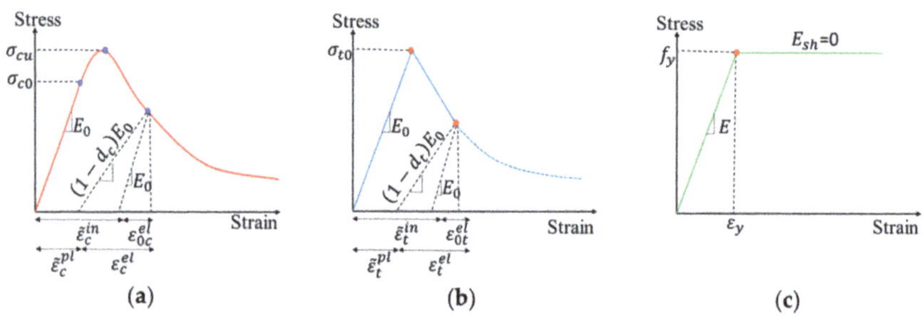

Figure 4. Illustration of the constitutive relationships of (**a**) concrete in compression; (**b**) concrete in tension; and (**c**) steel bars.

Table 2. Parameters used in the CDP model.

Parameter	Dilation Angle	Eccentricity Parameter	Viscosity Parameter	Shape of the Yielding Surface	Stress Ratio
Value	36°	0.1	0.001	0.667	1.05

3. Results

3.1. Characteristics of Biochar

The physicochemical characteristics of the biochar are listed in Table 3. The yield of the vetiver root biochar was 53.76% with a surface area of 308.15 m^2/g. The pH was about 11. This biochar had a high carbon concentration. According to the acid digestion method [31], the biochar also had trace amounts of inorganic components likely due to the presence of the engineered soil used to grow the plants in the greenhouse. The molar H/C, O/C, and N/C ratios were 0.04, 0.25, and 0.3, respectively. The H/C and O/C ratios indicate that the biochar is more aromatic than polar. Previous research also indicated that the high pH and high liming value of this biochar make it a potential liming agent for the reclamation of acid mine drainage-impacted soils [30].

Table 3. Physicochemical characteristics of biochar.

Parameters	Results
Pyrolysis temperature	500 °C
Pyrolysis time	60 min
Yield	53.76%
BET surface area	308.2 m^2/g
pH value	10.98 ± 0.11
Electric conductivity	162.07 ± 13.13 µS/cm
Ash content	26.4%
Cation exchange capacity	82 cmol/kg
Bulk density	0.62 gm/mL
Carbon	68.58%
Hydrogen	2.78%
Oxygen	1.75%
Nitrogen	20.7%
Hydrogen/Carbon	0.041
Oxygen/Carbon	0.026
Nitrogen/Carbon	0.302

The efficiency of using biochar in removing heavy metals through multi-metal batch adsorption experiments is shown in Figure 5. In the adsorption experiments, 100 mg/L of each metal was used to react with the biochar. After 24 h of reaction, the percentage of the removal of heavy metals was 99.99% for Pb, 91.67% for Cu, 78.17% for Zn, and 72.00% for Cu. The contaminated biochar was compared with pristine biochar in producing concrete

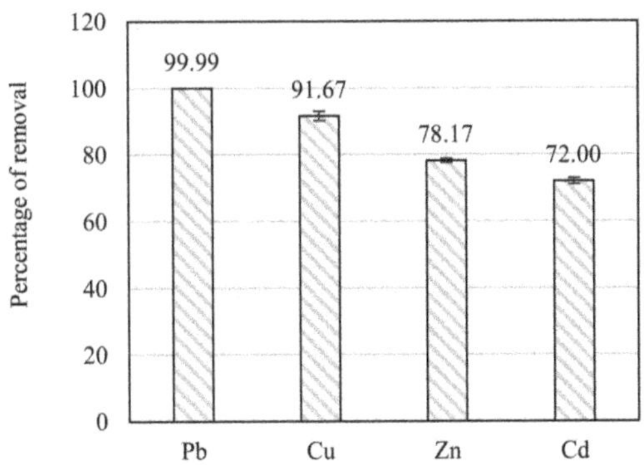

Figure 5. Results of the removal percentage of the heavy metals using biochar.

3.2. Characteristics of Concrete

3.2.1. Workability

The effect of biochar contents on the workability of fresh concrete mixtures is shown in Figure 6. When the HRWR content was the same in all the investigated concrete mixtures, the workability of concrete decreased with the increase of the biochar content, meaning that the use of biochar compromises the workability of concrete. Such results are consistent with the findings from prior research [53–56]. The main reason for the reduction of the workability of concrete is that the porous biochar particles have a large specific surface area and can absorb the mixing water when they are not pre-soaked.

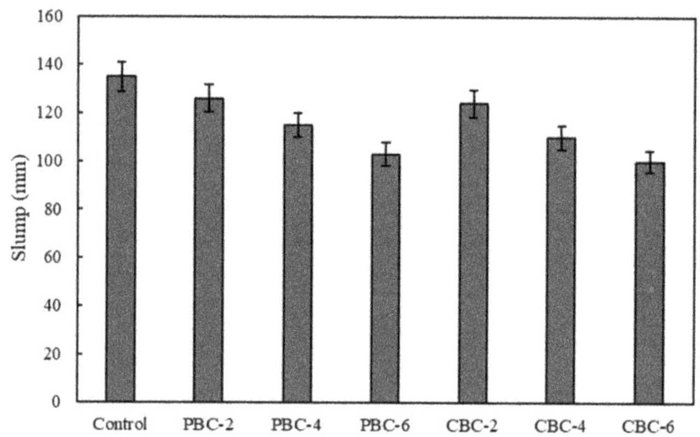

Figure 6. Results of the slump for evaluating the workability of the concrete mixtures.

In this research, the biochar was not pre-soaked because pre-soaking biochar with additional water would increase the total water content and, thus, the water-to-cement ratio of the concrete mixtures. It is known that the increase in the water-to-cement ratio tends to reduce the mechanical strength and durability of concrete. In practice, if the workability is sufficient for construction, it is often effective to increase the HRWR content [57]. The test results also show that the presence of heavy metals adsorbed by biochar does not affect the workability of concrete.

3.2.2. Compressive Strengths

The compressive strengths of the concrete mixtures at 7 days and 28 days are shown in Figure 7. When pristine biochar was used, with the biochar content varying from 0 to 6%, the compressive strengths of concrete at 7 days and 28 days followed the same trend: The compressive strengths first increased and then decreased. The highest compressive strengths were achieved when the pristine biochar content was 2%.

Such results are caused by two competing effects. On one hand, biochar provides filler effects to increase the compressive strength of concrete. On the other hand, biochar introduces defects such as pores and interfacial transition zones to decrease the compressive strength of concrete. When the biochar content was lower than 2%, the filler effect was dominant. When the biochar content was higher than 2%, the weakening effects became dominant. These results are consistent with previous research [15,58]. The results show that the heavy metals in the biochar reduce the compressive strength of concrete. When the contaminated biochar content was 2%, the compressive strengths at 7 days and 28 days were comparable with those of the control concrete.

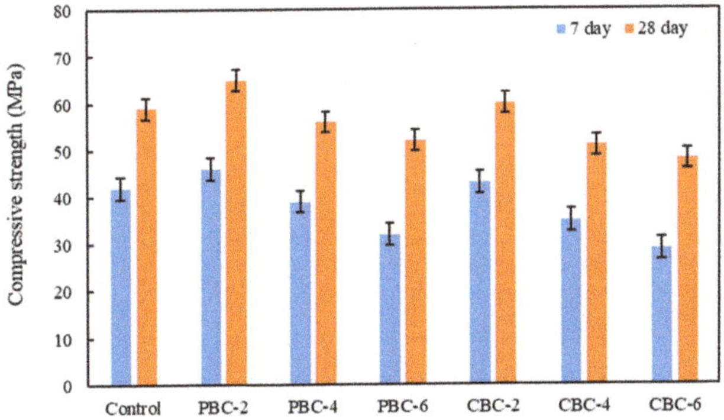

Figure 7. Results of the compressive strengths of the mixtures at 7 days and 28 days.

3.2.3. Heat of Hydration

The results of the heat of hydration are shown in Figure 8. When the pristine biochar was used, the biochar content increased from 0 to 6%. The peak heat flow was delayed from 11.9 h to 13.3 h. With the increase in contaminated biochar from 0 to 6%, the peak heat flow was delayed from 11.9 h to 14.0 h. These results reveal that the use of biochar retarded the hydration reactions of cement.

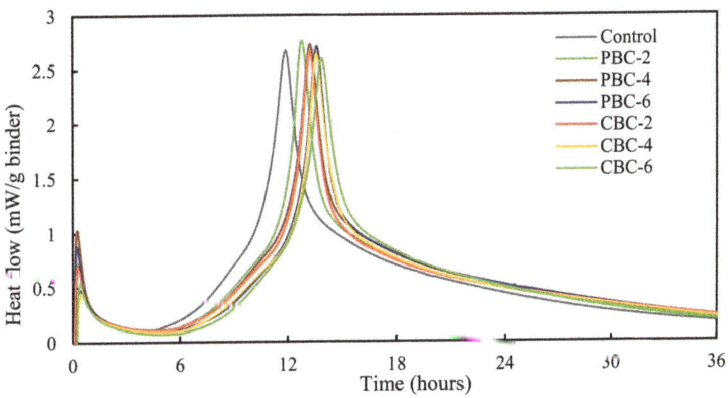

Figure 8. Results of the heat of hydration normalized by the mass of cement.

According to reference [59], the hydration kinetics of cement paste, mortar, and concrete is related to the presence of impurities and saccharides in biochar and the pretreatment of biochar. The addition of biochar may either accelerate or delay the hydration of cement. For instance, the biochar prepared from sorghum retarded the hydration heat (approximately 2 h) and decreased the peak magnitude [59]. The underlying reason is that the pyrolyzing sorghum at 500 °C decomposes the cellulose and hemicellulose, which are the key components that hinder the hydration of cement [60]. In addition, Gupta et al. [54] found that, compared with the cement paste with dry biochar, the addition of pre-soaked biochar further promoted the hydration of cement because the pre-soaked biochar released extra water to the cement matrix.

In this study, the roots were pyrolyzed at 500 °C for 60 min to produce biochar. The cellulose and hemicellulose were decomposed during the production process to retard the cement hydration. Moreover, the dry biochar powder without pre-soak treatment was directly used in this study. Since the mechanical strengths of concrete are gained through the hydration reactions, the results of the heat of hydration explain why the excessive addition of biochar reduced the compressive strengths of concrete. The results also show that the presence of heavy metals further retarded the hydration reactions of cement, thus further reducing the compressive strengths of concrete.

3.2.4. Autogenous Shrinkage

The results of the autogenous shrinkage of the investigated concrete mixtures are shown in Figure 9. The use of biochar efficiently reduced the autogenous shrinkage of the concrete. With the increase in pristine biochar content from 0 to 6%, the autogenous shrinkage at 28 days was reduced from 264 με to 126 με (by 50%). With the increase in contaminated biochar content from 0 to 6%, the autogenous shrinkage at 28 days was reduced from 264 με to 112 με (by 58%).

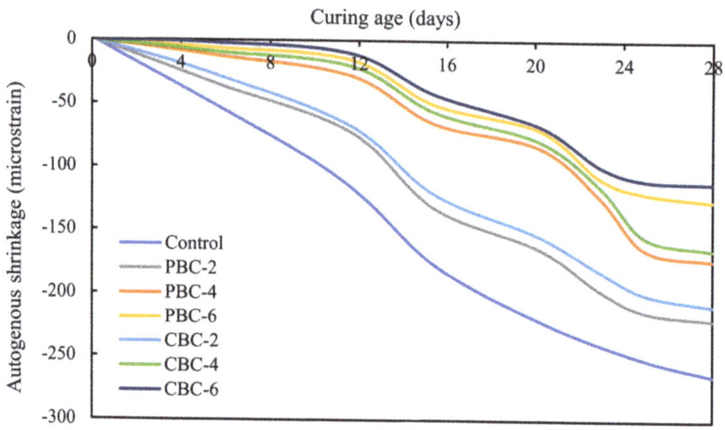

Figure 9. Results of the autogenous shrinkage of the concrete mixtures at different ages.

There are two main underlying mechanisms: (1) porous biochar particles absorbed water and released the internal curing water during the hardening process to delay the drop of the internal relative humidity inside the concrete, thus reducing the autogenous shrinkage. (2) The biochar retarded the hydration of cement, thus alleviating the autogenous shrinkage.

3.3. *Leaching Potential*

The concentrations of the heavy metals in the leachates from the SPLP test are listed in Table S1 in the Supplementary Material. The concentrations of the heavy metals are evaluated in accordance with US EPA requirements. The concentrations of Hg, As, Se, Pb, Cd, Ag, and Ba are far below the EPA limits. EPA has not set limits for Cu and Zn. Many of the results were below the detection limit, suggesting that the leaching potential of the biochar and concrete was low. In other words, there is no concern about the leaching of heavy metals when contaminated biochar is used to produce concrete.

The concentrations of the heavy metals in the leachates from the standard TCLP test are listed in Table S2. The values of the concentrations of the heavy metals are different from those from the SPLP test but are consistent with the observations that the concentrations of the heavy metals are far below the EPA limits. Many of the results were below the detection limit, corroborating that there is no concern about the leaching of the heavy metals when the contaminated biochar is utilized to produce concrete.

The concentrations of the heavy metals of the progressive TCLP test are listed in Tables S3–S5. Again, the results are consistent with the results from the standard TCLP test. The progressive TCLP test further verifies that the concentrations of the heavy metals are far below the EPA limits. There were incremental increases in the concentrations of the heavy metals as the test continued progressively with time. However, the low concentrations show that such concrete will not leach out heavy metal contamination despite being in contact with acidic environmental factors such as acid rain.

The pH values of the leachate in the progressive TCLP test are plotted in Figure 10. The pH values at the completion of each treatment were much higher than the pH value of the original extraction solution, which was 2.88. The increase in the pH value can be attributed to the high pH of the concrete. The pH value dropped as the TCLP extraction progressed, indicating that the biochar can function as a buffer. The pH values are not very sensitive to the change in the biochar content or the change in the concrete age.

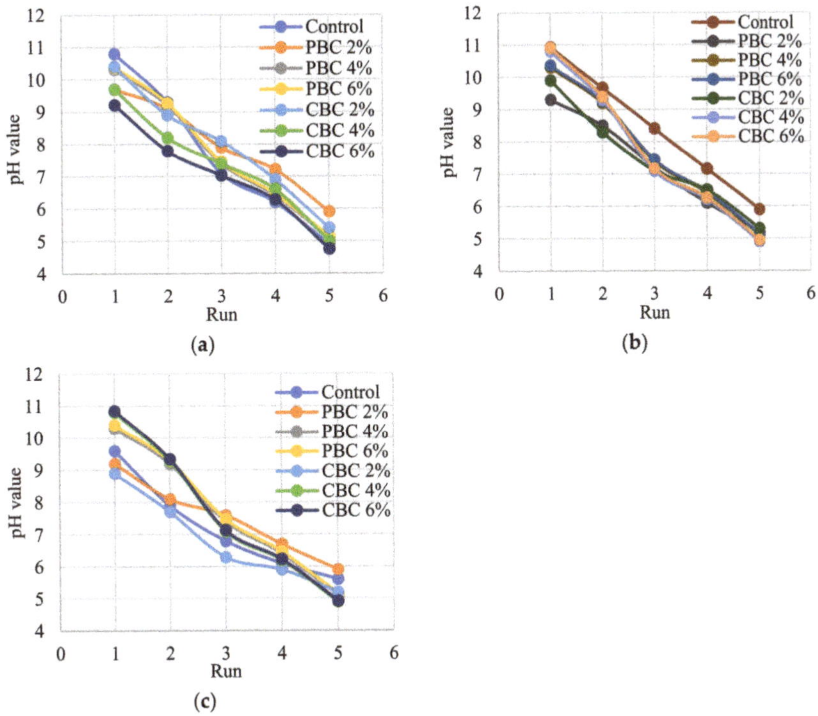

Figure 10. Results of the pH values in the progressive TCLP tests of the different concrete mixtures at different ages: (**a**) 1 day, (**b**) 14 days, and (**c**) 28 days.

3.4. Finite Element Analysis

Figure 11 shows the load–displacement curves obtained from finite element analysis. The load increased linearly with displacement until the first crack in the concrete appeared. Then, the slope of the curves decreased due to the development of the cracks in concrete. The beams could resist a higher load after the concrete was cracked because of the use of steel bars.

The differences in peak load between Beams 1, 4, and 5 were small. For instance, the difference in peak load between Beams 4 and 5 was 4 MPa. Given the vertical axis in Figure 11, the difference between the peak loads is unclear. A clearer representation of the difference in peak load and energy dissipation is given in Section 4.1.

Previous studies revealed that the biochar addition does not significantly alter the modulus of elasticity of mortar [58,61]. For instance, an addition of 8% biochar (by mass) reduced the elastic modulus of mortar by approximately 10%. This study mainly investigated the amendment of cement mortar with different biochar ratios. Without the fiber reinforcement, the cement mortar would show brittle rupture. In this case, the tensile behaviors of the investigated cement mortar were not satisfactory. Thus, further experimental tests to characterize the tensile behavior of cement mortar with different biochar contents in this study were not conducted.

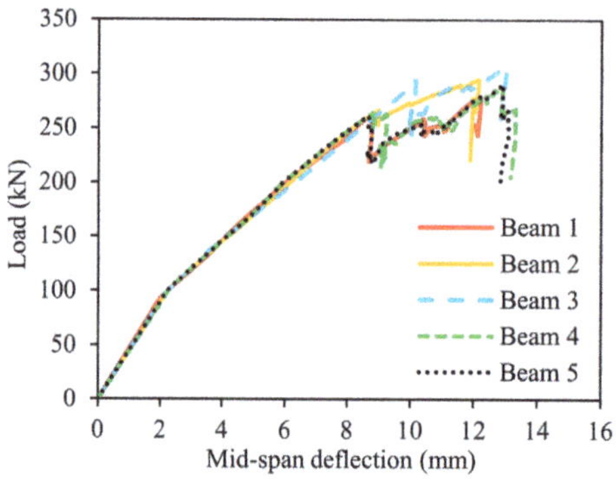

Figure 11. Results of the load–displacement curves of the five reinforced concrete beams.

The cracks in the concrete are quantitatively represented by a tension damage index, which is denoted by DAMAGET. DAMAGET is in the range of 0 to 1: 0 means there is no tension damage (i.e., crack), and 1 means the concrete is completely cracked. Figure 12 shows the finite element analysis results of the development of cracks in Beam 3 with the increase of the displacement applied to the beam. With the increase of the displacement, the cracks in concrete are initiated from the mid-span section and then developed toward the two ends of the beam. The cracks are mainly flexural cracks located within the flexural span. After the flexural cracks were developed to a certain level, diagonal shear cracks were generated in the shear spans.

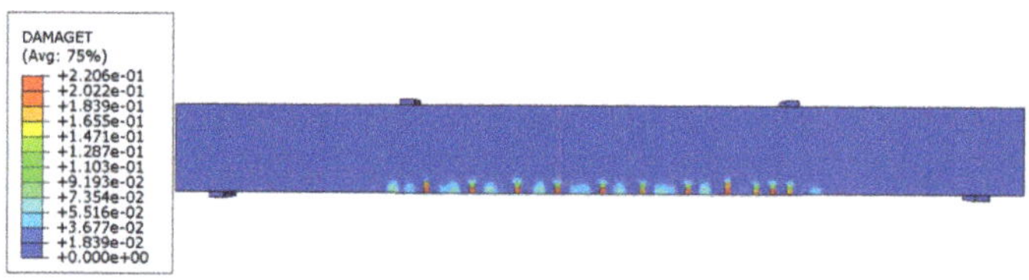

(**a**)

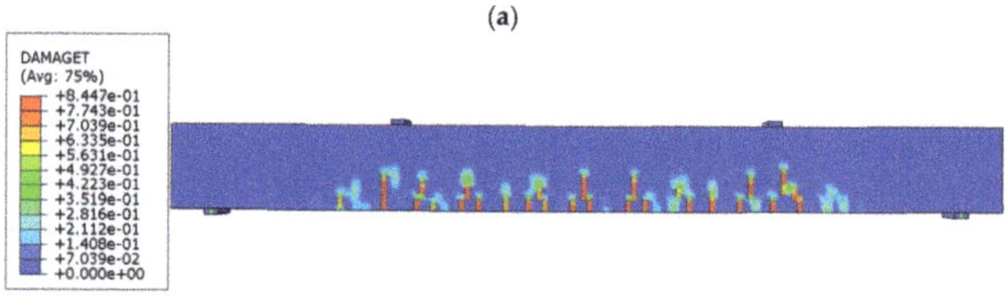

(**b**)

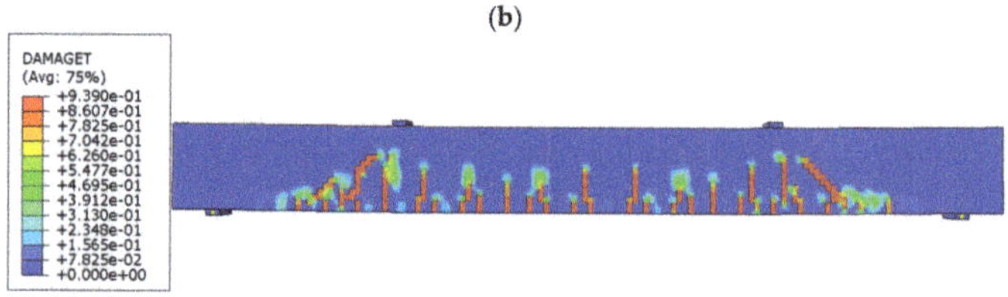

(**c**)

Figure 12. Visualization of the development of cracks in Beam 3 under increasing displacements: (**a**) 1.34 mm, (**b**) 5.53 mm, and (**c**) 10.22 mm.

4. Discussion

4.1. Constructional and Mechanical Performance

The workability of concrete is relevant to the construction quality as poor workability often causes problems with the placement and consolidation of concrete in the construction of structures such as bridges and buildings [62]. The reason is that poor consolidation in turn produces voids, which compromise not only the mechanical properties such as the compressive strengths but also the durability of concrete. Figure 6 shows that as the biochar content increases from 0 to 6%, the slump of concrete decreases from 135 mm to 103 mm. The reduction of the slump adversely affects the construction quality of the reinforced concrete beams. Therefore, in structural applications, it might be necessary to use a water reducer to increase the workability of concrete with biochar.

The results of finite element analysis show that utilizing biochar can improve the mechanical performance of the investigated beams. Figure 13a shows the load capacity results of the five reinforced concrete beams. The use of pristine and contaminated biochar improved the flexural performance of the reinforced concrete beams. The load capacities of Beam 2, Beam 3, Beam 4, and Beam 5 increased by 6%, 11%, 3%, and 5%, respectively, compared with the load capacity of Beam 1. The load capacity of Beam 3 with PBC-2, PBC-4, and PBC-6 is 11% higher than that of Beam 1.

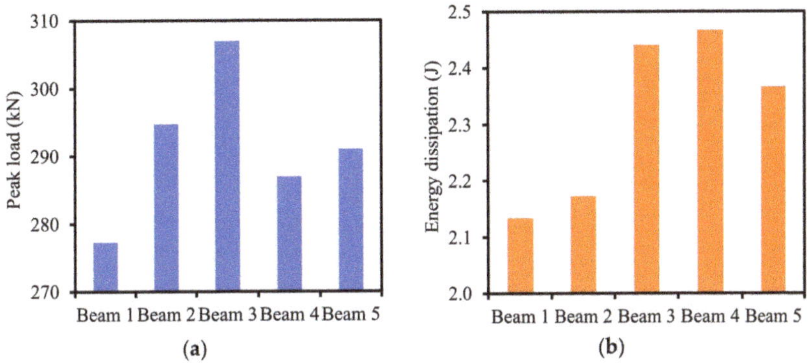

Figure 13. The comparison of the flexural behaviors of the investigated beams: (**a**) peak load, and (**b**) energy dissipation.

The results of energy dissipation of the investigated beams are shown in Figure 13b. Energy dissipation was calculated by the area under the load–deflection curve. The use of biochar in Beam 2, Beam 3, Beam 4, and Beam 5 increased the energy dissipation by 7%, 14% 16%, and 14%, respectively, compared with Beam 1.

4.2. Economic and Environmental Performance

The economic and environmental performance of the beams were evaluated using the data in Table 4. The unit material cost and unit carbon emission of the different ingredients were reported in previous research [63,64]. Biochar was regarded as a waste material [65]. The material cost and carbon emission of each beam were calculated by Equation (2) and Equation (3), respectively:

$$M = \sum_{i=1}^{n} m_i r_i \quad (2)$$

where M is the unit cost of a concrete mixture per cubic meter (unit: $\$/m^3$); m_i is the unit cost (unit: $/kg) of the ith ingredient of the mixture (i = 1, 2, 3, . . . , n, and n = 5), as listed in Table 4; and r_i is the mass of the ith ingredient of the mixture (unit: kg/m^3), as listed in Table 1.

$$C = \sum_{i=1}^{n} c_i r_i \quad (3)$$

where C is the carbon footprint of a concrete mixture; c_i is the unit carbon emission of the ith ingredient of the mixture (i = 1, 2, 3, . . . , n, and n = 5), as listed in Table 4; and r_i is the mass of the ith ingredient of the mixture, as listed in Table 1.

Table 4. Inventory data of unit material cost and unit carbon emission.

Material	Unit Cost ($/kg)	Unit Carbon Emission (kg/kg)
Portland cement	0.11 [63]	0.83 [66]
Biochar	0 [67]	0 [67]
River sand	0.02 [68]	0.01 [69]
HRWR	3.60 [70]	0.72 [64]
Water	0 [71]	0 [66]
Steel rebar	0.68 [72]	1.38 [73]

The calculation results are plotted in Figure 14. The results show that the use of biochar in the concrete reduces the cost and the carbon footprint of the beams. When the beam is made of multiple layers of concrete with different biochar contents, the cost and the carbon footprint of the beam can be further reduced. Beam 3 and beam 5 achieved the lowest cost and lowest carbon footprint.

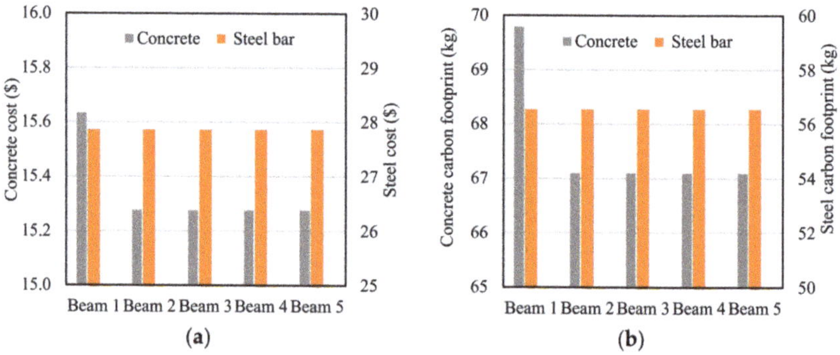

Figure 14. Results of the cost and carbon footprint of the reinforced concrete beams: (**a**) cost and (**b**) carbon footprint.

5. Conclusions

This study investigates the effects of using biochar in low-cost low-carbon concrete on the mechanical, economic, and environmental performance of reinforced concrete beams through comprehensive laboratory experiments of the concrete and finite element analysis of the reinforced concrete beams. The following conclusions can be drawn from the above investigations:

- The use of biochar was able to increase the compressive strength of concrete. When biochar was used to replace 2% of Portland cement in the concrete, the compressive strengths of the concrete were increased. When the replacement percentage was higher than 2%, the use of biochar reduced the compressive strengths of concrete. Compared with pristine biochar, the contamination of biochar by heavy metals reduced the compressive strengths due to the delay of hydration of cement.
- The use of biochar decreased the shrinkage of concrete. When the biochar was used to replace 6% of Portland cement in the concrete, the autogenous shrinkage of concrete was reduced from 264 με to 112 με. The main mechanism was the internal curing effect of porous biochar. Compared with pristine biochar, the contamination of biochar by heavy metals reduced the autogenous shrinkage due to the delay of hydration of cement.

- The use of biochar decreased the workability of concrete. When biochar was used to replace 6% of Portland cement in the concrete, the slump of the concrete was reduced from 135 mm to 103 mm. The main mechanism was the water absorption effect of porous biochar. Compared with pristine biochar, the contamination of biochar by heavy metals did not affect the workability of concrete.
- The use of biochar in concrete did not cause a leaching problem. The concentrations of the heavy metals were well below the EPA limits, indicating that the biochar that was used to absorb heavy metals was not a hazardous waste. In other words, it is promising to use biochar to immobilize the heavy metals and then produce cost-effective eco-friendly concrete.
- The use of concrete blended with biochar was able to improve the mechanical properties, economic, and environmental performance of reinforced concrete beams. Using concrete with different biochar contents in layers was able to further improve the economic and environmental performance of reinforced concrete beams while retaining the flexural properties of the beams.

Future research is needed to understand the influence of contaminated biochar on the durability of concrete and reinforced concrete beams. It is also important to evaluate the effect of biochar on the demand for water reducers to ensure adequate flowability and constructional performance.

Supplementary Materials: The following supporting information can be downloaded at: https://www.mdpi.com/article/10.3390/ma16062522/s1, Table S1: Results of the SPLP test (unit: μg/L). Table S2: Results of the TCLP test (unit: μg/L). Table S3: Progressive TCLP data at 1 day (unit: μg/L). Table S4: Progressive TCLP data at 14 days (unit: μg/L). Table S5: Progressive TCLP data at 28 days (unit: μg/L).

Author Contributions: Conceptualization, W.M., Y.B. and D.S.; methodology, S.N., J.D. and Y.B.; software, S.N., J.D. and R.B.; validation, W.M., Y.B. and D.S.; investigation, S.N., J.D. and R.B.; resources, supervision, project administration, funding acquisition, W.M., Y.B. and D.S.; data curation, S.N. and J.D.; writing—original draft preparation, S.N., J.D. and Y.B.; writing—review and editing, W.M., Y.B. and D.S.; visualization, S.N., J.D. and R.B. All authors have read and agreed to the published version of the manuscript.

Funding: This research received no external funding.

Institutional Review Board Statement: Not applicable.

Informed Consent Statement: Not applicable.

Data Availability Statement: Data presented in this study are available on request from the corresponding author.

Acknowledgments: The authors would like to acknowledge Lehigh for the generous donation of Portland cement used in this study.

Conflicts of Interest: The authors declare no conflict of interest. The funders had no role in the design of the study; in the collection, analyses, or interpretation of data; in the writing of the manuscript; or in the decision to publish the results.

References

1. Barcelo, L.; Kline, J.; Walenta, G.; Gartner, E. Cement and carbon emissions. *Mater. Struct.* **2013**, *47*, 1055–1065. [CrossRef]
2. Andrew, R.M. Global CO_2 emissions from cement production, 1928–2018. *Earth Syst. Sci. Data* **2019**, *11*, 1675–1710. [CrossRef]
3. Meng, W.; Valipour, M.; Khayat, K.H. Optimization and performance of cost-effective ultra-high performance concrete. *Mater. Struct.* **2017**, *50*, 29. [CrossRef]
4. Du, J.; Liu, Z.; Christodoulatos, C.; Conway, M.; Bao, Y.; Meng, W. Utilization of off-specification fly ash in preparing ultra-high-performance concrete (UHPC): Mixture design, characterization, and life-cycle assessment. *Resour. Conserv. Recycl.* **2022**, *180*, 106136. [CrossRef]
5. Li, X.; Lv, X.; Zhou, X.; Meng, W.; Bao, Y. Upcycling of waste concrete in eco-friendly strain-hardening cementitious composites: Mixture design, structural performance, and life-cycle assessment. *J. Clean. Prod.* **2021**, *330*, 129911. [CrossRef]
6. Wong, C.L.; Mo, K.H.; Yap, S.P.; Alengaram, U.J.; Ling, T.-C. Potential use of brick waste as alternate concrete-making materials: A review. *J. Clean. Prod.* **2018**, *195*, 226–239. [CrossRef]

7. Guo, P.; Meng, W.; Nassif, H.; Gou, H.; Bao, Y. New perspectives on recycling waste glass in manufacturing concrete for sustainable civil infrastructure. *Constr. Build. Mater.* **2020**, *257*, 119579. [CrossRef]
8. Sharma, R.; Bansal, P.P. Use of different forms of waste plastic in concrete—A review. *J. Clean. Prod.* **2016**, *112*, 473–482. [CrossRef]
9. Mahjoubi, S.; Barhemat, R.; Meng, W.; Bao, Y. AI-guided auto-discovery of low-carbon cost-effective ultra-high performance concrete (UHPC). *Resour. Conserv. Recycl.* **2023**, *189*, 106741. [CrossRef]
10. Du, J.; Meng, W.; Khayat, K.H.; Bao, Y.; Guo, P.; Lyu, Z.; Abu-Obeidah, A.; Nassif, H.; Wang, H. New development of ultra-high-performance concrete (UHPC). *Compos. Part B Eng.* **2021**, *224*, 109220. [CrossRef]
11. Spiesz, P.; Rouvas, S.; Brouwers, H. Utilization of waste glass in translucent and photocatalytic concrete. *Constr. Build. Mater.* **2016**, *128*, 436–448. [CrossRef]
12. Torres, A.; Bartlett, L.; Pilgrim, C. Effect of foundry waste on the mechanical properties of Portland Cement Concrete. *Constr. Build. Mater.* **2017**, *135*, 674–681. [CrossRef]
13. Ling, T.C.; Nor, H.M. Granulated waste tyres in concrete paving block. In Proceedings of the 6th Asia-Pacific Structural Engineering and Construction Conference, Kuala Lumpur, Malaysia, 5–6 September 2006.
14. Modarres, A.; Hesami, S.; Soltaninejad, M.; Madani, H. Application of coal waste in sustainable roller compacted concrete pavement-environmental and technical assessment. *Int. J. Pavement Eng.* **2016**, *19*, 748–761. [CrossRef]
15. Gupta, S.; Kua, H.W.; Koh, H.J. Application of biochar from food and wood waste as green admixture for cement mortar. *Sci. Total Environ.* **2018**, *619–620*, 419–435. [CrossRef]
16. Khushnood, R.A.; Ahmad, S.; Restuccia, L.; Spoto, C.; Jagdale, P.V.; Tulliani, J.-M.; Ferro, G.A. Carbonized nano/microparticles for enhanced mechanical properties and electromagnetic interference shielding of cementitious materials. *Front. Struct. Civ. Eng.* **2016**, *10*, 209–213. [CrossRef]
17. Gupta, S.; Kua, H.W.; Pang, S.D. Effect of biochar on mechanical and permeability properties of concrete exposed to elevated temperature. *Constr. Build. Mater.* **2019**, *234*, 117338. [CrossRef]
18. Tan, K.; Qin, Y.; Wang, J. Evaluation of the properties and carbon sequestration potential of biochar-modified pervious concrete. *Constr. Build. Mater.* **2022**, *314*, 125648. [CrossRef]
19. Tan, K.; Wang, J. Substrate modified with biochar improves the hydrothermal properties of green roofs. *Environ. Res.* **2023**, *216*, 114405. [CrossRef]
20. Aziz, M.A.; Zubair, M.; Saleem, M.; Alharthi, Y.M.; Ashraf, N.; Alotaibi, K.S.; Aga, O.; Al Eid, A.A.A. Mechanical, non-destructive, and thermal characterization of biochar-based mortar composite. *Biomass Convers. Biorefinery* **2023**, 1–14. [CrossRef]
21. Mo, L.; Fang, J.; Huang, B.; Wang, A.; Deng, M. Combined effects of biochar and MgO expansive additive on the autogenous shrinkage, internal relative humidity and compressive strength of cement pastes. *Constr. Build. Mater.* **2019**, *229*, 116877. [CrossRef]
22. Meng, W.; Khayat, K. Effects of saturated lightweight sand content on key characteristics of ultra-high-performance concrete. *Cem. Concr. Res.* **2017**, *101*, 46–54. [CrossRef]
23. Lyu, Z.; Shen, A.; Meng, W. Properties, mechanism, and optimization of superabsorbent polymers and basalt fibers modified cementitious composite. *Constr. Build. Mater.* **2021**, *276*, 122212. [CrossRef]
24. Chen, L.; Zhang, Y.; Wang, L.; Ruan, S.; Chen, J.; Li, H.; Yang, J.; Mechtcherine, V.; Tsang, D.C. Biochar-augmented carbon-negative concrete. *Chem. Eng. J.* **2021**, *431*, 133946. [CrossRef]
25. Wang, L.; Chen, L.; Tsang, D.C.; Guo, B.; Yang, J.; Shen, Z.; Hou, D.; Ok, Y.S.; Poon, C.S. Biochar as green additives in cement-based composites with carbon dioxide curing. *J. Clean. Prod.* **2020**, *258*, 120678. [CrossRef]
26. Godwin, P.M.; Pan, Y.; Xiao, H.; Afzal, M.T. Progress in preparation and application of modified biochar for improving heavy metal ion removal from wastewater. *J. Bioresour. Bioprod.* **2019**, *4*, 31–42. [CrossRef]
27. Wang, L.; Wang, Y.; Ma, F.; Tankpa, V.; Bai, S.; Guo, X.; Wang, X. Mechanisms and reutilization of modified biochar used for removal of heavy metals from wastewater: A review. *Sci. Total Environ.* **2019**, *668*, 1298–1309. [CrossRef]
28. Chen, Q.Y.; Tyrer, M.; Hills, C.D.; Yang, X.M.; Carey, P. Immobilisation of heavy metal in cement-based solidification/stabilisation: A review. *Waste Manag.* **2009**, *29*, 390–403. [CrossRef]
29. Neve, S.; Sarkar, D.; Zhang, Z.; Datta, R. Optimized Production of Second-Generation Bioethanol from a Spent C4 Grass: Vetiver (*Chrysopogon zizanioides*). *Energies* **2022**, *15*, 9597. [CrossRef]
30. Neve, S.; Sarkar, D.; Datta, R. Effects of Pyrolysis Temperature and Residence Time on Physicochemical Properties of Biochar Derived from Spent Vetiver Roots. In Proceedings of the ASA, CSSA, SSSA International Annual Meeting, Baltimore, MD, USA, 8 November 2022.
31. Kimbrough, D.E.; Wakakuwa, J.R. Acid digestion for sediments, sludges, soils, and solid wastes. A proposed alternative to EPA SW 846 Method 3050. *Environ. Sci. Technol.* **1989**, *23*, 898–900. [CrossRef]
32. Feng, J.; Yang, Z.; Zeng, G.; Huang, J.; Xu, H.; Zhang, Y.; Wei, S.; Wang, L. The adsorption behavior and mechanism investigation of Pb(II) removal by flocculation using microbial flocculant GA1. *Bioresour. Technol.* **2013**, *148*, 414–421. [CrossRef]
33. *ASTM C143*; Standard Test Method for Slump of Hydraulic-Cement Concrete. ASTM: West Conshohocken, PA, USA, 2020.
34. *ASTM C109*; Standard Test Method for Compressive Strength of Hydraulic Cement Mortars. ASTM: West Conshohocken, PA, USA, 2021.
35. *ASTM C1698*; Standard Test Method for Autogenous Strain of Cement Paste and Mortar. ASTM: West Conshohocken, PA, USA, 2020.

36. Montour, M.R.; Hageman, P.L.; Meier, A.L.; Theodorakos, P.; Briggs, P.H. *EPA Method 1312 (Synthetic Precipitation Leaching Procedure); Leachate Chemistry Data for Solid Mine Waste Composite Samples from Silverton and Leadville, Colorado*; US Geological Survey: Reston, Virginia, 2022; pp. 98–624.
37. EPA, U. S. *Toxicity Characteristics Leaching Procedure, Method 1311. Test Methods for the Evaluation of Solid Waste*; EPA, U. S.: Washington, DC, USA, 1992.
38. Li, X.D.; Poon, C.S.; Sun, H.; Lo, I.M.C.; Kirk, D.W. Heavy metal speciation and leaching behaviors in cement based solidified/stabilized waste materials. *J. Hazard. Mater.* **2001**, *82*, 215–230. [CrossRef]
39. *OriginPro, Version 2022b*; OriginLab Corporation: Northampton, MA, USA, 2022.
40. Liu, Y.; Bao, Y.; Deng, L.; Zhang, Q. Experimental and finite element investigations on shear behaviors of stud connectors embedded in Engineered Cementitious Composite (ECC). *Eng. Struct.* **2023**, *277*, 115438. [CrossRef]
41. Meng, W.; Khayat, K.H. Experimental and numerical studies on flexural behavior of ultrahigh-performance concrete panels reinforced with embedded glass fiber-reinforced polymer grids. *Transp. Res. Rec.* **2016**, *2592*, 38–44. [CrossRef]
42. Li, X.; Lu, X.; Qi, J.; Bao, Y. Flexural behavior of fire-damaged concrete beams repaired with strain-hardening cementitious composite. *Eng. Struct.* **2022**, *261*, 114305. [CrossRef]
43. Li, X.; Li, Y.; Yan, M.; Meng, W.; Lu, X.; Chen, K.; Bao, Y. Cyclic behavior of joints assembled using prefabricated beams and columns with Engineered Cementitious Composite (ECC). *Eng. Struct.* **2021**, *247*, 113115. [CrossRef]
44. ABAQUS/CAE. *Software for Technical Computation*; Version 2017; Simulia Corp.: Providence, RI, USA, 2017.
45. Vellaipandian, K.; Mydeen, M.S.S.M.; Periasamy, R.P.; Soosaimarian, J.J. Effects of metakaolin on the shear capacity of EBFRP RC beams: An experimental and numerical investigation. *Constr. Build. Mater.* **2023**, *367*, 130291. [CrossRef]
46. Obaidat, A.T. Flexural behavior of reinforced concrete beam using CFRP hybrid system. *Eur. J. Environ. Civ. Eng.* **2021**, *26*, 6165–6187. [CrossRef]
47. Long, X.; Lee, C. Modelling of Two Dimensional Reinforced Concrete Beam-Column Joints Subjected to Monotonic Loading. *Adv. Struct. Eng.* **2015**, *18*, 1461–1474. [CrossRef]
48. *ACI 318-19*; Building Code Requirements for Structural Concrete (ACI 318-19) and Commentary. American Concrete Institute: Farmington Hills, MI, USA, 2019.
49. Sumer, Y.; Aktaş, M. Defining parameters for concrete damage plasticity model. *Chall. J. Struct. Mech.* **2015**, *1*, 149–155. [CrossRef]
50. Hafezolghorani, M.; Hejazi, F.; Vaghei, R.; Jaafar, M.S.B.; Karimzade, K. Simplified Damage Plasticity Model for Concrete. *Struct. Eng. Int.* **2017**, *27*, 68–78. [CrossRef]
51. Salmon, C.G.; Johnson, J.E.; Malhas, F.A. *Steel Structures Design and Behavior*, 5th ed.; Pearson: London, UK, 2008.
52. Asadnia, M.; Roddis, W.K. Modeling out-of-flatness and residual stresses in steel plate girders. In Proceedings of the Annual Stability Conference Structural Stability Research Council, Baltimore, MD, USA, 10–13 April 2018.
53. Choi, W.C.; Yun, H.D.; Lee, J.Y. Mechanical properties of mortar containing bio-char from pyrolysis. *J. Korea Inst. Struct. Maint. Insp.* **2012**, *16*, 67–74.
54. Gupta, S.; Kua, H.W. Effect of water entrainment by pre-soaked biochar particles on strength and permeability of cement mortar. *Constr. Build. Mater.* **2018**, *159*, 107–125. [CrossRef]
55. Muthukrishnan, S.; Gupta, S.; Kua, H.W. Application of rice husk biochar and thermally treated low silica rice husk ash to improve physical properties of cement mortar. *Theor. Appl. Fract. Mech.* **2019**, *104*, 102376. [CrossRef]
56. Sirico, A.; Bernardi, P.; Belletti, B.; Malcevschi, A.; Dalcanale, E.; Domenichelli, I.; Fornoni, P.; Moretti, E. Mechanical characterization of cement-based materials containing biochar from gasification. *Constr. Build. Mater.* **2020**, *246*, 118490. [CrossRef]
57. Meng, W.; Khayat, K.H. Effect of hybrid fibers on fresh properties, mechanical properties, and autogenous shrink-age of cost-effective UHPC. *J. Mater. Civ. Eng.* **2018**, *30*, 04018030. [CrossRef]
58. Maljaee, H.; Madadi, R.; Paiva, H.; Tarelho, L.; Ferreira, V.M. Incorporation of biochar in cementitious materials: A roadmap of biochar selection. *Constr. Build. Mater.* **2021**, *283*, 122757. [CrossRef]
59. Tan, K.H.; Wang, T.Y.; Zhou, Z.H.; Qin, Y.H. Biochar as a partial cement replacement material for developing sustainable concrete: An overview. *J. Mater. Civ. Eng.* **2021**, *33*, 03121001. [CrossRef]
60. Gupta, S.; Kua, H.W. Combination of biochar and silica fume as partial cement replacement in mortar: Performance evaluation under normal and elevated temperature. *Waste Biomass Valoriz.* **2020**, *11*, 2807–2824. [CrossRef]
61. Gupta, S.; Kua, H.W.; Pang, S.D. Biochar-mortar composite: Manufacturing, evaluation of physical properties and economic viability. *Constr. Build. Mater.* **2018**, *167*, 874–889. [CrossRef]
62. Tattersall, G.H. *Workability and Quality Control of Concrete*; CRC Press: Boca Raton, FL, USA, 1991.
63. Wille, K.; Boisvert-Cotulio, C. Material efficiency in the design of ultra-high performance concrete. *Constr. Build. Mater.* **2015**, *86*, 33–43. [CrossRef]
64. Chiaia, B.; Fantilli, A.P.; Guerini, A.; Volpatti, G.; Zampini, D. Eco-mechanical index for structural concrete. *Constr. Build. Mater.* **2014**, *67*, 386–392. [CrossRef]
65. Dixit, A.; Gupta, S.; Pang, S.D.; Kua, H.W. Waste Valorisation using biochar for cement replacement and internal curing in ultra-high performance concrete. *J. Clean. Prod.* **2019**, *238*, 117876. [CrossRef]
66. Long, G.; Gao, Y.; Xie, Y. Designing more sustainable and greener self-compacting concrete. *Constr. Build. Mater.* **2015**, *84*, 301–306. [CrossRef]

67. Aman, A.M.N.; Selvarajoo, A.; Lau, T.L.; Chen, W.H. Biochar as cement replacement to enhance concrete composite properties: A review. *Energies* **2022**, *15*, 7662. [CrossRef]
68. Zhang, D.; Jaworska, B.; Zhu, H.; Dahlquist, K.; Li, V.C. Engineered Cementitious Composites (ECC) with lime-stone calcined clay cement (LC3). *Cem. Concr. Compos.* **2020**, *114*, 103766. [CrossRef]
69. Shi, Y.; Long, G.; Ma, C.; Xie, Y.; He, J. Design and preparation of ultra-high performance concrete with low environmental impact. *J. Clean. Prod.* **2019**, *214*, 633–643. [CrossRef]
70. Alsalman, A.; Dang, C.N.; Martí-Vargas, J.R.; Hale, W.M. Mixture-proportioning of economical UHPC mixtures. *J. Build. Eng.* **2019**, *27*, 100970. [CrossRef]
71. Adamu, M.; Mohammed, B.S.; Liew, M.S. Mechanical properties and performance of high volume fly ash roller compacted concrete containing crumb rubber and nano silica. *Constr. Build. Mater.* **2018**, *171*, 521–538. [CrossRef]
72. Available online: https://www.dailymetalprice.com/metalpricecharts.php?c=st&u=mt&d=240 (accessed on 17 February 2023).
73. Fantilli, A.; Mancinelli, O.; Chiaia, B. The carbon footprint of normal and high-strength concrete used in low-rise and high-rise buildings. *Case Stud. Constr. Mater.* **2019**, *11*, e00296. [CrossRef]

materials

MDPI

Article

Effect of Through-the-Thickness Delamination Position on the R-Curve Behavior of Plain-Woven ENF Specimens

Mazaher Salamat-Talab [1], Alireza Akhavan-Safar [2], Ali Zeinolabedin-Beygi [1], Ricardo J. C. Carbas [2,*] and Lucas F. M. da Silva [3]

[1] Department of Mechanical Engineering, Arak University of Technology, Arak 38181-41167, Iran
[2] Institute of Science and Innovation in Mechanical and Industrial Engineering (INEGI), Rua Dr. Roberto Frias 400, 4200-465 Porto, Portugal
[3] Departamento de Engenharia Mecânica, Faculdade de Engenharia, Universidade do Porto, Rua Dr. Roberto Frias, 4200-465 Porto, Portugal
* Correspondence: rcarbas@fe.up.pt

Abstract: In this study, the effect of through-the-thickness delamination plane position on the R-curve behavior of end-notch-flexure (ENF) specimens was investigated using experimental and numerical procedures. From the experimental point of view, plain-woven E-glass/epoxy ENF specimens with two different delamination planes, i.e., $[0_{12}//0_{12}]$ and $[0_{17}//0_{7}]$, were manufactured by hand lay-up method. Afterward, fracture tests were conducted on the specimens by aiding ASTM standards. The main three parameters of R-curves, including the initiation and propagation of mode II interlaminar fracture toughness and the fracture process zone length, were analyzed. The experimental results revealed that changing the delamination position in ENF specimen has a negligible effect on the initiation and steady steady-state toughness values of delamination. In the numerical part, the virtual crack closure technique (VCCT) was used in order to analyze the imitation delamination toughness as well as the contribution of another mode on the obtained delamination toughness. The numerical results indicated that by choosing an appropriate value of cohesive parameters, the trilinear cohesive zone model (CZM) is capable of predicting the initiation as well as propagation of the ENF specimens. Finally, the damage mechanisms at the delaminated interface were investigated with microscopic images taken using a scanning electron microscope.

Keywords: delamination; plain-woven composites; mode II interlaminar fracture toughness; cohesive zone model (CZM); virtual crack closure technique

check for updates

Citation: Salamat-Talab, M.; Akhavan-Safar, A.; Zeinolabedin-Beygi, A.; Carbas, R.J.C.; da Silva, L.F.M. Effect of Through-the-Thickness Delamination Position on the R-Curve Behavior of Plain-Woven ENF Specimens. *Materials* **2023**, *16*, 1811. https://doi.org/10.3390/ma16051811

Academic Editor: Ayman El-Zohairy

Received: 20 January 2023
Revised: 14 February 2023
Accepted: 17 February 2023
Published: 22 February 2023

1. Introduction

Polymer matrix composites (PMCs) are currently of interest to industry and academia—especially in the automotive, aerospace, wind turbine, military, and aerospace sectors—due to their superior properties, such as high strength-to-weight ratio and good mechanical performance. The use of glass-fiber-reinforced polymer (GFRP) [1,2] and carbon-fiber-reinforced polymer (CFRP) [3] has become more established because of their good mechanical properties. One of the most important mechanisms leading to damage in these materials is delamination. Delamination refers to a crack that propagates between two adjacent layers in laminated composites causing a decrease in the strength and stiffness of the structure, often leading to sudden and catastrophic failures. Investigating the delamination behavior of laminated composites in different failure modes has attracted the attention of many researchers [4–9]. One of the key parameters in the delamination analysis of composites is fracture toughness, which has been studied by several researchers [10–16]. Various samples, such as end-notched flexure (ENF) specimens, 4ENF, end-loaded split (ELS), and over notched flexure (ONF) are used to investigate the mode II interlaminar fracture toughness of laminated composites.

Morais and Pereira [17] investigated the mode II interlaminar fracture in ENF and 4ENF carbon/epoxy samples with unidirectional fibers using the effective crack method. Parameters including the beam thickness, width, crack position, and stacking sequence affect the mode II fracture toughness. Wang et al. [18] investigated the mode II interlaminar fracture toughness using four types of ENF, ELS, 4ENF, and ONF test specimens for unidirectional carbon/epoxy composites. The results showed that the initial fracture toughness value obtained from the ENF test has a smaller scatter in the obtained experimental data. Additionally, the fracture toughness obtained from the ENF test was almost 2% lower than the 4ENF test. Andersons and Konig [19] comprehensively investigated the effect of different parameters on the fracture toughness of laminated composites. They observed that the traditional fracture toughness characterization through various tests on unidirectional composites (according to the interface lay-up and loading mode) can lead to over- or underestimation of crack growth resistance. Yashiro et al. [20] characterized the behavior of the R-curve under mode II loading condition. For this purpose, they used double-end notched tension samples. The results showed that the mode II fracture toughness corresponds well with the results obtained from ENF samples. Chai [21] investigated the effect of the interface fiber angle and the properties of the matrix on the fracture toughness of laminated composites with carbon fibers. However, in this study, he did not report the resistance curve of ENF samples. Hosseini et al. [22] investigated the effect of the mat layer between the woven fiber layers on the mode II fracture toughness of glass/epoxy composites. They showed that the presence of the mat layer between the layers reduces the initiation and propagation values of fracture toughness by 59% and 35%, respectively. Mollón et al. [23] analyzed the effect of crack plane position on the critical strain energy release rate in carbon/epoxy ENF samples considering unidirectional fibers. They found that the deviation of the crack plane from the middle surface of the sample has no effect on the strain energy release rate of mode II. These researchers used a numerical method based on the virtual crack-closure technique (VCCT) to obtain fracture toughness. The effect of interfacial angles $0/\theta$ and $\theta/-\theta$ in multidirectional samples on the mode II fracture toughness was studied by Pereira et al. [24]. The results showed that increasing the angle θ in the studied delamination interface angles causes an increase in the interlaminar fracture toughness. Akhavan-Safar et al. [25] studied the effect of natural date palm tree fibers on the resistance curve of ENF samples. They showed that adding 5 wt% of petiole fibers increased the initial fracture toughness of glass/epoxy laminated composites by 61%. Additionally, they found that adding 5 wt% of bunch fibers improved the propagation of mode II interlaminar fracture toughness by 73%.

Numerical modeling of the delamination phenomenon can be conducted using two common methods: cohesive zone model (CZM) and VCCT. CZM is a suitable choice for crack initiation and propagation analysis in various materials, such as metals, adhesives, and laminated composites, whose main advantage is the elimination of the singularity at the crack tip.

The traction–separation model is considered as a key parameter for CZM [26,27]. Many researchers have used CZM to numerically simulate the load–displacement behavior of ENF specimens [28–33]. Moura and Morais [34] used a bilinear CZM based on 3D interlayer finite element modeling for ENF and ELS samples. Chandra et al. [35] compared the exponential and bilinear CZM shapes. They showed that the shape of the interface laws has a significant effect on the numerical simulation results. Williams and Hadavinia [36] analytically investigated the initiation and propagation of delamination in composites. Anyfantis and Tsouvalis [37] studied the resistance curve behavior of unidirectional ENF samples by using different data reduction methods, including corrected beam theory (CBT), the compliance calibration method (CCM), and the compliance-based beam method (CBBM). In addition, they used different CZM shapes, such as trapezoidal, linear-exponential, and bilinear. They found that the CBBM in relation to the trapezoidal CZM can estimate the experimental data more precisely. In another study, five CZM shapes, including the equivalent linear elastic, equivalent constant stress, cubic, bilinear, and exponential models with the assumption of interface shear strength and mode II fracture toughness, were analyzed by Ouyang et al. [38]. It was found that the CZM shape has a great effect on the critical load and the load–displacement curve for short cracks. Salamat-Talab et al. [39] used bilinear, trapezoidal, exponential, and linear-exponential models to predict the load–displacement behavior in unidirectional laminated composite under mode II loading. The results showed that the use of linear–exponential softening behavior provides a suitable answer for estimating the load-displacement. Heidari-Rarani and Ghasemi [40] developed a trilinear CZM for ENF samples by using the effect of the R-curve. They found the cohesive parameters using a semi-analytical method and showed the effectiveness of their model for analyzing the propagation behavior. Dourado et al. [41] presented a bilinear model for unidirectional specimens with a large fracture process zone. The model parameters were determined by fitting the load-displacement data obtained from finite element analysis (FEA) with the relevant experimental data.

As indicated above, despite the extensive analysis of mode II fracture toughness of composites, however, the effect of through-the-thickness delamination position on the mode II interlaminar fracture toughness and on the cohesive modelling of plain-woven laminated composites has not been well explored. Accordingly, the present study deals with this topic. To achieve this, in the first step, the delamination behavior of symmetric and asymmetric ENF samples manufactured by E-glass/epoxy plain woven laminated composite was experimentally investigated. Then, the distribution of the strain energy release rate across the sample was obtained numerically using VCCT. In addition, the suitable CZM shape for predicting the initiation and propagation of delamination was also investigated and the most accurate cohesive parameters were presented. Numerical and experimental results were also compared.

2. Properties and Fabrication of Samples

The composite samples used in this study consist of 24 layers made using the hand lay-up method. The fibers used are woven glass fibers of 200 g/m^2 with a harness of 2. EPL 1012 resin and RH 112 hardener with a weight ratio of 100 to 12 were used to make the plain-woven composites. In order to create pre-cracks within the laminates, a Teflon tape with a thickness of 13 microns was used. In addition, to change the position of the crack in the samples, the stacking sequences of $[0_{12}//0_{12}]$ and $[0_{17}//0_{7}]$—where "//" indicates the position of the crack between the layers—were used. In the stacking sequences of $[0_{12}//0_{12}]$, the crack was placed between Layers 12 and 13, and in the stacking sequences of $[0_{17}//0_{7}]$, the crack was placed between Layers 17 and 18. Considering that the fabric is used with harness 2, so the layer shown with zero angle has 90-degree fibers equal to zero fibers. The samples were placed at ambient temperature for 24 h for the initial curing, and after that, in order to post-cure, they were subjected to a temperature of 60 °C for 4 h (according to the manufacturer's recommendation) to improve their mechanical properties. ENF samples were cut to a width of 25 mm and a total length of 140 mm, as shown in Figure 1. The geometrical properties of the symmetric and asymmetric ENFs are summarized in Table 1. After cutting, the edges of the sample were polished with soft sandpaper to better show the crack during growth. It should be noted that for the repeatability of the test results, at least three specimens were manufactured and tested for each stacking sequence. Tensile test samples and three-point bending tests were performed to determine the mechanical properties of plain-woven composites such as in-plane shear modulus and flexural modulus according to ASTM D3518 [42] and ASTM D790 [43] standards, respectively. Additionally, in accordance with the ASTM D3039 [44] standard, the longitudinal and transverse elastic modulus were determined, as shown in Table 2.

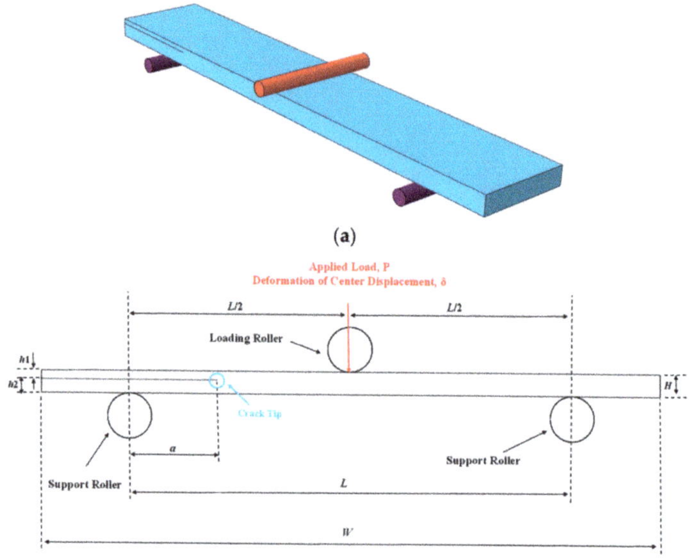

Figure 1. Schematic and geometrical configuration of ENF specimen (**a**) 3D scheme (**b**) dimensions and boundary condition.

Table 1. Geometrical properties of ENF specimen with symmetrical and asymmetrical delamination plane.

Geometrical Parameters	ENF Specimen with Symmetrical Delamination Plane (S-ENF) $[0_{12}//0_{12}]$	ENF Specimen with Asymmetrical Delamination Plane (As-ENF) $[0_{17}//0_{7}]$
Total length of the beam, *W* (mm)	140	140
Span length, *L* (mm)	100	100
Beam width, *b* (mm)	25	25
Initial crack length, *a* (mm)	30	30
The thickness of arm-1, h_1 (mm)	2.06	1.22
The thickness of arm-2, h_2 (mm)	2.06	2.94
Total thickness of the beam, *H* (mm)	4.12	4.16

Table 2. The mechanical properties of E-glass/epoxy laminated composites.

Properties	Value
Longitudinal elastic modulus (E_1)	19.5 GPa
Transverse elastic modulus (E_2)	19.5 GPa
In-plane shear modulus (G_{12})	3.1 GPa
Flexural modulus (E_b)	18.1 GPa
Poisson's ratio (ν_{12})	0.29

Fracture Test and Data Reduction Method

A universal testing machine (Santam-STM-150, SANTAM Engineering Design Company, Tehran, Iran) and a Bongshin load cell with a capacity of 2000 Kgf were used to conduct the fracture tests and to obtain the load–displacement curves. For this purpose, to determine the mode II fracture toughness according to the ASTM D7905 standard [45], which proposes the compliance calibration approach, three-point bending tests were first carried out on the ENF samples along the cracks of 20, 25, 35, and 40 mm lengths to obtain the compliance of the specimens. It should be noted that the crack length is measured from the center of the left roller support. Then, the fracture tests were conducted on specimens with a crack length of 30 mm. The crosshead speed for the test was considered to be 0.5 mm/min. In addition, based on the above standard, the mode II interlaminar fracture toughness was obtained using Equation (1):

$$G_{IIc} = \frac{3P^2ma^2}{2b} \tag{1}$$

In this equation, G_{IIC} is the mode II interlaminar fracture toughness, a is the crack length, m is the calibration parameter, P is the applied load along the crack length, and b is the sample width. To obtain the calibration parameter, it is necessary to draw the compliance diagram in terms of the third power of the crack length. The slope of the curve fitted to this plot indicates the m value. To measure the length of the crack, according to Figure 2, a sticky paper ruler was placed on the side of the samples, and the length of the crack was measured during the test. It should be noted that in order to obtain the interlaminar fracture toughness, the load, the load line displacement, and the length of the crack should be recorded simultaneously. The load and displacement values were recorded at 10 s intervals, and to measure the crack length, images of the sample were taken at the same time interval with a suitable quality (4000 × 6000 pixels) using the Canon D80 digital camera (Canon Inc., Tokyo, Japan).

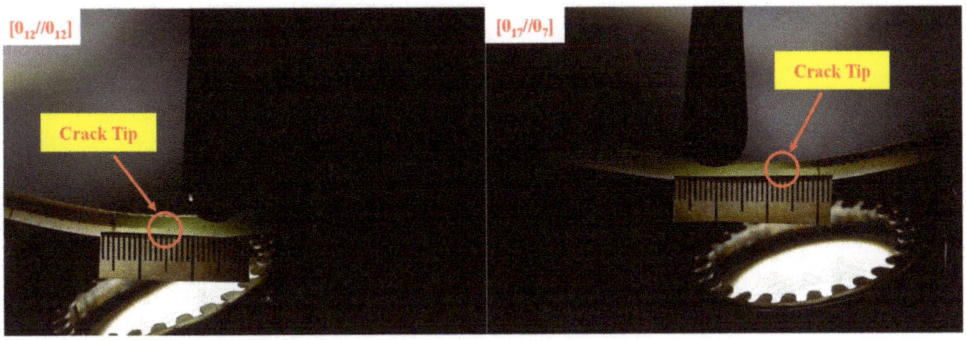

Figure 2. ENF specimens under mode II loading.

3. Results and Discussion

The load–displacement curves for the ENF samples obtained from the fracture test are illustrated in Figure 3. The behavior of these graphs is initially linear due to the elastic behavior of these materials, and then it changes to a nonlinear state, which actually indicates initial crack growth in composite samples. These data were used to calculate the initial fracture toughness of the samples. As the graphs become nonlinear, the stiffness decreases but the load continues to increase, and after it has reached its maximum value, the fracture toughness fluctuates around a certain value, which corresponds to the propagation value or steady-state value of fracture toughness. It should be noted that after the load reaches its maximum value, with the increase in the delamination length, the load starts to decrease. According to Figure 3, it is clear that the maximum load in the stacking sequence of $[0_{17}//0_7]$ is higher than that of the $[0_{12}//0_{12}]$ because of the higher stiffness of samples with the $[0_{17}//0_7]$ stacking sequence [46]. In addition, the behavior of the load–displacement curve indicates that the crack growth in plain-woven composites has occurred in a stable manner. The symmetry (S-ENF) and asymmetry (As-ENF) results in Figure 3 correspond to samples with $[0_{12}//0_{12}]$ and $[0_{17}//0_7]$ stacking sequence, respectively.

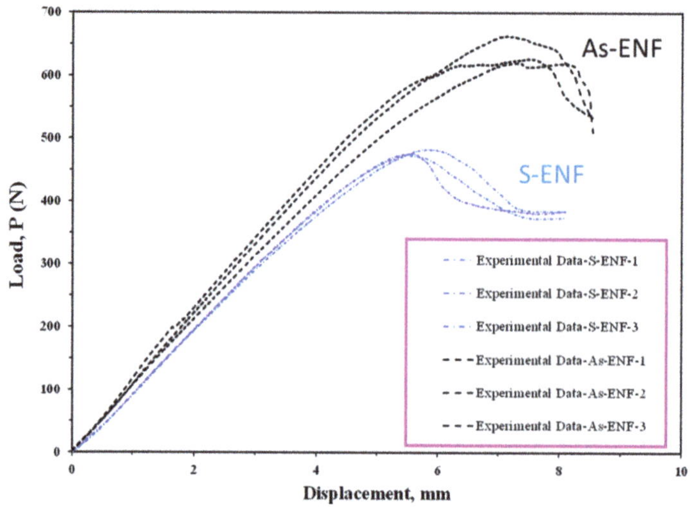

Figure 3. Load–displacement curves of ENF specimens with symmetric (S-ENF) and asymmetric (As-ENF) delamination positions.

3.1. Fracture Toughness and R-Curve

Figure 4 shows the R-curves (resistance curve that denotes the strain energy release rate (G_{II}) vs. the delamination length) of ENF samples with symmetric and asymmetric delamination planes. Comparing the results for two symmetric and asymmetric samples shows that the initiation and propagation of mode II interlaminar fracture toughness for ENF specimens with asymmetric delamination positions through the thickness increases by 4.85% and 9.28%, respectively, which is not a significant increase. In other words, two samples have almost identical R-curves. Maximum load, fracture process zone length, and initiation and propagation values of fracture toughness are listed in Table 3. It should be mentioned that the length at which the crack starts to grow to reach the steady state fracture toughness is called steady state fracture process zone (FPZ) length, l_{FPZ}^{ss}.

Table 3. Average values of maximum load, steady-state FPZ length, initiation, and propagation values fracture toughness for ENF specimens with different delamination planes.

Delamination Plane	Maximum Load, P (N)	Initial Fracture Toughness, G_{IIC}^{i} (J/m^2)	Propagation Fracture Toughness, $G_{IIC}^{Prop.}$ (J/m^2)	FPZ Length, L_{FPZ}^{ss} (mm)
$[0_{12}//0_{12}]$	477.60 ± 4.48	226.82 ± 85.96	1905.85 ± 34.74	8.67 ± 0.29
$[0_{17}//0_{7}]$	636.87 ± 23.03	237.83 ± 77.50	2082.71 ± 63.96	11.50 ± 1.80

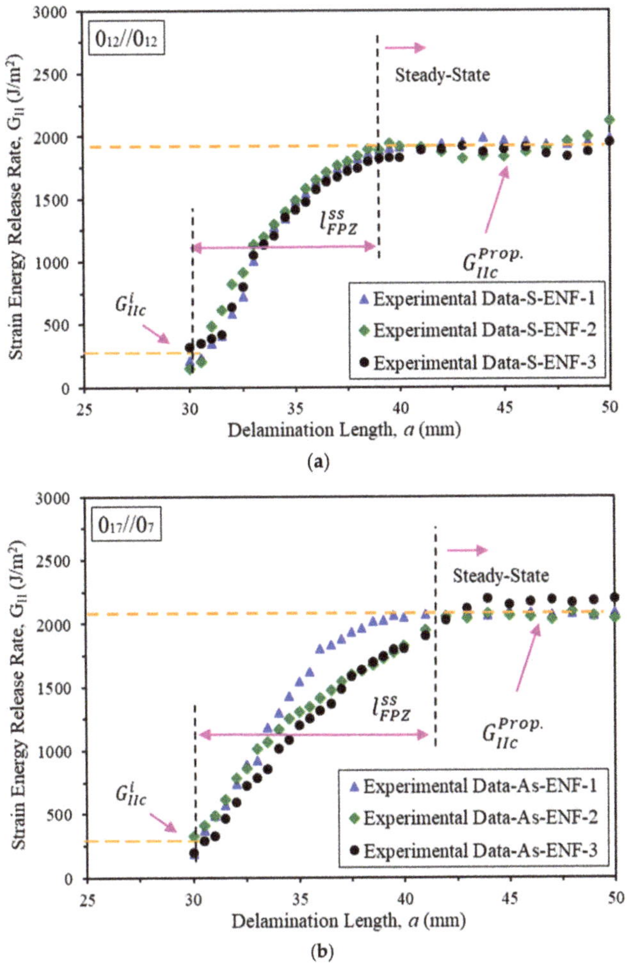

Figure 4. R-curves of ENF specimens with (**a**) symmetric delamination planes (S-ENF) and (**b**) asymmetric delamination planes (As-ENF).

3.2. Morphology

In order to better understand the mechanisms of crack initiation and propagation, images of the interface of ENF samples with different delamination planes were prepared by the scanning electron microscope (SEM) method. As a significant result in Figure 5, it can be concluded that the main mechanism of crack growth in both samples is shear huckles, which occur due to shear stresses at the interface. Additionally, fiber traces can be mentioned as a mechanism of crack growth, which shows the effect of fiber residue on the resin. Therefore, since the damage mechanisms at the interface of the samples are similar, there is no significant difference in their fracture toughness.

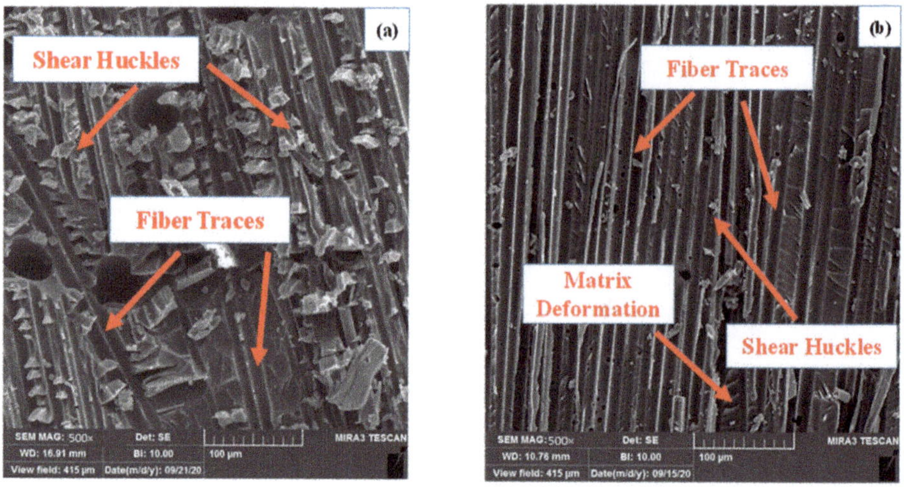

Figure 5. SEM images of delamination interfaces of (**a**) S-ENF and (**b**) As-ENF samples.

4. Finite Element Simulation of the Delamination Growth

4.1. Simulation of the Delamination Using CZM

In this section, the numerical simulation of delamination growth in ENF samples using Abaqus software is discussed. One of the important goals in this study is the evaluation and efficiency of the cohesive model in simulating the fracture process. It should be noted that for the numerical simulation, the dimensions of the samples considered for fracture tests were taken into account.

In CZM, the crack growth can be modeled as a gradual decrease in the stiffness of the cohesive element, which is based on the softening of the elements at the crack tip region. When the stress reaches the maximum value, the gradual decrease in the stiffness of the cohesive element begins, and when it reaches zero, the element completely fails. In this research, three different CZMs have been used, as shown in Figure 6a–c. In the bilinear CZM, the stress value increases linearly and after reaching the maximum cohesive shear stress, the damage starts in the element. The softening zone in this type of model is linear as shown in Figure 6a. However, in this model, the process of increasing the damage parameter is nonlinear. In the second model, which is a trapezoidal model, after the initial linear and elastic region, the constant shear stress region is experienced in the cohesive zone. In other words, when the amount of slipping increases in the delamination area, the amount of shear stress is constant, but damage occurs in the element. It should also be noted that the area under the CZM curve up to point s_1^* is equal to the initial fracture toughness. After this part, the linear softening area of the cohesive element is created until finally, the tolerable traction value in the element reaches zero. In the third model, which is the trilinear model, after the elastic region, there are two linear softening regions with unequal slopes. The first softening zone continues until the area under the curve in this zone reaches the value of the initial fracture toughness. Then, the softening region continues with another slope and linearly, and the amount of traction reaches zero at the end of the damage. The area under the mentioned curve is also equal to the difference between the initial and propagation fracture toughness. Unlike other models, in the bilinear model, there are not any steps that separates the initiation and propagation areas. Therefore, the entire area under the curve is considered to be propagation. In the following, the performance of each of these models in predicting the delamination behavior of symmetric and asymmetric ENF samples is examined. k_p is the stiffness of the linear–elastic region of the traction–separation law. In this study, for interfacial stiffness, values between $10^4 \frac{N}{mm^3}$ to $10^6 \frac{N}{mm^3}$ were used, which has an insignificant effect in the numerical results [39]. The s_f^* is the final point of damage and after this point, no traction exists between the two crack surfaces. This parameter in cohesive laws can be calculated using the other defined parameters.

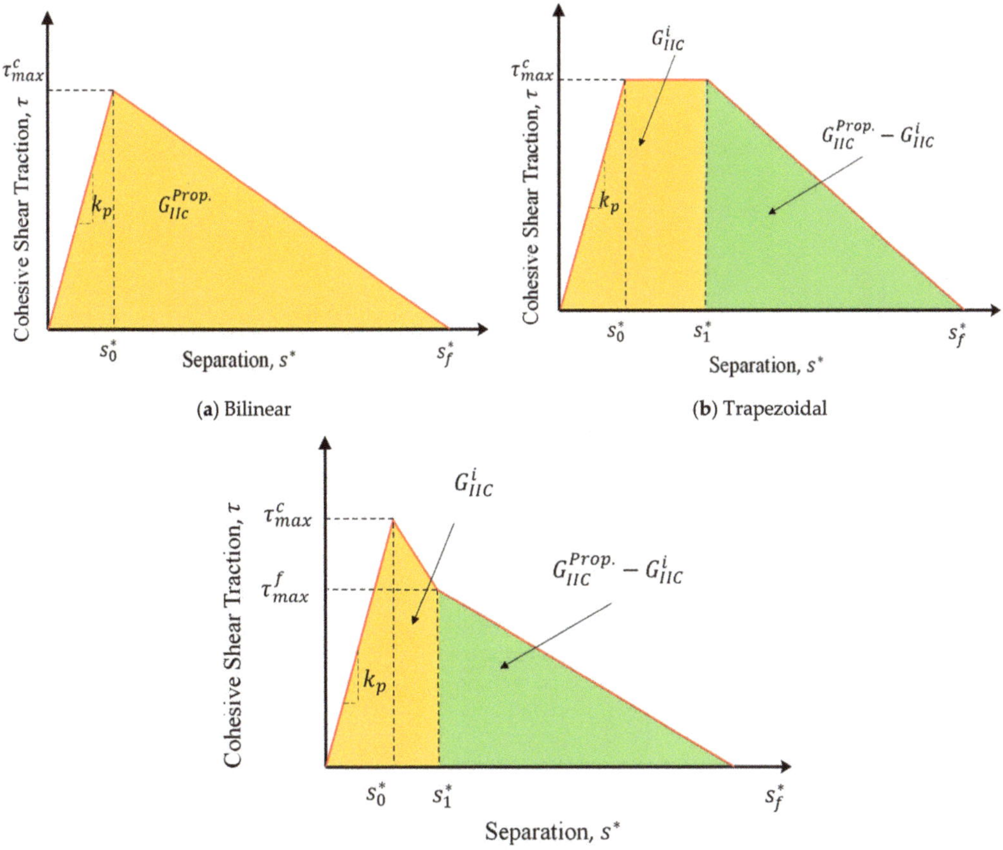

Figure 6. Different types of cohesive shear traction versus separation.

In order to model the delamination area and the possible path of crack growth, eight-node three-dimensional cohesive (COH3D8) elements with a thickness of 0.01 mm were used. In addition, eight-node linear brick (C3D8R) elements were used for the whole ENF sample. For more accurate numerical results and to prevent divergence of finite element analysis, as well as to shorten the analysis time, in the middle section of the beam along the longitudinal direction, after the crack tip, the mesh was refined (using 10 elements for each 1 mm), while in the other sections 2 elements were used per 1 mm. Additionally, in this section, 10 elements were used along the width of the beam (8 elements per 1 mm), and for the other sections of the beam, 2 elements per 1 mm were used through the longitudinal direction. In the transverse direction, 10 elements are considered (Figure 7). The simply supported boundary conditions are considered for the right and left nodes of the ENF beam, and the vertical displacement is taken for the node regions of the upper arm of the beam at the loading point.

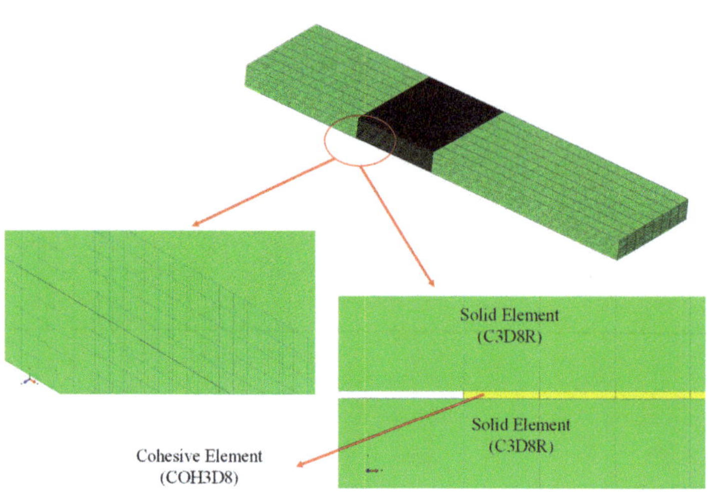

Figure 7. Mesh generated for the ENF specimens.

Maximum cohesive shear traction is one of the important and effective parameters in CZM for load–displacement prediction in FEA. In order to determine the value of the maximum cohesive shear traction, first—using the bilinear model and different values of the maximum stress—the starting point of delamination is predicted. By considering the numerical results provided in Figure 8, the stiffness and maximum load increase with increasing the maximum cohesive shear stress, τ^{c}_{Max}. In addition, it can be seen that this increase is stopped when the maximum traction reaches 30 MPa. Therefore, in this study, the maximum stress value in the cohesive area is considered to be 30 MPa in all numerical models. It should be noted that considering the results shown in Figure 8, using the initial fracture toughness the delamination behavior of the symmetric ENF sample cannot be predicted well, and therefore, it is necessary to use the propagation fracture toughness in FEA in order to correctly model the delamination behavior of these samples. In Figure 9, the finite element results are compared to experimental ones. Numerical results show that the bilinear and trapezoidal models are not able to predict the delamination behavior of the sample when the delamination growth starts. Additionally, the predicted maximum load in these CZM types (bilinear and trapezoidal) is much higher than its experimental values. In order to complete and correct these numerical results, a trilinear model has been used to predict the load–displacement curves. As can be seen in Figure 10, the value of τ^{f}_{Max} has a great influence on the nonlinear region when the delamination growth starts. It also influences the numerically obtained maximum load. Additionally, the maximum load predicted in the finite element model increases with this stress. By using the trial and error method, the value of this stress has been considered to be approximately 15 MPa. The numerical results for this stress indicated a good compliance of the load-displacement curve of the finite element model with its experimental results. Now, according to the stresses obtained, the trilinear model with the obtained stresses was used to model the initiation and propagation of the delamination in the asymmetric ENF samples. The numerical results are compared with the experimental data in Figure 11. Numerical results show that the trilinear model was able to predict more precisely the load–displacement behavior of the asymmetric sample. In other words, it can be said that by extracting the CZM in the symmetric ENF sample, the obtained cohesive model parameters can be used to predict the delamination behavior of the samples that are also asymmetric and under mode II loading.

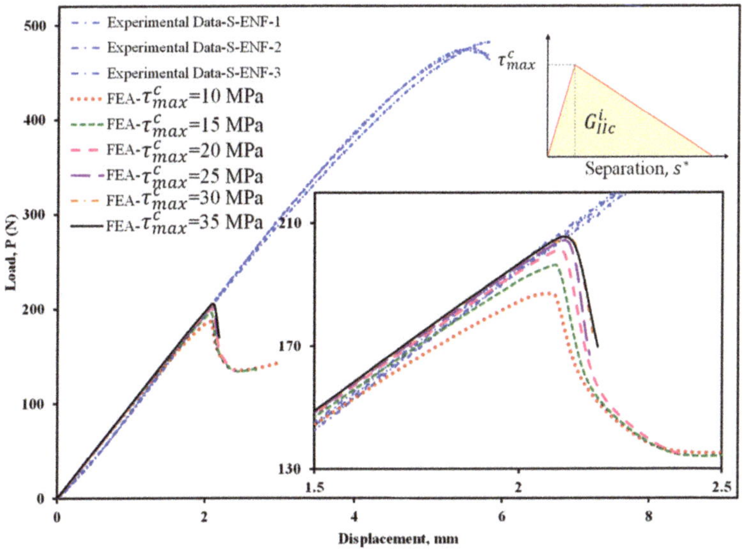

Figure 8. Prediction of initiation point of load–displacement curves by FEA using bilinear CZM and initiation fracture toughness.

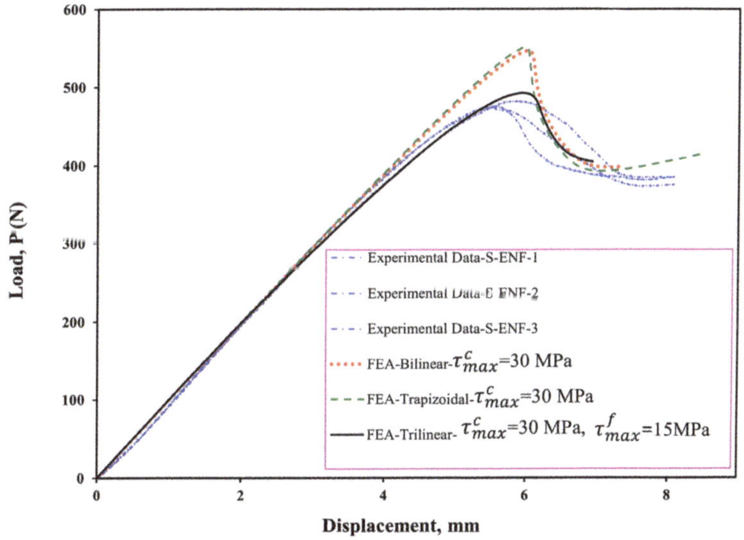

Figure 9. A comparison of the load–displacement curves of experiments and estimates using FEA for ENF specimen with symmetric delamination plane (S-ENF).

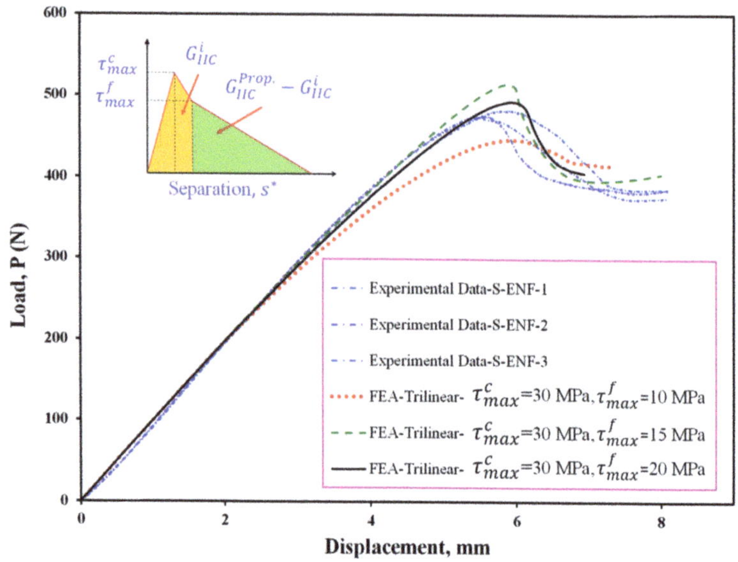

Figure 10. Effect of trilinear CZM parameters on the predicted load–displacement curve for ENF specimen with symmetric delamination plane (S-ENF).

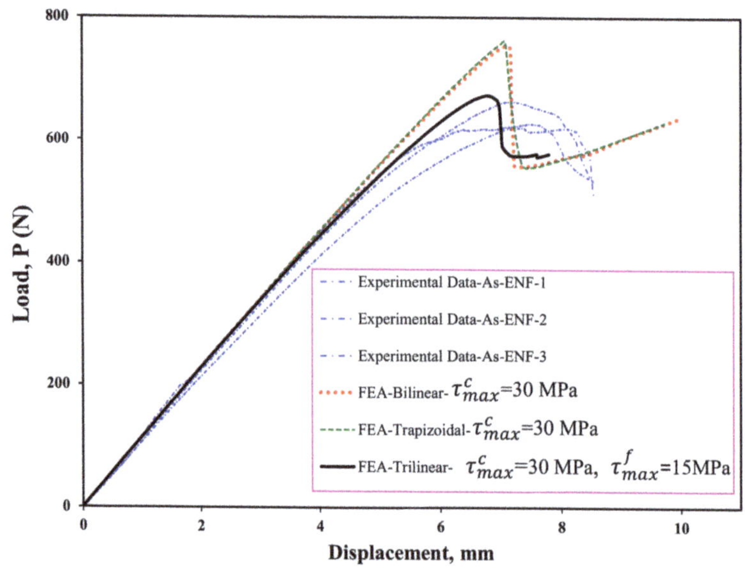

Figure 11. A comparison of the load–displacement curves of experiments and estimated by FEA for ENF specimen with asymmetric delamination plane (As-ENF).

4.2. *Effect Delamination Position and Mode Mixity on the Obtained Fracture Toughness*

In order to check the distribution of strain energy release rate in symmetric and asymmetric ENF samples, the VCCT method has been used by authors [47–49]. In order to numerically model these samples, a three-dimensional model according to Figure 12 was used. In Figure 12, Δa and Δb are the crack extension along the longitudinal direction and width wise of the delaminated beam, respectively. The boundary conditions in this modeling are the same as in the previous case. The strain energy release rate is highly dependent on the considered crack extension value (Δa). In other words, the length of the element considered at the crack tip. In this research, by considering different lengths for the elements near the crack tip, the sensitivity of the results to the size of the elements was investigated. Numerical results showed that the value of Δa = 0.1 mm is a suitable length for the element near the crack tip. It should be also noted that 25 elements were considered in the width of the beam, and 5 elements were considered for each of the arms of the beam along its thickness. The numerical results of the strain energy release rate distribution across the width of the beam for two ENF samples with symmetric and asymmetric delamination planes are shown in Figures 13 and 14, respectively. Accordingly, the mode II strain energy release rate has a uniform distribution along the beam width except in the vicinity of the two edges. In addition, the mode III strain energy release rate contributed to the obtained results near both edges of the ENF beam due to the free edge effect [49]. Additonaly, mode I strain energy release rate is approximately zero along with the beam width. It can be concluded that the mode mixity has a negligible effect on the extracted strain energy release rate. In other words, pure mode II occurs in both symmetric and asymmetric samples. It should be noted that in Figures 13 and 14, G_I, G_{II}, and G_{III} are the strain energy release rates in mode I, mode II, and mode III, respectively.

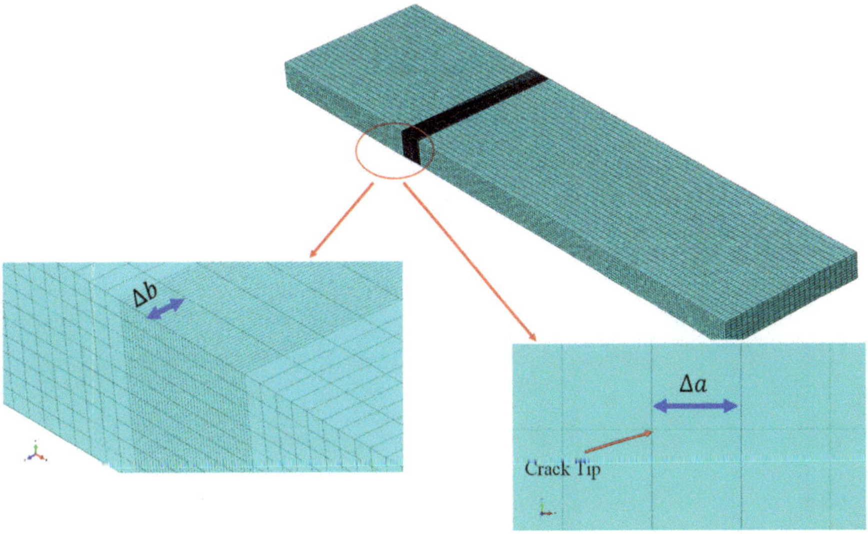

Figure 12. Meshing style of the specimen and the magnified view of the crack tip of ENF specimen.

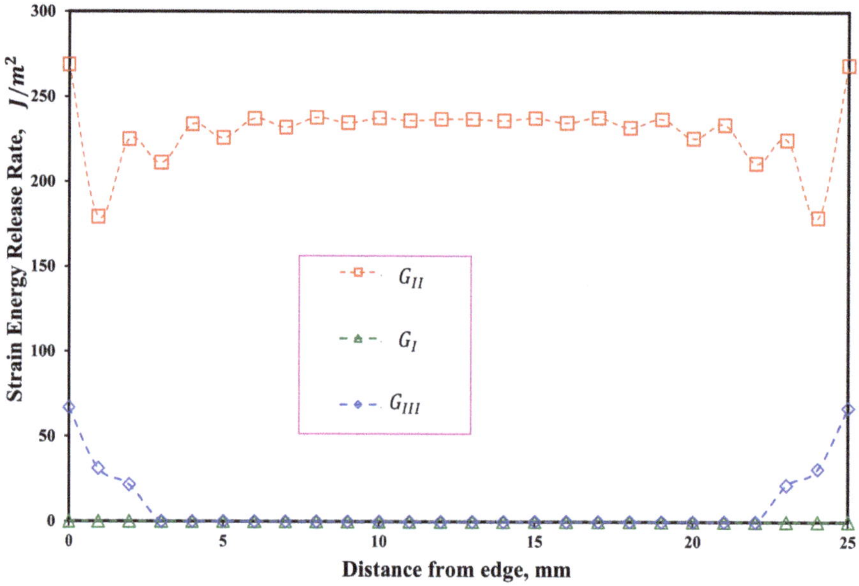

Figure 13. Distribution of the strain energy release rate along the width of ENF specimens with symmetric delamination plane (S-ENF).

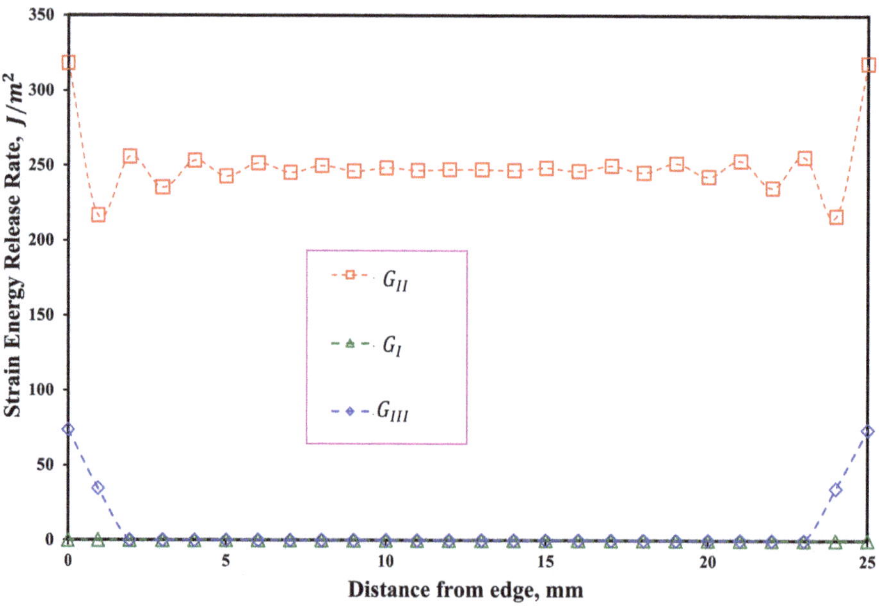

Figure 14. Distribution of the strain energy release rate along the width of ENF specimens with asymmetric delamination plane (As-ENF).

5. Conclusions

In the first part of this study, the plain-woven glass/epoxy ENF samples were fabricated with symmetry and asymmetry using the hand lay-up method. Then, according to the ASTM standards, mode II fracture tests were performed. The delamination behavior of the ENF samples was investigated and the R-curves were extracted. The experimental results showed that the initiation and propagation values of fracture toughness through the thickness delamination position of laminated composites increased by 4.85% and 9.28%, respectively, which is not a significant increase. This means that the symmetry and asymmetry samples have almost the same R-curves. In addition, to better understand the mechanisms of delamination initiation and growth, images of the interface of ENF samples were produced using SEM. Shear huckles and fiber traces can be mentioned as the crack growth mechanism. In the second part of this study, using the finite element method, the delamination growth was modeled. In the numerical analysis, CZM was used to predict the initiation and propagation of delamination. Therefore, three different models—including bilinear, trapezoidal, and trilinear—were used, and the performance of each model was investigated. The numerical results indicated that by choosing an appropriate value for cohesive parameters, the trilinear CZM is capable of predicting the initiation as well as propagation of the delamination phenomenon for both the symmetric and asymmetric ENF specimens. In addition, the distribution of strain energy release rate obtained by the VCCT across the beam width showed that due to the free edge effect, the mode III strain energy release rate contributes only in the vicinity of two edges of the beam width. In other words, the contribution of mode III to the obtained results is negligible.

Author Contributions: Conceptualization, M.S.-T.; Data curation, M.S.-T.; Formal analysis, A.A.-S.; Investigation, M.S.-T. and A.Z.-B.; Methodology, M.S.-T. and A.A.-S., Project administration, L.F.M.d.S., Supervision, L.F.M.d.S. and R.J.C.C., Validation, M.S.-T.; Writing—original draft, M.S.-T. and A.Z.-B.; Writing—review & editing A.A.-S., L.F.M.d.S. and R.J.C.C. All authors have read and agreed to the published version of the manuscript.

Funding: This research received no external funding.

Institutional Review Board Statement: Not applicable.

Informed Consent Statement: Not applicable.

Data Availability Statement: Not applicable.

Conflicts of Interest: The authors declare no conflict of interest.

References

1. Moazzami, M.; Ayatollahi, M.; Akhavan-Safar, A.; De Freitas, S.T.; Poulis, J.; Da Silva, L. Effect of Cyclic Aging on Mode I Fracture Energy of Dissimilar Metal/Composite DCB Adhesive Joints. *Eng. Fract. Mech.* **2022**, *271*, 108675. [CrossRef]
2. Moazzami, M.; Ayatollahi, M.; Akhavan-Safar, A.; De Freitas, S.T.; Poulis, J.; Da Silva, L. Influence of Cyclic Aging on Adhesive Mode Mixity in Dissimilar Composite/Metal Double Cantilever Beam Joints. *Proc. IME J. Mater. Des. Appl.* **2022**, *236*, 1476–1488. [CrossRef]
3. Akhavan-Safar, A.; Eisaabadi Bozchaloei, G.; Jalali, S.; Beygi, R.; Ayatollahi, M.; Da Silva, L. Impact Fatigue Life of Adhesively Bonded Composite-Steel Joints Enhanced with the Bi-Adhesive Technique. *Materials* **2023**, *16*, 419. [CrossRef] [PubMed]
4. Shokrieh, M.; Heidari-Rarani, M.; Ayatollahi, M. Delamination R-curve as a Material Property of Unidirectional Glass/Epoxy Composites. *Mater. Des.* **2012**, *34*, 211–218. [CrossRef]
5. Shokrieh, M.; Zeinedini, A.; Ghoreishi, S. On the Mixed Mode I/II Delamination R-curve of E-Glass/Epoxy Laminated Composites. *Compos. Struct.* **2017**, *171*, 19–31. [CrossRef]
6. Gong, Y.; Jiang, L.; Li, L.; Zhao, J. An Experimental and Numerical Study of the Influence of Temperature on Mode II Fracture of a T800/Epoxy Unidirectional Laminate. *Materials* **2022**, *15*, 8108. [CrossRef] [PubMed]
7. Huo, X.; Jiang, Z.; Luo, Q.; Li, Q.; Sun, G. Mechanical Characterization and Numerical Modeling on the Yield and Fracture Behaviors of Polymethacrylimide (PMI) Foam Materials. *Int. J. Mech. Sci.* **2022**, *218*, 107033. [CrossRef]
8. Ahmadi-Moghadam, B.; Taheri, F. Influence of Graphene Nanoplatelets on Modes I, II and III Interlaminar Fracture Toughness of Fiber-Reinforced Polymer Composites. *Eng. Fract. Mech.* **2015**, *143*, 97–107. [CrossRef]
9. Zeinolabedin Beygi, A.; Salamat-talab, M.; Farrokhabadi, F.; Moslemi Naeini, H. Experimental Investigation of the Effect of Natural Microfibers on the Mode I Fracture Toughness of Plain-Woven Laminated Composites. *Modares Mech. Eng.* **2022**, *22*, 271–279. Available online: https://mme.modares.ac.ir/article-15-53223-en.html (accessed on 30 November 2022). (In Persian).
10. Almansour, F.; Dhakal, H.; Zhang, Z.Y. Investigation into Mode II Interlaminar Fracture Toughness Characteristics of Flax/Basalt Reinforced Vinyl Ester Hybrid Composites. *Compos. Sci. Technol.* **2018**, *154*, 117–127. [CrossRef]

11. Akhavan-Safar, A.; Salamat-Talab, M.; Delzendehrooy, F.; Barbosa, A.; Da Silva, L. Mode II Fracture Energy of Laminated Composites Enhanced with Micro-Cork Particles. *J. Braz. Soc. Mech. Sci. Eng.* **2021**, *43*, 490. [CrossRef]
12. Shin, Y.C.; Lee, W.I.; Kim, H.S. Mode II Interlaminar Fracture Toughness of Carbon Nanotubes/Epoxy Film-Interleaved Carbon Fiber Composites. *Compos. Struct.* **2020**, *236*, 111808. [CrossRef]
13. Srivastava, V.K.; Gries, T.; Veit, D.; Quadflieg, T.; Mohr, B.; Kolloch, M. Effect of Nanomaterial on Mode I and Mode II Interlaminar Fracture Toughness of Woven Carbon Fabric Reinforced Polymer Composites. *Eng. Fract. Mech.* **2017**, *180*, 73–86. [CrossRef]
14. Ekeoseye, W.S.; Kolasangiani, K.; Oguamanam, D.C.; Bougherara, H. Mode II Interlaminar Fracture Toughness of Flax/Glass/Epoxy Hybrid Composite Materials: An Experimental and Numerical Study. *J. Nat. Fibers* **2020**, *19*, 4286–4300. [CrossRef]
15. Lan, B.; Liu, Y.; Mo, S.; He, M.; Zhai, L.; Fan, L. Interlaminar Fracture Behavior of Carbon Fiber/Polyimide Composites Toughened by Interleaving Thermoplastic Polyimide Fiber Veils. *Materials* **2021**, *14*, 2695. [CrossRef] [PubMed]
16. Salamat-talab, M.; Zeinolabedin Beygi, A.; Seyyednejad, M. Experimental Investigation of the Effect of Interface Fiber Angle on the Fracture Toughness of Woven Laminated Composites Under Mode II Loading. *Modares Mech. Eng.* **2021**, *21*, 225–233. Available online: http://mme.modares.ac.ir/article-15-45627-en.html (accessed on 25 March 2021). (In Persian).
17. De Morais, A.; Pereira, A. Application of the Effective Crack Method to Mode I and Mode II Interlaminar Fracture of Carbon/Epoxy Unidirectional Laminates. *Compos. Part A* **2007**, *38*, 785–794. [CrossRef]
18. Wang, W.X.; Nakata, M.; Takao, Y.; Matsubara, T. Experimental Investigation on Test Methods for Mode II Interlaminar Fracture Testing of Carbon Fiber Reinforced Composites. *Compos. Part A* **2009**, *40*, 1447–1455. [CrossRef]
19. Andersons, J.; König, M. Dependence of Fracture Toughness of Composite Laminates on Interface Ply Orientations and Delamination Growth Direction. *Compos. Sci. Technol.* **2004**, *64*, 2139–2152. [CrossRef]
20. Yashiro, S.; Agata, T.; Yoshimura, A. A New Approach for Evaluating Crack Growth Resistance Curve of Mode II Delamination by Doubly End-Notched Tension Tests. *Adv. Compos. Mater.* **2018**, *27*, 119–133. [CrossRef]
21. Chai, H. Interlaminar Shear Fracture of Laminated Composites. *Int. J. Fract.* **1990**, *43*, 117–131. [CrossRef]
22. Hosseini, M.R.; Taheri-Behrooz, F.; Salamat-Talab, M. Mode II Interlaminar Fracture Toughness of Woven E-Glass/Epoxy Composites in the Presence of Mat Interleaves. *Int. J. Adhes. Adhes.* **2020**, *98*, 102523. [CrossRef]
23. Mollón, V.; Bonhomme, J.; Argüelles, A.; Viña, J. Influence of the Crack Plane Asymmetry Over G_{II} Results in Carbon Epoxy ENF Specimens. *Compos. Struct.* **2012**, *94*, 1187–1191. [CrossRef]
24. Pereira, A.; De Morais, A.; Marques, A.; De Castro, P. Mode II Interlaminar Fracture of Carbon/Epoxy Multidirectional Laminates. *Compos. Sci. Technol.* **2004**, *64*, 1653–1659. [CrossRef]
25. Akhavan-Safar, A.; Salamat-Talab, M.; Delzendehrooy, F.; Zeinolabedin-Beygi, A.; Da Silva, L. Effects of Natural Date Palm Tree Fibers on Mode II Fracture Energy of E-Gass/Epoxy Plain-Woven Laminated Composites. *J. Braz. Soc. Mech. Sci. Eng.* **2022**, *44*, 457. [CrossRef]
26. Akhavan-Safar, A.; Marques, E.A.; Carbas, R.J.; Da Silva, F. Cohesive Zone Modelling-CZM. In *Cohesive Zone Modelling for Fatigue Life Analysis of Adhesive Joints*; Springer: Berlin/Heidelberg, Germany, 2022; pp. 19–42. [CrossRef]
27. Ebadi-Rajoli, J.; Akhavan-Safar, A.; Hosseini-Toudeshky, H.; Da Silva, L. Progressive damage modeling of composite materials subjected to mixed mode cyclic loading using cohesive zone model. *Mech. Mater.* **2020**, *143*, 103322. [CrossRef]
28. Giuliese, G.; Palazzetti, R.; Moroni, F.; Zucchelli, A.; Pirondi, A. Cohesive Zone Modelling of Delamination Response of a Composite Laminate with Interleaved Nylon 6, 6 Nanofibers. *Compos. Part B* **2015**, *78*, 384–392. [CrossRef]
29. Saghafi, H.; Ghaffarian, S.; Salimi-Majd, D.; Saghafi, H.A. Investigation of Interleaf Sequence Effects on Impact Delamination of Nano-Modified Woven Composite Laminates Using Cohesive Zone Model. *Compos. Struct.* **2017**, *166*, 49–56. [CrossRef]
30. Khaled, B.M.; Shyamsunder, L.; Holt, N.; Hoover, C.G.; Rajan, S.D.; Blankenhorn, G. Enhancing the Predictive Capabilities of a Composite Plasticity Model Using Cohesive Zone Modeling. *Compos. Part A* **2019**, *121*, 1–17. [CrossRef]
31. Arrese, A.; Insausti, N.; Mujika, F.; Perez-Galmés, M.; Renart, J. A Novel Experimental Procedure to Determine the Cohesive Law in ENF Tests. *Compos. Sci. Technol.* **2019**, *170*, 42–50. [CrossRef]
32. Garg, N.; Prusty, B.G.; Ooi, E.T.; Song, C.; Pearce, G.; Phillips, A.W. Application of Scaled Boundary Finite Element Method for Delamination Analysis of Composite Laminates Using Cohesive Zone Modelling. *Compos. Struct.* **2020**, *253*, 112773. [CrossRef]
33. Sioutis, I.; Tserpes, K. Development and Numerical Implementation of a Modified Mixed-Mode Traction–Separation Law for the Simulation of Interlaminar Fracture of Co-Consolidated Thermoplastic Laminates Considering the Effect of Fiber Bridging. *Materials* **2022**, *15*, 5108. [CrossRef] [PubMed]
34. De Moura, M.; De Morais, A. Equivalent Crack Based Analyses of ENF and ELS Tests. *Eng. Fract. Mech.* **2008**, *75*, 2584–2596. [CrossRef]
35. Chandra, N.; Li, H.; Shet, C.; Ghonem, H. Some Issues in the Application of Cohesive Zone Models for Metal–Ceramic Interfaces. *Int. J. Solids Struct.* **2002**, *39*, 2827–2855. [CrossRef]
36. Williams, J.; Hadavinia, H. Analytical Solutions for Cohesive Zone Models. *J. Mech. Phys. Solids* **2002**, *50*, 809–825. [CrossRef]
37. Anyfantis, K.N.; Tsouvalis, N.G. Experimental and Numerical Investigation of Mode II Fracture in Fibrous Reinforced Composites. *J. Reinf. Plast. Compos.* **2011**, *30*, 473–487. [CrossRef]
38. Ouyang, Z.; Li, G. Nonlinear Interface Shear Fracture of End Notched Flexure Specimens. *Int. J. Solids Struct.* **2009**, *46*, 2659–2668. [CrossRef]

39. Salamat-Talab, M.; Shokrieh, M.; Mohaghegh, M. On the R-curve and Cohesive Law of Glass/Epoxy End-Notch Flexure Specimens with 0//θ Interface Fiber Angles. *Polym. Test.* **2021**, *93*, 106992. [CrossRef]
40. Heidari-Rarani, M.; Ghasemi, A. Appropriate Shape of Cohesive Zone Model for Delamination Propagation in ENF Specimens with R-Curve Effects. *Theor. Appl. Fract. Mech.* **2017**, *90*, 174–181. [CrossRef]
41. Dourado, N.; De Moura, M.; De Morais, A.; Pereira, A. Bilinear Approximations to the Mode II Delamination Cohesive Law Using an Inverse Method. *Mech. Mater.* **2012**, *49*, 42–50. [CrossRef]
42. *ASTM D3518*; Standard Test Method for In-Plane Shear Response of Polymer Matrix Composite Materials by Tensile Test of a±45° Laminate. A. ASTM International: West Conshohocken, PA, USA, 2001.
43. *ASTM D790*; Standard Test Methods for Flexural Properties of Unreinforced and Reinforced Plastics and Electrical Insulating Materials. Annual book of ASTM Standards. S. ASTM International: West Conshohocken, PA, USA, 1997.
44. *ASTM D3039/DM*; Standard Test Method for Tensile Properties of Polymer Matrix Composite Materials. A. ASTM International: West Conshohocken, PA, USA, 2008.
45. *ASTM D7905 D7905M*; Standard Test Method for Mode II Interlaminar Fracture Toughness of Undirectional Fiber-reinforced Polymer Matrix Composites. ASTM International: West Conshohocken, PA, USA, 2014.
46. Carlsson, L.A.; Kardomateas, G.A. *Structural and Failure Mechanics of Sandwich Composites*; Springer Science & Business Media: Berlin/Heidelberg, Germany, 2011. [CrossRef]
47. Krueger, R. Virtual Crack Closure Technique: History, Approach, and Applications. *Appl. Mech. Rev.* **2004**, *57*, 109–143. [CrossRef]
48. Orifici, A.C.; Krueger, R. Benchmark Assessment of Automated Delamination Propagation Capabilities in Finite Element Codes for Static Loading. *Finite. Elem. Anal. Des.* **2012**, *54*, 28–36. [CrossRef]
49. Davidson, B.D.; Kruger, R.; König, M. Three Dimensional Analysis and Resulting Design Recommendations for Unidirectional and Multidirectional End-Notched Flexure Tests. *J. Compos. Mater.* **1995**, *29*, 2108–2133. [CrossRef]

materials MDPI

Article

Development of Concrete Mixture for Spun-Cast Full-Scale Precast Concrete Pipes Incorporating Bundled Steel and Polypropylene Fibers

Adeel Faisal [1], Safeer Abbas [1,*], Syed Minhaj Saleem Kazmi [2,*] and Muhammad Junaid Munir [2,*]

1 Civil Engineering Department, University of Engineering and Technology, Lahore 54890, Pakistan
2 School of Engineering, RMIT University, Melbourne, VIC 3001, Australia
* Correspondence: safeer.abbas@uet.edu.pk (S.A.); minhajkazmi17@gmail.com or syed.kazmi@rmit.edu.au (S.M.S.K.); muhammad.munir@rmit.edu.au (M.J.M.)

Abstract: Spin casting is the oldest method of manufacturing precast concrete pipes among all existing methods. While improved concrete mixtures incorporating fibers for other methods of concrete pipe manufacturing, such as the vibration method and roller compaction method, have been developed, no such concrete mixture has yet been developed for spun-cast concrete pipes. This study was designed to explore the possibility of incorporating locally manufactured steel fibers and commercially available polypropylene fibers to develop an improved concrete mixture for use in the manufacturing of full-scale spun-cast concrete pipes. The used steel fibers were of two types, i.e., straight and bundled steel fibers, manufactured by cutting locally available long straight and bundled steel wires, respectively. Various dosages of steel fibers (i.e., 20, 30, 40, and 50 kg/m^3) and polypropylene fibers (i.e., 5, 10, 15, and 20 kg/m^3) were used in mono and hybrid (steel and polypropylene) forms. The properties in the fresh state and mechanical properties of the test mixtures were investigated. Full-scale spun-cast concrete pipes having a 450 mm internal diameter were manufactured and tested using the three-edge bearing test. The compressive strength of the mixtures was largely insensitive to the dosage of the fibers. The splitting tensile strength of all fiber-reinforced concrete mixtures was higher than that of the reference mixture without fibers, with a 24% increase recorded for the concrete mixture incorporating 50 kg/m^3 of bundled steel fibers relative to the reference mixture with no fibers. The flexural performance of the fiber-reinforced concrete mixtures was superior to that of the reference mixture without fibers in terms of flexural strength, toughness, residual strength, and crack control, with up to 28% higher flexural strength relative to the reference mixture without fibers. The three-edge bearing tests on full-scale spun-cast pipes incorporating steel fibers showed that the use of fibers is a promising alternative to the traditional steel cage in spun-cast concrete pipes.

Keywords: spun-cast method; bundled fiber; three-edge bearing test; crack control

Citation: Faisal, A.; Abbas, S.; Kazmi, S.M.S.; Munir, M.J. Development of Concrete Mixture for Spun-Cast Full-Scale Precast Concrete Pipes Incorporating Bundled Steel and Polypropylene Fibers. *Materials* **2023**, *16*, 512. https://doi.org/10.3390/ma16020512

Academic Editor: Yuri Ribakov

Received: 12 December 2022
Revised: 26 December 2022
Accepted: 31 December 2022
Published: 5 January 2023

1. Introduction

Water is a basic necessity of life, and access to safe water is recognized as a basic human right [1]. Most consumed water becomes wastewater and requires safe disposal [2]. The demand for water supply and consequently the amount of wastewater generated are rising due to growth in population, rise in demand per capita, and expansion in economic activities [3,4]. It is estimated that about 85 billion m^3 of wastewater is generated annually in North America alone [5]. The worldwide annual generation of wastewater was estimated to be about 380 billion m^3 in 2020 and is expected to have increased by 24% in 2030 and by 51% in 2050 [6]. Wastewater, if not properly disposed of, poses threat to the environment, as it contains pathogens, which cause many infectious diseases such as diarrhea, typhoid, hepatitis, and cholera [7]. The safe disposal of wastewater involves collection, conveyance, and treatment of the wastewater before releasing it into water bodies [8]. Concrete pipes are important components of wastewater collection and conveyance systems.

Concrete pipes are manufactured by various methods. The most adopted methods are vibration, roller compaction, and the spinning or centrifugal method [9]. These methods differ mainly in the mechanism of consolidating the concrete and the type of concrete used. In vibration method, a vertical pipe mold with an annular space is fed with dry-mix concrete. The concrete is consolidated through intense vibrations and hydraulic pressing from the top. The vertical casting position and dry-mix nature of the concrete allow the pipe to be demolded immediately after casting, as it can support its self-weight [10]. In the roller compaction method, a mold with an internal diameter equal to the required external diameter of the pipe is used. Dry-mix concrete is dropped into the mold. The dry-mix concrete lands on the top of a computer-controlled rotating roller head, equipped with a set of small rotors. The centrifugal action of the small rotors spreads the concrete towards the wall of the mold. The rotating roller head mechanically compacts the concrete against the wall of the mold to form the internal surface of the pipe. The rotating roller head gradually ascends to the top to make the required length of the pipe. The pipe is demolded immediately, and the mold is reused for another pipe [11]. In the centrifugal method, a pipe mold, fitted with end rings, is supported on two rollers. A variable-speed electrical motor is used to rotate the rollers. The rotating rollers rotate the pipe mold. Concrete with an initial water/cementitious materials ratio ranging between 0.40 and 0.50 (depending on whether a water-reducing admixture is used) is fed uniformly into the rotating mold throughout the length, either manually or by a concrete feeder [12]. The concrete mixture must be wet enough to allow the concrete to stick to the walls of the rotating mold. The required wall thickness is controlled by the end rings. When enough concrete has been poured to form the required wall thickness, the speed of rotation is increased. The centrifugal action of rotation pushes the concrete against the wall of the mold, resulting in the extraction of water due to its lower specific gravity than the remaining concrete constituents. Thus, the final water/cementitious materials ratio is decreased. However, the laitance formed on the inner surface of the pipe must be cleaned with the help of a brush. It is important that the initial concrete mixture is not too wet, as otherwise the extraction of excess water will take too long, which not only reduces the efficiency of the process but also results in the drainage of cement paste from the surface of the aggregates, leaving behind weak and porous concrete. The durability of concrete is adversely affected by increased porosity [13,14]. At the end of the rotation, the concrete pipe, along with the mold, is lifted and placed in the curing area. The pipe is demolded after 24 h and cured by ponding or showers.

The manufacturing of concrete pipes by the vibration method and roller compaction method is performed using computer-controlled automatic plants. These methods have the advantages of high efficiency, better quality control due to automation, and high concrete strength due to a low water/cementitious materials ratio. These methods have the additional advantage of producing concrete pipe with uniform concrete strength throughout the wall thickness. Due to these advantages, these methods are popular in North America and Europe. Manufacturing by the centrifugal method is performed either manually or using an automated plant. Due to the high initial investment and technical expertise required for an automated plant, spun-cast pipes are still manufactured manually in developing countries, including Pakistan. It is known that the centrifugal method of manufacturing concrete pipes results in the segregation of concrete ingredients, with larger particles tending to move towards the outer side and the smaller particles towards the inner side [15]. This sets up a gradient in the strength of the concrete over the thickness of the pipe wall such that the strength is higher on the outer side relative to the inner side of the wall [16].

The tensile strength of plain concrete is low, and hence concrete pipes are vulnerable to cracking [17]. As concrete pipes are buried structures, the cracks often go undetected [18]. The cracks pave the way for the corrosion of conventional steel bars by allowing chlorides to reach the surfaces of the bars [19–21]. The cracking of a concrete pipe can potentially lead to the final failure of the pipe and associated problems such as road accidents and inconvenience to the public (Figure 1).

Figure 1. Road collapse due to an underlying pipe failure.

Structural fibers such as steel fibers and polypropylene fibers are known to enhance the tensile strength, toughness, and resistance of concrete against cracking [22–26]. Due to the ongoing improvement in the properties of synthetic fibers by the polymer industry, their application in concrete is increasing. For instance, Wang et al. [27] explored the impact of polyoxymethylene fiber on the mechanical performance of seawater–sea-sand concrete at various ages. The incorporation of the polyoxymethylene fibers into the seawater–sea-sand concrete resulted in the improvement of the compressive, splitting tensile, and flexural strength of the concrete, especially at the age of 28 days. Xue et al. [28] suggested that the seawater–sea-sand concrete can be utilized to conserve river sand and fresh water by using polyoxymethylene fibers for improved resistance to early age cracking and enhanced mechanical properties. Fang et al. [29] studied the effects of steel fibers having various geometries and shapes on the mechanical performance of ultra-high-performance concrete (UHPC). It was found that hook-ended steel fibers resulted in better improvement in the tensile performance compared with straight steel fibers. Khan et al. [30] found that aligning the steel fibers in the desired direction using magnetic field while the concrete is still in a fresh state resulted in the significant enhancement of the cracking and ultimate loads of concrete. Abed et al. [31] reported improvement in the flexure and shear performance of self-compacting concrete incorporating steel and polypropylene fibers with increasing dosage of the fibers. The workability reduced with increasing fiber content, although it was still in the permissible range. Wolfel et al. [32] concluded from their study on various types of polypropylene fibers that the efficiency of the fibers could be enhanced by use of higher strength polypropylene fibers with increased surface roughness. Wang et al. [33] studied the effect of various types of fibers on the crack resistance of concrete and reported improved resistance of the polypropylene fiber-reinforced concrete to cracking induced by restrained shrinkage.

Several researchers have explored the use of fibers in concrete pipes produced by either the vibration method or the roller compaction method [34–48]. The use of fibers in concrete pipes manufactured by the vibration method or the roller compaction method is relatively straightforward, considering that the aforementioned studies did not report any segregation of fibers from the concrete matrix. Moreover, as mentioned earlier, these methods utilize dry-cast concrete with a low water/cementitious materials (w/cm) ratio. For instance, Peyvandi et al. [44] and Mohamed et al. [43] used w/cm ratios of 0.32 and 0.38, respectively. On the other hand, the spun-cast technique of manufacturing precast concrete pipes involves concrete with a w/cm ratio between 0.4 and 0.5. Studies on inclusion of fibers in spun-cast concrete products are scant. Raju et al. [49] investigated the impact of the addition of steel fibers on the shear resistance of spun-cast concrete piles. The research concluded that the steel fibers could be used to replace conventional shear reinforcement in spun-cast concrete piles. No comprehensive study could be found by the authors in the published literature on the use of fibers in spun-cast concrete pipes. This study is aimed at comprehensively investigating the mechanical performance of concrete incorporating locally manufactured straight steel fibers, novel bundled steel fibers, commercially available polypropylene fibers, and hybrid fibers. Full-scale spun-cast concrete pipes with inside diameters of 450 mm were also manufactured in a local pipe factory with fibers completely or partially replacing the conventional steel cage reinforcement and subjected to the three-edge bearing test (TEBT) in the laboratory.

2. Research Significance

The application of fiber-reinforced concrete in manufacturing of precast concrete pipes has been a subject of research for the last two decades. The potential of using steel fibers or polypropylene fibers to replace conventional steel cages for the production of precast concrete pipes has been recognized in the EN standard BS EN 1916 [50], ASTM C1765 [51], and ASTM C1818 [52]. However, the current standards on the subject are still in the developing stage. Unlike the standards for conventionally reinforced concrete pipes such as ASTM C76 [53] which provide detailed guidance for the amount of steel reinforcement required for each class of pipe, the current standards on fiber-reinforced concrete pipes do not provide any detailed guidance such as the dosage of fibers required for a particular class of pipe. Due to the lack of such guidelines, commercial concrete pipe manufacturers have shown reluctance toward adopting fiber-reinforced concrete for the production of precast concrete pipes. Moreover, the current standards for fiber-reinforced concrete pipes have been developed mostly based on studies conducted on the two types of fibers, i.e., steel fibers and polypropylene fibers in mono form. Very scant literature is available on the application of steel and polypropylene fibers in the hybrid form for manufacturing precast concrete pipes. Furthermore, the biggest limitation of the previous studies lies in the manufacturing technique used for the production of the concrete pipes. All previous studies on fiber-reinforced concrete pipes have been conducted on the concrete pipes produced by either the vibration method or the roller compaction method. The spun-cast method for the manufacturing of precast concrete pipes is very commonly adopted in developing countries due to the easy and economical installation of its plants. However, limited study has been reported so far on the application of fiber-reinforced concrete in spun-cast concrete pipes. The potential of incorporating fibers into spun-cast concrete pipes needs to be explored because spun-cast concrete is known to exhibit segregation of concrete ingredients based on their specific gravities [16]. It is important to study the behavior of fibers in spun-cast concrete pipes, as the fibers could segregate from the concrete matrix due to the difference in their specific gravities. There is a further need to investigate other fiber types which were not investigated in previous studies to allow for the incorporation of locally available fibers into precast concrete pipes. Therefore, this study was planned to investigate the effects of various proportions of fibers on the mechanical properties of concrete to be used in the spun-cast method. The studied fibers were locally available straight steel fibers, novel bundled steel fibers, and polypropylene (PP) fibers. Furthermore, hybrid steel and PP fibers were also investigated.

3. Materials

The materials used by a local concrete pipe manufacturer for the manufacturing of conventional spun-cast concrete pipes were employed in the current study. Ordinary Portland cement was used in all concrete mixtures. River sand having a fineness modulus of 2.15 and coarse aggregate having a nominal maximum size of 12.5 mm from a local crusher plant were used. The control concrete mixture was designed according to the guidelines of ACI 211.3R [54] with a target compressive strength of 36 MPa and a desired slump of 0 to 25 mm. The selection of the target compressive strength was based on the minimum compressive strength of 27.6 MPa required by ASTM C76 [53] for conventional concrete pipes. The desired slump was chosen to mimic the workability of concrete used for the local manufacturing of spun-cast concrete pipes. The concrete mixture proportions employed in this study are presented in Table 1.

Table 1. Concrete mixture design.

Ingredient	Mass/Cement Mass
Cement	1.00
Fine aggregate	1.50
Coarse aggregate	2.92
Water	0.47

Locally manufactured steel fibers of two types and commercially available polypropylene fibers were used (Figure 2). One type of steel fiber was manufactured by cutting locally available straight steel wire having a smooth surface, and the other was manufactured by cutting locally available bundled steel wire into straight lengths of 25 mm. The bundled steel wire (Figure 3) consisted of a strand of 12 interwound thin steel wires wrapped by a steel wire in the form a spiral (Figure 3). Therefore, the bundled steel fiber had a rough surface, suitable for improved bonding with concrete. The fibers employed in this study had the properties given in Table 2. Relatively low dosages of steel fibers and PP fibers were used in this study to avoid the severe reduction in workability of concrete resulting from high dosages of fibers, which could translate into reduced mechanical performance due to the difficulty in achieving the consolidation of concrete [27,28]. The steel fibers were used at dosages of 20, 30, 40, and 50 kg/m^3 of concrete. The polypropylene fibers were added at dosages of 5, 10, 15, and 20 kg/m^3.

Table 2. Properties of used fibers.

Fiber Type	Shape	Length (mm)	Diameter (mm)	Aspect Ratio	Tensile Strength (MPa)	Modulus of Elasticity (GPa)
Steel fibers	Solid fiber with smooth surface and straight ends	25	0.68	36	1569	200
Steel fibers	Bundled fiber and straight ends	25	0.79 *	32	1428	200
PP fibers	Embossed surface	55	0.91 *	60	460	9

PP = Polypropylene fiber; * Equivalent diameter.

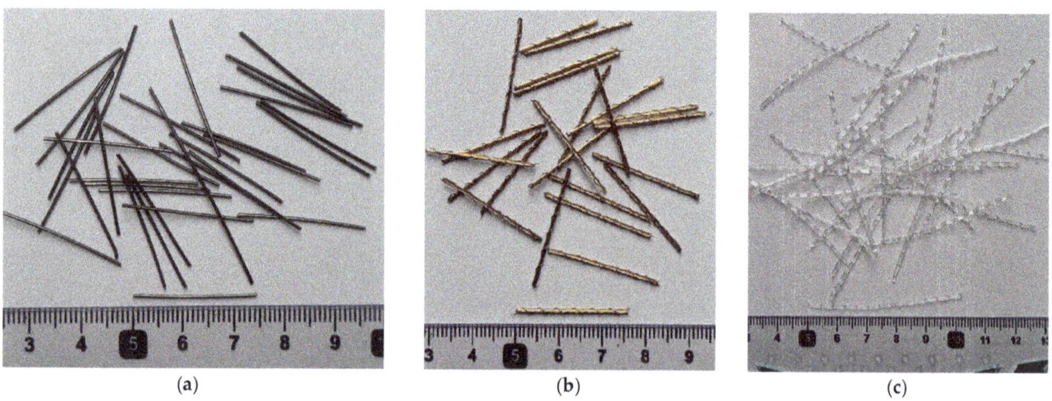

Figure 2. Fibers used: (**a**) steel fibers, (**b**) bundled steel fibers, (**c**) PP fibers.

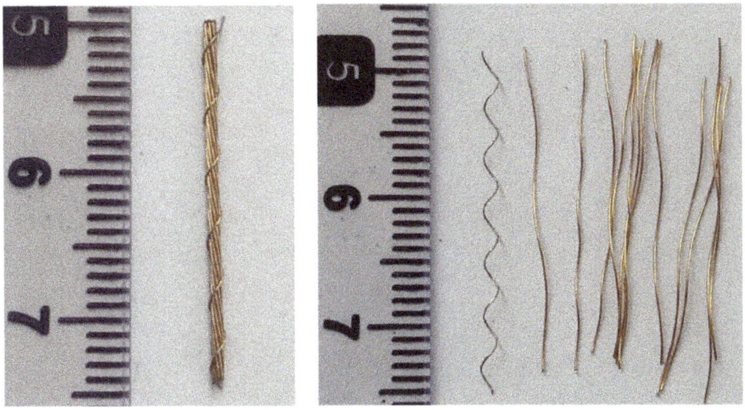

(**a**) (**b**)

Figure 3. Structure of bundled steel fiber: (**a**) bundled steel fiber, (**b**) unwound steel wires.

4. Experimental Methodologies

A pan-type mixer with capacity of 56 L was used for mixing the concrete constituents, (Figure 4). Coarse aggregate, sand, and cement were initially dry-mixed for three minutes. Next, the mixing water was added gradually. The mixing was continued for another three to five minutes. Lastly, fibers were added so as to avoid fiber balling, and the mixing was continued until the fibers were completely dispersed in the concrete.

Figure 4. Pan-type mixer used for mixing of concrete.

The slump and the fresh-state density of concrete were measured immediately after the preparation of the mixture in accordance with ASTM C143 [55] and ASTM C138 [56], respectively. For the density of fresh concrete, a 9.3 L capacity cylindrical measure was placed on a vibrating table and filled in two layers. Each layer was given external vibrations for consolidation. Cylindrical specimens with dimensions of 150 × 300 mm and prismatic specimens with dimensions of 100 × 100 × 350 mm were prepared from each mixture. The consolidation of all mixtures was achieved in two layers using a vibrating table. The specimens were demolded after 24 h and cured for 28 days through water ponding to simulate the field curing of small-diameter concrete pipes.

The 150 × 300 mm cylindrical specimens were tested in compression for compressive strength according to ASTM C39 [57]. Plaster of Paris was used to cap the specimens before testing (Figure 5a). Continuous loading at a rate of 0.25 MPa/s was applied to the specimens. Three identical cylinders for compression testing were molded from each mixture to represent the average behavior.

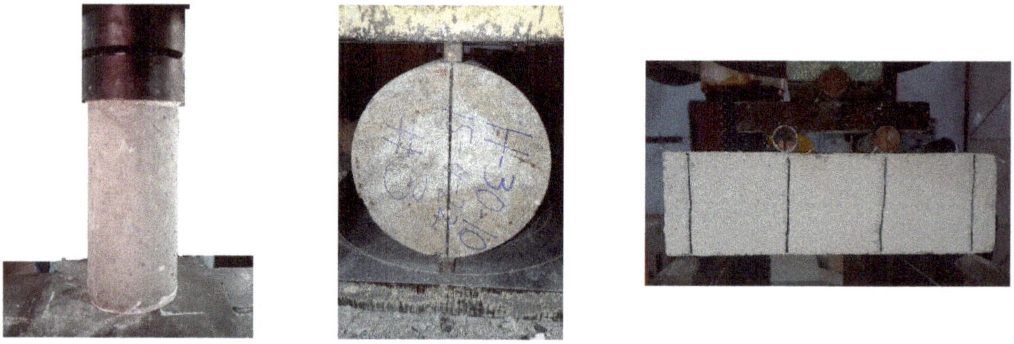

(**a**) (**b**) (**c**)

Figure 5. Setups for various tests: (**a**) compressive strength, (**b**) splitting tensile strength, (**c**) flexural strength.

Splitting tensile strength testing was conducted on the 150 × 300 mm cylindrical specimens according to ASTM C496 [58] (Figure 5b). Each specimen was loaded at a rate of 1 MPa/s. Three identical specimens were tested for each mixture.

The flexural performance of the concrete mixtures was assessed in accordance with ASTM C1609 [59] on the 100 × 100 × 350 mm prismatic specimens, (Figure 5c). Displacement-controlled loading at a rate of 0.05 mm/min was applied. The crack width was monitored using leaf gages having various thicknesses, similar to the crack-measuring leaf gauge used in the three-edge bearing test on concrete pipes. Table 3 shows the test matrix.

Table 3. Test matrix.

Sr. No.	Mixture Designation	Fiber Dosage (kg/m^3)		Mixture Type	Remarks
		Steel	PP		
1	PC	0	0	Plain concrete	Control mix
2	S20	20	0	Steel fiber-reinforced concrete	Straight steel fibers
3	S30	30	0		
4	S40	40	0		
5	S50	50	0		
6	S20B	20	0		Bundled steel fibers
7	S30B	30	0		
8	S40B	40	0		
9	S50B	50	0		
10	P5	0	5	PP fiber-reinforced concrete	Polypropylene fibers with embossed surface
11	P10	0	10		
12	P15	0	15		
13	P20	0	20		
14	H20/5	20	5	Hybrid fiber-reinforced concrete	Hybrid of bundled steel fibers and PP fibers
15	H20/10	20	10		
16	H30/10	30	10		

PC = Control concrete mixture without fibers; S20 = Concrete mixture having 20 kg/m^3 of straight steel fibers; S30B = Concrete mixture having 30 kg/m^3 of bundled steel fibers; P5 = Concrete mixture having 5 kg/m^3 of polypropylene fibers; H20/5 = Concrete mixture having hybrid bundled steel fiber (20 kg/m^3) and PP fibers (5 kg/m^3).

5. Results and Discussion

5.1. Material Characterization of Concrete Mixtures Incorporating Fibers

5.1.1. Fresh Properties

Although the workability of fiber-reinforced concrete mixtures is best determined by vibration-based tests such as Vebe time test, the workability of the concrete mixtures tested in this study was evaluated using the slump test due to the simplicity of the test and its widespread use in the industry. The slump and fresh density results of the tested mixtures are given in Table 4. The slump of the concrete slightly decreased with the increasing dosage of the fibers. For instance, the slump of the mixture with 50 kg/m^3 of straight steel fibers was 5 mm as compared to 20 mm for the plain concrete mixture. However, the concrete mixture incorporating 20 kg/m^3 of PP fibers showed zero slump. Similarly, the fresh density of the concrete also marginally decreased as the dosage of the fibers increased. For instance, the fresh density of the concrete mixture incorporating 50 kg/m^3 of straight steel fibers was about 2% lower than that of the plain concrete mixture. This shows that the slight reduction in the slump resulting from the addition of the fibers in the dosages employed in this study did not significantly affect the level of consolidation of most of the fiber-reinforced concrete mixtures. However, the fresh density of the concrete mixture incorporating 20 kg/m^3 of PP fibers was about 6% less than that of the plain concrete mixture without fibers. This was because the mixture had a zero slump and could not be consolidated sufficiently.

Table 4. Results of various tests.

Sr. No.	Mixture Designation	Slump (mm)	Fresh Density (kg/m^3)	Compressive Strength (MPa)	Splitting Tensile Strength (Mpa)
1	PC	20	2513	32.6	3.21
2	S20	15	2502	34.5	3.27
3	S30	10	2493	35.2	3.47
4	S40	5	2486	36.5	3.73
5	S50	5	2474	37.1	3.92
6	S20B	15	2498	33.9	3.32
7	S30B	15	2491	34.4	3.49
8	S40B	10	2482	36.8	3.78
9	S50B	10	2477	38.4	3.98
10	P5	10	2498	32.9	3.29
11	P10	5	2488	31.8	3.35
12	P15	5	2476	31.5	3.48
13	P20	0	2412	30.8	3.57
14	H20/5	10	2495	32.4	3.29
15	H20/10	5	2484	33.1	3.37
16	H30/10	5	2473	34.1	3.55

5.1.2. Compressive Strength

The results of the compressive strength tests are given in Table 4. Different trends were observed for the different varieties of fibers employed in this research. In general, the compressive strength of the concrete mixtures incorporating steel fibers (straight or bundled) moderately improved with the addition of the fibers. For instance, the compressive strength of the concrete mixture increased from 34.5 Mpa to 37.1 Mpa when the dosage of the bundled steel fibers was raised from 20 kg/m^3 to 50 kg/m^3. A similar trend was observed for the straight steel fibers. The increase in the dosage of the steel fibers increases the likelihood that cracks will be arrested by fibers crossing the cracks, thus leading to somewhat higher compressive strength [60] (Figure 6a). The compressive strength of the mixtures incorporating PP fibers seemed to be insensitive to the dosage of the PP fibers. However, the compressive strength of the mixture having 20 kg/m^3 of PP fibers was 6% lower as compared to the plain concrete mixture without fibers. This is ascribed to the insufficient consolidation of the mixture due to a loss in workability [61] (slump decreased to zero when the PP fiber content was 20 kg/m^3).

(a) (b)

Figure 6. Tested fiber-reinforced cylinders: (**a**) compression test, (**b**) splitting tensile strength test.

The dosage of the fibers in the concrete mixtures incorporating hybrid fibers negligibly affected their compressive strength.

5.1.3. Splitting Tensile Strength

The results of the splitting tensile strength tests are listed in Table 4. The splitting tensile strength increased as the dosage of the fibers increased. This was true for all types of fibers employed in this study. For instance, the splitting tensile strength increased from 3.27 MPa to 3.92 MPa (20% increase) when the dosage of the straight steel fibers increased from 20 kg/m^3 to 50 kg/m^3. This is ascribed to the fact that raising the fiber dosage increases the number of fibers that can possibly intersect and hold together cracks, thus postponing the splitting of the cylinders [62] (Figure 6b). The bundled steel fibers seemed somewhat more effective as compared to the straight steel fibers. However, the difference in their splitting tensile strengths was only marginal at the same dosage. For instance, the splitting tensile strength of the mixture having the straight steel fibers was 3.73 MPa as compared to 3.78 MPa for the mixture incorporating the bundled steel fibers. This shows that the splitting tensile strength is more sensitive to the dosage of the steel fibers than to the type of the steel fibers used. The splitting tensile strength of the mixtures incorporating PP fibers and those incorporating hybrid fibers (bundled steel fibers and PP fibers) ranged from 1.02 to 1.11 times the splitting tensile strength of the plain concrete mixture without fibers. In comparison, the mixtures incorporating the steel fibers showed splitting tensile strengths ranging from 1.02 to 1.24 times that of the plain concrete mixture without fibers. This shows that the steel fibers were more efficient than the PP fibers and the hybrid fibers in increasing the splitting tensile strength.

5.1.4. Flexural Strength

The results of the flexural tests are presented in Table 5. The peak flexural strength of the fiber-reinforced concrete mixtures was higher than that of the plain concrete mixture, with the increase depending on the dosage of the fibers. This is attributed to the arresting of micro cracks by the fibers, thus delaying the formation of macro cracks [63]. The use of bundled steel fibers resulted in higher flexural strength compared with that of straight steel fibers at the same dosage. For instance, the peak flexural strength increased by 12% and 28% for straight steel fibers and bundled steel fibers, respectively, at the same dosage of 50 kg/m^3, compared with the plain concrete specimens without fibers. This is possibly because of the relatively roughened surface of the bundled steel fibers formed as the thin individual fibers are interwound to form the bundle, leading to better bonding with the concrete. The PP fibers were found to have a negligible effect on the peak flexural strength with an increase of about 10% at a fiber dosage of 20 kg/m^3, in agreement with the results of the previous studies [64,65].

Table 5. Results of flexure tests.

Sr. No.	Mixture Designation	First-Peak Strength (f_1) and Deflection (δ_1)		Residual Strength (MPa) at Deflection of		Toughness (Joule)
		f_1 (MPa)	δ_1 (mm)	L/600	L/150	
1	PC	5.11	0.05	0	0	1.2
2	S20	5.24	0.05	0.78	0.47	5.2
3	S30	5.48	0.05	1.46	0.73	8.6
4	S40	5.6	0.05	1.73	1.12	10.5
5	S50	5.74	0.05	1.96	1.38	12
6	S20B	5.65	0.05	0.62	0.2	3.5
7	S30B	6.06	0.05	1.79	1.18	11.2
8	S40B	6.39	0.06	2.36	1.43	14.1
9	S50B	6.56	0.06	2.69	1.87	16.4
10	P5	5.14	0.05	1.2	1.34	9.1
11	P10	5.27	0.05	1.65	1.83	12.1
12	P15	5.4	0.05	2.09	2.31	14.7
13	P20	5.61	0.05	2.35	2.63	16.6
14	H20/5	5.74	0.05	2.73	2.25	17.4
15	H20/10	5.87	0.05	3.08	2.47	19.9
16	H30/10	6.32	0.06	4.39	3.89	27.6

The concrete mixture incorporating both PP fibers and bundled steel fibers showed better or similar improvement in peak flexural strength compared with the cumulative increase due to the individual fibers. Comparison with the concrete mixtures incorporating either the bundled steel fibers or the PP fibers shows that the enhancement in the peak flexural strength of the hybrid fiber-reinforced concrete mixtures was dependent mainly on the dosage of the bundled steel fibers.

Load-Deflection Response

Figure 7 illustrates the load-deflection curves of the prismatic specimens under third-point loading. Initially, as expected, the load increased linearly with the deflection up to the peak load, irrespective of the fiber type and dosage. Following the peak load, all the tested specimens showed a drop in the load with the increasing deflection. The occurrence of the drop in the load coincided with the appearance of a crack in the middle third of the span. The plain concrete specimen abruptly lost its load capacity following the peak load, and the test was terminated. In the case of the fiber-reinforced specimens, the fibers bridged the crack, and the specimens continued to resist the load. However, the specimens incorporating low dosages of fibers (20 kg/m^3 of the steel fibers or 5 kg/m^3 of the PP fibers) showed sharp drops in load-carrying capacity after the peak load. This is due to the insufficient number of fibers available to bridge the crack [66].

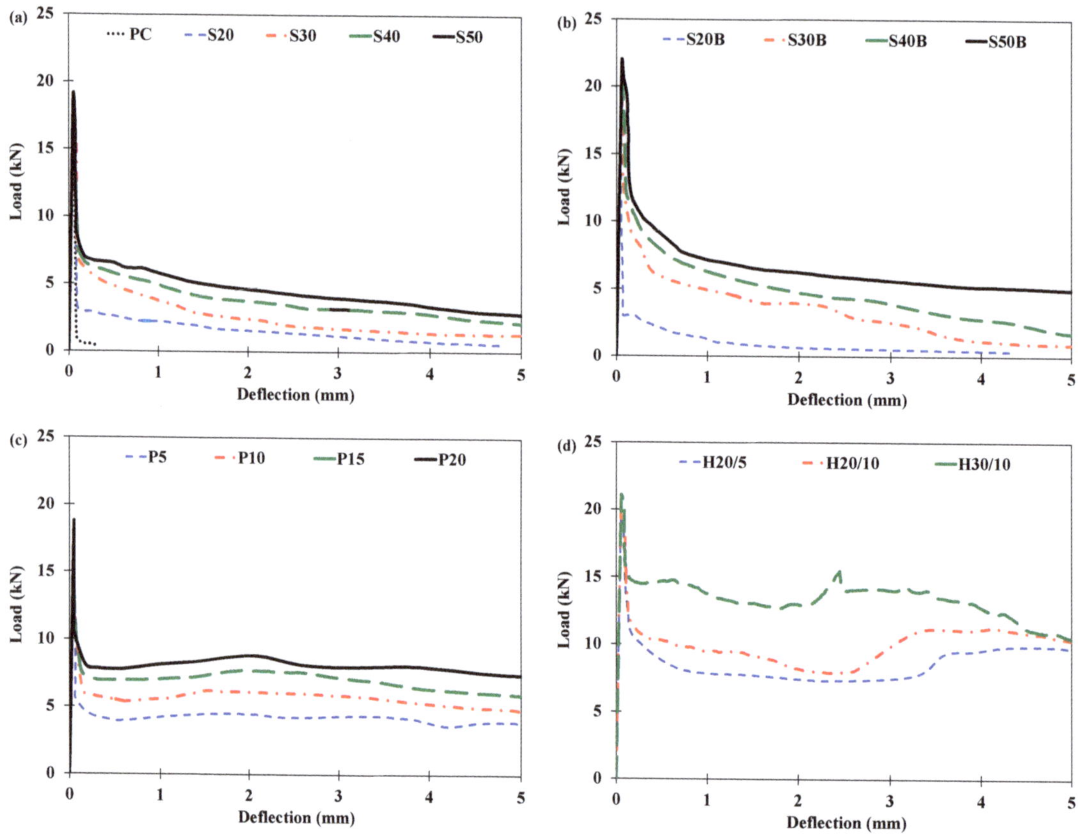

Figure 7. Load-deflection curves in flexure for specimens incorporating (**a**) straight steel fibers, (**b**) bundled steel fibers, (**c**) PP fibers, (**d**) hybrid fibers (bundled steel fibers and PP fibers).

The post-cracking behavior of fiber-reinforced concrete depends on the material, geometry, and dosage of the fibers used [67]. The immediate drop in the load-carrying capacity after the peak load depended on the type of the fibers used. For instance, the load-carrying capacities of specimens incorporating straight steel fibers (50 kg/m^3), those with bundled steel fibers (50 kg/m^3), and those with PP fibers (20 kg/m^3) dropped to 37%, 53%, and 43% of the respective peak loads at a deflection of about 0.17 mm. The specimens incorporating hybrid fibers (30 kg/m^3 of bundled steel fibers and 10 kg/m^3 of PP fibers) had a residual strength of about 70% of their peak load at the same deflection of 0.17 mm.

Furthermore, the specimens incorporating straight steel fibers had the steepest rate of loss in residual strength, followed by the specimens incorporating bundled steel fibers. This is possibly because both kinds of steel fibers had a straight geometry with no hooks at the ends, which led to the pullout of the steel fibers from the concrete matrix at low bond stresses [68]. By contrast, the specimens incorporating only PP fibers maintained their residual load-carrying capacities with increasing deflection up to about 5 mm. This was due to the continuously embossed surface of the PP fibers, leading to improved bond performance with the concrete matrix. The specimens incorporating hybrid fibers (PP and bundled steel fibers) initially maintained nearly constant residual strength as the imposed deflection increased, followed by second peaks in the load. The hybrid fiber-reinforced concrete specimens were the only specimens tested in this study which showed second peaks in the load. The occurrence of the second peak was dependent on the dosage of the fibers used in the hybrid fiber system. The increase in the dosage of the fibers resulted in the occurrence of the second peak at lower deflection. All the second peaks in the load were lower than the first peaks, and therefore, the first-peak strength also represented the flexural strength of the specimens.

A comparison of the residual strengths corresponding to deflections of L/600 (0.5 mm) and L/150 (2.0 mm) is given in Figure 8. In general, the residual strength increased with the increase in the dosage of the fibers. However, the type of the fibers used had the most pronounced impact on the residual strength. The bundled steel fibers had higher residual strength as compared to the straight steel fibers. For instance, the residual strengths at L/600 (0.5 mm) and L/150 (2.0 mm) were 1.96 MPa and 1.38 MPa, respectively, for the straight steel fibers, and 2.69 MPa and 1.87 MPa, respectively, for the bundled steel fibers at the same dosage of 50 kg/m^3. The residual strength of the bundled steel fibers was generally higher at smaller deflection (L/600 = 0.5 mm) as compared to the PP fibers, but lower at larger deflection (L/150 = 0.5 mm). The use of the hybrid fibers (bundled steel fibers and PP fibers) resulted in higher residual strengths at both smaller and larger deflections. For example, the specimens incorporating hybrid fibers (30 kg/m^3 of bundled steel fibers and 10 kg/m^3 of PP fibers) resulted in residual strengths of 4.39 MPa and 3.89 MPa at deflections of L/600 (0.5 mm) and L/150 (0.5 mm), respectively.

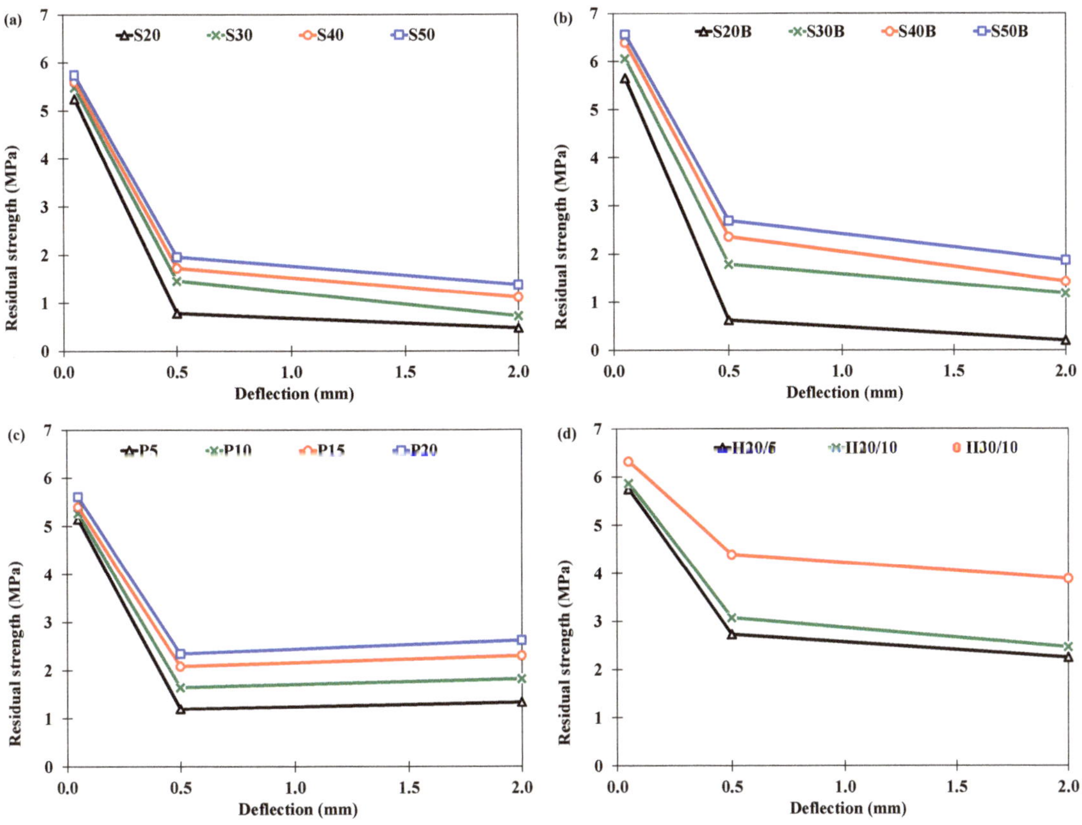

Figure 8. Residual strengths at deflections of L/600 (0.5 mm) and at L/150 (2.0 mm) for specimens incorporating (**a**) straight steel fibers, (**b**) bundled steel fibers, (**c**) PP fibers, (**d**) hybrid fibers (bundled steel fibers and PP fibers).

Toughness

Toughness indicates the ability to maintain the load after cracking of the concrete matrix [69]. The toughness was computed by determining the area under the load-deflection curve up to a deflection of L/150 (2.0 mm), following the guidelines of [59]. The toughness of the tested prismatic specimens in flexure is given in Table 5. The toughness increased with the increase in the dosage of the fibers. For instance, the toughness of the bundled steel fibers increased from 3.5 J to 14.1 J when the fiber dosage was increased from 20 kg/m^3 to 40 kg/m^3. The straight steel fibers generally had lower toughness as compared to the bundled steel fibers at the same dosage. For instance, the toughness values of the straight steel fibers and bundled steel fibers were 12.0 J and 16.4 J, respectively, at the same dosage of 50 kg/m^3. This was attributed to the superior post-peak performance of the bundled steel fibers. The PP fibers and the bundled steel fibers showed comparable toughness values at similar levels of dosage of the fibers, except that the bundled steel fibers had much lower toughness (3.5 J) at the first level of dosage used (that is, 20 kg/m^3). This shows that the dosage of 20 kg/m^3 of steel fibers is inadequate to improve the post-peak behavior of concrete [43]. The hybrid fibers (bundled steel fibers and PP fibers) outperformed the rest of the fibers used in this study. For instance, the specimen incorporating hybrid fibers (30 kg/m^3 of bundled steel fibers and 10 kg/m^3 of PP fibers) had a toughness of 27.6 J, which is the highest among all the specimens tested in this study.

Crack and Failure Pattern

All the tested specimens showed no cracking during the initial loading phase in which the load increased linearly with the deflection. At the end of the linear elastic phase of the loading, all the tested specimens showed a crack in the central third of the span. As the imposed deflection increased, the crack width increased, and the crack propagated towards the compression face. Nearly all the tested specimens developed a single crack, and multiple cracking was not observed (Figure 9). Only the specimen reinforced with 20 kg/m^3 of PP fibers showed a secondary crack at a spacing of about 25 mm from the primary crack (Figure 10). The secondary crack appeared when the residual load was about 37% of the peak load. The formation of single cracks in the tested specimens is due to the relatively low fiber dosage employed in this study [70].

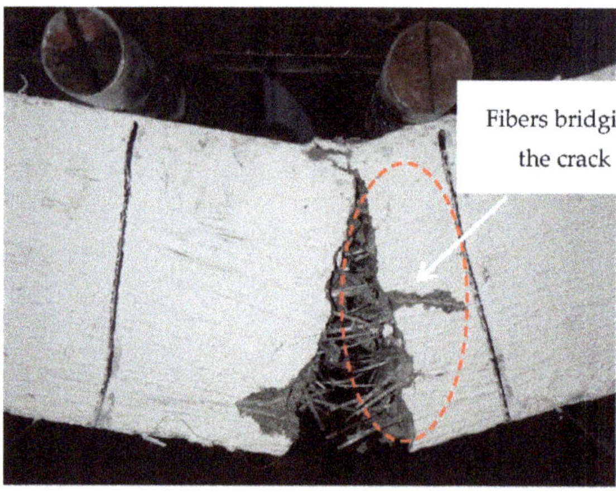

Figure 9. Fibers bridging the crack that appeared in the central third of the specimen tested in flexure.

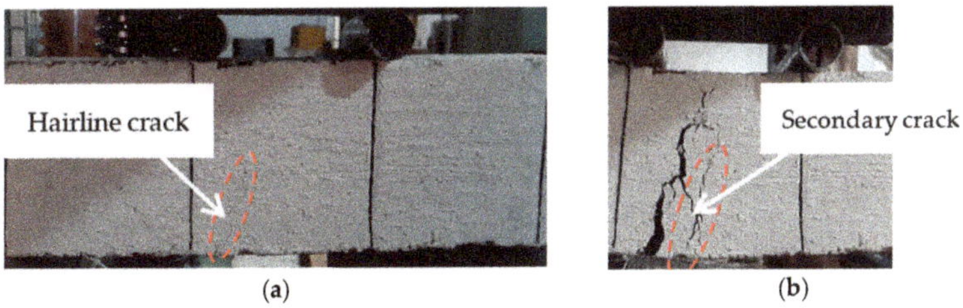

Figure 10. Cracking in the specimen incorporating 20 kg/m^3 of PP fibers: (**a**) first hairline crack, (**b**) secondary crack.

Considering the intended application of the concrete mixtures for manufacturing spun-cast full-scale concrete pipes, the crack width was measured at discrete points during loading and plotted against the load (Figure 11). The crack propagation and widening consisted of two phases. In the first phase, unstable crack propagation and widening took place immediately after the initiation of the crack. The second phase was characterized by stable crack propagation and widening [71]. It is worth mentioning that the specimens without fibers and those with low fiber dosages (20 kg/m^3 of steel fibers or 5 kg/m^3 of PP fibers) displayed a prominent unstable crack propagation phase, and a loud tick sound was heard when the crack appeared and quickly propagated towards the compression face. The increase in the dosage of the fibers enhanced the ability of the specimen to hold the crack, and thus, mostly stable crack propagation was observed. The first point of the load vs. crack width curves corresponds to the peak load at which the crack is barely visible. The next points on the load vs. crack width curves depict the ability of the specimen to resist crack opening. In general, the load vs. crack width curves confirm the earlier findings on the basis of the load vs. deflection curves. The bundled steel fibers were more efficient in holding a crack as compared to the straight steel fibers. For instance, the loads corresponding to 0.26 mm crack width were 8.51 kN and 6.72 kN for the bundled steel fibers and the straight steel fibers, respectively. The specimens incorporating hybrid fibers (bundled steel fibers and PP fibers) were the most effective in controlling the crack width. The load at 0.26 mm crack width was 14.38 kN for the specimen incorporating hybrid fibers (30 kg/m^3 of bundled steel fibers and 10 kg/m^3 of PP fibers).

Furthermore, visual examination of the fracture surfaces of the specimens indicated that both the straight and the bundled steel fibers predominantly failed by fiber pullout, (Figure 12a,b). This is perhaps due to the high strength of the steel fibers and the lack of end anchorage. By contrast, the failure of the PP fibers was primarily due to fiber rupture (Figure 12c). This could be due to the relatively lower strength of the PP fibers and improved bonding with concrete matrix due to their continuously embossed surface.

5.2. *Manufacturing and Testing of Full-Scale Spun-Cast Concrete Pipes*

To study the potential of incorporating the fibers into the manufacturing of full-scale spun-cast concrete pipes, full-scale pipes, having a 450 mm inside diameter and a 63 mm wall thickness, were manufactured at a local facility using the spinning technique. Concrete pipes are traditionally reinforced with a steel cage consisting of circumferential and longitudinal bars. For this study, initially, concrete pipe without any conventional cage was manufactured, incorporating straight steel fibers at a dosage of 20 kg/m^3. The objective of this initial trial was to observe the behavior of the steel fibers in the spinning process. Accordingly, the concrete ingredients were mixed in a drum-type mixture according to the proportions given in Table 1. After dry-mixing of coarse aggregate, fine aggregate, and cement for two to three minutes, the mixing water was introduced, and wet mixing was performed for another three to five minutes. The straight steel fibers were then gradually added to the mixer so as to avoid fiber balling. The mixing was continued until the steel fibers were dispersed in the concrete mixture. The prepared concrete mixture was unloaded onto the casting yard, from where it was manually poured into the pipe mold with shovels.

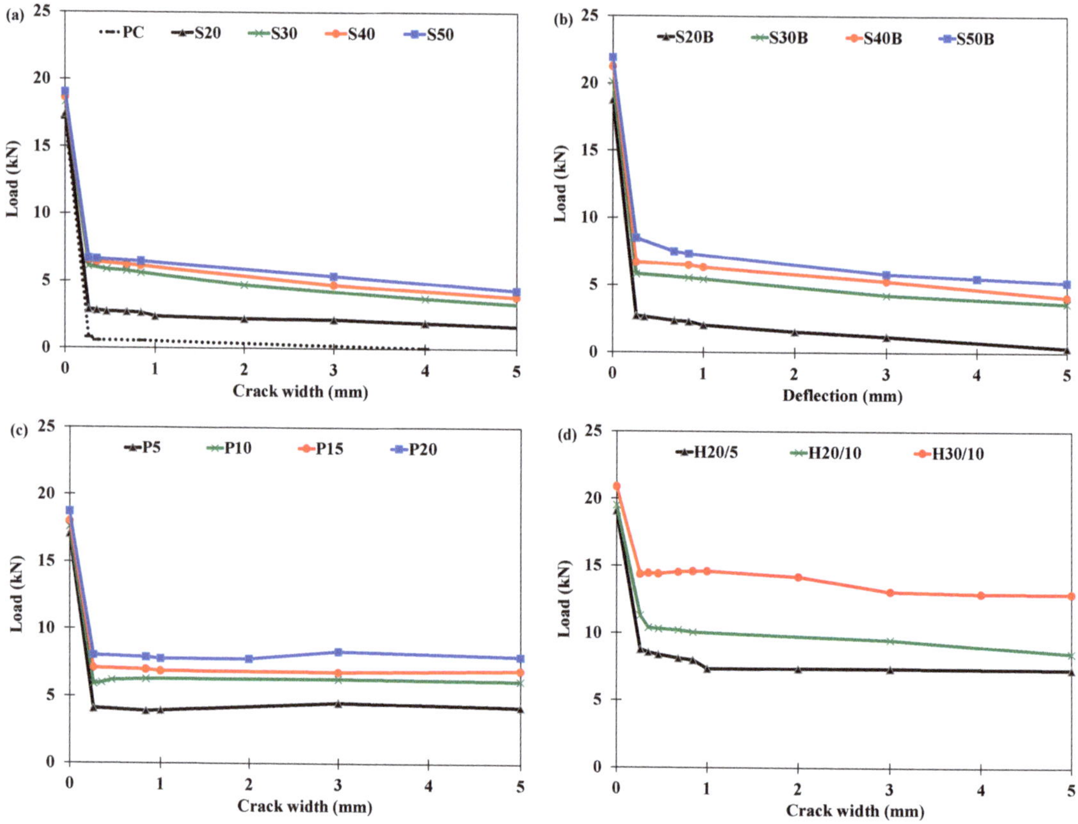

Figure 11. Load vs. crack width recorded at discrete points during the testing for specimen incorporating (**a**) straight steel fibers, (**b**) bundled steel fibers, (**c**) PP fibers, (**d**) hybrid fibers (bundled steel fibers and PP fibers).

The pipe mold was mounted on two rollers which induced rotating motion to the mold (Figure 13). Initially, the mold was set into rotation at a speed of 160 rpm to 180 rpm. In a conventional concrete pipe, a reinforcing steel cage is inserted into the mold, which helps retain the fresh concrete as it is poured into the spinning mold. The exclusion of the conventional steel cage from the pipe for this study presented a challenge in ensuring the fresh concrete would stick to the wall of the mold. As the fresh concrete was poured into the mold, some of the concrete stuck to the wall of the mold, while some of it fell off. The problem was overcome by reducing the initial spinning speed to 130 rpm to 140 rpm, so as to allow the concrete to stick to the wall of the mold. As the thickness of the pipe wall built up with more concrete poured into the mold, the concrete could bear higher spinning speed without falling off. Therefore, the spinning speed was increased back to 160 rpm to 180 rpm. The pouring of the concrete was stopped when the desired wall thickness was attained. Finally, the spinning speed of the mold was increased to 230 rpm to 250 rpm. It is the centrifugal forces resulting from the spinning process that result in the consolidation of concrete. The water, being the lightest of all the ingredients of concrete, was forced out of the concrete. The water was then drained out with the help of a brush. The expulsion of water from the concrete has the advantage of a reduced water/cementitious material (w/cm) ratio in the final concrete. The spinning was continued until no more water could be extracted from the concrete.

Figure 12. Fracture surfaces of specimens incorporating various fibers: (**a**) straight steel fibers, (**b**) bundled steel fibers, (**c**) PP fibers, (**d**) hybrid fibers.

The centrifugal forces presented the biggest challenge to maintaining the uniform dispersion of the steel fibers in the spinning concrete. The visual inspection of the freshly cast concrete pipe revealed that the steel fibers, having much higher specific gravity (7.85), segregated towards the outer side of the wall of the pipe and that no steel fiber could be seen on the inside of the wall of the pipe (Figure 14).

In the second phase, four additional concrete pipes having an inside diameter of 450 mm and a wall thickness of 63 mm were manufactured. Plain concrete pipe without any reinforcement, conventional reinforced concrete pipe, concrete pipe with a conventional steel cage as well as 20 kg/m^3 of the straight steel fibers, and concrete pipe with a conventional steel cage as well as 40 kg/m^3 of the straight steel fibers were cast. The conventional steel reinforcement comprised six longitudinal bars of 4 mm diameter and circumferential rebar of 4 mm diameter spaced at 82 mm.

Figure 13. Pipe mold on rotating wheels.

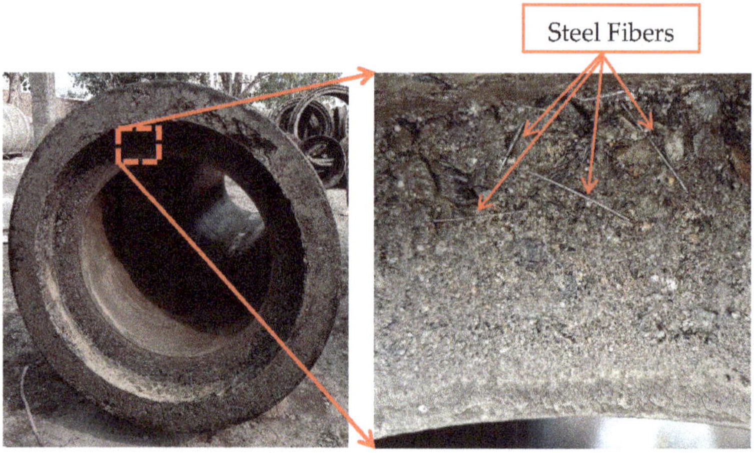

Figure 14. Distribution of steel fibers seen in the freshly cast concrete pipe.

The pipes were tested in three-edge bearing loading (Figures 15 and 16), following the guidelines of ASTM C497 [72]. The D-loads corresponding to 0.3 mm crack width and the ultimate load recorded for the tested pipes are given in Table 6. The D-load is the load per unit length per unit diameter of the pipe. Figure 17 shows the typical load-deflection curve for the tested pipes. The inclusion of the steel fibers in the concrete pipes improved both the cracking load (0.3 mm crack width) and the ultimate load. This was due to the ability of the steel fibers to bridge a crack formed along the springlines, (Figure 18a). However, no such crack bridging was observed for the cracks long the invert and the crown of the pipe, (Figure 18b). This was because the steel fibers had relocated towards the exterior side of the wall of the pipe during the spinning process. This was further confirmed by visual inspection of the fracture surface of the pipe (Figure 19). It can be seen that steel fibers were present closer to the outer side of the wall.

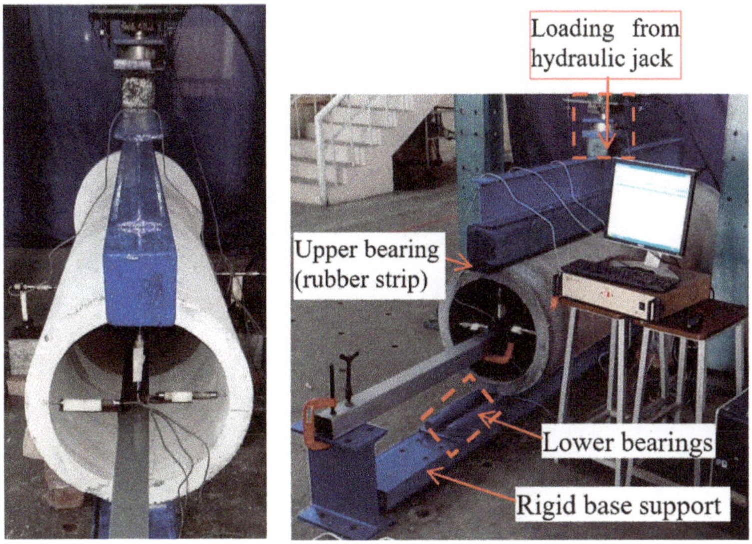

(**a**) (**b**)

Figure 15. Setup of three-edge bearing test: (**a**) side view, (**b**) 3D view.

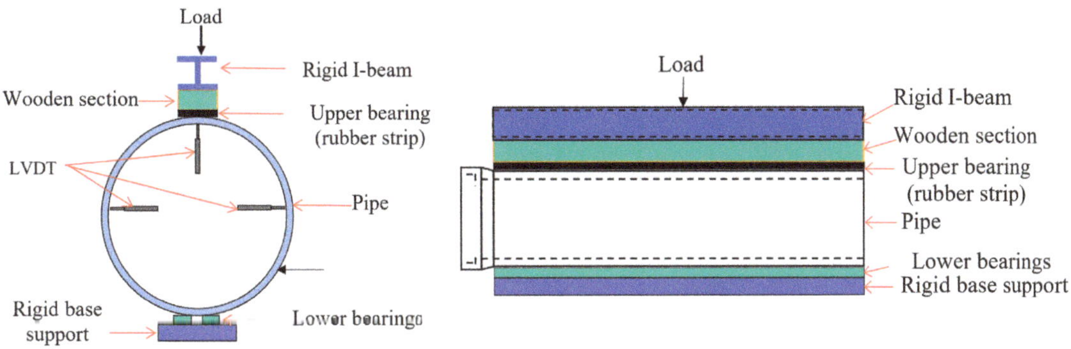

(**a**) (**b**)

Figure 16. Schematic of three-edge bearing test: (**a**) cross-sectional view, (**b**) longitudinal view.

Table 6. Results of three-edge bearing tests.

Pipe	Reinforcement Details	Steel Fibers (kg/m^3)	D-Load$_{0.3mm\ crack}$ (N/m/mm)	D-Load$_{ult}$ (N/m/mm)
1	-	-	60	60
2	-	20	72	72
3	Conventional steel rebar	-	105	140
4	Conventional steel rebar	20	114	155
5	Conventional steel rebar	40	130	185

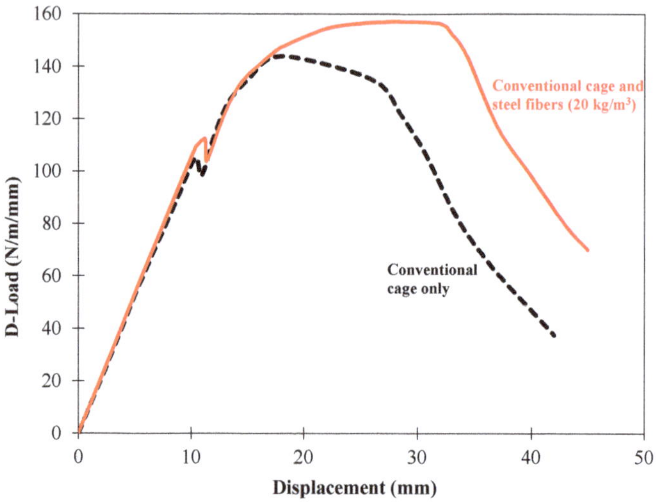

Figure 17. Load-deflection curves of the pipes tested under the three-edge bearing condition.

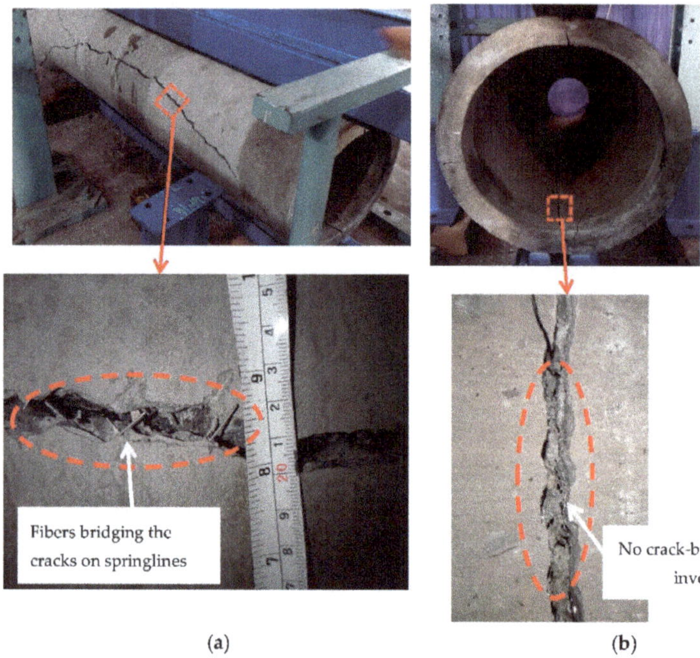

Figure 18. Cracking pattern of the tested concrete pipe incorporating steel fibers: (**a**) crack along springlines, (**b**) crack at invert.

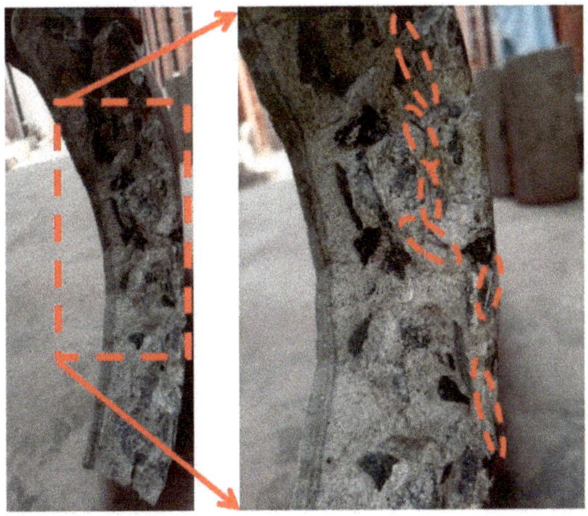

Figure 19. Fracture surface of the tested concrete pipe incorporating steel fibers (Location of steel fibers mentioned at outer edge enclosed in red dashed lines).

6. Conclusions

This study explored the mechanical properties of various concrete mixtures incorporating steel and PP fibers in mono and hybrid forms at varying dosages with the aim to incorporate the fibers into concrete mixtures for the manufacturing of full-scale spun-cast concrete pipes. The following specific conclusions can be drawn from this study:

- The dosage and type of fibers generally showed little or no effect on the compressive strength of the fiber-reinforced concrete mixtures. However, some improvement in the compressive strength of up to 18% was observed for the concrete mixture incorporating 50 kg/m^3 of bundled steel fibers, as compared to the plain concrete mixture without fibers.
- The splitting tensile strength and flexural strength of all fiber-reinforced concrete mixtures were consistently higher as compared to the plain concrete mixture without fibers. The concrete mixture incorporating hybrid fibers (30 kg/m^3 of bundled steel fibers and 10 kg/m^3 PP fibers) exhibited much better improvement in flexural strength (24% higher), residual flexural strength at small and large deflections (4.39 MPa at L/600 and 3.89 MPa at L/150, as compared to nil for plain concrete mixture), and toughness (23 times higher) in comparison with the plain concrete mixture without fibers.
- The fibers can be incorporated into concrete mixtures for the manufacturing of full-scale spun-cast concrete pipes without significant changes in the manufacturing process. The tests on full-scale pipes showed that the use of fibers as partial or complete replacements for the conventional steel cage is a viable option to enhance the crack resistance of concrete pipes and eliminate or reduce the time-consuming fabrication of conventional steel cages.
- The use of hybrid fibers consisting of high-specific-gravity fibers (such as steel fibers) and low-specific-gravity fibers (such as PP fibers) can be an effective strategy to deal with the tensile stresses on both the inner and outer sides of the wall of a concrete pipe a with thickness too small to accommodate two layers of conventional rebar.

Author Contributions: Conceptualization, S.A. and A.F.; experimentation and methodology, A.F. and S.A.; validation, S.A., S.M.S.K., and M.J.M.; analysis of results, S.A., A.F., S.M.S.K., and M.J.M.; resources, S.A. and A.F.; data curation, A.F.; manuscript writing, S.A. and A.F.; review and editing of manuscript, S.A., S.M.S.K. and M.J.M.; project supervision administration, S.A.; funding acquisition, S.A. and A.F. All authors have read and agreed to the published version of the manuscript.

Funding: This research received no external funding.

Institutional Review Board Statement: Not applicable.

Informed Consent Statement: Not applicable.

Data Availability Statement: Not applicable.

Acknowledgments: The authors would like to thank the undergraduate students and technical lab staff of the Civil Engineering Department at UET Lahore for assisting during the performance of laboratory tests.

Conflicts of Interest: The authors declare no conflict of interest.

References

1. United Nations. *The Human Rights to Safe Drinking Water and Sanitation*; United Nations General Assembly: New York, NY, USA, 2016; Volume 660.
2. Qadir, M.; Wichelns, D.; Raschid-Sally, L.; McCornick, P.G.; Drechsel, P.; Bahri, A.; Minhas, P.S. The Challenges of Wastewater Irrigation in Developing Countries. *Agric. Water Manag.* **2010**, *97*, 561–568. [CrossRef]
3. Mateo-Sagasta, J.; Raschid-Sally, L.; Thebo, A. Global Wastewater and Sludge Production, Treatment and Use. In *Wastewater*; Springer: Dordrecht, The Netherlands, 2015; pp. 15–38.
4. Sato, T.; Qadir, M.; Yamamoto, S.; Endo, T.; Zahoor, A. Global, Regional, and Country Level Need for Data on Wastewater Generation, Treatment, and Use. *Agric. Water Manag.* **2013**, *130*, 1–13. [CrossRef]
5. United Nations. *Rising Reuse of Wastewater in Forecast but World Lacks Data on "Massive Potential Resource"*; United Nations University: Tokyo, Japan, 2013.
6. Qadir, M.; Drechsel, P.; Jiménez Cisneros, B.; Kim, Y.; Pramanik, A.; Mehta, P.; Olaniyan, O. Global and Regional Potential of Wastewater as a Water, Nutrient and Energy Source. *Nat. Resour. Forum* **2020**, *44*, 40–51. [CrossRef]
7. World Health Organization. *Water Quality: Guidelines, Standards, and Health: Assessment of Risk and Risk Management for Water-Related Infectious Disease*; World Health Organization: Geneva, Switzerland, 2001.
8. Zhang, J.; Cao, X.S.; Meng, X.Z. Sustainable Urban Sewerage System and Its Application in China. *Resour. Conserv. Recycl.* **2007**, *51*, 284–293. [CrossRef]
9. Humes Concrete Pipe Brochure. Available online: https://www.humes.co.nz/assets/Uploads/9516411875/Humes-Concrete-Pipe-Brochure-2020.pdf (accessed on 26 April 2022).
10. Nedal, M. Experimental and Numerical Study on Full-Scale Precast Steel Fibre-Reinforced Concrete Pipes. Ph.D. Thesis, Western University, London, ON, Canada, 2015.
11. Mohsen, A. Evaluation Of The Performance Of Steel Fiber Reinforced Concrete Pipes Produced By Packerhead Method. Master's Thesis, University of Texas at Arlington, Arlington, TX, USA, 2014.
12. Joshi, N.G. *Concrete Pipes and Pipelines*; Narosa Publishing House Pvt. Ltd.: New Delhi, India, 2015.
13. Abbas, S.; Kazmi, S.M.S.; Munir, M.J. Potential of Rice Husk Ash for Mitigating the Alkali-Silica Reaction in Mortar Bars Incorporating Reactive Aggregates. *Constr. Build. Mater.* **2017**, *132*, 61–70. [CrossRef]
14. Gencel, O.; Kazmi, S.M.S.; Munir, M.J.; Kaplan, G.; Bayraktar, O.Y.; Yarar, D.O.; Karimipour, A.; Ahmad, M.R. Influence of Bottom Ash and Polypropylene Fibers on the Physico-Mechanical, Durability and Thermal Performance of Foam Concrete: An Experimental Investigation. *Constr. Build. Mater.* **2021**, *306*, 124887. [CrossRef]
15. Heger, F.J.; McGrath, T.J. Radial Tension Strength of Pipe and Other Curved Flexural Members. *ACI J. Proc.* **1983**, *80*, 33–39. [CrossRef]
16. Wen, Q.-J.; Jing, H.-W.; Sanda, S.; Zhuang, S.-S. Experimental Investigation of Mechanical Properties of Centrifugal Concrete in Circular Pipes. *J. Mater. Civ. Eng.* **2017**, *29*, 04016251. [CrossRef]
17. Feng, Q.; Kong, Q.; Huo, L.; Song, G. Crack Detection and Leakage Monitoring on Reinforced Concrete Pipe. *Smart Mater. Struct.* **2015**, *24*, 115020. [CrossRef]
18. Dirksen, J.; Clemens, F.H.L.R.; Korving, H.; Cherqui, F.; le Gauffre, P.; Ertl, T.; Plihal, H.; Müller, K.; Snaterse, C.T.M. The Consistency of Visual Sewer Inspection Data. *Struct. Infrastruct. Eng.* **2013**, *9*, 214–228. [CrossRef]
19. Wang, F.; Xue, X.; Hua, J.; Chen, Z.; Huang, L.; Wang, N.; Jin, J. Non-Uniform Corrosion Influences on Mechanical Performances of Stainless-Clad Bimetallic Steel Bars. *Mar. Struct.* **2022**, *86*, 103276. [CrossRef]
20. Hua, J.; Fan, H.; Xue, X.; Wang, F.; Chen, Z.; Huang, L.; Wang, N. Tensile and Low-Cycle Fatigue Performance of Bimetallic Steel Bars with Corrosion. *J. Build. Eng.* **2021**, *43*, 103188. [CrossRef]
21. Hua, J.; Wang, F.; Huang, L.; Wang, N.; Xue, X. Experimental Study on Mechanical Properties of Corroded Stainless-Clad Bimetallic Steel Bars. *Constr. Build. Mater.* **2021**, *287*, 123019. [CrossRef]
22. Abbas, S.; Soliman, A.M.; Nehdi, M.L. Exploring Mechanical and Durability Properties of Ultra-High Performance Concrete Incorporating Various Steel Fiber Lengths and Dosages. *Constr. Build. Mater.* **2015**, *75*, 429–441. [CrossRef]
23. Abbas, S.; Ishaq, M.A.A.; Kazmi, S.M.S.; Munir, M.J.; Ali, S. Investigating the Behavior of Waste Alumina Powder and Nylon Fibers for Eco-Friendly Production of Self-Compacting Concrete. *Materials* **2022**, *15*, 4515. [CrossRef]
24. Kazmi, S.M.S.; Munir, M.J.; Wu, Y.F.; Patnaikuni, I. Effect of Macro-Synthetic Fibers on the Fracture Energy and Mechanical Behavior of Recycled Aggregate Concrete. *Constr. Build. Mater.* **2018**, *189*, 857–868. [CrossRef]

25. Kytinou, V.K.; Chalioris, C.E.; Karayannis, C.G. Analysis of Residual Flexural Stiffness of Steel Fiber-Reinforced Concrete Beams with Steel Reinforcement. *Materials* **2020**, *13*, 2698. [CrossRef]
26. Guo, H.; Jiang, L.; Tao, J.; Chen, Y.; Zheng, Z.; Jia, B. Influence of a Hybrid Combination of Steel and Polypropylene Fibers on Concrete Toughness. *Constr. Build. Mater.* **2021**, *275*, 122132. [CrossRef]
27. Wang, F.; Hua, J.; Xue, X.; Wang, N.; Yao, Y. Effects of Polyoxymethylene Fiber on Mechanical Properties of Seawater Sea-Sand Concrete with Different Ages. *Polymers* **2022**, *14*, 3472. [CrossRef]
28. Xue, X.; Wang, F.; Hua, J.; Wang, N.; Huang, L.; Chen, Z.; Yao, Y. Effects of Polyoxymethylene Fiber on Fresh and Hardened Properties of Seawater Sea-Sand Concrete. *Polymers* **2022**, *14*, 4969. [CrossRef]
29. Fang, H.; Gu, M.; Zhang, S.; Jiang, H.; Fang, Z.; Hu, J. Effects of Steel Fiber and Specimen Geometric Dimensions on the Mechanical Properties of Ultra-High-Performance Concrete. *Materials* **2022**, *15*, 3027. [CrossRef]
30. Khan, S.; Qing, L.; Ahmad, I.; Mu, R.; Bi, M. Investigation on Fracture Behavior of Cementitious Composites Reinforced with Aligned Hooked-End Steel Fibers. *Materials* **2022**, *15*, 542. [CrossRef]
31. Abed, M.A.; Fořt, J.; Naoulo, A.; Essa, A. Influence of Polypropylene and Steel Fibers on the Performance and Crack Repair of Self-Compacting Concrete. *Materials* **2021**, *14*, 5506. [CrossRef]
32. Wölfel, E.; Brünig, H.; Curosu, I.; Mechtcherine, V.; Scheffler, C. Dynamic Single-Fiber Pull-Out of Polypropylene Fibers Produced with Different Mechanical and Surface Properties for Concrete Reinforcement. *Materials* **2021**, *14*, 722. [CrossRef]
33. Wang, L.; He, T.; Zhou, Y.; Tang, S.; Tan, J.; Liu, Z.; Su, J. The Influence of Fiber Type and Length on the Cracking Resistance, Durability and Pore Structure of Face Slab Concrete. *Constr. Build. Mater.* **2021**, *282*, 122706. [CrossRef]
34. Haktanir, T.; Ari, K.; Altun, F.; Karahan, O. A Comparative Experimental Investigation of Concrete, Reinforced-Concrete and Steel-Fibre Concrete Pipes under Three-Edge-Bearing Test. *Constr. Build. Mater.* **2007**, *21*, 1702–1708. [CrossRef]
35. Wilson, A.; Abolmaali, A. Performance of Synthetic Fiber-Reinforced Concrete Pipes. *J. Pipeline Syst. Eng. Pract.* **2014**, *5*, 04014002. [CrossRef]
36. Lee, S.; Park, Y.; Abolmaali, A. Investigation of Flexural Toughness for Steel-and-Synthetic-Fiber-Reinforced Concrete Pipes. *Structures* **2019**, *19*, 203–211. [CrossRef]
37. Park, Y.; Abolmaali, A.; Mohammadagha, M.; Lee, S. Structural Performance of Dry-Cast Rubberized Concrete Pipes with Steel and Synthetic Fibers. *Constr. Build. Mater.* **2015**, *77*, 218–226. [CrossRef]
38. Abolmaali, A.; Mikhaylova, A.; Wilson, A.; Lundy, J. Performance of Steel Fiber–Reinforced Concrete Pipes. *Transp. Res. Rec.* **2012**, *2313*, 168–177. [CrossRef]
39. Park, Y.; Abolmaali, A.; Beakley, J.; Attiogbe, E. Thin-Walled Flexible Concrete Pipes with Synthetic Fibers and Reduced Traditional Steel Cage. *Eng. Struct.* **2015**, *100*, 731–741. [CrossRef]
40. Mohamed, N.; Nehdi, M.L. Rational Finite Element Assisted Design of Precast Steel Fibre Reinforced Concrete Pipes. *Eng. Struct.* **2016**, *124*, 196–206. [CrossRef]
41. Nehdi, M.L.; Mohamed, N.; Soliman, A.M. Investigation of Buried Full-Scale SFRC Pipes under Live Loads. *Constr. Build. Mater.* **2016**, *102*, 733–742. [CrossRef]
42. Mohamed, N.; Soliman, A.M.; Nehdi, M.L. Mechanical Performance of Full-Scale Precast Steel Fibre-Reinforced Concrete Pipes. *Eng. Struct.* **2015**, *84*, 287–299. [CrossRef]
43. Mohamed, N.; Soliman, A.M.; Nehdi, M.L. Full-Scale Pipes Using Dry-Cast Steel Fibre-Reinforced Concrete. *Constr. Build. Mater.* **2014**, *72*, 411–422. [CrossRef]
44. Peyvandi, A.; Soroushian, P.; Jahangirnejad, S. Enhancement of the Structural Efficiency and Performance of Concrete Pipes through Fiber Reinforcement. *Constr. Build. Mater.* **2013**, *45*, 36–44. [CrossRef]
45. Deng, Z.; Liu, X.; Chen, P.; de la Fuente, A.; Zhou, X.; Liang, N.; Han, Y.; Du, L. Basalt-Polypropylene Fiber Reinforced Concrete for Durable and Sustainable Pipe Production. Part 1: Experimental Program. *Struct. Concr.* **2021**, *23*, 311–327. [CrossRef]
46. al Rikabi, F.T.; Sargand, S.M.; Kurdziel, J. Evaluation of Synthetic Fiber Reinforced Concrete Pipe Performance Using Three-Edge Bearing Test. *J. Test Eval.* **2018**, *47*, 942–958. [CrossRef]
47. al Rikabi, F.T.; Sargand, S.M.; Kurdziel, J.; Khoury, I. Performance of Thin-Wall Synthetic Fiber–Reinforced Concrete Pipes under Short and Long-Term Loading. *J. Test Eval.* **2020**, *48*, 3713–3733. [CrossRef]
48. Fouad, T.; Kurdziel, A.R.J.; Husam, H.; Hussein, S.M.S. Experimental Investigation of Thin-Wall Synthetic Fiber- Reinforced Concrete Pipes. *ACI Struct. J.* **2018**, *115*, 1671–1681. [CrossRef]
49. Raju, R.A.; Akiyama, M.; Lim, S.; Kakegawa, T.; Hosono, Y. A Novel Casting Procedure for SFRC Piles without Shear Reinforcement Using the Centrifugal Forming Technique to Manipulate the Fiber Orientation and Distribution. *Constr. Build. Mater.* **2021**, *303*, 124232. [CrossRef]
50. *BS EN 1916*; Concrete Pipes and Fittings, Unreinforced, Steel Fibre and Reinforced. British Standards Institution: London, UK, 2002.
51. *ASTM C1765*; Standard Specification for Steel Fiber Reinforced Concrete Culvert, Storm Drain, and Sewer Pipe. American Society for Testing and Materials: West Conshohocken, PA, USA, 2019.
52. *ASTM C1818*; Standard Specification for Synthetic Fiber Reinforced Concrete Culvert, Storm Drain, and Sewer Pipe. American Society for Testing and Materials: West Conshohocken, PA, USA, 2022.
53. *ASTM C76*; Standard Specification for Reinforced Concrete Culvert, Storm Drain, and Sewer Pipe. American Society for Testing and Materials: West Conshohocken, PA, USA, 2022.

54. *ACI Committee 211*; Guide for Selecting Proportions for No-Slump Concrete (ACI 211.3R-02). American Concrete Institute: Farmington Hills, MI, USA, 2009.
55. *ASTM C143*; Standard Test Method for Slump of Hydraulic-Cement Concrete. American Society for Testing and Materials: West Conshohocken, PA, USA, 2020.
56. *ASTM C138*; Standard Test Method for Density (Unit Weight), Yield, and Air Content (Gravimetric) of Concrete. American Society for Testing and Materials: West Conshohocken, PA, USA, 2017.
57. *ASTM C39*; Standard Test Method for Compressive Strength of Cylindrical Concrete Specimens. American Society for Testing and Materials: West Conshohocken, PA, USA, 2021.
58. *ASTM C496*; Standard Test Method for Splitting Tensile Strength of Cylindrical Concrete Specimens. American Society for Testing and Materials: West Conshohocken, PA, USA, 2017.
59. *ASTM C1609*; Standard Test Method for Flexural Performance of Fiber-Reinforced Concrete (Using Beam With Third-Point Loading). American Society for Testing and Materials: West Conshohocken, PA, USA, 2019.
60. Yang, K.-H. Tests on Concrete Reinforced with Hybrid or Monolithic Steel and Polyvinyl Alcohol Fibers. *ACI Mater. J.* **2011**, *108*, 664–672.
61. Ramakrishnan, V.; George, Y.W.; Hosalli, G. Flexural Fatigue Strength, Endurance Limit, and Impact Strength of Fiber Reinforced Concrete. In Proceedings of the International Symposium on Recent Developments in Concrete Fiber Composites; Transportation Research Board: Washington, DC, USA, 1989; pp. 17–24.
62. Abbass, W.; Khan, M.I.; Mourad, S. Evaluation of Mechanical Properties of Steel Fiber Reinforced Concrete with Different Strengths of Concrete. *Constr. Build. Mater.* **2018**, *168*, 556–569. [CrossRef]
63. Li, B.; Chi, Y.; Xu, L.; Shi, Y.; Li, C. Experimental Investigation on the Flexural Behavior of Steel-Polypropylene Hybrid Fiber Reinforced Concrete. *Constr. Build. Mater.* **2018**, *191*, 80–94. [CrossRef]
64. Banthia, N.; Soleimani, S.M. Flexural Response of Hybrid Fiber-Reinforced Cementitious Composites. *ACI Mater. J.* **2005**, *102*, 382–389. [CrossRef]
65. Wang, J.; Dai, Q.; Si, R.; Guo, S. Mechanical, Durability, and Microstructural Properties of Macro Synthetic Polypropylene (PP) Fiber-Reinforced Rubber Concrete. *J. Clean. Prod.* **2019**, *234*, 1351–1364. [CrossRef]
66. Yao, W.; Li, J.; Wu, K. Mechanical Properties of Hybrid Fiber-Reinforced Concrete at Low Fiber Volume Fraction. *Cem. Concr. Res.* **2003**, *33*, 27–30. [CrossRef]
67. *ACI Committee 544*; Guide to Design with Fiber-Reinforced Concrete (544.4R-18). American Concrete Institute: Farmington Hills, MI, USA, 2018.
68. Afroughsabet, V.; Ozbakkaloglu, T. Mechanical and Durability Properties of High-Strength Concrete Containing Steel and Polypropylene Fibers. *Constr. Build. Mater.* **2015**, *94*, 73–82. [CrossRef]
69. *ACI Committee 544*; Report on Measuring Mechanical Properties of Hardened Fiber-Reinforced Concrete (544.9R-17). American Concrete Institute: Farmington Hills, MI, USA, 2017.
70. Dopko, M.; Najimi, M.; Shafei, B.; Wang, X.; Taylor, P.; Phares, B.M. Flexural Performance Evaluation of Fiber-Reinforced Concrete Incorporating Multiple Macro-Synthetic Fibers. *Transp. Res. Rec.* **2018**, *2672*, 1–12. [CrossRef]
71. *ACI Committee 544*; Report on Design and Construction of Steel Fiber-Reinforced Concrete Elevated Slabs (544.6R-15). American Concrete Institute: Farmington Hills, MI, USA, 2015.
72. *ASTM C497*; Standard Test Methods for Concrete Pipe, Concrete Box Sections, Manhole Sections, or Tile. American Society for Testing and Materials: West Conshohocken, PA, USA, 2020.

materials MDPI

Article

Effect of the Partial Replacement of Cement with Waste Granite Powder on the Properties of Fresh and Hardened Mortars for Masonry Applications

Zuzanna Zofia Woźniak , Adrian Chajec and Łukasz Sadowski *

Department of Materials Engineering and Construction Processes, Wrocław University of Science and Technology, Wybrzeże Wyspiańskiego 27, 50-370 Wroclaw, Poland
* Correspondence: lukasz.sadowski@pwr.edu.pl

Abstract: Granite is a well-known building and decorative material, and, therefore, the amount of produced waste in the form of granite powder is a problem. Granite powder affects the health of people living near landfills. Dust particles floating in the air, which are blown by gusts of wind, can lead to lung silicosis and eye infections, and can also affect the immune system. To find an application for this kind of waste material, it was decided to study the effect of partially replacing cement with waste granite powder on the properties of fresh and hardened mortars intended for masonry applications. The authors planned to replace 5%, 10%, and 15% of cement with waste material. Series of mortar with the addition of granite powder achieved 50% to 70% of the compressive strength of the reference series, and 60% to 76% of the bending strength of the reference series. The partial replacement of cement with the granite powder significantly increased the water sorption coefficient. The consistency of the fresh mortar, and its density and water absorption also increased when compared to the reference series. Therefore, Granite powder can be used as a partial replacement of cement in masonry mortars.

Keywords: granite powder; waste; cement mortar

check for updates

Citation: Woźniak, Z.Z.; Chajec, A.; Sadowski, Ł. Effect of the Partial Replacement of Cement with Waste Granite Powder on the Properties of Fresh and Hardened Mortars for Masonry Applications. *Materials* **2022**, *15*, 9066. https://doi.org/10.3390/ma15249066

Academic Editor: Ayman El-Zohairy

Received: 15 November 2022
Accepted: 15 December 2022
Published: 19 December 2022

1. Introduction

The constantly developing construction sector results in a variety of masonry structures, among others, being made [1]; therefore, there is a need to modify traditional masonry mortars. One of the factors influencing the development of the construction industry is the market demand for finding a use for waste materials. Another factor that affects the development of the construction industry is the desire to carry out work as quickly and correctly as possible [2]. Additionally, there is also a need to develop materials that have a variety of properties. Despite the rapid development of the construction industry and the creation of new building materials, there is also a problem with regards to the correct execution of cement mortars [3].

The main purpose of masonry mortar is to permanently connect masonry elements. The mortar must have the correct consistency, and a good compressive and bending strength. A mortar should be selected according to the masonry element and the weather conditions in which the masonry works are conducted. The masonry mortar affects the capillary rise in the wall in which it is used; therefore, the intended use of the brick object is also important. The choice of components of masonry mortars, which are partially replaced with waste materials, is not accidental. In many studies, waste materials are added to mortars as a substitute for cement or aggregate, which is a solution to the global problem associated with these two materials. Using waste materials as substitute materials for cement reduces the costs related to, for example, the storage of such waste materials, and is an appropriate method of their disposal [4,5]. Many articles over the past few years have been devoted to the study of carbon dioxide emissions. The current challenge, due to the continuous emission of carbon dioxide into the atmosphere, is to take care of the environment. Cement production produces carbon dioxide CO_2, which is responsible for 5% of the total anthropogenic carbon dioxide emissions [4], with some sources even giving a value of 7% [6] or 8% [7]. Over the past few years, cement plants have been modernized in order to reduce the CO_2 emissions resulting from cement production. However, the problem has not been completely resolved. Therefore, scientists are constantly conducting more studies that aim to reduce or optimize the amount of used cement. In cement mortars, it has been found to be possible to replace cement with various waste materials. Even a small replacement of cement with waste material can have a positive impact on the environment [8,9].

Many countries have a problem finding a way to use waste materials in the construction industry [10]. There is an interest among scientists in waste materials, such as granite powder, which shows that there is a problem related to the use of this material [11–13]. Granite is an igneous rock, the hardness of which is estimated to be 7 on the Mohs scale. This material, as a rock, is a very durable material that does not dissolve in water. Therefore, granite powder is a mineral material of natural origin. This material is found on virtually every continent, even in Antarctica. In Poland, the biggest granite deposits are located in Lower Silesia [14]. The extraction of granite in Europe is counted in millions of tons per year. Granite and marble are the second most frequently mined non-metallic minerals [15], and are the most frequently used decorative materials [16]. They are used for the production of tombstones, road surfaces, sidewalks, and fences. The processing of these materials generates by-products, such as granite powder and coarse aggregate. Granite powder has sharp edges due to cutting processes. The shape of granite powder particles differs from the shape of cement particles, therefore such added waste material fills in the aggregate skeleton. The experimental part of this work is to investigate the effect of granite powder on the selected properties of mortars. Currently, granite powder is not a commonly used material in construction, but it is used, for example, as a fertilizer for plants. It can also be used instead of sand to fill paving stones. However, the amounts of granite powder waste indicate that there is an insufficient use of this waste. There is a problem with the disposal of such granite powder, because it is a non-biodegradable material [17], and often ends up in landfills. Piles of material pose a threat to human life due to gusts of wind that carry away their particles. A person exposed to a long-term contact with crystalline silicon dioxide SiO_2 is at risk of lung silicosis. People staying or living in the vicinity of dust dumps containing SiO_2 have problems with their immune systems, and are exposed to all kinds of lung changes [18]. Granite powder also causes eye infections due to air pollution [17].

This article aims to investigate the effect of granite powder on selected properties of cement mortars. The focus was on properties, such as wall capillary suction, water absorption, temperature, and strength. All these properties affect the integrity of the joint between the mortar and the structural element. These properties correlate with the factors to which the mortar is exposed in its normal working conditions. In their research, the authors tried to find a possibility of using waste material, such as granite powder, in mortars, which would result in less cement being produced for this purpose. This will in turn contribute to the reduction of carbon dioxide emissions. The replacement of a few percent of cement with another material may have a small effect in reducing the amount of carbon dioxide that is emitted into the atmosphere. However, when taking into consideration the entire production of cement that is used in mortars, such a replacement can significantly reduce the amount of carbon dioxide emissions. The conducted research is very important due to the possibility of improving the quality of life of people and the environment. The article describes an innovative approach to environmental protection.

2. Materials and Methods

This point has been divided into several points. In Section 2.1, the materials from which the mortar series were created are described.

2.1. Materials Used in the Research

In the research, Portland Cement type II 32.5R (Odra in Opole, Poland), aggregate (Górażdże, Poland), granite powder (Strzegom, Poland), and tap water were used. The particle size distribution of the used granite powder is shown in Table 1. The chemical compositions are provided in Table 2. The grain size of the aggregate was 0–1.8 mm. The aggregate was dried for one week at a temperature of 25 °C.

Table 1. Graining curve of the waste material used in the research.

Sieve Size (mm)	<0.02	0.02	0.032	0.063	0.1	0.14
Passes through a sieve (%)	0	5	51.9	82.9	90.2	100

Table 2. Chemical composition of the used granite powder (%).

Chemical Compound	SiO_2	Al_2O_3	Fe_2O_3	CaO	MgO	FeO	NaO	K_2O
Granite powder (%)	54	20	2	8	6	4	3	3

The chemical compositions are provided in Table 2. The grain size of the aggregate was 0–1.8 mm. The aggregate was dried for one week at a temperature of 25 °C.

Four cementitious mortar compositions were investigated. The binder of the reference series was made just with Portland Cement type II 32.5R. Three series were made by replacing 5%, 10%, and 15%, respectively, of the mass of the cement with granite powder. The components of the mix used for the tests came from the same batch of products. The materials were stored at a constant temperature. The cement, aggregate, and granite did not have any lumps. The prepared proportions of the components are presented in Table 3.

Table 3. The proportions of the components of the designed cement mortar mixes.

Series (-)	Portland Cement (%)	Granite Powder (%)	Aggregate (%)	Water (%)	Water/Cement Ratio	Water/Binder Ratio
REF	21.45	-	64.34	14.21	0.66	0.66
GP5	20.38	1.07	64.34	14.21	0.70	0.66
GP10	19.30	2.14	64.34	14.21	0.74	0.66
GP15	18.23	3.22	64.34	14.21	0.78	0.66

Mixing took place using a mortar drill, and began with placing the aggregate, cement, and granite powder in measured amounts into a mixer. After about 30 s, the ingredients were mixed manually to eliminate fractionation. Then, three-quarters of the total mass of water was added while mixing continuously with a mortar drill for approximately 30 s. Afterwards, manual mixing was conducted in order to clear the bottom of the container of leftover unmixed ingredients. One-quarter of the total water was added during the mixing. The final mixing was for about one minute.

2.2. Properties of the Fresh Mortars

Immediately after the completion of the mixing process, the temperature of the fresh mortar was examined. A pin thermometer was used to test the temperature of the mix. The thermometer measures the temperature within the range from −50 °C to +300 °C, with an accuracy of ±1 °C, and a resolution of 0.1 °C. The temperature was measured every half an hour, starting from immediately after mixing to 90 min after the end of the mixing process. The consistency of the evaluated series was measured using a flow table in accordance with PN-EN 1015-3 [19]. The test equipment includes a flow table, truncated cones, and a compactor. The test involved determining the spreading diameter of the mortar on a flow table. The bulk density was evaluated by measuring the weight of a 1000 cm^3 container filled with mortar in two layers. Each layer was vibrated in order to remove any air bubbles.

2.3. *Properties of the Hardened Cement Mortars*

The following sub-sections describe how the tests were conducted on hardened samples of the masonry mortars.

2.3.1. Bending Strength and Compressive Strength

After determining the bulk density of the fresh mix, samples with dimensions of 40 × 40 × 160 mm were taken for the bending strength tests. Before the filling moulds, their inner surface was covered with a release agent. The hardened series were demoulded after 3 days, and kept in an air temperature of 20 ∓ 2 °C. The bending strength tests were conducted in accordance with PN-EN 196-1 [20] after 28 days of the samples maturing. The value of the bending strength was obtained based on the arithmetic mean of the three evaluated samples, which was rounded up to 0.01 MPa.

After the bending strength tests, the compressive strength tests were conducted in accordance with PN-EN 1015-11 on six halves of 40 × 40 × 160 mm beams, which were obtained after the bending strength tests [21].

2.3.2. Water Absorption

Water absorption is a measure of the amount of water being absorbed into a material, and has an influence on the capillary suction of a wall [20]. Moisture mainly migrates up a masonry wall through mortar. Mass water absorption is determined from Equation (1):

$$n_w = \frac{m_n - m_s}{m_s} 100, \ (\%) \tag{1}$$

where:

n_w—mass water absorption (%);
m_n—mass of the material sample when saturated with water (kg);
m_s—mass of a sample of the material in its dry state (kg).

First, the samples were marked, and a scale was drawn on them (a line in the middle of the sample). The samples were then weighed with an accuracy of 1 g and placed on washers in a vessel so that they would not touch the bottom of the vessel. The washer consisted of two layers of 1 mm thick plastic mesh. The vessel was filled with water to half the height of the samples. The water temperature was 18 ± 2 °C. After 24 h, the samples were wiped and then weighed on scales with an accuracy of 1 g. After that, water was added to the level of 10 mm above the surface of the sample. After 24 h, the weight of the samples was evaluated until the weight of the saturated samples was constant.

2.3.3. Capillary Suction

Capillary suction of a wall leads to an increase in the moisture content of the wall, and thus to imperfections in the appearance of the building [22]. It affects the thermal insulation of the wall and the durability of wooden elements. The capillary suction tests involved recording the capillary rate by measuring the mass of the samples. In the first period of time, the amount of absorbed water by the samples was greater, which was also shown in the studies of Halina Garbalińska et al. [23]. The measurements continued until a constant weight of all the series was achieved-with a deviation of 1%. Therefore, it was decided to introduce five phases of measuring the mass of the samples. The time distances between individual measurements were as follows:

- Phase I lasted 5 h and measurements were taken every hour;
- Phase II lasted 8 h and measurements were taken every 2 h;
- Phase III lasted 24 h and measurements were taken every 4 h;
- Phase IV lasted 16 h and measurements were taken every 8 h;

- Phase IV lasted 36 h and measurements were taken every 12 h.

The research was conducted until a constant weight was obtained for each series. The main aim of the test was to calculate the water sorption coefficient using Equation (2):

$$A = \frac{\Delta m_t}{F\,\Delta\sqrt{t}} \tag{2}$$

where:

A—water sorption coefficient [$kg/(m^2h^{1/2})$];
Δm_t—increase in the sample's mass (kg);
F—suction surface (m^2);
$\Delta\sqrt{t}$—increment of the root of time ($h^{1/2}$).

A plastic grate was placed on the bottom of a transparent plastic container. The samples were weighed and numbered, then placed on a 3.5 cm thick plastic grate. Water was gradually added so that the samples were immersed to a height of 5 mm. The water level was kept constant up to the immersion height of the samples of 5 mm. The samples were weighed until moisture equilibrium was reached.

3. Results

3.1. Temperature

The temperature of the mixes affects the time of their transport, the properties of the obtained hardened mortars [24], and the maintenance of the properties of the fresh mortars, e.g., their workability. A comparison of the temperature test results for the GP series and the REF series is presented in Figure 1.

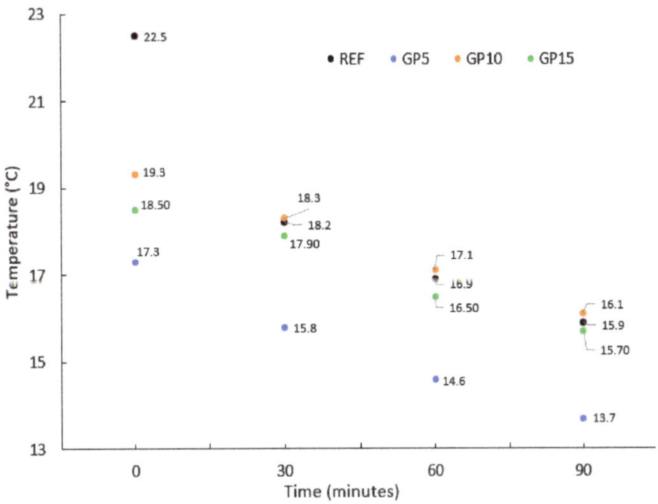

Figure 1. The temperature of the series of the cement mortar evaluated half an hour after the end of the mixing process and up to 2 h after the end of the mixing process.

The temperature for all the cases decreases over time. The highest temperature difference during 90 min was achieved by the reference series, and amounted to 6.4 °C. The GP5 series had a difference of 3.6 °C. The GP10 and GP15 series obtained the smallest temperature differences in the evaluated time, amounting to 3.2 °C and 2.8 °C, respectively. It can be concluded that with an increasing amount of granite powder added to the mix, the temperature difference, measured at 90 min intervals, decreases. Immediately after the end of the mixing process, the temperature for the series with the addition of granite powder varied by several degrees. The largest temperature difference in the first measurement was over 5 °C between the REF and GP5 series, with the reference series having the highest temperature of 22 °C in the first measurement. The biggest difference in the last measurement was 2.4 °C between the GP10 and GP5 series. The results of the last measurement were similar for the REF, GP10, and GP15 series. The difference between the GP15 and REF series and the GP15 and REF series was 0.2 °C. The differences in temperature for these series after 90 min from the end of the mixing process did not show big differences. This means that replacing the cement with granite powder in the amount of 10% and 15% of the cement mass will not significantly affect the temperature of the mix within 30 min to 90 min from the end of the mixing process. The exception was the GP5 series, for which the mix temperature was the lowest. Such a temperature may have a positive effect on concreting in summer conditions. The decrease in the temperature is due to the reduction in the amount of pozzolanic material. Pozzolanic material reacts as soon as it comes into contact with water. Replacing cement with a material with different pozzolanic properties causes the temperature to drop. Z. Woźniak et al. [24] noticed a similar trend regarding a drop in temperature for the series with the addition of granite powder when compared to the reference series. The authors measured the temperature immediately after mixing, then again 30 min after the end of the mixing process. They examined mortar series with a 10%, 15%, and 20%, respectively, replacement of cement with granite powder. In the case of the second measurement (after 30 min), there was a clear relationship between the increasing amount of granite powder in the binder and the decreasing temperature. Another mineral material that influences the hydration temperature is limestone powder. Grzeszczyk et al. [25] claimed that the addition of lime powder in mortar reduces the heat of hydration.

3.2. *Consistency of the Mix and Its Bulk Density*

Results of the consistency tests based on two measurements for each series are presented in Figure 2a, and results of bulk density are presented in Figure 2b. The results of the series with a partial replacement of cement with granite powder (green columns) were compared with the results obtained for the reference series (grey column). Error bars indicate the standard deviation from the mean value for the evaluated series.

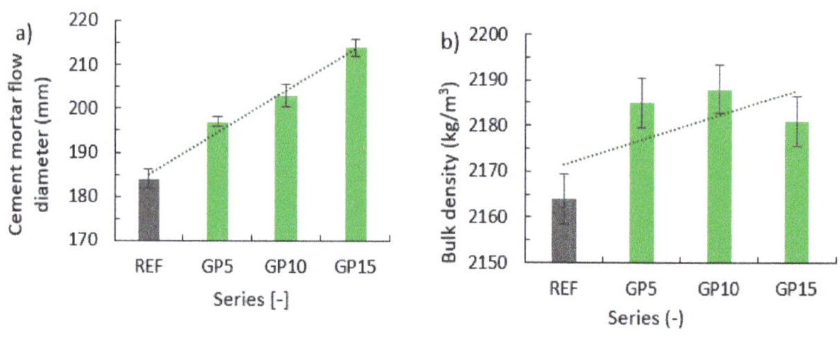

Figure 2. Properties of the fresh cement mortars with a partial replacement of cement with the granite powder: (**a**) cement mortar flow diameter; (**b**) bulk density.

The addition of the granite powder waste material significantly increased the consistency of the mix when compared to the reference series. As the amount of waste material in a mix increases, the flow increases. The largest flow was achieved for the GP15 series, and it was 30 mm larger than that of the REF series. This difference is about 16% of the flow of the REF series. Prokopski et al. [26] examined the consistency of concrete mixes with the addition of granite powder using the slump method. He achieved a similar effect-the consistency of mixes increased with the addition of granite powder. The authors suggest that this effect is related to the addition of material with a smaller grain size than that of cement and sand. The addition of granite powder increased the packing density and decreased the porosity of the cement. Granite powder particles in the form of dust surround cement particles and cause the filling in of voids and the reduction in the friction between the particles.

The partial replacement of cement with granite powder resulted in an increase in density when compared to the reference series. As the amount of granite powder increases, the bulk density increases. The highest density among all the evaluated series was achieved by the GP10 series with a 10% replacement of cement with granite powder. Granite has a smaller fraction than the used cement. Replacing cement with material with smaller fractions will increase the packing density, which will in turn affect the bulk density. The addition of the granite powder caused the filling of voids between the sand and cement, which increased the bulk density. Similar results were obtained by the authors of article [24], in which granite powder caused a higher volume density of the mix when compared to the series without the additive. Woźniak and Chajec [24] observed that it may be related to the filling in of the aggregate skeleton with granite powder, which has less particles than cement. Article [11] also investigated the effect of adding granite powder to the cement mass on the bulk density, and it was observed that the partial replacement of cement with granite powder increased the bulk density of the mortar. Four series were made, in which 10%, 20%, and 30%, respectively, of the cement was replaced with granite powder. The highest density was achieved by the series with the 10% replacement of cement. It was observed that with the increasing amount of granite powder in the series, the bulk density of the mix increased.

3.3. Bending Strength and Compressive Strength

Figure 3a presents the bending strength, and Figure 3b shows the compressive strength. The error bars refer to the standard deviation from the average values for the results obtained by the series.

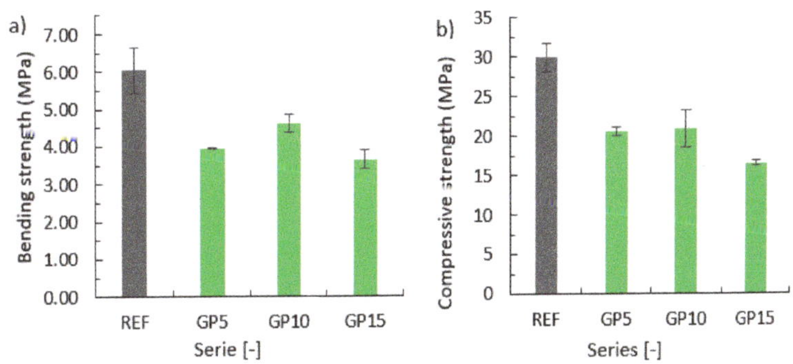

Figure 3. Properties of the hardened cementitious mortars with a partial replacement of cement with the granite powder: (**a**) bending strength; (**b**) compressive strength.

Based on Figure 3a, it can be concluded that the partial replacement of cement with granite powder reduces the bending strength. The reference series achieved the highest bending strength. The second-highest compressive strength was achieved by the GP10 series. The lowest values were achieved by the GP15 series. For the evaluated series, there is no relationship between the amount of cement replaced with a given waste material and the bending strength. However, it can be concluded that the partial replacement of cement with the granite powder significantly reduced the bending strength. The most optimal amount of replaced cement with the materials used is 10% by weight of cement.

Similar values were achieved by Woźniak et al. [24], where the relationship between a decreasing bending strength and an increasing amount of granite powder in the binder was obtained. The authors evaluated the bending strength after 28 days of the maturation of mortars, which were prepared with a 10%, 20%, and 30%, respectively, replacement of cement with granite powder, and were named GP10, GP20, and GP30, respectively. The difference between the GP15 and the REF series amounted to 44.78% of the compressive strength of the REF series. In the case of the GP10 series, this difference was equal to 30.27% of the REF series. A similar result, but lower than that of the GP10 series, was achieved by the GP5 series, with its compressive strength being equal to 20.54 MPa. Ramadji et al. [13] also investigated the effect of the compressive strength of mortars prepared with the replacement of cement with granite powder in amounts of 10%, 15%, 20%, respectively, of the weight of CEM I cement. It was noticed that with the increase in the amount of granite powder, the compressive strength decreased after 7, 14, and 28 days. After 90 days, the difference between the series with the 10% replacement of cement with granite powder, and the REF series was only 0.1 MPa. This difference was equal to 0.2% of the strength value of the REF series. Nascimento et al. [27] evaluated the compressive and tensile strength of mortars with a partial replacement of cement or aggregate with granite powder. The series with the cement replacement achieved lower strengths than the reference series, which they claim is due to the reduction of the amount of the binder. Higher strengths were obtained for the series with sand replacement, which is due to an increase in the packing density, an increase in the specific weight, and greater resistance.

3.4. Water Absorption

Figure 4 presents the results of water absorption tests for the reference series (grey column), and the series with a partial replacement of cement with granite powder (green columns). Error bars indicate the standard deviation from the mean value for the evaluated series.

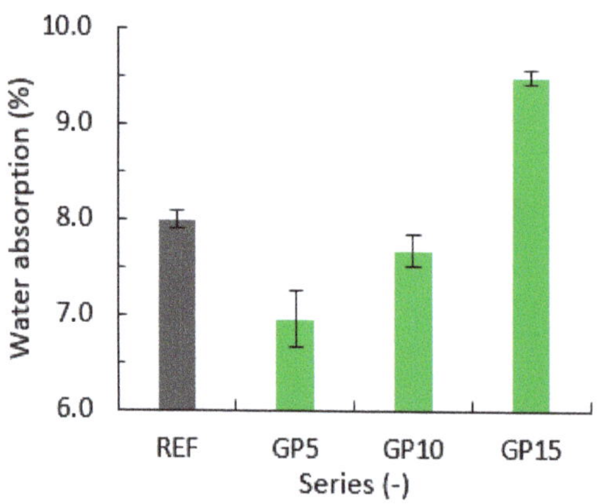

Figure 4. Water absorption for the reference series and series with a partial replacement of cement with granite powder (GP).

There is a tendency of increased water absorption with an increased amount of waste material in the mortar. The GP5 series had the lowest water absorption. The series with a 15% replacement of cement with granite powder had the highest water absorption value, which was higher than that of the REF series by 1.51%. Gupta et al. [17] claim that water absorption is inversely proportional to mechanical resistance. Therefore, the highest water absorption was achieved for the series with the lowest compressive strength (GP15).

Similar results were obtained by Chajec et al. [12], who evaluated the absorbability of cement screeds placed on a concrete substrate. This type of flooring is mainly used in housing construction. He evaluated four series of samples, where granite powder was used as a cement substitute in amounts of 10% (GP10), 20% (GP20), and 30% (GP30), respectively, of the cement mass. The values for the series with additives were compared to the REF series. The samples were subjected to two different curing conditions. Some of the samples were stored in air at 20 °C and 50% humidity, and some were subjected to wet curing, where the temperature was 24 °C and the humidity was 85%. The authors noted that the curing conditions had a significant effect on the water absorption, which was evaluated after 90 days of maturation. For the series hardened in the air, there was a tendency of increased water absorption with the addition of granite powder. In turn, there was a reverse tendency for the series cured in wet conditions. It was found that the partial replacement of cement with granite powder did not reduce water absorption if appropriate hardening conditions were maintained.

A similar Influence of granite powder on water absorption in cement mortar was also observed by Alok Damare et al. [28]. They used granite powder as a partial replacement of sand in amounts of 2.5%, 5%, 7%, 10%, 12.5%, 15%, 17.5%, and 20%, respectively, of the sand's weight. Dry samples were placed in a water container for 48 h, then the amount of absorbed water was measured in percent by weight. The series with the 5% and 7.5% sand replacement with granite powder showed a reduced water absorption. The 2.5% sand replacement series showed the same absorbance as the reference series. The remaining series showed a tendency of increased water absorption with increasing amounts of granite powder in the mortar.

3.5. Capillary Suction

Measurements of the mass of the samples, which were planned at equal time intervals divided into five phases, allowed for the creation of graphs showing the process and speed of capillary rising for each of the series. The planned time of the measurements simulates the mortar being exposed for several days to atmospheric factors such as rain, snow, dew, and high air humidity. This situation also reflects the threat to walls caused by flooding or an increase in the level of groundwater. Figure 5 shows the dependence of the change of Δmt/F over time for the reference series, and for the series with the partial replacement of cement with granite powder.

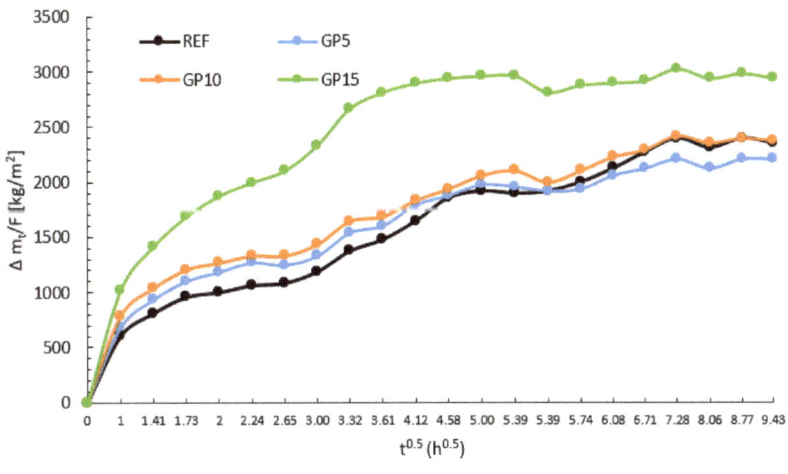

Figure 5. Diagram of Δmt/F as a function of time root $t^{0.5}$ for the series with the partial replacement of cement with granite powder and the reference series.

Based on the mass measurements during the test, it can be concluded that for the period up to 29 h, there is a tendency for higher water absorption with an increasing amount of granite powder in the mortar. The differences between the REF, GP10, and GP10 series are slight, while the GP15 series is significantly different from the other series. The differences between the GP15 series and the other series reached about 1300 kg/m^2. On the basis of Figure 6, the water sorption coefficient "A" was determined, which is the slope of the straight section of the function $f(t^{0.5}) = \Delta m_t/F$. These rectilinear sections are presented in the graphs below. The same time intervals were selected (with three measurements), and, based on them, linear trend lines with equation descriptions were created in Figure 6. Error bars indicate the standard deviation from the mean value for the evaluated series.

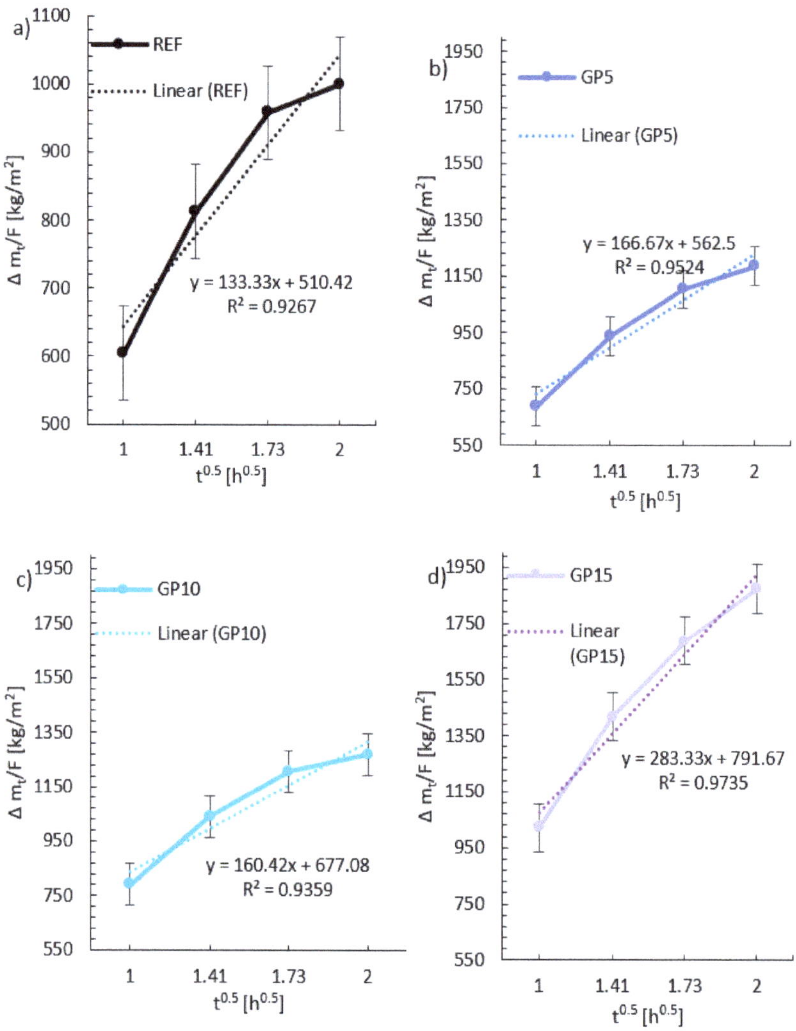

Figure 6. Graphs of capillary suction for the three measuring points with a linear trend line for: (**a**) the reference series; (**b**) the series with the 5% replacement of cement with granite powder; (**c**) the series with the 10% replacement of cement with granite powder; (**d**) the series with the 15% replacement of cement with granite powder.

Based on the directional coefficients of the linear trend lines, Figure 7 was created in order to show the water sorption coefficient.

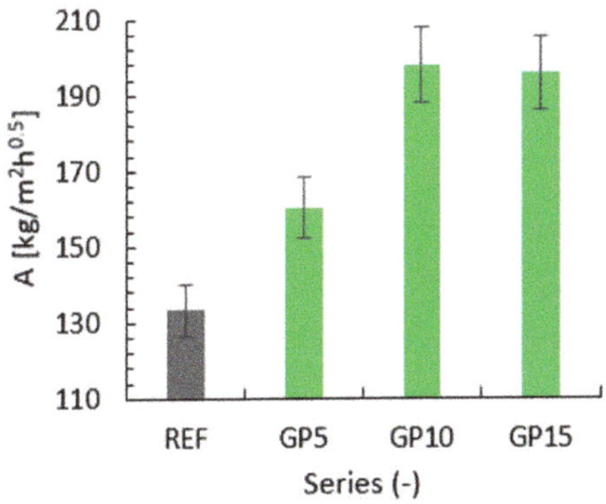

Figure 7. Water sorption coefficients for the reference series (REF) and the series with the partial replacement of cement with granite powder (GP).

The capillary suction varies significantly between the evaluated series. The series with a partial replacement of cement with granite powder had higher water sorption coefficients than the reference series. The highest coefficients out of all the GP series were achieved by the GP10 and GP15 series. Their results differed by only 2.09 $kg/m^2h^{0.5}$. The high sorption coefficients of the GP series may be due to the fact that the surface of the cement particles is very rough, and the particles of granite powder, which are smaller, penetrate into the uneven surface of the cement particles. Prokopski et al. [26] tests were conducted on cubic samples with dimensions of $10 \times 10 \times 10$ cm. They found that the samples made on the basis of CEM I 32.5 R cement had a higher water absorption than the series based on CEM I 42.5 R. They also showed that the addition of granite reduces the wicking capacity. The difference may result from the used test method, because the water was fed at a pressure of 500 kPa, and the penetration was measured in mm. The mass measuring method used in the conducted tests is more accurate due to the fact that it is the amount of water that is absorbed into the sample that is measured. Due to this, the total amount of water, not only the water contained on the surface, can be determined. Additionally, in the case of the samples with dimensions of $4 \times 4 \times 16$ cm, the differences in the absorbed water across their width may be smaller than in the case of the wider samples with dimensions of $10 \times 10 \times 10$ cm. The level of water contained in the centre of a larger sample may be much lower than for smaller samples. The surface of the samples is also important because the capillary suction is different in the case of sample that have surfaces of different roughness.

4. The Importance of the Research

The cost of cement mortar with addition of granite powder, could be reduced. Another advantage of using this granite powder could be a reduction in the cost of storing this material. The quality of the storage could also be improved, which would in turn reduce the amount of materials exposed to wind that can transport it to neighbouring settlements. Replacing cement with a different material could reduce the price of mortar. Such a mortar could be used, for example, in dry places due to its increased sorption when compared to traditional mortar. Mortar with the addition of granite powder could be used in walls that are not subjected to significant bending and compressive loads. However, further studies are needed to verify the potential application of such a mortar.

5. Conclusions

The conducted studies filled the gaps in the knowledge related to the dependence of Δ mt/F. The positive effect of partially replacing cement with granite powder on the temperature of fresh cement mortar, and its water absorption, was described. Based on the presented study, the following main conclusions can be drawn:

Reducing the amount of pozzolanic material resulted in a decrease in the temperature of the fresh mortar. This property could enable masonry works to be conducted in summer conditions. The addition of a material with a different shape of grains causes an increase in the flow of a mix and changes in its specific density, which in turn affects the properties of hardened mortars.

The creation of innovative masonry mortars may be associated with the deterioration of their mechanical properties-replacing cement with granite powder for the evaluated series led to a decrease in compressive strength and bending strength. Among the evaluated series, the optimal value of replacing cement with granite powder is 5% to 10% of the weight of cement. Changes in the mechanical properties for these series are acceptable due to the positive aspects associated with the use of granite powder.

To sum up, an appropriate selection of the optimal amount of granite powder in mortar may lead to such a mortar being used. The desired properties of mortars depend on the type of construction material and their exposure to weather conditions during setting and use. Future research should focus on investigating other properties, such as the setting speed with regards to mortar temperature, or the mortar's adhesion to bricks. In addition, further research is needed in order to estimate the exact amount of cement that can be replaced with granite powder, which in turn would contribute to the development of the practical application of the newly designed mortar. Moreover, subsequent research would bring scientists closer to both the implementation of the proposed solution and the creation of a ready-made mix of mortar ingredients.

Author Contributions: Writing—original draft, Z.Z.W., A.C. and Ł.S. All authors have read and agreed to the published version of the manuscript.

Funding: The authors received funding from the project supported by the National Centre for Research and Development, Poland (grant no. LIDER/35/0130/L-11/19/NCBR/2020 "The use of granite powder waste for the production of selected construction products").

Institutional Review Board Statement: Not applicable.

Informed Consent Statement: Not applicable.

Data Availability Statement: The data presented in this article are available within the article.

Conflicts of Interest: The authors declare no conflict of interest.

References

1. Teja, P.R.R.; Sahu, S.; Sarkar, P.; Davis, R. Compressive Strength Prediction Models for Fly-Ash Brick Masonry. *Pract. Period. Struct. Des. Constr.* **2022**, *27*, 04022014. [CrossRef]
2. Drobiec, Ł. Stan obecny i perspektywy rozwoju konstrukcji murowych. *Inżynieria I Bud.* **2018**, *74*, 7–8.
3. Hoła, J.; Sadowski, Ł.; Hoła, A. The effect of failure to comply with technological and technical requirements on the condition of newly built cement mortar floors. *SAGE J.* **2018**, *233*, 268–275. [CrossRef]
4. Mannan, M.A.; Ganapathy, C. Concrete from an agricultural waste-oil palm shell (OPS). *Build. Environ.* **2004**, *39*, 441–448. [CrossRef]
5. Amin, M.N.; Khan, K.; Saleem, M.U.; Khurram, N.; Niazi, M.U.K. Aging and Curing Temperature Effects on Compressive Strength of Mortar Containing Lime Stone Quarry Dust and Industrial Granite Sludge. *Materials* **2017**, *10*, 642. [CrossRef] [PubMed]
6. Strunge, T.; Renforth, P.; van der Spek, M. Towards a business case for CO_2 mineralisation in the cement industry. *Commun. Earth Environ.* **2022**, *3*, 59. [CrossRef]
7. Winnefeld, F.; Leemann, A.; German, A.; Lothenbach, B. CO_2 storage in cement and concrete by mineral carbonation. *Curr. Opin. Green Sustain. Chem.* **2022**, *38*, 100672. [CrossRef]
8. Li, L.G.; Wang, Y.M.; Tan, Y.P.; Kwan, A.K.H. Filler technology of adding granite dust to reduce cement content and increase strength of mortar. *Powder Technol.* **2019**, *342*, 388–396. [CrossRef]
9. AAbulencia, B.; Villoria, M.B.D.; Libre, R.G.D.; Quiațchon, P.R.J.; Dollente, I.J.R.; Guades, E.J.; Promentilla, M.A.B.; Garciano, L.E.O.; Ongpeng, J.M.O. Geopolymers as Sustainable Material for Strengthening and Restoring Unreinforced Masonry Structures: A Review. *Buildings* **2021**, *11*, 532. [CrossRef]
10. Shah, M.I.; Amin, M.N.; Khan, K.; Niazi, M.S.K.; Aslam, F.; Alyousef, R.; Javed, M.F.; Mosavi, A. Performance Evaluation of Soft Computing for Modeling the Strength Properties of Waste Substitute Green Concrete. *Materials* **2021**, [illegible]. [CrossRef]

11. Chajec, A. Granite powder vs. Fly ash for the sustainable production of air-cured cementitious mortars. *Materials* **2021**, *14*, 1208. [CrossRef] [PubMed]
12. Chajec, A.; Sadowski, Ł.; Moj, M. The adhesive and functional properties of cementitious overlays modified with granite powder. *Int. J. Adhes. Adhes.* **2021**, *117*, 103008. [CrossRef]
13. Ramadji, C.; Messan, A.; Prud'homme, E. Influence of granite powder on physico-mechanical and durability properties of mortar. *Materials* **2020**, *13*, 5406. [CrossRef]
14. Charakterystyka Granitu Jak Powstał Granit?—Luxstone Granity. Available online: https://luxstone-granity.pl/charakterystyka-granitu/ (accessed on 25 May 2022).
15. Ostrowski, K.; Stefaniuk, D.; Sadowski, Ł.; Krzywiński, K.; Gicala, M.; Różańska, M. Potential use of granite waste sourced from rock processing for the application as coarse aggregate in high-performance self-compacting concrete. *Constr. Build. Mater.* **2020**, *238*, 117794. [CrossRef]
16. Gencel, O.; Ozel, C.; Koksal, F.; Erdogmus, E.; Martínez-Barrera, G.; Brostow, W. Properties of concrete paving blocks made with waste marble. *J. Clean. Prod.* **2012**, *21*, 62–70. [CrossRef]
17. Gupta, L.K.; Vyas, A.K. Impact on mechanical properties of cement sand mortar containing waste granite powder. *Constr. Build. Mater.* **2018**, *191*, 155–164. [CrossRef]
18. Konopacka, Ż.; Nędza, Z.; Stopa, M. Oznaczenie wolnej krystalicznej krzemionki w pyłach emitowanych na stanowiskach pracy w górnictwie skalnym. *Min. Sci. Miner. Aggreg.* **2015**, *22*, 75–81.
19. *PN-EN 1015-3*; Methods of Test for Mortar for Masonry. Determination of Consistence of Fresh Mortar (by Flow Table). Polish Committee for Standardization: Warsaw, Poland, 2015.
20. *PN-EN 196-1:2016-07*; Methods of Testing Cement—Part 1: Determination of Strength. Polish Committee for Standardization: Warsaw, Poland, 2016.
21. *PN-EN 1015-11:202-04*; Methods of Test for Mortar for Masonry—Determination of Flexural and Compressive Strength of Hardened Mortar. Polish Committee for Standardization: Warsaw, Poland, 2020.
22. Rirsch, E.; Zhang, Z. Rising damp in masonry walls and the importance of mortar properties. *Constr. Build. Mater.* **2010**, *24*, 1815–1820. Available online: https://daneshyari.com/en/article/260364 (accessed on 14 December 2022). [CrossRef]
23. Garbalińska, H.; Kosmaczewska-Możejko, A. Badanie współczynnika sorpcji kapilarnej betonów komórkowych klas gęstości 400, 500, 600 i 700. *Builder* **2020**, *275*, 16–18. [CrossRef]
24. Woźniak, Z.; Chajec, A. Wpływ mączki granitowej na wybrane właściwości zapraw cementowych. *Mater. Bud.* **2022**, *1*, 60–63. [CrossRef]
25. Grzeszczyk, S.; Janowska-Renkas, E. The Influence of Limestone Microfiler in Cement on the Heat of Hardening. *Fizyka Budowli w Teorii I Praktyce* **2005**, *1*, 111–117. Available online: https://yadda.icm.edu.pl/baztech/element/bwmeta1.element.baztech-article-LOD4-0003-0013/c/Grzeszczyk.pdf (accessed on 14 December 2022).
26. Prokopski, G.; Marchuk, V.; Huts, A. The effect of using granite dust as a component of concrete mixture. *Case Stud. Constr. Mater.* **2020**, *13*, e00349. [CrossRef]
27. Nascimento, A.S.; Pereira dos Santos, C.; Cavalvante de Melo, F.M.; Iliviera, V.G.A.; Oliveira, R.M.P.B.; Macedo, Z.S.; Alves de Oliveira, H. Production of plaster mortar with incorporation of granite cutting wastes. *J. Clean. Prod.* **2020**, *265*, 121808. [CrossRef]
28. Damare, A.; Gupta, P.; Baweja, R. Utilization of granite powder in cement mortar. *Int. J. Res. Publ. Eng. Technol.* **2017**, *3*, 2454–7875.

materials

Article

Performance of Sustainable Insulated Wall Panels with Geopolymer Concrete

Balamurali Kanagaraj [1], Tattukolla Kiran [1], Jayakumar Gunasekaran [2], Anand Nammalvar [1,*], Prince Arulraj [1], Beulah Gnana Ananthi Gurupatham [3] and Krishanu Roy [4,*]

1 Department of Civil Engineering, Karunya Institute of Technology and Sciences, Coimbatore 641114, India
2 Department of Civil Engineering, Sri Krishna College of Technology, Coimbatore 641042, India
3 Division of Structural Engineering, College of Engineering Guindy Campus, Anna University, Chennai 600025, India
4 School of Engineering, The University of Waikato, Hamilton 3216, New Zealand
* Correspondence: nanand@karunya.edu (A.N.); krishanu.roy@waikato.ac.nz (K.R.)

Abstract: The increase in the population creates an increased demand for construction activities with eco-friendly, sustainable, and high-performance materials. Insulated concrete form (ICF) is an emerging technology that satisfies the sustainability demands of the construction sector. ICF is a composite material (a combination of expanded polystyrene (EPS) and geopolymer concrete (GPC)) that enhances the performance of concrete (such as thermal insulation and mechanical properties). To investigate the axial strength performance, five different types of prototypes were created and tested. Type I (without reinforcement): (a) hollow EPS without concrete, (b) alternative cells of EPS filled with concrete, (c) and all the cells of EPS filled with concrete; and Type II (with reinforcement): (d) alternative cells of EPS filled with concrete; (e) and all the cells of EPS filled with concrete. Amongst all the five prototypes, two grades of GPC were employed. M15 and M20 grades are used to examine the effectiveness in terms of cost. For comparing the test results, a reference masonry unit was constructed with conventional clay bricks. The main aim of the investigation is to examine the physical and mechanical performance of sandwich-type ICFs. The presence of polystyrene in ICF changes the failure pattern from brittle to ductile. The result from the study reveals that the Type II prototype, i.e., the specimen with all the cells of EPS filled with concrete and reinforcement, possesses a maximum load-carrying capacity greater than the reference masonry unit. Therefore, the proposed ICF is recommended to replace the conventional load-bearing system and non-load-bearing walls.

Keywords: expanded polystyrene; geopolymer concrete; sustainability; masonry; axial strength

check for updates

Citation: Kanagaraj, B.; Kiran, T.; Gunasekaran, J.; Nammalvar, A.; Arulraj, P.; Gurupatham, B.G.A.; Roy, K. Performance of Sustainable Insulated Wall Panels with Geopolymer Concrete. *Materials* **2022**, *15*, 8801. https://doi.org/10.3390/ma15248801

Academic Editor: Ayman El-Zohairy

Received: 31 October 2022
Accepted: 7 December 2022
Published: 9 December 2022

1. Introduction

Expanded polystyrene (EPS) is considered one of the common materials used instead of aggregates to reduce weight and for improving thermal comfort [1]. EPS is a kind of stable foam with low density, consisting of discrete air voids in a polymer matrix [2]. As a lightweight artificial aggregate, EPS is commercially available and can be incorporated in mortar or concrete to produce lightweight insulating concrete [3]. From the engineering point of view, the advantage of using EPS among different types of LWAs is the low water absorption due to its lower porosity. In addition, the low thermal conductivity of EPS (0.03–0.04 W/mK) is one of the main reasons for its use in the construction industry, particularly for insulting purposes [4]. Previous research has promoted the applications of EPS-based concrete in construction and building products. Currently, EPS lightweight concrete is used in various structural and non-structural elements such as precast concrete panels, cladding panels, composite flooring systems, subbase materials in pavements, floating marine structures, and insulating building elements [5]. However, replacing the normal-weight aggregate with EPS lightweight aggregate increases the drying shrinkage strain [6,7].

Due to the over-exploitation of virgin construction and building materials, there is a need to develop a suitable alternative eco-friendly building material to satisfy the demand for infrastructural activities. EPS is found to be one of the alternatives fabricated of recycled materials; since EPS is a non-degradable material, it can be recycled after its first life, second life, and so on [8]. EPS is recycled either in the form of beads or modified with heat treatment [9]. Recycling and using such materials in the construction sector protects the environment by preventing the emission of harmful gases when incinerated.

Quality construction practices are gaining demand with the increase in the population [10]. ICF is found to be a promising construction practice. ICF walls are fabricated by placing concrete in the individual cells of EPS foam. Construction of ICF structures has many advantages over conventional structures based on cost, ease of construction practices, thermal insulation, etc. Some other benefits are heatproof, soundproof, faster construction, lower maintenance, and resistance against insects, wind, and disasters [11]. In the United States (US), nearly 3% of houses are constructed using ICF systems [12]. ICF is one of the sustainable, eco-friendly alternatives to conventional carbon-emitting cement-based construction practices. ICF's superior performance is mainly due to the EPS [13]. ICFs are used as construction materials in the construction of schools, colleges, hospital buildings, etc. ICF claims several advantages over conventional building materials [14]. ICF has been a boon to the construction sectors in India since 2013; ICF-based structures have been accepted in the US, Germany, Japan, Canada, and Mexico. Moreover, the construction cost of ICF structures is 5–10% less than conventional practices. ICF construction practices require more awareness to accept this new system of the construction sectors to meet the current requirements of the infrastructure industry.

EPS employed for producing ICF should have a density of 20 kg/m^3 to 40 kg/m^3 and a thickness of 50 mm to 100 mm. ICF panels are generally employed to construct a wall system to portion the functional area with a structural wall system. The wall systems are mainly subjected to axial compression due to the load transfer from the floor system. The main intention of the axial load test is to determine the ultimate load-carrying capacity, failure pattern, and stress–strain relationship, which helps to design the ICF panels. Past studies have reported that axial compression tests are performed on different wall panels, such as foamed wall panels, composite panels, reinforced panels, precast sandwich panels, precast foamed panels, panels with openings, and panels with shear connectors. Similarly, a four-point bending test was performed on bridge deck panels, Ferro-cement panels, lightweight panels, fiber-reinforced sandwich panels, and hollow concrete wall panels.

However, OPC production requires argillaceous and calcareous materials and is energy-intensive. The main reasons for the emission of greenhouse gas during the production of OPC are calcination and fossil-fuel combustion [15–17]. The production of conventional Portland cement increases yearly by about 8% to 10%. These cement types account for 8% of global carbon dioxide emissions (CO_2-e) [18]. To reduce global CO_2-e, an alternative binder must be used. Supplementary cementitious materials (SCMs) such as sugarcane bagasse ash, Metakaolin (MK), red mud, and Ground Granulated Blast Furnace Slag (GGBS) are industrial by-products. SCMs have cementitious properties and are employed as a cement alternative [19,20]. A durable structure with less greenhouse gas emission and with less energy can be obtained by the addition of fly ash to the concrete [21–24]. Investigations are carried out on SCMs for producing sustainable, eco-friendly concrete. The outcomes of using SCMs as partial cement replacements are gaining economic impact in reducing CO_2-e during its production.

In 1978, a novel binding medium was named geopolymer, which employs 0% cement in its production. These geopolymers are a suitable alternative that completely replaces conventional Portland cement concrete. The early age strength of geopolymer is higher than that of traditional cement concrete and possesses higher fire resistance. Moreover, these SCMs can replace cement by 20% to 30% by weight, beyond which the strength reduction is more significant. The dissolution makes the production of geopolymer concrete (GPC) of source material possible with the alkaline solution (combination of NaOH and Na_2SiO_3) [25]. The source material can be either from geological origin or industrial by-products such as GGBS, MK, Palm Oil Fuel Shell Ash (POFA), Rice Husk Ash (RHA), Fly Ash (FA), and so on [26]. Previous research has shown that GPC concrete has superior performance compared with OPC concrete in terms of mechanical and durability characteristics [25,27].

A combination of GPC and ICF can form a sustainable construction practice. The proposed system's significant advantages are lightweight, ease of placement, faster construction, optimized cost, strength, and eco-friendly. Reduction in cost and CO_2 emissions are some of the indicators that might be helpful to achieve sustainability goals. The effective utilization of GPC in the ICF system can improve the load-withstanding capacity of the wall panel. With this background, the current study aims to analyze the load-deformation behavior of GPC-ICF systems subjected to axial compressive loading.

Insulated concrete form (ICF) panels are structural wall panels fabricated by pouring concrete in interlocked expanded polystyrene (EPS) that hold the concrete together during the curing process. The EPS form is a permanent part of a wall panel and provides thermal insulation to the building, whereas the reinforced concrete affords a structural system to the building [12]. Applications of ICFs are extended to a wide range of building constructions, including residences, theatres, schools, and hospitals [13]. EPS is a by-product of the petroleum industry and is derived by the styrene hydrocarbon polymer (polystyrene) expansion using pentane gas. An EPS bead consists of 2% raw material and 98% of air, which is chemically composed of two elements: carbon and hydrogen [11]. Generally, EPS sheets have been used in various applications, including impact mitigation packaging, protective helmet, expansion joints, construction filling material, false ceilings, and food packaging material. Diverse structural and geotechnical applications of EPS are also found in the literature, namely structural insulated panels [28], composite structural insulated panels [5], insulated concrete sandwich panels [14], lightweight concrete sandwich panels [29], and thermal insulators [2].

The past study found that the axial compression test was performed on various panels of fibers, concrete, and EPS sheets. Research shows that the wall panel characterization focused on load-deflection behavior and load-carrying capacity of the structural element. Moreover, fewer types of research are available on the ICF wall panels subjected to axial compressive loads. In the present work, the ICF wall panels are constructed by creating a hollow outer shell and filling the hollow core with concrete. The study's main aim is to examine the ultimate load-carrying capacity of ICF, load-deflection behavior, and failure pattern of ICF wall panels with GPC filling. The findings may be helpful for future work with large-scale models.

2. Materials and Methods

2.1. Binder

Class F Fly Ash (FA) and GGBS were the source materials obtained from the local market. The chemical composition of the binder material is illustrated in Table 1 [30].

Table 1. The oxide composition of FA and GGBS.

Oxides	FA (%)	GGBFS (%)
SiO_2	61.2	42.4
Al_2O_3	26.9	13.2
Fe_2O_3	6.21	1.12
CaO	1.91	41.2
MgO	0.29	1.3
Na_2O	0.58	0.3
K_2O	1.21	0.7
Ti	0.4	-

2.1.1. Filler

Locally available manufactured sand (M-sand) and crushed granite chips were employed as the filler materials (fine and coarse aggregates) for the production of GPC. The specific gravity, water absorption, and fineness modulus of M-sand were found to be 2.75, 1.37%, and 2.59, respectively. The specific gravity and water absorption of coarse aggregate were found to be 2.81 and 0.92%, respectively. The size of the M-sand employed in the present investigation falls between 4.75 mm and below, whereas coarse aggregate is between 12.5 mm and 20 mm, confirming IS 383 [31]. Figure 1 illustrates the semi-log graph of M-sand and coarse aggregate.

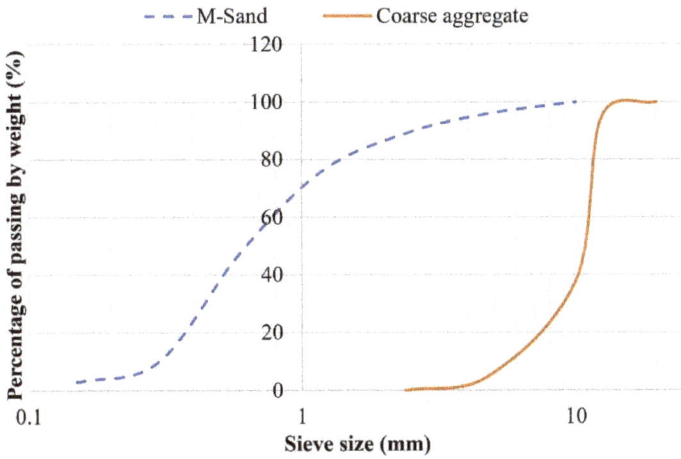

Figure 1. Particle size distribution of M-Sand and coarse aggregates.

2.1.2. Activator Solution (AS)

A synthesis between an alumino-silica source and an alkaline activator solution [13], yields an alumino-silica source and an alkaline activator solution (AS). AS is a combination of NaOH (NH) and Na_2SiO_3 (NS) [32,33]. NH and NS are obtained from local dealers in Coimbatore. In general, the NH is available in the form of flakes or pellets, and it must be made into a solution by dissolving the required amount of NH pellets or flakes in water. NS is available in the form of liquid and can be used directly.

2.1.3. Expanded Polystyrene (EPS)

EPS is a combination of air and styrene, and the whole wall shell system is fabricated of 98% air and 2% styrene. Due to the advantage of the EPS shell, the ICF wall panels retain their shape even after failure. Table 2 illustrates the material properties of an ICF wall.

Table 2. Properties of the ICF wall.

Specification	Material Properties
Size of the ICF wall	20.32 cm (8 inches) × 30.50 cm (12 inches) × 121.00 cm (48 inches)
Water vapor Permeability	0.0020 ng·s·m^2·Pa
Moisture Absorption (humidity):% mass	0.4417
Moisture Absorption (humidity):% volume	0.0099

2.1.4. Mix Design

As mentioned above, AS is a combination of NH and NS. The NH concentration employed in the present investigation is four molarity, and the ratio of AS used is 1:1.5 [34]. The pellets or flakes of NH are to be made into a solution before concrete production [35].

The AS was made one day before the casting to initiate the polymerization reaction [25]. The mix design carried out in the study was entirely based on the trial-and-error method. Table 3 illustrates the recipe for the mix involved in the current investigation.

Table 3. Mix ratios employed in the current investigation.

Type of Mix	Binder (kg/m^3)		Aggregate (kg/m^3)		Alkaline Activator (kg/m^3)		Molarity (M)
	FA	GGBS	M-Sand	Coarse	NH	NS	
M15	180	20	450	650	28.96	43.44	4
M20	100	100	450	650	28.96	43.44	4

2.2. Methods

2.2.1. Specimen Preparation

The filler materials, fine and coarse aggregates, were mixed in the rotary-type mixer drum for 1–2 min. Then binder materials FA and GGBS were added to the mixer drum, and the drum was rotated for 2–3 min to achieve uniformity in the mix. After attaining uniformity in the mix, the prepared AS mix was added; then the mixture was rotated for another 2 min to achieve homogeneity [35]. After preparing the mix, fresh properties were tested on the prepared mix to check its workability. After checking the slump test, the freshly prepared mix was poured into the EPS shell's hollow portion. The mix was set at room temperature for 24 h to initiate the polymerization reaction process. After setting, the ICF (combination of EPS and GPC) was cured at a room temperature of 27 ± 2 °C for 28 days to achieve the target strength. At the end of 28 days of the curing period, the ICF was tested to examine the load-carrying capacity. Table 4 shows the details of specimens prepared for the study. Figure 2 represents the outer shell of the EPS form. Figures 3–5 represent the casting of ICF panels and masonry reference units. For better performance with respect to durability of the steel reinforcement stainless steel of various types (Ferritic, Duplex, and Austenitic) can also be used [36].

Figure 2. Structure of the EPS shell.

Table 4. Details of the ICF specimens.

Specimen Type	Reinforcement Type	Specimen Description	Type of Mix
S1	(Without reinforcement)	Hollow EPS without concrete	-
S2		Alternative cells of EPS were filled with concrete	M15
S3		All the cells of EPS were filled with concrete	M15
S4		Alternative cells of EPS were filled with concrete	M20
S5		All the cells of EPS were filled with concrete	M20
S6	(With reinforcement)	Alternative cells of EPS were filled with concrete	M15
S7		All the cells of EPS were filled with concrete	M15
S8		Alternative cells of EPS were filled with concrete	M20
S9		All the cells of EPS were filled with concrete	M20
S10		Masonry (reference)	-

Figure 3. Reinforcing ICF wall.

Figure 4. The casting of the ICF.

2.2.2. Workability Test

The freshly prepared mix was tested for its workability using a slump cone. A slump test was performed to assess the fresh property of the GPC. A 300 mm high slump cone was employed in the present investigation to assess the fresh property. Waste lubricant was applied on the inner side of the cone initially to prevent the sticking of concrete with the cone. Then, the mold was filled with concrete in three layers with the compaction of 25 rods for each layer, after filling the cone with the freshly prepared concrete. The excessive layer at the top of the mold was trimmed, and slowly, the mold was lifted. The difference in GPC height and the mold is represented as the slump value [37].

Figure 5. Construction of the conventional brick masonry.

2.2.3. Test Setup for Axial Compression Test on ICF

The axial compression test on the ICF was conducted with a Servo hydraulic UTM of 100 T capacity (MTS). The MTS system is operated based on displacement control; the displacement rate is controlled during the loading. The ICF's maximum load-carrying capacity and load-deformation behavior are recorded for every 0.01 mm displacement. The ICF panels are positioned in the vertical direction of the MTS system. A 20 mm thick plate and the spacer bars are provided at the top of the wall panel to distribute the load uniformly throughout the specimen. Above that, a two-point loading platform is provided to transfer the load from the MTS system to the 20 mm thick plate as shown in Figure 6.

Figure 6. Load testing setup of ICF in MTS (axial compression tests).

3. Results and Discussion

3.1. Workability

As per ASTM C 143 [37], the fresh property test on freshly prepared concrete is conducted. A slump cone test was conducted immediately after preparing the mix to determine the workability of freshly prepared geopolymer concrete. It is found from the test results that the M15 mix possesses a target slump of 78 mm, and the M20 mix retains an 85 mm slump.

3.2. Compressive Strength (CS)

The CS test was performed on a hollow EPS shell and EPS shell (with M15 and M20 concrete) using a computerized UTM of 100 T capacity following IS 516:2004 [38], as shown in Figure 7. The CS test was performed on the 28-day cured GPC specimens. The reference specimen (EPS shell) possesses a maximum average CS of 9.56 MPa. The EPS shell with M15 geopolymer concrete contains an average CS of 20.6 MPa, and the EPS shell with M20 geopolymer concrete possesses an average CS of 29 MPa. Higher strength development of geopolymer concrete was attributed to the presence of GGBS content. From Figure 8 it is clear that an increase in the GGBS content increases the CS of the geopolymer concrete. The curing of geopolymer concrete does not require water, so the curing of geopolymer mix is initiated at room temperature conditions. The strength development of the mix is mainly attributed to the polymerization reaction of the alumina-silica source with the alkaline solutions. The load was applied at the top of the geopolymer concrete blended ICF system. Initially, the geopolymer concrete present in the core of the ICF system reaches its ultimate strength and tends to collapse completely by splitting the concrete into pieces. The concrete was no longer strong enough to withstand the additional load, but the EPS outer periphery contributed to unite the system and prevent wall failure.

Figure 7. Compressive strength test setup of EPS shell and composite unit (EPS with concrete).

3.3. Axial Load Deformation of ICF

The axial load test was performed on five types of ICF wall panels. The testing of ICF wall panels is depicted in Figure 6. From the test, it was found that (S1) hollow EPS without concrete failed by crushing, and no sign of buckling was seen in this failure type; and when the load was released, the EPS shell regained its original shape, similar results were found by the researchers [12]. All the other types of panels possess a similar kind of failure; during the test, it was ensured that no load was applied eccentrically on the top of the panel. The original shape of the ICF is not degraded on loading the ICF and under deformation; while releasing the load, it regains its original shape. The elastic nature of the outer shell adds advantages to the concrete present in the core of the wall panel. On the contrary, the failure pattern of the concrete was not visible due to the outer shell of the EPS.

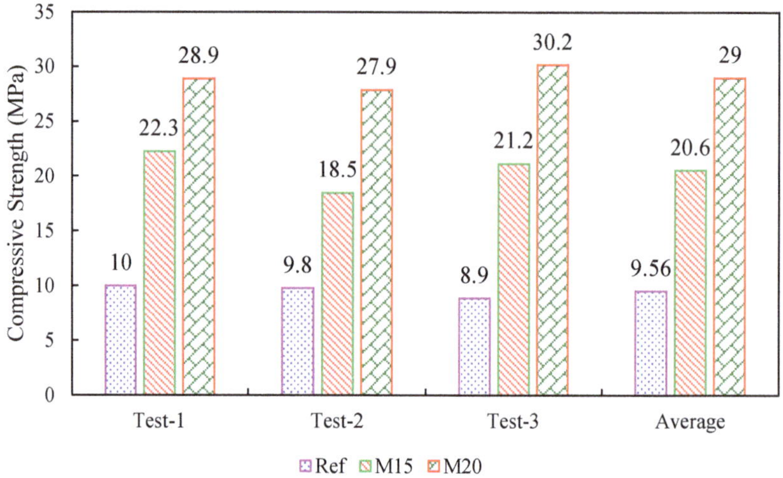

Figure 8. Compressive strength results of EPS with M15 and M20 concrete mixes.

Figure 9 depicts the load-deformation graph of ICF wall panels and masonry units. Specimen S1 (hollow EPS without concrete) exhibits the lowest load of 53 kN with a deformation of 19 mm. Specimen S2 (alternative cells of EPS were filled with concrete without reinforcement, M15) shows a load of 149 kN with a deformation of 15 mm. Specimen S3 (all the cells of EPS were filled with concrete without reinforcement, M15) exhibits a load of 176 kN with a deformation of 20 mm. To understand the loading capacity of the EPS panel under axial compressive loading, the concrete is filled in the voids of the EPS. This is also to optimize the use of concrete in the EPS cells. Lower-grade concrete is preferred to reduce the preparation cost. Specimen S4 (alternative cells of EPS were filled with concrete without reinforcement, M20) demonstrated a load of 230 kN with deformation of 10 mm. Specimen S5 (all the cells of EPS were filled with concrete without reinforcement, M20) showed a load of 260 kN with a deformation of 20 mm. It is observed that the EPS with M20 grade has shown a higher load with less deformation. This is seen in the specimen with the GPC filled in all cells. The reinforcements in the wall system improve the compressive strength and bending capacity. The reinforcements were added to increase the strength and stability of the structural wall system. Four vertical bars of 8 mm diameter and five ties of 6 mm diameter were employed in the present investigation. The ties were provided at a constant interval of 150 mm. The reinforcement was provided in each cell of the wall panel to measure the maximum load-carrying capacity of the reinforced ICF-based wall panels. This also connects the EPS panels as an anchorage to form continuity. The details of load-carrying capacity and axial deformation of EPS panels with GPC and reinforcements are as follows: Specimen S6 (alternative cells of EPS were filled with concrete, M15) exhibits a load of 226 kN with deformation of 20 mm. Specimen S7 (all the cells of EPS were filled with concrete, M15) indicates a load of 300 kN with deformation of 11 mm. Specimen S8 (alternative cells of EPS were filled with concrete, M20) shows a load of 350 kN with a deformation of 8 mm. Specimen S9 (all the cells of EPS were filled with concrete, M20) exhibits a load of 360 kN with a deformation of 11 mm. Specimen S10 (reference masonry unit) shows a load of 150 kN with deformation of 16 mm.

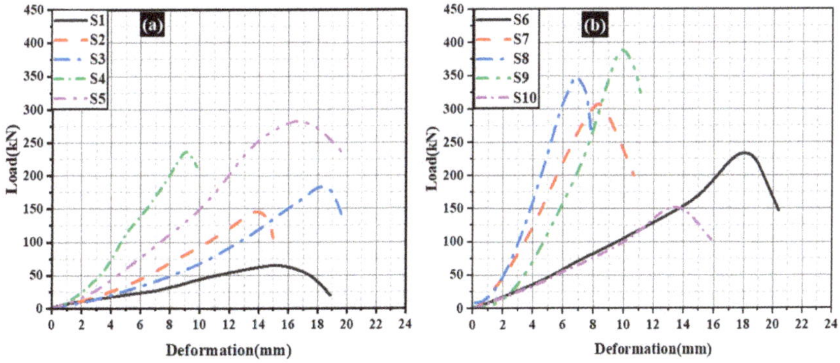

(**a**) Without reinforcement (**b**) With reinforcement

Figure 9. Load–axial-deformation behavior of ICF wall panels and masonry units.

The above test results show that the specimens without reinforcement possess higher deformation than those with reinforcement. Moreover, an increase in the grade of concrete increases the ultimate failure load and decreases the deformation of the ICF wall panels. Specimen S6 possesses a maximum deformation of 20.5 mm; this can be attributed to hollow, empty cells in the ICF wall panels. Moreover, rapid failure was seen in the brick masonry wall. When the ultimate load is reached, the brick masonry walls crumbles; this type of failure is avoided in ICF wall panels. The presence of an EPS shell can be attributed to this. The elastic nature of the EPS shell holds the GPC at the core even after failure. Sudden failure (brittle) was seen in the conventional brick wall. In the present investigation, ICF-based wall panels do not show brittle failure in all cases when they are subjected to loading.

The ultimate load of all the ICF wall panels is higher than that of conventional brick masonry. Furthermore, after reaching the ultimate load, the load is removed, and the EPS system holds the crushed concrete inside the core and prevents the failure mode (brittle) of GPC. The load vs. deflection behavior of conventional masonry and ICF-based walls is depicted in Figure 10.

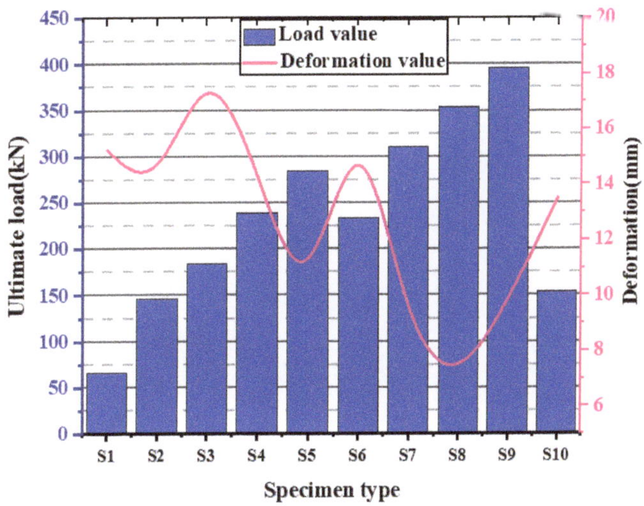

Figure 10. The ultimate load carried by the ICF wall panels, along with their deformation.

In general, concrete is strong in compression and weak in tension. Once concrete reaches its ultimate strength, the shape and structure of concrete tend to deteriorate and collapse completely. In the case of an ICF wall, the structure does not collapse even after it reaches its peak load. While loading, the ICF wall tends to compress and regain its original shape after the load release; this might be attributed to the presence of the EPS periphery (which comprises 98% air and 2% polystyrene). In conclusion, "the encapsulation of concrete in the EPS system changes the failure type from brittle to ductile."

In the case of both M15 and M20 mixes, cracks developed at the top of the specimen, and the specimens failed due to crushing. Under uniform loading, all the ICF panels exhibited significant resistance to compressive stress. The projected cells of panels have shown their failure under compressive loading. In both cases (M15 and M20 mix), the initial crack started at a load of 25 kN, and further, the cracks tend to develop at the consequent loads till failure. Figure 11, shows the axial stiffness of the ICF wall. In addition to the load-deflection parameter, load to weight ratio, was estimated and is shown in Figure 12. The weight of the individual ICF wall panel was found in terms of kg and concerted into Newtons for better assessment. It is found that S4 and S8 possess the maximum value; this shows that S4 and S8 have higher efficiency compared with all other panel types.

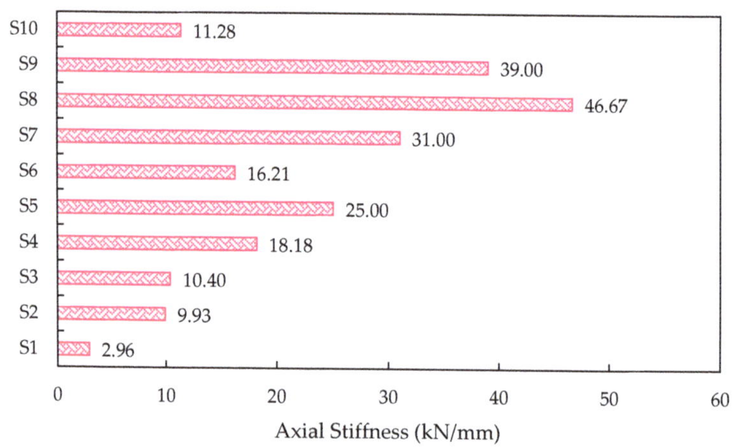

Figure 11. The axial stiffness of the ICF walls.

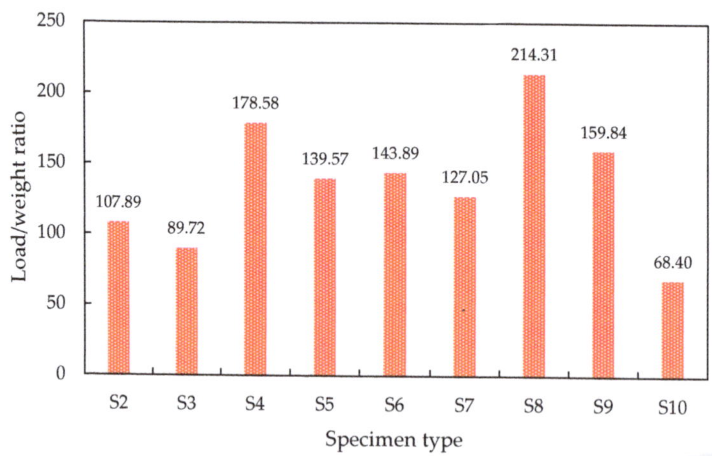

Figure 12. The load-to-weight ratio of the ICF wall panels.

3.4. Strain Energy

The strain energy is the energy absorbed by a body while undergoing deformation. The strain energy is an indirect indicator used to evaluate ductility. The area under the load-displacement characteristic curve is used to evaluate the ductility. The absorption of energy is related to the ductility of the material. Because the ICF wall panel absorbs more energy than the reference panel, it implicitly reflects ICF's superior ductility. Equation (1) may be used to determine the energy absorption (EA) of a wall panel [12].

$$\text{Energy Absorption capacity (EAC)} = V\int_0^{\varepsilon_c} \sigma d\varepsilon \quad (1)$$

where V is the specimen volume in cubic meters, ε_c is compressive strain, and σ is compressive stress. The compressive load-displacement curve may also be used to calculate energy absorption. The area represents the absorbed strain energy by the wall panel under the compressive load-displacement curve [12]. The ICF wall panels' high strain energy indicates their high ductility compared with the reference panels. Figure 13 illustrates the energy-absorbing capacity of the wall panels.

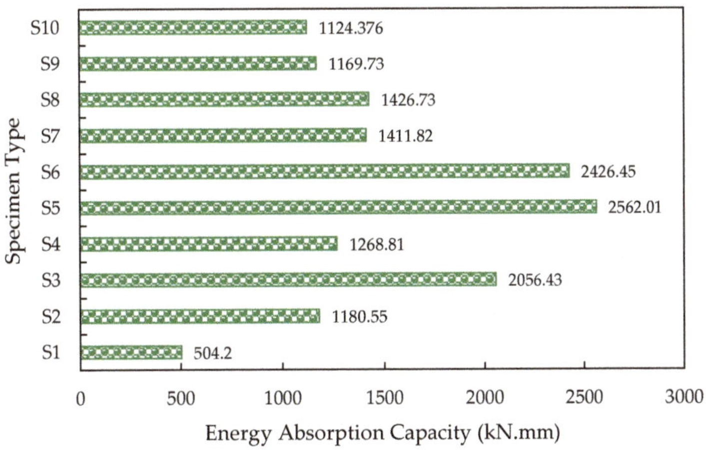

Figure 13. Energy-absorbing capacity of the ICF-based wall panels.

3.5. Crack Pattern

Due to the elastic nature of the EPS shell, cracks were not observed in all the ICF-based wall panels. Many minor cracks were seen on the bottom side of the ICF wall panels, then the top surface of the wall panels, as seen in Figures 14 and 15. One or two major cracks were observed in the top surface of the ICF wall panels. Almost all the ICF-based wall panels possess similar failure and crack patterns; the presence of an EPS shell could be attributed to this. Notably, the degradation in load capacity is mainly due to the crushing of concrete in the cells, but the shielding of EPS can enhance the load-carrying capacity further. The final reduction in the load-carrying capacity of panels is due to the tearing of EPS at the bottom phase.

It is observed that the ICF wall panels exhibit the ability to resist the load even after the failure of the concrete core. However, a sudden failure was observed in the reference panel. The reference panel broke into pieces after failure, which is presented in the ICF wall panels because of the elastic nature of the EPS sheets enveloping them. This elastic property of EPS holds the concrete core in position even after its failure. Brittle failure was observed in the reference panel. ICF does not exhibit any such failure; instead, it displays an elastic nature when the axial compressive load is applied. Moreover, after removing the load, EPS in ICF holds together the failed concrete and prevents the collapse of failed concrete when it reaches its peak load.

Figure 14. Failure caused while loading.

Figure 15. Failure pattern of ICF panels.

3.6. Cost Efficiency

ICF and brick walls' cost efficiency is measured by taking the strength-to-cost ratio [39]. The cost involved in this analysis is considered based on the materials' purchase cost. All the raw ingredients are obtained from local dealers in Coimbatore. The cost of the raw materials is found to be: 130 USD/MT for cement; 67.88 USD/MT for GGBS; 7.31 USD/MT for M-sand; 9.14 USD/MT for coarse aggregate; and 13 USD/MT for FA. The production cost of ICF panels and brick walls is marginal. The early age of GPC is relatively higher than conventional cement-based concrete. The strength-to-cost ratio of the ICF and brick wall is illustrated in Table 5. The higher the value of cost efficiency, the higher the effectiveness in its production. M20 grade concrete mix possesses higher strength than M15 grade mix.

This can be attributed to the higher percentages of GGBS content in it. The production cost of the ICF panels and their counterpart brick walls is marginal; this can be attributed to alkaline activators, sodium hydroxide and sodium silicate solutions. Figure 16 illustrates the cost-efficiency analysis of M15 and M20 grade GPC specimens.

Table 5. Cost-efficiency analysis of ICF and brick walls.

S.No	Material	ICF					Brick Wall	
			M15		M20			
		Rate	QTY	Cost	QTY	Cost	QTY	Cost
		(USD/MT)	(1 Wall)	(USD/Wall)	(1 Wall)	(USD/Wall)	(1 Wall)	(USD/Wall)
1	Cement	134.91	-	-	-	-	30	4.05
2	FA	13.49	180	2.43	100	1.35	-	-
3	GGBS	78.6	20	1.57	100	7.86	-	-
4	CA	9.44	650	6.14	350	3.30	-	-
5	MS	7.55	450	3.40	450	3.40	60	0.45
6	SS	121.41	43.44	5.27	43.44	5.27	-	-
7	SH	303.33	28.96	8.78	28.96	8.78	-	-
8	Bricks	0.15/brick	-	-	-	-	275	41.25
9	ICF panel	10/panel	2	20	2	20	-	-
10	Total Cost USD/wall			47.59		49.97		45.75
11	28 days compressive strength (MPa)			20.60		29.00		6.20
12	Cost efficiency (MPa/USD/wall)			0.43		0.58		0.14

Note: SH: Sodium hydroxide; SS: Sodium silicate; QTY: Quantity.

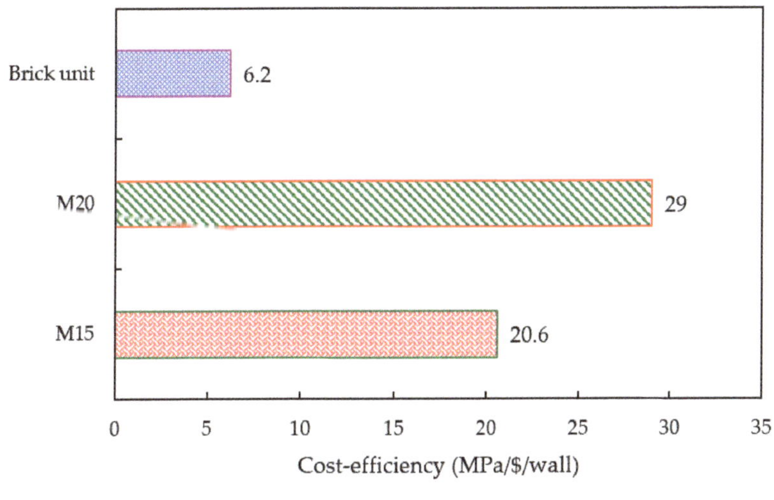

Figure 16. Cost-efficiency analysis of ICF and brick walls.

The cost-efficiency analysis shows that the ICF-based system provides higher effectiveness in terms of its cost of production. It is concluded that ICF wall panels can be used as alternatives to conventional brick-based walls to enhance the construction speed and lower the production cost.

3.7. Advantages of ICF over Masonry Walls

ICF does not require skilled construction labor, so the labor cost is considerably lower. The ICF construction time of the ICF wall is reduced by 50%, and an increase in the construction unit decreases the labor cost. Due to the interconnectivity of EPS with the aid of reinforcement, ICF has higher resistance against static loads. ICF construction in hot and cold regions can reduce power consumption costs due to high heat insulation. Plumbing and electrical ducts can be installed more easily in ICF than in conventional masonry units. ICF can reduce noise as a sound barrier between rooms and create a better living environment. ICF walls are more cost efficient than traditional masonry units. EPS is a recycled material; therefore, the construction of structures using ICF can be considered under green construction practices.

4. Conclusions

The performance of encapsulated concrete EPS wall panels under axial compression was studied in the current investigation. The major conclusions of the findings are reported below:

- The encapsulation of concrete in the EPS system changes the failure type from brittle to ductile, and the elastic property of the EPS holds the encapsulated concrete in the core even after the failure;
- The result from the study reveals that Specimen S9 (all the cells of EPS were filled with concrete, M20) exhibits a load of 360 kN with deformation of 11 mm more than the reference masonry unit and all other models;
- In comparison with the reference wall panel, Specimen S2, the load-carrying capacity of Specimen S2 was 3.33% less, whereas the Specimens S3, S4, S5, S6, S7, S8, and S9 gained an increase of 60%, 87%, 57%, 107%, 133%, and 160%, respectively;
- Due to the EPS outer shell, the wall panels possess higher deformations than the conventional brick masonry wall;
- The brick masonry walls exhibit brittle failure; without any sign, the masonry wall fails into pieces, whereas the ICF wall panels retain their shape even after failure. Therefore, ICF can be recommended to replace the conventional load-bearing and non-load-bearing walls;
- Research on the carbon and energy-effective mix design proportion of GPC-based ICF to attain target strength, workability, durability, and sustainability may be carried out. Further investigations are required to elaborately explain the failure pattern, loading-deflection behavior, material characterization of EPS, and the infilling materials so that national standards/guidelines can be developed.

Author Contributions: B.K., J.G., A.N., P.A.: Investigation, Methodology, Formal analysis, Writing—original draft, Writing—review & editing. A.N., P.A.: Resources, Supervised the research as well as the analysis of results. B.K., T.K., J.G., A.N.: Project administration, Visualization, Review and Editing. A.N., P.A., B.G.A.G., K.R.: Validation, suggested and chose the journal for submission. A.N., B.G.A.G., K.R.: Collaborated, Co-ordinated the research, Reviewed for submission. All authors have read and agreed to the published version of the manuscript.

Funding: This research received no external funding.

Institutional Review Board Statement: Not applicable.

Informed Consent Statement: Not applicable.

Data Availability Statement: The data presented in this study are available on request from the corresponding author.

Conflicts of Interest: This manuscript has not been submitted to, nor is it under review by, another journal or other publishing venue. The authors have no affiliation with any organization with a direct or indirect financial interest in the subject matter discussed in the manuscript. The authors declare no conflict of interest.

References

1. Assaad, J.J.; Mikhael, C.; Hanna, R. Recycling of Waste Expanded Polystyrene Concrete in Lightweight Sandwich Panels and Structural Applications. *Clean. Mater.* **2022**, *4*, 100095. [CrossRef]
2. Meddage, D.P.P.; Chadee, A.; Jayasinghe, M.T.R.; Rathnayake, U. Exploring the Applicability of Expanded Polystyrene (EPS) Based Concrete Panels as Roof Slab Insulation in the Tropics. *Case Stud. Constr. Mater.* **2022**, *17*, e01361. [CrossRef]

3. Souza, T.B.; Lima, V.M.E.; Araújo, F.W.C.; Miranda, L.F.R.; Melo Neto, A.A. Alkali-Activated Slag Cellular Concrete with Expanded Polystyrene (EPS)—Physical, Mechanical, and Mineralogical Properties. *J. Build. Eng.* **2021**, *44*, 103387. [CrossRef]
4. Garhwal, A.; Sharma, S.; Roy, A.B.D. Performance of Expanded Polystyrene (EPS) Sandwiched Concrete Panels Subjected to Accelerated Corrosion. *Structures* **2022**, *43*, 1057–1072. [CrossRef]
5. Jiang, T.; Wang, Y.; Shi, S.; Yuan, N.; Ma, R.; Wu, X.; Shi, D.; Sun, K.; Zhao, Y.; Li, W.; et al. Compressive Behavior of Lightweight Concrete Using Aerogel-Reinforced Expanded Polystyrene Foams. *Case Stud. Constr. Mater.* **2022**, *17*, e01557. [CrossRef]
6. Maghfouri, M.; Alimohammadi, V.; Gupta, R.; Siberian, M.; Azaria, P.; Hashemi, M.; Asadi, I.; Roychand, R. Drying Shrinkage Properties of Expanded Polystyrene (EPS) Lightweight Aggregate Concrete: A Review. *Case Stud. Constr. Mater.* **2022**, *16*, e00919. [CrossRef]
7. Khatib, J.; Jahami, A.; Baalbaki, O. Flexural Characteristics of Reinforced Concrete Beams Containing Lightweight Aggregate in the Tensile Zone. *Sustain. Constr. Mater. Technol.* **2019**, 1–9. [CrossRef]
8. Jiang, P.; Chen, Y.; Li, N.; Zhou, L.; Pu, S.; Wang, W. Strength Properties and Microscopic Mechanism of Lime and Fly Ash Modified Expandable Poly Styrene Lightweight Soil Reinforced by Polypropylene Fiber. *Case Stud. Constr. Mater.* **2022**, *17*, e01250. [CrossRef]
9. Mochalov, A.M.; Akulova, M.V.; Sokolova, J.A.; Sokolova, A.G. New Flame-Retardant Composition for Lowering Contribution of Expanded Poly Styrene to the Propagation of Fire. *IOP Conf. Ser. Mater. Sci. Eng.* **2021**, *1015*, 012064. [CrossRef]
10. Ding, F.; Wu, X.; Xiang, P.; Yu, Z. New Damage Ratio Strength Criterion for Concrete and Lightweight Aggregate Concrete. *ACI Struct. J.* **2021**, *118*, 165–178. [CrossRef]
11. Ekrami, N.; Garat, A.; Fung, A.S. Thermal Analysis of Insulated Concrete Form (ICF) Walls. *Energy Procedia* **2015**, *75*, 2150–2156. [CrossRef]
12. Arun Solomon, A.; Hemalatha, G. Characteristics of Expanded Polystyrene (EPS) and Its Impact on Mechanical and Thermal Performance of Insulated Concrete Form (ICF) System. *Structures* **2020**, *23*, 204–213. [CrossRef]
13. Lopez, A.; Bazaez, R.; Leiva, G.; Loyola, R.; Gómez, M. Experimental Study of In-Plane Flexural Behavior of Screen-Grid Insulated Concrete Form Rectangular and T-Shaped Walls. *Eng. Struct.* **2021**, *247*, 113128. [CrossRef]
14. Arun Solomon, A.; Hemalatha, G.; Sudheer, G.; Shelton, J.; Carmichael, J. Exploring the Impact of EPS Incorporation on Insulated Concrete Form (ICF) Wall Panels under Axial Compression and Flexure. *J. King Saud Univ.-Eng. Sci.* **2022**, *in press*. [CrossRef]
15. Lowe, D.; Roy, K.; Das, R.; Clifton, C.; Lim, J. Full-scale experiments on splitting behaviour of concrete slabs in steel-concrete composite beams with shear stud connection. *Structures* **2020**, *23*, 126–138. [CrossRef]
16. Rajamony Laila, L.; Gurupatham, B.G.A.; Roy, K.; Lim, J.B.P. Effect of Super Absorbent Polymer on Microstructural and Mechanical Properties of Concrete Blends Using Granite Pulver. *Struct. Concr.* **2020**, *22*, E898–E915. [CrossRef]
17. Rajamony Laila, L.; Gurupatham, B.G.A.; Roy, K.; Lim, J.B.P. Influence of Super Absorbent Polymer on Mechanical, Rheological, Durability, and Microstructural Properties of Self-Compacting Concrete Using Non-Biodegradable Granite Pulver. *Struct. Concr.* **2020**, *22*, E1093–E1116. [CrossRef]
18. Moro, C.; Francioso, V.; Velay-Lizancos, M. Impact of Nano-TiO_2 Addition on the Reduction of Net CO_2 Emissions of Cement Pastes after CO_2 Curing. *Cem. Concr. Compos.* **2021**, *123*, 104160. [CrossRef]
19. Ndahirwa, D.; Zmamou, H.; Lenormand, H.; Leblanc, N. The Role of Supplementary Cementitious Materials in Hydration, Durability and Shrinkage of Cement-Based Materials, Their Environmental and Economic Benefits: A Review. *Clean. Mater.* **2022**, *5*, 100123. [CrossRef]
20. Kanagaraj, B.; Anand, N.; Andrushia, A.D.; Lubloy, E. Investigation on Engineering Properties and Micro-Structure Characteristics of Low Strength and High Strength Geopolymer Composites Subjected to Standard Temperature Exposure. *Case Stud. Constr. Mater.* **2022**, *17*, e01608. [CrossRef]
21. Madan, C.S.; Munuswamy, S.; Joanna, P.S.; Gurupatham, B.G.A.; Roy, K. Comparison of the Flexural Behavior of High-Volume Fly Ash Based Concrete Slab Reinforced with GFRP Bars and Steel Bars. *J. Compos. Sci.* **2022**, *6*, 157. [CrossRef]
22. Madan, C.S.; Panchapakesan, K.; Anil Reddy, P.V.; Joanna, P.S.; Rooby, J.; Gurupatham, B.G.A.; Roy, K. Influence on the Flexural Behaviour of High-Volume Fly-Ash-Based Concrete Slab Reinforced with Sustainable Glass-Fibre-Reinforced Polymer Sheets. *J. Compos. Sci.* **2022**, *6*, 169. [CrossRef]
23. Chinnasamy, Y.; Joanna, P.S.; Kothanda, K.; Gurupatham BG, A.; Roy, K. Behavior of Pultruded Glass-Fiber-Reinforced Polymer Beam-Columns Infilled with Engineered Cementitious Composites under Cyclic Loading. *J. Compos. Sci.* **2022**, *6*, 338. [CrossRef]
24. Sivanantham, P.; Gurupatham, B.G.A.; Roy, K.; Rajendiran, K.; Pugazhlendi, D. Plastic Hinge Length Mechanism of Steel-Fiber-Reinforced Concrete Slab under Repeated Loading. *J. Compos. Sci.* **2022**, *6*, 164. [CrossRef]
25. Kanagaraj, B.; Anand, N.; Lubloy, E. Performance Evaluation of Sodium Silicate Waste as a Replacement for Conventional Sand in Geopolymer Concrete. *J. Clean. Prod.* **2022**, *375*, 134172. [CrossRef]
26. Prusty, J.K.; Pradhan, B. Investigation on Effect of Precursor Materials and Sand-to-Binder Ratio on Strength Development, Acid Resistance and Microstructure Evolution of Geopolymer Mortar. *Constr. Build. Mater.* **2022**, *346*, 128501. [CrossRef]
27. Qaidi, S.M.A.; Sulaiman Atrushi, D.; Mohammed, A.S.; Unis Ahmed, H.; Faraj, R.H.; Emad, W.; Tayeh, B.A.; Mohammed Najm, H. Ultra-High-Performance Geopolymer Concrete: A Review. *Constr. Build. Mater.* **2022**, *346*, 128495. [CrossRef]
28. Wu, H.C.; Sun, P. New Building Materials from Fly Ash-Based Lightweight Inorganic Polymer. *Constr. Build. Mater.* **2007**, *21*, 211–217. [CrossRef]
29. Ahmad, A.; Singh, Y. Flexural Behavior of Expanded Polystyrene Core Reinforced Concrete Sandwich Panels with Different Construction Methods and End Conditions. *Structures* **2021**, *34*, 2900–2911. [CrossRef]

30. Kanagaraj, B.; Anand, N.; Johnson Alengaram, U.; Samuel Raj, R.; Kiran, T. Exemplification of Sustainable Sodium Silicate Waste Sediments as Coarse Aggregates in the Performance Evaluation of Geopolymer Concrete. *Constr. Build. Mater.* **2022**, *330*, 127135. [CrossRef]
31. *IS 383*; Coarse and Fine Aggregate for Concrete-Specification. Bureau of Indian Standards: New Delhi, India, 2016.
32. Kanagaraj, B.; Kiran, T.; Anand, N.; Al Jabri, K.; Justin, S. Development and Strength Assessment of Eco-Friendly Geopolymer Concrete Made with Natural and Recycled Aggregates. *Constr. Innov.* **2022**. *ahead-of-print*. [CrossRef]
33. Kanagaraj, B.N.A.; Kiran, T.; Prince, A.S.J. Experimental Investigation on Fresh and Hardened Properties of Geopolymer Concrete Blended with Recycled Concrete Aggregate. *Indian Concr. J.* **2022**, *96*, 29–41.
34. Kanagaraj, B.; Anand, N.; Samuel Raj, R.; Lubloy, E. Performance Evaluation on Engineering Properties of Sodium Silicate Binder as a Precursor Material for the Development of Cement-Free Concrete. *Dev. Built Environ.* **2022**, *12*, 100092. [CrossRef]
35. Kanagaraj, B.; Anand, N.; Alengaram, U.J.; Praveen, B.; Tattukolla, K. Performance Evaluation on Engineering Properties and Sustainability Analysis of High Strength Geopolymer Concrete. *J. Build. Eng.* **2022**, *60*, 105147. [CrossRef]
36. Fang, Z.; Roy, K.; Chi, Y.; Chen, B.; Lim, J.B.P. Finite element analysis and proposed design rules for cold-formed stainless steel channels with web holes under end-one-flange loading. *Structures* **2021**, *34*, 2876–2899. [CrossRef]
37. *ASTM C 143*; Standard Test Method for Slump of Hydraulic-Cement Concrete. ASTM International: West Conshohocken, PA, USA, 2014. [CrossRef]
38. *IS 516*; Method of Tests for Strength of Concrete-Specifications. Bureau of Indian Standards: New Delhi, India, 2004.
39. Darvish, P.; Johnson Alengaram, U.; Soon Poh, Y.; Ibrahim, S.; Yusoff, S. Performance Evaluation of Palm Oil Clinker Sand as Replacement for Conventional Sand in Geopolymer Mortar. *Constr. Build. Mater.* **2020**, *258*, 120352. [CrossRef]

materials

MDPI

Article

Durability of Reinforced Concrete with Additions of Natural Pozzolans of Volcanic Origin

Juan J. Santana [1,*], Natalia Rodríguez-Brito [2], Concepción Blanco-Peñalver [2], Vicente F. Mena [1,3] and Ricardo M. Souto [3,4,*]

1 Department of Process Engineering, University of Las Palmas de Gran Canaria, 35017 Las Palmas de Gran Canaria, Spain
2 Consejería de Obras Públicas y Transportes, Gobierno de Canarias, 38270 San Cristóbal de La Laguna, Spain
3 Department of Chemistry, Universidad de La Laguna, P.O. Box 456, 38200 La Laguna, Spain
4 Institute of Material Science and Nanotechnology, Universidad de La Laguna, 38200 La Laguna, Spain
* Correspondence: juan.santana@ulpgc.es (J.J.S.); rsouto@ull.es (R.M.S.)

Abstract: In this work, the properties of concrete modified with dosages of natural pozzolans (NP) in substitution of cement or superfine aggregates were evaluated. Proportions of 20/80 pozzolan/cement or pozzolan/superfine aggregates were selected for the additions of quarry and tuff pozzolans. Pozzolanic activity, durability, compressive strength, characteristic resistance, settling consistency, density, electrical resistivity, depth of water penetration, accessible porosity, and carbonation and chloride penetration were determined for the resulting concrete mixtures, and they were subsequently compared to the values obtained for the reference concrete batches without additions. The results of the cementitious mixtures supplemented with tuff (PZT) and quarry (PZQ) pozzolans, expressed in mmol/L, are consistent with the pozzolanism test, with $[Ca(OH)_2]/[OH^-]$ ratios at 7 days are 6.03/60.19 for PZQ and 1.78/92.78 PZT. In addition to the pozzolanic activity at these dosages, the characteristic resistance and durability parameters required by EHE-08 were verified. Particular attention was given to the determination of the diffusion of chloride ions, introducing an instrumental modification of the accelerated integral method. The modification provides values of diffusion coefficients similar to those obtained by the other methods with the advantage of greater stability and quality of the measurement.

Keywords: reinforced concrete; chloride penetration; carbonation; durability; useful life prediction; integral method; multi-regime method; modeling

Citation: Santana, J.J.; Rodríguez-Brito, N.; Blanco-Peñalver, C.; Mena, V.F.; Souto, R.M. Durability of Reinforced Concrete with Additions of Natural Pozzolans of Volcanic Origin. *Materials* **2022**, *15*, 8352. https://doi.org/10.3390/ma15238352

Academic Editor: Ayman El-Zohairy

Received: 20 October 2022
Accepted: 22 November 2022
Published: 24 November 2022

1. Introduction

The reduction in CO_2 emissions into the atmosphere has become a subject of great importance in recent decades in order to reduce the greenhouse effect [1]. Among the various types of industries, the cement sector is one of the largest contributors to CO_2, estimated at around 8% of total anthropogenic greenhouse gas (GHG) emissions [2–8]. One of the ways to reduce this impact is to reduce cement consumption by partially replacing it with supplementary cementitious materials (SCMs), which can be by-products from industry or natural pozzolans (NP), such as fly ash (FA), ground granulated blast furnace slag (GGBS), silica fume (SF), metakaolin (MK), limestone (L), fine glass powder (GP), rice husk ash (RHA), and volcanic tuffs (natural pozzolans) [5,6,9–20]. These latter SCMs include both pozzolanic tuffs and those from the extraction of quarry aggregates. Indeed, the abundance of these materials in certain geographic areas, such as volcanic regions, makes them attractive candidates as a partial substitute for cement, mortar, and concrete [1,10,21–25]. These materials are often disposed of in landfills, thus constituting a major environmental problem [5]. Therefore, its reuse would reduce the environmental impact, apart from CO_2 emissions, the reduction in the cost of concrete manufacturing, and the improvement of mechanical properties [17]. On the other hand, it is difficult to draw general conclusions on the behavior or to predict the reactivity of a given pozzolan, thus arousing a growing interest in the study of these tuffs or zeolitic deposits [26–31].

The effect of pozzolan additions to mortars and concretes varies depending on the type of pozzolan and the percentage of cement or aggregate substituted, among other factors. Generally speaking, pozzolan additions have been reported to increase durability [5,13–15,21,26,32–38], decrease the heat of hydration [1,38], increase resistance to sulfate attack [1,39], increase setting time, prevent water penetration [1,40], and reduce energy cost per unit of cement [37]. Specifically, the effect these additions have on mechanical strength depends on many factors, and conflicting results can often be found in the literature.

When natural pozzolans were used as a cement substitute in the range of 10–30%, Xia et al. [14] reported that the mechanical strength varied differently depending on the type of pozzolan, a fact corroborated by other authors [29,41]. Najimi et al. [1] found that the compressive strength was only affected slightly (actually less than 5%) by substituting about 25% of the Portland cement with natural tuff pozzolan. Costafreda et al. [11] recommend the industrial manufacturing of pozzolanic cement with the proportions 75:25% and 70:30% with natural pozzolans since they provide practically the same mechanical strength. Çullu et al. [42] observed that increasing the percentage of various natural pozzolans at the expense of cement produced a decrease in compressive strength, proposing that the optimum dose may be 10%. Turanli et al. [43] found that the compressive strength of mortar containing large amounts of volcanic ash was lower than that of Portland cement at any age less than 90 days. Khan et al. [18] found that improved fineness of volcanic ash up to 30% mass replacement of cement in mortar demonstrated compressive strength comparable to control and reference fly ash mortars at all ages. In recent work, these authors [19] demonstrated that the incorporation of those ultrafine volcanic ashes in percentages of 10% to 30% improved the compressive strength after 91 days of curing. These results, therefore, do not make it possible to extract a clear trend as to the effect exerted by the additions of natural pozzolans.

Regarding the penetration of aggressive substances, the addition of pozzolans leads to an improvement in the reduction in chloride ion penetration [1], but not in the carbonation process. Thus, Xia et al. [14] found that the addition of natural pozzolans caused a decrease in the resistance to carbonation and a decrease in the diffusion coefficients of chlorides, parameters related to the corrosion of reinforced concrete, chloride being the main cause of reinforcement corrosion [44,45]. The corrosive process has arisen interest in the last decades because it has been identified as the main process related to the durability of reinforced concrete, which leads to great economic losses and, on the other hand, a huge environmental cost due to increased CO_2 emissions originated by the consumption of raw materials in the repair or replacement of damaged structures [46,47]. Although authors such as Kaid et al. [48] and Fajardo et al. [49] found that concrete mixed with natural pozzolan showed excellent corrosion resistance, quantification is necessary when these concretes are going to be used in areas with a high chloride content in the environment. For this reason, the penetration of aggressive agents is the basis for quantifying the durability of a structure, although there is no rigorous standardized methodology for predicting its useful life [50]. The first proposals for systematizing the notion of useful life were introduced in the 1970s [51,52] from the observation of the first degradations. These initiatives were by Tuutti reviewed these initiatives [53], and their use among researchers is currently widespread, even if their concretization in codes or recommendations is still very nascent [54].

In the present study, the behavior of concrete samples dosed with additions of natural pozzolans present in the Canary Islands (i.e., tuff pozzolan and quarry pozzolan) in substitution of cement or fine aggregates was tested by physico-chemical and mechanical assays according toto the regulations in force. The pozzolanic activity of the samples and the penetration of aggressive agents were evaluated and compared to samples without addition. The study of the penetration of chlorides in samples of reinforced concrete dosed with these additions stands out for its importance in the corrosion of the reinforcements, using the pool immersion method, the multi-regime test, and the integral method with an instrumental modification. In this way, the diffusion coefficients of the chlorides in the samples were determined. The durability of the tested mixtures was determined from the carbonation and chloride penetration models included in the Spanish regulation EHE-08 [54]. The feasibility of the additions was established based on the results obtained from mechanical and durability tests.

2. Materials and Methods

2.1. Materials and Sample Preparation

Two existing pozzolans in the study area, namely tuff (PZT) and quarry (PZQ) pozzolans, were investigated as possible additives for reinforced concrete. PZT comes from the crushing and grinding of rock classified as unwelded Ignimbrite lithotype (IGNS) [55]. They are made up of a chaotic mix of pumice fragments, ashes, rock fragments, and crystals. They have a trachytic or phonolitic composition, and petrographic microscopy reveals crystals of anorthoclase or sanidine, vitreous fragments of irregular morphology, fragments of aegirine, hornblende, and opaque minerals, all in a microcrystalline or vitreous matrix. PZT is a residue that comes from the clearing of an aggregate quarry located in the area (SATOCAN quarry in the municipality of Arico, Tenerife, Canary Islands). It is a powder with a real density or a specific weight of 2550 kg/m^3, which is produced as waste during the exploitation in search of the optimal layers for the extraction of aggregates. Particle size analysis was performed for both pozzolans, and XFR analysis of PZT was also performed.

In this work, an II/A-P/42.5R MR-type cement was used, and Table 1 lists its chemical, physical and mechanical properties. Reinforced concretes were dosed in the laboratory with characteristic resistances 25, 30, and 35 N/mm^2 (HA25, HA30, and HA 35, respectively) using this cement and additions of either PZT or PZQ. The concrete samples were assayed using the Fuller [56] method, and their main characteristics are given in Table 2. Pozzolans were introduced in a percentage of 20% by subtracting the amount of cement in the D3 samples, while in the D4 samples, the pozzolans were included as superfine aggregates by subtracting their weight from the basalt sand and not from cement. All proposed dosages and mixtures have been manufactured with expressed water/cement ratios (w/c), in which the moisture of the aggregates has been subtracted from the amount of water.

The pozzolanic activity index with lime (PAI) of the concretes with pozzolan contents was determined according to the ASTM C311/C311M-13 standard [57] and further established in the UNE 80303-2 standard [58] in item 7.2.b), and according to the test method UNE EN 196-5 [59] after 7 days.

The aggregates used were gravel and sand from a local quarry with CE marking. The basaltic aggregate, coded A, was selected from the typical quarry materials of the Canary Islands for the manufacture of concrete. These are gravel particles with an approximate size of 20 mm, chosen from the storage material of the assays.

Table 1. Chemical, physical, and mechanical properties of the cement (type II/A-P/42.5R MR).

	II/A-P/42.5R MR
Loss on ignition (LOI)	3.4
Insoluble residue	10.1
Chloride (Cl^-, %)	0.01
Sulphates (SO_3, %)	3.38
Initial setting time (min)	150
Pozzolan content (%)	19
Clinker content (%)	76
Gypsum content (%)	5
End of hardening (EH, min)	215
Compressive strength at 7 days (N/mm^2)	37.0
Compressive strength at 28 days (N/mm^2)	51.6

Table 2. Identification, classification and characteristics of all concrete samples tested. All were made with type II/A-P/42.5R MR cement.

Code	D1P	D3P	D3PZT	D3PZQ	D4P	D4PZT	D4PZQ
Type of Aggregates	**Reference**	**Reference**	**Tuff Pozzolan**	**Quarry Pozzolan**	**Reference**	**Tuff Pozzolan**	**Quarry Pozzolan**
Characteristic resistance (N/mm^2)	HA-35	HA-30	HA-30	HA-30	HA-25	HA-25	HA-25
Cement content (kg/m^3)	325	300	240	240	300	300	300
Pozzolan content (kg/m^3)	-	-	60	60	-	60	60
Basalt gravel (kg/m^3)	806.19	945.75	945.75	945.75	945.75	945.75	945.75
Basalt sand (kg/m^3)	688.08	502	502	502	502	442	442
African sand (kg/m^3)	250.23	296.75	296.75	296.75	296.75	296.75	296.75
Water content (L/m^3)	156	167	167	167	198	195	195
Cement ratio (%)	0.51	0.56	0.57	0.57	0.76	0.73	0.73

The specimens of each sample were manufactured and stored in accordance with the UNE-EN 12390-1:2001 [60] and UNE-EN 12390-2:2009 [61] standards and according to the geometry established for each particular test. Thus, cylindrical specimens with a diameter of 15 cm and a height of 30 cm were fabricated for the tests of bending and compression, water penetration, chloride penetration by the pond method, direct resistivity, and Wenner. Next, cubic specimens with 10 cm edges were fabricated for chloride penetration tests using the integral method and accessible porosity. Sections 15 cm in diameter and 2.5 cm thick were employed in the multi-regime method. Finally, prismatic specimens of dimensions 15 cm × 15 cm × 60 cm were used in the natural and accelerated carbonation tests. For the integral method, a corrugated steel bar was embedded, such as armor, in each of the specimens. These bars were insulated and then laid out so that the exposed surface was 6 cm long, fully embedded in the concrete. In all cases, once the concrete was properly vibrated, it was hardened in a humid chamber for 28 days. After the hardening period, the specimens were subjected to various tests.

The samples were coded according to a letter-number-letter (*DXY*) sequence as described below, according to its definition according to EHE-08 [54], and taking into account their main characteristics, such as the grade and type of cement, the water/cement ratio and type of addition. The letter *D* indicates that the samples were mixed concrete made in the laboratory, whereas the digit *X* indicates the number of the mix, up to a total of 3 types of concrete dosages. They were followed by letters or acronyms, *Y*, which indicated the modifications that were made, depending on whether or not it included an addition (e.g., the letter *P*, to codify the concrete reference mixes, without addition; the letters PZT, for concrete mixes with the addition of pozzolan extracted from a block of tuff, and the letters PZQ, for concrete mixes with the addition of quarry pozzolan). Table 2 also shows sample coding and identification.

2.2. Tests and Methods

The concrete specimens were subjected to various tests as follows. Compressive strength at 28 days, according to UNE-EN 12390-3 [62], and resistance, according to UNE-EN 196-1:2005 [63]. Determination of fresh or settling consistency using the Abrams cone, according to UNE EN 12350-2 [64]. The density of hardened concrete UNE-EN 12390-7 [65]. Electrical resistivity using the direct method (Direct Resistivity) according to UNE 83988-1 [66] and by the four points or Wenner method (Indirect Resistivity) according to UNE 83988-2 [67]. Depth of penetration of water under pressure according to UNE 83309-90 [68] and UNE-EN 12390-8 [69]. Accessible porosity, according to UNE-EN 1936 [70] and according to UNE 83980:2014 [71]. Measurement of carbonation by the natural method at 90 days according to UNE 112011-94 [72], UNE 112011:2011 [73], and UNE 83993-1 [74], as well as by the accelerated method according to UNE 83993-2 [75]. Measurement of chloride penetration through the "Pond" arrangement according to PrUNE 83986 [76] and CEN/TS 12390-11 [77].

Determination of chloride content in hardened concrete according to UNE-EN 14629 [78]. The multi-regime test was performed according to UNE 83987 [79], whereas the integral method is described in UNE 83992-2 [80], and the polarization resistance was measured according to UNE 112072 [81]. Finally, the pozzolanic activity with lime (PAI) was determined according to ASTM C311/C311M-13 [57], UNE 80303-2 [58], and UNE EN 196-5 [59].

2.3. Modification of the Accelerated Integral Method

The assembly of the experimental cell was carried out in accordance with the specifications in standard UNE 83992-2 [80]. Once the concrete specimen was prepared, PVC cylinders were attached to its top side to serve as containers for a 0.6 M NaCl + 0.4 M $CuCl_2$ solution, and a Cu electrode was introduced to serve as the cathode of an electrochemical cell. The union of the plastic cylinder with the surface of the concrete specimen was sealed using neutral silicone, and the whole assembly is shown on the left side of Figure 1.

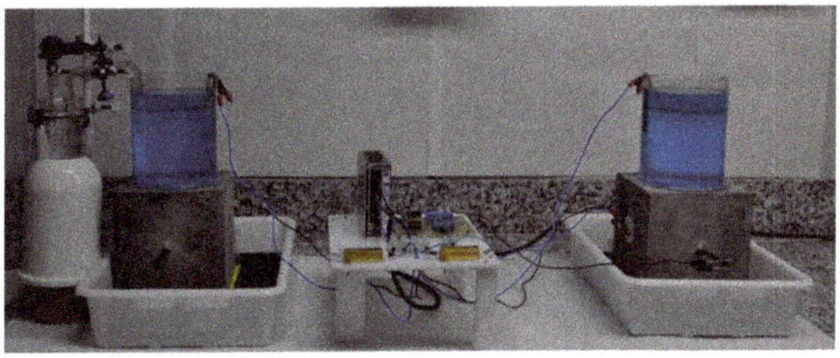

Figure 1. An assembly comprising two samples in parallel is used in the accelerated integral method.

Along with this setup, an instrumental modification is proposed here, which, although it does not affect the electrochemical nature of the assay, would improve the quality of the data for correct interpretation. Although the standard method indicates that a conventional reference electrode (calomel or Ag/AgCl, for instance) can be introduced into the attached electrochemical cell to determine the potential values of the embedded rods, we instead propose to include a fixed electrode along the plane perpendicular to the location of the embedded bar. This arrangement is shown schematically in Figure 2 and can be found for the specimen on the right of Figure 1.

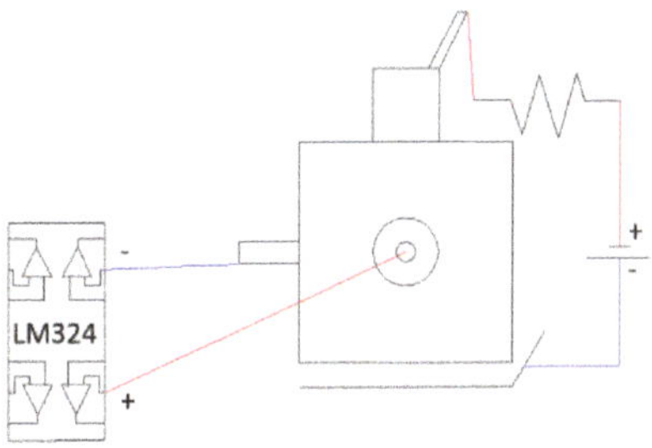

Figure 2. Wiring diagram of the implemented system. An electrode is fixed to the free side of the specimen on a plane perpendicular to that of the embedded bar.

While the standard method indicates a conventional reference electrode (either calomel or Ag/AgCl, etc.) can be introduced in the attached electrochemical cell in order to determine the potential values of the embedded bars, we propose instead to include a fixed electrode along the plane perpendicular to the location of the embedded bar. This arrangement is schematically described in Figure 2, and it can be observed for the specimen on the right in Figure 1. Thus, to measure a reference potential, the negative pole of the voltmeter is connected to the embedded steel, and the positive pole is to the electrode attached to the concrete. In this way, a measurement without fluctuations is obtained by having the constant reference of the external electrode. The two assemblies were set in parallel and with the same power supply potential in order to analyze any eventual differences arising. It was found that a voltage drop occurred after connecting the multimeter to the sample using the standard arrangement of the method. To overcome this problem, and taking into account that the values obtained at the time of connection and just before the voltage drop occurred were identical to those observed with the assembly proposed in the standard, an electrical circuit was included in the assembly, which would allow stabilization of the signal, although without affecting its value. The modification consists of introducing in the measuring clamp a control component powered by an external DC power supply, thus making it possible to measure data at any time without the drawbacks of the aforementioned instabilities.

As can be seen in Figures 2 and 3, the control element is an operational model LM324 which, powered by a TDK-Lambda power supply, model LS75-12, delivers a stable reading. The time drift and characteristic ripple of the power supply were characterized using a Pintek oscilloscope, model PS-605. A 0.1 Ω resistor in series with the potential source allowed measuring the current entering each probe without the need to disconnect the circuit. According to Ohm's law, the intensity of the current circulating through a component of known resistance can be determined if we know the potential that is established across it.

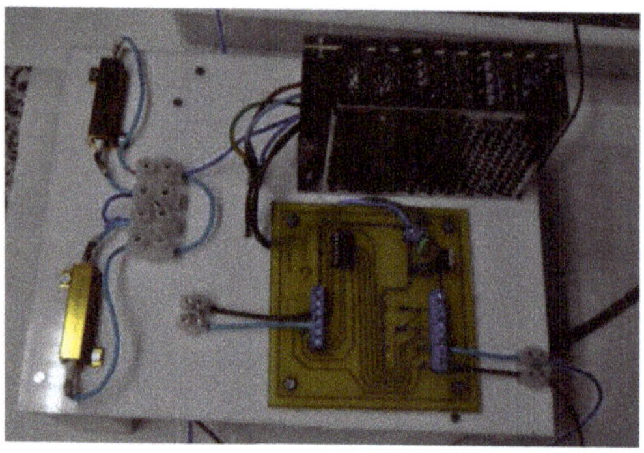

Figure 3. Voltage source and data acquisition system.

2.4. Determination of the Durability

In order to assess the feasibility of the proposed dosages, the durability of the mixtures was determined using the carbonation and the chloride penetration models given in Annex 9 of EHE-08 [82]. Thus, the time required for carbonation at a distance *d* from the concrete surface was determined using the carbonation model:

$$t = \left(\frac{d_p}{K_c}\right)^2 \quad (1)$$

where d_p is the carbonation depth in mm and t is the time in years. The carbonation factor K_c was determined using:

$$K_c = c_{env} \cdot c_{air} \cdot a' \cdot f_{cm}^{b'} \quad (2)$$

where f_{cm} is the compressive strength of the concrete, in N/mm^2, which can be estimated from the specified characteristic strength (f_{ck}) as ($f_{cm} = f_{ck} + 8$). c_{env} is the environmental factor, c_{air} is the air-entraining factor, and a' and b' are parameters which depend on the type of binder (see [82] for more details).

On the other hand, the time required for a certain chloride concentration C_{th} to occur at a distance d from the concrete surface is estimated according to the chloride penetration model by:

$$t = \left(\frac{d_p}{K_{Cl}}\right)^2 \tag{3}$$

where d_p is the depth in mm, and t is the time in years. The chloride penetration coefficient K_{Cl} is given by:

$$K_{Cl} = \alpha\sqrt{12D(t)}\left(1 - \sqrt{\frac{C_{th} - C_b}{C_s - C_b}}\right) \tag{4}$$

where α is a unit conversion factor equal to 56,157; $D(t)$ is the chloride effective diffusion coefficient, for age t, expressed in cm^2/s; C_{th} is the chloride concentration on the surface of the concrete, expressed in % of cement weight; C_b is the content of chloride from materials (aggregates, cement, water, etc.), when the concrete mix is prepared. The diffusion coefficient of chloride varies with the age of the concrete according to:

$$D(t) = D(t_0)\left(\frac{t_0}{t}\right)^n \tag{5}$$

where $D(t_0)$ is the chloride diffusion coefficient at age t_0, $D(t)$ is the coefficient at age t, and n is the age factor. The latter can be taken as equal to 0.5 in the absence of precise values determined by tests on the concrete considered. The value of $D(t_0)$ comes from EHE-08 [82], depending on the type of cement and the w/c ratio.

In order to predict the useful life in the design phase from the calculation of the initiation and propagation models described above, the following considerations were made for the starting data of certain parameters, which were kept fixed in all cases. The considered structure was classified as a tenement, whose minimum required value of the useful life in the project (t_g) is 50 years. This implies a calculated useful life (t_d) of 55 years according to Equation (6):

$$t_d = \gamma_t \cdot t_g \tag{6}$$

The diameter of the steel bar (generic) was taken as Ø = 12 mm. Accordingly, on the basis of the EHE-08, the covers of the steel bar (d_r) would be: (1) 20 mm for an IIa environment, with cement, additions, and a characteristic resistance between 25 and 40 N/mm^2; (2) 25 mm for an IIb environment, with cement, additives and characteristic resistance between 25 and 40 N/mm^2; or (3) 45 mm for an IIIa environment and our type of cement and projected useful life of 50 years. Regarding the environmental coefficient, the most unfavorable condition was selected, which corresponds to an "environment protected from the rain," adopting the value c_{env} = 1. The aeration coefficient takes the value c_{air} = 1 for entrained air below 4.5%. The chloride content provided by the raw materials was set as C_b = 0. Although the critical concentration (C_{th}, given as a percentage by weight of cement) should be 0.6% according to EHE-08, it was 0.4% for the samples tested using the pond method. Other parameters of the tested samples that are necessary for the calculation of the models (i.e., type of concrete, environment, water/cement ratio, and amount of cement) are introduced in the following Sections as employed.

The working life due to the corrosion of reinforcements was determined by adding together the initiation period and the corrosion propagation period, that in the case of corrosion by carbonation, is given by:

$$t_L = \left(\frac{d_p}{K_c}\right)^2 + \frac{80}{\phi}\frac{d_r}{V_{corr}} \tag{7}$$

In the case of corrosion by chlorides, this will be:

$$t_L = \left(\frac{d_P}{K_{Cl}}\right)^2 + \frac{80}{\phi}\frac{d_r}{V_{corr}} \tag{8}$$

3. Results and Discussion

3.1. Characterization of the Components and the Concrete

Table 3 shows the particle size distribution of the tuff and quarry pozzolans used as additives in this study, whereas Figure 4 shows their particle size distributions. In addition, the XRF analysis of the PZT is given in Table 4. Next, Table 5 summarizes the most relevant characteristics of the aggregates used.

Table 3. Particle size analysis of additives by wet sieving.

	Tuff Pozzolan	Quarry Pozzolan
UNE Sieves	% Passing Through	
0.158	16.02	16.02
0.112	14.38	14.38
0.080	12.75	12.75
0.051	11.11	11.11
0.027	7.85	7.85
0.021	6.21	6.21
0.015	6.21	6.21
0.011	6.21	6.21
0.008	4.58	4.58
0.003	2.94	2.94

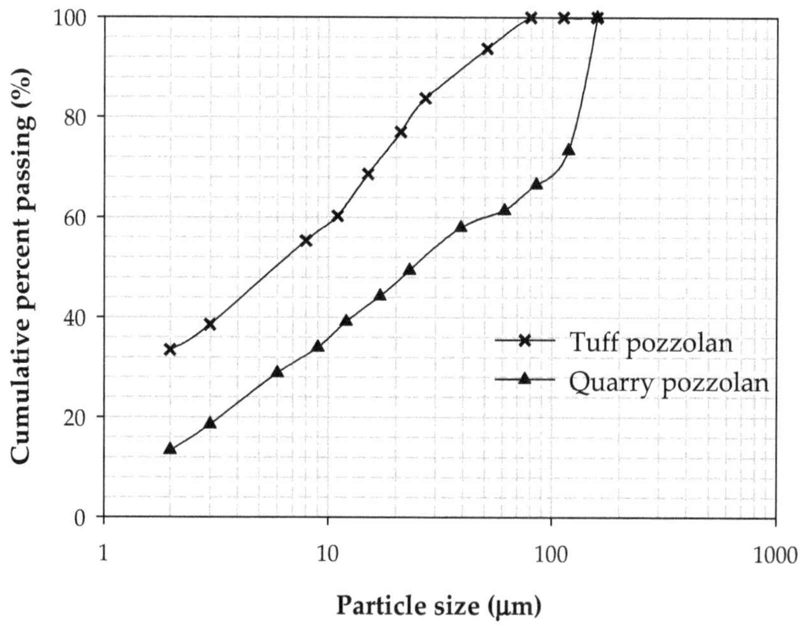

Figure 4. Particle size distribution of the used tuff and quarry pozzolans.

Table 4. XRF analysis of the tuff pozzolan.

SiO_2	Al_2O_3	Fe_2O_3	CaO	MgO	Na_2O	K_2O	MnO	P_2O_5	TiO_2	LOI	SO_4
50.19	17.35	3.08	1.45	1.31	6.42	4.94	0.15	0.10	0.64	14.38	0.21

Table 5. Origin, size, and absorption coefficient of the aggregates.

Aggregate	Origin	Size	Absorption Coefficient
Sand	Siliceous	0/2	1.50
Sand	Basaltic	0/4	3.95
Gravel	Basaltic	10/20	3.65

Particle size distribution of the used tuff pozzolan and quarry pozzolan is shown in Figure 4. It can be seen that the tuff pozzolan had finer particles than the quarry pozzolan used in this study. The median particle size (50% passing size) of tuff pozzolan was almost 6 μm, while quarry pozzolan showed a particle size of 24 μm.

Table 6 presents the results of the flexural strength, compressive strength, and resistance activity index [58] of the additions with the reference cement. According to the test method of the UNE EN 196-1 standard [63], with a 75/25 mixture by weight, the compressive strength determined after 28 days must be equal to or greater than 75% of that corresponding to the standard cement I 42.5 R/S.

Table 6. Flexural strength, compressive strength, and resistance activity index of the additives in the reference cement.

Sample	Aggregate	Flexural Strength (N/mm^2)	Compressive Strength (N/mm^2)	RAI (%)
Cement (I/42.5 R/SR)	Reference	10.1	5.4	100
75% Cement + 25% PZT	Tuff pozzolan	9.1	5.4	100
75% Cement + 25% PZQ	Quarry pozzolan	8.7	4.8	88.9

The reference samples, as well as the various additions, were characterized by means of the tests indicated in Section 2.3. Table 7 presents the results both from the tests carried out and from the calculation processes described in the standard that governs each of the tests

The chemical analysis carried out for PZQ, according to ASTM C618 [83], shows a $SiO_2 + Al_2O_3 + Fe_2O_3$ composition of 70.62% (type N pozzolan). This value is greater than the 70% set by the standard. Furthermore, the SO_3 value is below 3%. On the other hand, the LOI value is greater than the 12% set by the standard. This abnormal value is justified by the late geochemical alteration of the sample due to corrosion, in particular of the pumice stone in the simple. In fact, analysis of samples taken near the quarry from which the material was obtained showed LOI values below 4% [47]. Both types of pozzolans showed pozzolanic activity according to the lime pozzolanic activity index (PAI). The results of the cement mixtures with the additions of PZT and PZQ, expressed in mmol/L, are consistent with the pozzolanism test (see Figure 5). The $[Ca(OH)_2]/[OH^-]$ ratios at 7 days are 6.03/60.19 for PZQ and 1.78/92.78 PZT, with values under the saturation curve.

The additions of PZT and PZQ give concretes whose strengths correspond well to those required for concrete HA-25 (dosages D4) and HA-30 (dosages D3). When these are added to replace superfine aggregate, the pozzolan tuff shows increases of 22% and pozzolan quarry of 5% over the reference mix. When added as a replacement for cement, they cause a small decrease in resistance. All mixes exceed the required minimum characteristic strengths of 25 N/mm^2 and 30 N/mm^2, respectively. The dosages with the two pozzolans far exceed the limit of 75% of the resistant activity index with respect to the reference cement according to the UNE EN 196-1 standard. These results are in agreement with similar studies [1,47].

Table 7. Experimental results of the tests made on the concrete mixtures.

Code	D1P	D3P	D3PZT	D3PZQ	D4P	D4PZT	D4PZQ
Type of aggregates	53	52.5	54.7	54.7	52.5	54.7	54.7
Consistency (cm)	4	15	15	19	10	11.5	10.5
Fresh density (kg/m^3)	2399	2475	2379	2423	2335	2365	2365
Density after 24 h (kg/m^3)	2421	2449	2356	2408	2320	2343	2353
Density after 28 d (kg/m^3)	2428	2468	2391	2437	2339	2364	2374
Dry density (kg/m^3)	2324	2359	2255	2282	2110	2269	2278
Accumulated porosity (%)	14.33	15.5	18.4	20.34	22.5	15.59	15.4
RCS of concrete (N/mm^2)	57	52.79	51.34	50	38.14	46.54	43.72
RIM after 28 d (Ω m)	115.4	109.9	106.6	79.4	107.5	120	105.3
RIC after 28 d (Ω m)	109.5	111.9	104.1	80.7	99.8	120.6	95.3
RID after 28 d (Ω m)	112	121.8	109.8	86.8	104.8	118.3	103.1
Natural V_{CO2} (mm/year$^{\frac{1}{2}}$)	0	1.41	1.41	1.54	3.15	2.4	2.81
Accelerated V_{CO2} (mm/year$^{\frac{1}{2}}$)	0	7.55	14.05	15.6	17.67	14.64	16.48
Profile X_{Cl} (mm)	4.2	11.8	13.5	13	13.8	13	14.8
Profile V_{Cl} (mm/year$^{\frac{1}{2}}$)	8.46	23.66	27.04	26.04	27.85	26.18	29.7
Pond, $D_{ns} \times 10^{-8}$ (cm^2/s)	0.6329	4.1	6.363	7.789	7.205	4.492	5.841
Multi-regime, $D_{ns} \times 10^{-8}$ (cm^2/s)	3.0516	3.5094	2.3337	2.8636	4.8058	2.3694	2.4799
Initial/final intensity (mA)			5.2/9.45	5.72/15.71		4.89/8.33	5.62/10.17
Initial/final resistance (Ω)			2.307/1.270	2.098/0.764		2.045/1.440	2.135/1.180
Integral, $D_{ns} \times 10^{-8}$ (cm^2/s)			4.019	3.191		4.019	6.028
Depassivation time (s)			2,332,800	2,937,600		2,332,800	1,555,200

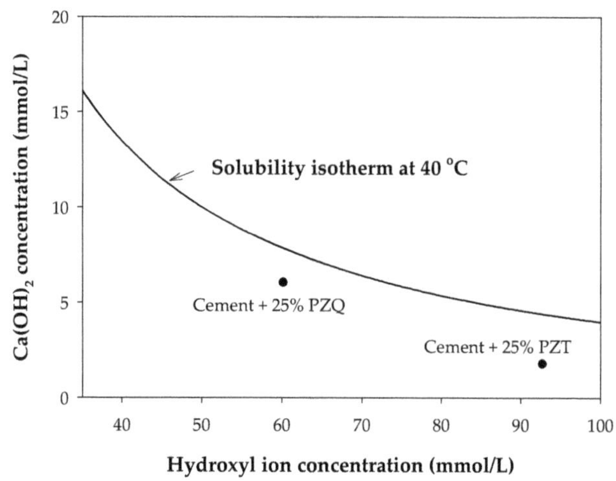

Figure 5. Variation of the pozzolanic behavior of the samples analyzed at 7 days.

The additions cause a change in the consistency of the concrete, making it more fluid. Thus, according to the classification of the Abrams cone, it varies from plastic and liquid (between 3 and >16 cm) to the fluid (10 to 15 cm) and liquid (≥16 cm) range. The density values ranged between 2100 and 2500 kg/m^3. In general, a decrease in the density is observed 24 h after demolding compared to fresh, followed by an increase in the density of the saturated specimens after 28 days of curing in a humid chamber, and finally, exhibits an expected decrease in the dry density, which varies between 3 and 10%. All of these increases and decreases in density are due to the increase or decrease in the amount of water in the sample.

The additions cause an appreciable change in the porosity of the samples. In the D3 dosages, with a lower water/cement ratio, the additions cause an increase in porosity that ranges from 31.23% in the case of tuff pozzolan to 18.71% in the case of quarry pozzolan. With regard to the D4 dosage, by maintaining the amount of cement and introducing the additions as superfine aggregate, the porosities were reduced in both cases by approximately 31%, thus improving the properties of the concrete. This variation in porosity complies with the results of water penetration required by EHE-08 for the durability control of concrete exposed in environments III and IV. In D3 concretes, water penetration results equal to or greater than the reference dosage are obtained, while for D4 dosages, in which the percentage of addition to the cement has not been subtracted, the results are not maintained but improved compared to the reference. The addition of pozzolan in the dosages with the highest water/cement ratio (i.e., 0.73%) effectively slows down the penetration of water.

It is also observed that the resistivity values obtained by the direct and the indirect (four-point) methods are close to 100 Ω m in all the batches. Pozzolanic additions cause a slight decrease in resistivity value when added to replace cement. However, when replacing the superfine aggregate, the tuff pozzolan causes a small increase in the resistivity values, while the quarry pozzolan causes almost no change in the measurement.

3.2. *Modified Integral Method*

As mentioned above, an instrumental modification was made to the integral method described in Section 2.3. From the measurement of the power supply ripple, a maximum of 15 mV AC was obtained at the 12 V DC output of the power supply. Note that, in the characteristics provided by the manufacturer, a maximum ripple of 120 mV is fixed for this model. On the other hand, we proceeded to obtain the drift over time of the source by performing two measurements at different times. The first was performed at time 0, which gave an output of 12.00 VDC. After 90 min, a second measurement was made, obtaining a value of 12.01 V on the screen, to obtain a value of 12.12 V after a month of testing.

After a month of testing had passed, the concrete specimen used in the experimental device validation experiment was broken. A colorimetric technique was used to determine the chloride penetration. By adding a solution of silver nitrate (0.1 M) and the corresponding reaction of the compound with Cl^- ions, it was possible to see with the naked eye a whitish coloration when the two species came into contact. It must be noted that over time, this color turned purple until it almost disappeared 2 weeks after its development. As can be seen in Figure 6, the most unfavorable penetration of the chloride species reached a depth of 0.5 cm. Table 7 shows the values of the diffusion coefficients, as well as the time until the depassivation of the different specimens.

Figure 6. Chloride penetration profile revealed. The chloride penetration front can be seen in the upper part of the broken sample.

3.3. Penetration of Aggressive Species

Regarding the natural carbonation tests, among the results obtained, dosage 1 (D1P) stands out from the rest because it was not affected by the aggressive process, unlike the rest. In the case of the evolution of the carbonation process, by adding the pozzolans to the detriment of the cement, the natural process is hardly affected, but in the case of the accelerated tests, the carbonation fronts increase by 86.20% when adding tuff pozzolan and 106.79% in the case of quarry pozzolan (cf. Table 7). However, by subtracting superfine aggregate (dosages D4), the samples with additions show penetration rate values lower than the reference (D4P) both for the natural test (namely, −23.77 for D4PZT and −10.80 for D4PZQ) and for the accelerated test (−17.13 for D4PZT, and −6.77% for D4PZQ).

When evaluating the penetration of chlorides, it is found that the non-steady state diffusion coefficients are all of the same order of magnitude, regardless of the method with which they were obtained. The multi-regime method is the one that presents the lowest values of non-stationary diffusion coefficients compared to the other two methods. The reference sample D1P is the one that exhibits the lowest value of the non-stationary diffusivity coefficient. It is found that the additions of pozzolans to the detriment of the superfine aggregate lead to an improvement in the behavior of the dosages with respect to the penetration of chloride ions. When introduced in the part of cement, the dosages improve compared to the reference when analyzed with the multi-regime method but not when analyzed by the pond method, where mixtures containing quarry pozzolan show an increase in 9% in chloride penetration compared to the reference dosage, and of 5.5% in the case of tuff pozzolan. The chloride profiles follow a linear trend when calculated as a percentage of the sample weight with respect to the depth of penetration (mm), with regression indices R^2 ranging from 0.891 (in the case of D2P) to 0.994 (for D4PZT).

According to the multi-regime method, mixes dosed with tuff pozzolan and quarry pozzolan have lower diffusion coefficients than their respective reference mixtures, whether they are introduced as either a cement or a superfine aggregate substitute. The greatest decreases in the coefficients were presented by the mixtures containing PZT, between 33 and 50%, and those containing PZQ, between 18 and 48%. These results coincide with those observed by other authors [17,84–88].

Finally, the modified integral method gives values similar to those obtained by the multi-regime method, with the advantage that this method provides information on the depassivation time of the reinforcement. A comparison with the reference dosages cannot be made as these have not been tested.

3.4. Estimation of Durability

Unfortunately, not all existing cement, according to RC-08 [89], are listed in the Durability Annex of the EHE-08 for the carbonation model. This was the case with the cement II/A-P/45.5 R/MR used in the tested samples. For this reason, this cement had to be considered in the model as Portland cement binder or Portland cement + 28% fly ash. On the other hand, an experimental comparison was made to observe the difference between using $f_{ck} + 8$ as the characteristic strength or, alternatively, the average compressive strength of cement, f_{cm}. Table 8 shows the results with the raw data provided by the EHE-08, whereas Table 9 shows the results obtained from the tests carried out. For the resistance data variable, the average compressive strength f_{cm} of each set was taken, and three options were considered to catalog the type of binder, namely, Portland Cement, Portland Cement + 28% fly ash, and Portland Cement + 9% silica fume.

To estimate the durability by applying the chloride penetration model (Equations (3) and (4)) and the data provided by EHE-08, the diffusion coefficient, the surface, and the critical chloride concentrations were introduced, together with the chloride penetration data obtained by means of the pond method. Table 10 shows the results obtained with the data provided by the EHE-08 when considering $D(t_0)$ for a CEM II/A-V cement according to its water/cement ratio. Therefore, the value of $D(t)$ is determined for t = 50 years (Equation (5)), while t_0 is taken at 28 days (i.e., 0.0767 years). For the calculation of K_{CL}, the parameter C_{th} by weight of cement is 0.6%, whereas the concentration of chlorides on the surface in weight of cement, C_s, was taken in all cases as the value corresponding to an IIIa environment (up to 500 m from the sea the shore) of 0.14%, although in reality only the reference sample D1 was manufactured for the IIIa environment. The lifetime values (t_L) thus determined were of the same order as those obtained using the carbonation model. All three dosages were found to be well above the design value (t_d) set at 55 years, confirming that the reference doses are adequate for use in real-world applications. It also predicts that dose D3 (HA-30/F/20/IIb) is the most durable according to the model, followed by D1 (HA-35/P/20/IIIa), although it should be noted that the V_{corr} of the latter is always higher due to the assumption of operation in an IIIa environment.

Table 8. Application of the carbonation model for the general calculation method for the useful life. Carbonation model according to EHE-08 data.

Sample	$f_{ck}+8$ (N/mm^2)	K_c	t_i (Year)	d_r (mm)	V_{corr} (μm/Year)	t_p (Year)	t_L (Year) ($t_L = t_i + t_p$)
		$f_{ck}+8$, and binder coefficients, $a' = 360$ and $b' = -1.2$					
D1	43	3.95	57.80	45	20	15	72.80
D3	38	4.58	42.97	25	2	83.33	126.30
D4	33	5.42	30.62	20	3	44.44	75.07
D1	43	3.95	57.80	45	20	15	72.80
		$f_{ck}+8$, and binder coefficients, $a' = 1800$ and $b' = -1.7$					
D1	43	3.01	99.42	45	20	15	114.42
D3	38	3.71	65.31	25	2	83.33	148.64
D4	33	4.72	40.42	20	3	44.44	84.87

Table 9. Application of the carbonation model for the general calculation method of the useful life. These values were obtained for each type of conglomerate using data extracted from the batches, and the average resistance of the concretes (f_{cm}).

Sample	f_{cm} (N/mm^2)	K_c	t_i (Year)	d_r (mm)	V_{corr} (μm/Year)	t_p (Years)	t_L (Year) ($t_L = t_i + t_p$)
		Type of binder, Portland cement ($a' = 1800$ and $b' = -1.7$)					
D1P	57.00	1.86	259.22	45	20	15	274.22
D3P	52.79	2.12	199.70	25	2	83.33	283.03
D3PZT	51.34	2.23	181.66	25	2	83.33	264.99
D3PZQ	50.00	2.33	166.03	25	2	83.33	249.37
D4P	38.14	3.69	66.13	20	3	44.44	110.57
D4PZT	46.54	2.63	130.11	20	3	44.44	174.55
D4PZQ	43.72	2.92	105.20	20	3	44.44	149.64
		Type of binder, Portland cement + 28% fly ash ($a' = 360$ and $b' = -1.2$)					
D1P	57	2.81	113.69	45	20	15	128.69
D3P	52.79	3.08	94.57	25	2	83.33	177.91
D3PZT	51.34	3.19	88.46	25	2	83.33	171.79
D3PZQ	50	3.29	83.02	25	2	83.33	166.35
D4P	38.14	4.56	43.35	20	3	44.44	87.79
D4PZT	46.54	3.59	69.89	20	3	44.44	114.34
D4PZQ	43.72	3.87	60.15	20	3	44.44	104.60

Table 11 presents the data for the calculation of the useful life of all the mixed samples, with the data provided by the EHE-08 and the value of $D(t_0)$ taken from each of the samples tested for chloride penetration from the pond method. For the derivation of $D(t)$, t was set as 50 years, and t_0 as 90 days (0.2466 years); and for the calculation of the chloride penetration coefficient (K_{CL}), the critical concentration (C_{th}) by weight of cement was 0.4%, and the concentration of chlorides at the surface by the weight of cement (C_s) corresponded to that of each sample.

Table 10. Application of the chloride penetration model for the general calculation method of the useful life according to data included in the EHE-08.

Sample	$D(t_0) \times 10^{-7}$ (cm^2/s)	$D(t = 50) \times 10^{-9}$ (cm^2/s)	K_{CL} (mm/Year$^{\frac{1}{2}}$)	t_i (Year)	d_r (mm)	V_{corr} (μm/Year)	t_p (Year)	t_L (Year) ($t_L = t_i + t_p$)
D1P	0.90	3.53	2.56	137.14	45	20	15.00	152.14
D3P	1.09	4.27	3.21	87.49	25	2	83.33	170.83
D4P	1.49	5.84	3.75	64.00	20	3	44.44	108.45

Table 11. Application of the chloride penetration model for the general calculation method of the useful life using data obtained from the pond immersion test.

Sample	$D(t_0) \times 10^{-8}$ (cm^2/s)	$D(t = 50) \times 10^{-9}$ (cm^2/s)	K_{CL} (mm/Year$^{\frac{1}{2}}$)	t_i (Year)	d_r (mm)	V_{corr} (μm/Year)	t_p (Year)	t_L (Year) ($t_L = t_i + t_p$)
D1P	0.63	0.45	2.15	194.47	45	20	15.00	209.47
D3P	4.10	2.88	6.35	22.32	25	2	83.33	105.66
D3PZT	6.36	4.47	7.49	16.06	25	2	83.33	99.39
D3PZQ	4.88	3.43	6.86	19.13	25	2	83.33	102.46
D4P	7.21	5.06	8.78	11.68	20	3	44.44	56.12
D4PZT	4.49	3.15	6.82	19.34	20	3	44.44	63.79
D4PZQ	5.84	4.10	7.72	15.12	20	3	44.44	59.56

By applying the durability model for corrosion processes considering carbonation as the relevant process determining the initiation period and using the data provided in the EHE-08, it is observed that the results greatly exceed the design value (t_d) established in 55 years (cf. Table 8). The three dosage types are suitable for use in fulfilling the conditions of the proposed construction element, being the dosage D3 (HA-30/20/F/IIb), the most durable according to the model, followed by D1 (HA-35/20/P /IIIa). This fact has been observed by other authors [13,33,35].

When introducing the test results of the dosages in the carbonation model (see Table 9), it is observed that the additions of tuff and quarry pozzolans cause an increase in the useful lifetime (t_L) which greatly exceeds the design value (t_d), of 55 years, as well as the value obtained for the reference mixes. Regarding the approach of cataloging the binder in the different batches, it seems that the choice of Portland cement +28% ash in the modeling is the most conservative, although it gives higher values than those of the useful life calculation.

When applying the chloride penetration model for the estimation of durability, the estimated useful life (t_L) values for the reference batches greatly exceed the design value (t_d) set in 55 years, which confirms that the reference dosages are suitable for the use and conditions of the proposed building element. Here again, dosage D3 (HA-30/F/20/IIb) appears to be the most durable according to the model, followed by D1 (HA-35/P/20/IIIa), although it should be clarified that the V_{corr} of the latter is always superior due to the premise of being an IIIa environment.

During the introduction of the pozzolanic additions, we observe that for the case of the D1P sample, the results obtained in the prediction from the experimental data largely exceed the estimate made by the EHE-08, while the additions decrease the service life considerably, although all exceed the design value of the useful life (i.e., t_d = 55 years).

4. Conclusions

All laboratory concrete mixes, both reference and added with tuff or quarry pozzolans, whether used as a substitute for cement or for the superfine aggregate in a percentage of 20%, exceed the minimum characteristic resistance and the durability values required by EHE-08 on the basis of the carbonation and chloride penetration models. Both types of pozzolans showed pozzolanic activity according to the lime pozzolanic activity index (PAI) with $[Ca(OH)_2]/[OH^-]$ ratios at 7 days are 6.03/60.19 for PZQ, and 1.78/92.78 for PZT.

In general, when replacing the cement, the additions cause an increase in porosity, a decrease in resistivity, and an increase in the depth of the carbonation front, while when replacing the superfine aggregates, they cause a decrease in porosity and a decrease in the carbonation front. The added mixes change their consistency with respect to the reference, so the introduction of the pozzolans influences the consistency of the mixes, making them more fluid.

The non-stationary diffusion coefficients obtained by the three methods are of the same order of magnitude, the highest values being those obtained by the pond method, followed by the integral method, and finally by the multi-regime method. In the pond immersion test, when the pozzolans replaced the cement, an increase in the penetration of chlorides compared to the reference dosage was observed. In the rest of the cases, a lesser diffusion of chlorides was observed when pozzolans replaced cement or superfine aggregates. There were decreases between 33 and 50% for the diffusion coefficients in the cementitious mixtures containing PZT, while those containing PZQ showed a more moderate decrease which varied between 18 and 48%.

Author Contributions: Conceptualization, J.J.S.; investigation, N.R.-B., C.B.-P. and V.F.M.; data curation, J.J.S. and N.R.-B.; methodology, J.J.S. and R.M.S.; supervision, J.J.S. and R.M.S.; writing—original draft preparation, J.J.S. and R.M.S.; writing—review and editing, J.J.S. and R.M.S.; visualization, J.J.S. and R.M.S.; funding acquisition, R.M.S. All authors have read and agreed to the published version of the manuscript.

Funding: The work was supported by the University of La Laguna and the Spanish Ministry of Science, Innovation, and Universities (Madrid, Spain) under contract No. 2022/0000586.

Institutional Review Board Statement: Not applicable.

Informed Consent Statement: Not applicable.

Data Availability Statement: The raw/processed data required to reproduce these findings cannot be shared at this time as the data also forms part of an ongoing study.

Conflicts of Interest: The authors declare no conflict of interest.

References

1. Najimi, M.; Jamshidi, M.; Pourkhorshidi, A. Durability of concretes containing natural pozzolan. *Proc. Inst. Civ. Eng. Constr. Mater.* **2008**, *161*, 113–118. [CrossRef]
2. Concrete needs to lose its colossal carbon footprint. *Nature* **2021**, *597*, 593–594. [CrossRef] [PubMed]
3. Miller, S.A.; Horvath, A.; Monteiro, P.J.M. Readily implementable techniques can cut annual CO_2 emissions from the production of concrete by over 20%. *Environ. Res. Lett.* **2016**, *11*, 074029. [CrossRef]
4. Fan, C.; Miller, S.A. Reducing greenhouse gas emissions for prescribed concrete compressive strength. *Constr. Build. Mater.* **2018**, *167*, 918–928. [CrossRef]
5. Tayeh, B.A.; Hamada, H.M.; Almeshal, I.; Bakar, B.H.A. Durability and mechanical properties of cement concrete comprising pozzolanic materials with alkali-activated binder: A comprehensive review. *Case Stud. Constr. Mater.* **2022**, *17*, e01429. [CrossRef]
6. Abiodun, Y.O.; Olanrewaju, O.A.; Gbenebor, O.P.; Ochulor, E.F.; Obasa, D.V.; Adeosun, S.O. Cutting cement industry CO_2 emissions through metakaolin use in construction. *Atmosphere* **2022**, *13*, 1494. [CrossRef]
7. Amran, M.; Makul, N.; Fediuk, R.; Lee, Y.H.; Vatin, N.I.; Lee, Y.Y.; Mohammed, K. Global carbon recoverability experiences from the cement industry. *Case Stud. Constr. Mater.* **2022**, *17*, e01439. [CrossRef]
8. Andrew, R.M. Global CO_2 Emissions from cement production, 1928–2018. *Earth Syst. Sci. Data* **2019**, *11*, 1675–1710. [CrossRef]
9. Qaidi, S.; Najm, H.M.; Abed, S.M.; Ahmed, H.U.; Al Dughaishi, H.; Al Lawati, J.; Sabri, M.M.; Alkhatib, F.; Milad, A. Fly ash-based geopolymer composites: A review of the compressive strength and microstructure analysis. *Materials* **2022**, *15*, 7098. [CrossRef] [PubMed]
10. Ababneh, A.; Matalkah, F. Potential use of jordanian volcanic tuffs as supplementary cementitious materials. *Case Stud. Constr. Mater.* **2018**, *8*, 193–202. [CrossRef]
11. Costafreda, J.L.; Martín, D.A.; Presa, L.; Parra, J.L. Altered volcanic tuffs from Los Frailes Caldera. A study of their pozzolanic properties. *Molecules* **2021**, *26*, 5348. [CrossRef]
12. Rakhimov, R.Z.; Rakhimova, N.R.; Gaifullin, A.R.; Morozov, V.P. Properties of portland cement pastes enriched with addition of calcined marl. *J. Build. Eng.* **2017**, *11*, 30–36. [CrossRef]

13. Sobol, K.; Markiv, T.; Terlyha, V.; Franus, W. Peculiarities of hydration processes of cements containing natural zeolite. *Bud. Archit.* **2015**, *14*, 105–113. [CrossRef]
14. Xia, J.; Guan, Q.; Zhou, Y.; Wang, J.; Gao, C.; He, Y.; Wang, Z.; Song, P. Use of natural pozzolans in high-performance concrete for the Mombasa–Nairobi railway. *Adv. Cem. Res.* **2021**, *33*, 318–330. [CrossRef]
15. Papadakis, V.G.; Tsimas, S. Supplementary cementing materials in concrete. Part I: Efficiency and design. *Cem. Concr. Res.* **2002**, *32*, 1525–1532. [CrossRef]
16. Wasim, M.; Abadel, A.; Abu Bakar, B.H.; Alshaikh, I.M.H. Future directions for the application of zero carbon concrete in civil engineering—A review. *Case Stud. Constr. Mater.* **2022**, *17*, e01318. [CrossRef]
17. Fazilati, M.; Mohammadi Golafshani, E. Durability properties of concrete containing amorphous silicate tuff as a type of natural cementitious material. *Constr. Build. Mater.* **2020**, *230*, 117087. [CrossRef]
18. Khan, K.; Amin, M.N. Influence of fineness of volcanic ash and its blends with quarry dust and slag on compressive strength of mortar under different curing temperatures. *Constr. Build. Mater.* **2017**, *154*, 514–528. [CrossRef]
19. Khan, K.; Amin, M.N.; Saleem, M.U.; Qureshi, H.J.; Al-Faiad, M.A.; Qadir, M.G. Effect of fineness of basaltic volcanic ash on pozzolanic reactivity, ASR expansion and drying shrinkage of blended cement mortars. *Materials* **2019**, *12*, 2603. [CrossRef] [PubMed]
20. Khan, K.; Amin, M.N.; Usman, M.; Imran, M.; Al-Faiad, M.A.; Shalabi, F.I. Effect of fineness and heat treatment on the pozzolanic activity of natural volcanic ash for its utilization as supplementary cementitious materials. *Crystals* **2022**, *12*, 302. [CrossRef]
21. Kushnir, A.R.L.; Heap, M.J.; Griffiths, L.; Wadsworth, F.B.; Langella, A.; Baud, P.; Reuschlé, T.; Kendrick, J.E.; Utley, J.E.P. The fire resistance of high-strength concrete containing natural zeolites. *Cem. Concr. Compos.* **2021**, *116*, 103897. [CrossRef]
22. Robayo-Salazar, R.A.; de Gutiérrez, M.; Puertas, F. Study of synergy between a natural volcanic pozzolan and a granulated blast furnace slag in the production of geopolymeric pastes and mortars. *Constr. Build. Mater.* **2017**, *157*, 151–160. [CrossRef]
23. Takahashi, T.; Shoji, S. Distribution and classification of volcanic ash soils. *Glob. Environ. Res.* **2002**, *6*, 83–97.
24. Lemougna, P.N.; MacKenzie, K.J.D.; Melo, U.F.C. Synthesis and thermal properties of inorganic polymers (geopolymers) for structural and refractory applications from volcanic ash. *Ceram. Int.* **2011**, *37*, 3011–3018. [CrossRef]
25. Kouamo Tchakoute, H.; Elimbi, A.; Diffo Kenne, B.B.; Mbey, J.A.; Njopwouo, D. Synthesis of geopolymers from volcanic ash via the alkaline fusion method: Effect of Al_2O_3/Na_2O molar ratio of soda–volcanic ash. *Ceram. Int.* **2013**, *39*, 269–276. [CrossRef]
26. Mertens, G.; Snellings, R.; Van Balen, K.; Bicer-Simsir, B.; Verlooy, P.; Elsen, J. Pozzolanic reactions of common natural zeolites with lime and parameters affecting their reactivity. *Cem. Concr. Res.* **2009**, *39*, 233–240. [CrossRef]
27. Kitsopoulos, K.P.; Dunham, A.C. Heulandite and mordenite-rich tuffs from Greece: A potential source for pozzolanic materials. *Miner. Depos.* **1996**, *31*, 576–583. [CrossRef]
28. Martínez-Ramírez, S.; Blanco-Varela, M.T.; Ereña, I.; Gener, M. Pozzolanic reactivity of zeolitic rocks from two different Cuban Deposits: Characterization of reaction products. *Appl. Clay Sci.* **2006**, *32*, 40–52. [CrossRef]
29. Fragoulis, D.; Chaniotakis, E.; Stamatakis, M.G. Zeolitic tuffs of Kimolos Island, Aegean Sea, Greece and their industrial potential. *Cem. Concr. Res.* **1997**, *27*, 889–905. [CrossRef]
30. Varela, M.T.B.; Ramírez, S.M.; Ereña, I.; Gener, M.; Carmona, P. Characterization and pozzolanicity of zeolitic rocks from two Cuban deposits. *Appl. Clay Sci.* **2006**, *33*, 149–159.
31. Liebig, E.; Althaus, E. Pozzolanic activity of volcanic tuff and Suevite: Effects of calcination. *Cem. Concr. Res.* **1998**, *28*, 567–575. [CrossRef]
32. Zhang, M.H.; Malhotra, V.M. Characteristics of a thermally activated alumino-silicate pozzolanic material and its use in concrete. *Cem. Concr. Res.* **1995**, *25*, 1713–1725. [CrossRef]
33. Çavdar, A.; Yetgin, Ş. Availability of tuffs from Northeast of Turkey as natural pozzolan on cement, some chemical and mechanical relationships. *Constr. Build. Mater.* **2007**, *21*, 2066–2071. [CrossRef]
34. Hunyak, O.; Sobol, K.; Markiv, T.; Bidos, V. The effect of natural pozzolans on properties of vibropressed interlocking concrete blocks in different curing conditions. *Prod. Eng. Arch.* **2019**, *22*, 3–6. [CrossRef]
35. Markiv, T.; Sobol, K.; Franus, M.; Franus, W. Mechanical and durability properties of concretes incorporating natural zeolite. *Arch. Civ. Mech. Eng.* **2016**, *16*, 554–562. [CrossRef]
36. Dedeloudis, C.; Zervaki, M.; Sideris, K.; Juenger, M.; Alderete, N.; Kamali-Bernard, S.; Villagrán, Y.; Snellings, R. Natural pozzolans. *RILEM State-Art Rep.* **2018**, *25*, 181–231.
37. Rojas, M.F.; Cabrera, J. The Effect of temperature on the hydration rate and stability of the hydration phases of Metakaolin–Lime–Water systems. *Cem. Concr. Res.* **2002**, *32*, 133–138. [CrossRef]
38. Shi, C. Pozzolanic reaction and microstructure of chemical activated lime-fly ash pastes. *ACI Mater. J.* **1998**, *95*, 537–545.
39. Janotka, I.; Krajči, L. Utilization of natural zeolite in Portland cement of increased sulphate resistance. *ACI Spec. Publ.* **2003**, *221*, 223–229.
40. Ahmadi, B.; Shekarchi, M. Use of natural zeolite as a supplementary cementitious material. *Cem. Concr. Compos.* **2010**, *32*, 134–141. [CrossRef]
41. Mostafa, N.Y.; El-Hemaly, S.A.S.; Al-Wakeel, E.I.; El-Korashy, S.A.; Brown, P.W. Characterization and evaluation of the pozzolanic activity of Egyptian industrial by-products: I: Silica fume and dealuminated Kaolin. *Cem. Concr. Res.* **2001**, *31*, 467–474. [CrossRef]
42. Çullu, M.; Bolat, H.; Vural, A.; Tuncer, E. Investigation of pozzolanic activity of volcanic rocks from the Northeast of the Black Sea. *Sci. Eng. Compos. Mater.* **2016**, *23*, 315–323. [CrossRef]

43. Turanli, L.; Uzal, B.; Bektas, F. Effect of large amounts of natural pozzolan addition on properties of blended cements. *Cem. Concr. Res.* **2005**, *35*, 1106–1111. [CrossRef]
44. Hu, J.Y.; Zhang, S.S.; Chen, E.; Li, W.G. A review on corrosion detection and protection of existing reinforced concrete (RC) structures. *Constr. Build. Mater.* **2022**, *325*, 126718. [CrossRef]
45. Bouteiller, V.; Tissier, Y.; Marie-Victoire, E.; Chaussadent, T.; Joiret, S. The application of electrochemical chloride extraction to reinforced concrete—A review. *Constr. Build. Mater.* **2022**, *351*, 128931. [CrossRef]
46. Syll, A.S.; Kanakubo, T. Impact of corrosion on the bond strength between concrete and rebar: A systematic review. *Materials* **2022**, *15*, 7016. [CrossRef]
47. Boschmann Käthler, C.; Poulsen, S.L.; Sørensen, H.E.; Angst, U.M. Investigations of accelerated methods for determination of chloride threshold for reinforcement corrosion in concrete. *Sustain. Resilient Infrastruct.* **2021**, *6*, 1–12. [CrossRef]
48. Kaid, N.; Cyr, M.; Julien, S.; Khelafi, H. Durability of Concrete containing a natural pozzolan as defined by a performance-based approach. *Constr. Build. Mater.* **2009**, *23*, 3457–3467. [CrossRef]
49. Fajardo, G.; Valdez, P.; Pacheco, J. Corrosion of steel rebar embedded in natural pozzolan based mortars exposed to chlorides. *Constr. Build. Mater.* **2009**, *23*, 768–774. [CrossRef]
50. Andrade, C.; Arriaga, L.Q. *Cursos Avanzados Eduardo Torroja: Durabilidad, Rehabilitación y Sostenibilidad*; Instituto de Ciencias de la Construcción Ediuardo Torroja: Madrid, Spain, 2014.
51. Valenta, O. General analysis of the methods of testing the durability of concrete. *Mater. Struct.* **1970**, *3*, 333–345.
52. Fagerlund, G. *Prediction of the Service Life of Concrete Exposed to Frost Action*; Swedish Cement and Concrete Research Institute: Stockholm, Sweden, 1979.
53. Tuutti, K. *Corrosion of Steel in Concrete*; Swedish Cement and Concrete Research Institute: Stockholm, Sweden, 1982.
54. Real Decreto 1247/2008, de 18 de Julio, por el que se aprueba la instrucción de hormigón estructural (EHE-08). 2008. Available online: https://www.boe.es/eli/es/rd/2008/07/18/1247 (accessed on 10 November 2022).
55. Hernández Gutiérrez, L.E. *Caracterización Geomecánica de las Rocas Volcánicas de las Islas Canarias*; Universidad de La Laguna: La Laguna, Spain, 2014.
56. Cánovas, M.F. *Hormigón*; Servicio de Publicaciones Colegio de Ingenieros de Caminos, Canales y Puertos: Madrid, Spain, 2002.
57. *C311/C311M−13*; Standard Test Methods for Sampling and testing Fly Ash or natural Pozzolans for Use in Portland-Cement Concrete. ASTM International: West Conshohocken, PA, USA, 2013.
58. *UNE 80303-2:2011*; Cements with Additional Characteristics. Part 2: Sea water Resisting Cements. AENOR, Asociación Española de Normalización y Certificación: Madrid, Spain, 2011.
59. *UNE-EN 196-5*; Methods of Testing Cement—Part 5: Pozzolanicity Test for Pozzolanic Cement. AENOR, Asociación Española de Normalización y Certificación: Madrid, Spain, 2011.
60. *UNE-EN 12390-1*; Testing Hardened Concrete—Part 1: Shape, Dimensions and other Requirements for Specimens and Moulds. AENOR, Asociación Española de Normalización y Certificación: Madrid, Spain, 2001.
61. *UNE-EN 12390-2*; Testing Hardened Concrete—Part 2: Making and curing Specimens for strength Test. AENOR, Asociación Española de Normalización y Certificación: Madrid, Spain, 2009.
62. *UNE-EN 12390-3*; Testing Hardened Concrete—Part 3: Compressive Strength of Test Specimens. AENOR, Asociación Española de Normalización y Certificación: Madrid, Spain, 2009.
63. *UNE-EN 196-1:2005*; Methods of Testing Cement—Part 1: Determination of Strength. AENOR, Asociación Española de Normalización y Certificación: Madrid, Spain, 2005.
64. *UNE-EN 12350-2*; Testing Fresh Concrete—Part 2: Slump Test. AENOR, Asociación Española de Normalización y Certificación: Madrid, Spain, 2009.
65. *UNE-EN 12390-7*; Testing Hardened Concrete—Part 7: Density of Hardened Concrete. AENOR, Asociación Española de Normalización y Certificación: Madrid, Spain, 2009.
66. *UNE 83988-1*; Concrete Durability. Test Methods. Determination of the Electrical Resistivity. Part 1: Direct Test (Reference Method). AENOR, Asociación Española de Normalización y Certificación: Madrid, Spain, 2008.
67. *UNE 83988-2:2008*; Concrete durability. Test methods. Determination of the Electrical Resistivity. Part 2: Four Points or Wenner Method. AENOR, Asociación Española de Normalización y Certificación: Madrid, Spain, 2008.
68. *UNE 83309:1990 EX*; Concrete Tests. Determination of the Depth of Penetration of Water under Pressure. AENOR, Asociación Española de Normalización y Certificación: Madrid, Spain, 1990.
69. *UNE-EN 12390-8:2001*; Testing Hardened Concrete—Part 8: Depth of Penetration of Water under Pressure. AENOR, Asociación Española de Normalización y Certificación: Madrid, Spain, 2001.
70. *UNE-EN 1936*; Natural Stone Test Methods—Determination of Real Density and Apparent Density, and of Total and Open Porosity. AENOR, Asociación Española de Normalización y Certificación: Madrid, Spain, 2007.
71. *UNE 83980:2014*; Concrete Durability. Test Methods. Determination of the Water Absorption, Density and Accessible Porosity for Water in Concrete. AENOR, Asociación Española de Normalización y Certificación: Madrid, Spain, 2014.
72. *UNE 112011:1994*; Assembly Corrosion. Determination of the Carbonatation Depth for in Service Concrete. AENOR, Asociación Española de Normalización y Certificación: Madrid, Spain, 1994.
73. *UNE 112011:2011*; Corrosion of Concrete Reinforcement Steel. Determination of the carbonatation Depth for In-Service Concrete. AENOR, Asociación Española de Normalización y Certificación: Madrid, Spain, 2011.

74. *UNE 83993-1:2013*; Durability of Concrete. Test Method. Measurement of Carbonation Penetration Rate in Hardened Concrete. Part 1: Natural Method. AENOR, Asociación Española de Normalización y Certificación: Madrid, Spain, 2013.
75. *UNE 83993-2:2013*; Durability of Concrete. Test Method. Measurement of Carbonation Penetration Rate in Hardened Concrete. Part 2: Accelerated Method. AENOR, Asociación Española de Normalización y Certificación: Madrid, Spain, 2013.
76. *PrNE 83986:2009*; Ensayo de Durabilidad del Hormigón. Determinación de la Difusión de Iones Cloruro. AENOR, Asociación Española de Normalización y Certificación: Madrid, Spain, 2009.
77. *CEN/TS 12390-11:2010*; Testing Hardened Concrete—Part 11: Determination of the Chloride Resistance of Concrete, Unidirectional Diffusion. ITeh Inc.: Wien, Austria, 2010.
78. *UNE-EN 14629:2007*; Products And Systems For The Protection And Repair Of concrete structures—Test methods—Determination of Chloride Content in Hardened Concrete. AENOR, Asociación Española de Normalización y Certificación: Madrid, Spain, 2007.
79. *UNE 83987:2014*; Concrete Durability. Test Methods. Measurement of Chloride Diffusion Coefficient in Hardened Concrete. Multiregime Method. AENOR, Asociación Española de Normalización y Certificación: Madrid, Spain, 2014.
80. *UNE 83992-2:2012 EX*; Durability of Concrete. Test Methods. Chloride Penetration Tests on Concrete. Part 2: Integral Accelerated Method. AENOR, Asociación Española de Normalización y Certificación: Madrid, Spain, 2012.
81. *UNE 112072:2011*; Laboratory measurement of Corrosion Speed Using the Polarization Resistance Technique. AENOR, Asociación Española de Normalización y Certificación: Madrid, Spain, 2011.
82. Annex 9—Additional Considerations on Durability According to the EHE-08. Available online: https://www.mitma.gob.es/recursos_mfom/annex_9.pdf (accessed on 10 November 2022).
83. *ASTM C618-22*; Standard Specification for Coal Fly Ash and Raw or Calcined Natural Pozzolan for Use in Concrete. ASTM International: West Conshohocken, PA, USA, 2022.
84. Limbachiya, M.; Meddah, M.S.; Ouchagour, Y. Use of recycled concrete aggregate in fly-ash concrete. *Constr. Build. Mater.* **2012**, *27*, 439–449. [CrossRef]
85. Kim, Y.; Hanif, A.; Kazmi, S.M.S.; Munir, M.J.; Park, C. Properties enhancement of recycled aggregate concrete through pretreatment of coarse aggregates—Comparative assessment of assorted techniques. *J. Clean. Prod.* **2018**, *191*, 339–349. [CrossRef]
86. Kim, T.; Davis, J.M.; Ley, M.T.; Kang, S.; Amrollahi, P. Fly ash particle characterization for predicting concrete compressive strength. *Constr. Build. Mater.* **2018**, *165*, 560–571. [CrossRef]
87. Lee, H.; Hanif, A.; Usman, M.; Sim, J.; Oh, H. Performance evaluation of concrete incorporating glass powder and glass sludge wastes as supplementary cementing material. *J. Clean. Prod.* **2018**, *170*, 683–693. [CrossRef]
88. Zhang, R.; Panesar, D.K. Mechanical properties and rapid chloride permeability of carbonated concrete containing reactive MgO. *Constr. Build. Mater.* **2018**, *172*, 77–85. [CrossRef]
89. *Real Decreto 956/2008*; Instrucción para la Recepción de Cementos RC-08. Government of Spain: Madrid, Spain, 2008.

materials

MDPI

Article

Influence of Adhesive Layer Thickness on the Effectiveness of Reinforcing Thin-Walled Steel Beams with CFRP Tapes—A Pilot Study

Ilona Szewczak [1,*], Malgorzata Snela [1] and Patryk Rozylo [2]

1 Faculty of Civil Engineering, Lublin University of Technology, 40 Nadbystrzycka Str, 20-618 Lublin, Poland
2 Faculty of Mechanical Engineering, Lublin University of Technology, Nadbystrzycka 36, 20-618 Lublin, Poland
* Correspondence: i.szewczak@pollub.pl

Abstract: When reinforcing thin-walled steel members with composite tapes, two issues often overlooked in published scientific papers should be considered, namely the correct thickness of the adhesive layer and the optimum bond length of the CFRP tape. In this article, the authors focused on the first of these issues. For this purpose, eight beams with a thin-walled box cross-section and a length of 3 m were subjected to bending in a four-point scheme. Six beams were reinforced with Sika CarboDur S512 composite tape, and two beams without reinforcement were tested as reference members. Three thicknesses of the adhesive layer (SikaDur-30) were analyzed: 0.6 mm, 1.3 mm and 1.75 mm. In addition to examining the effect of the thickness of the adhesive layer on displacements and deformations of thin-walled steel members, the load value at which the composite tape peeled off was also analyzed. Numerical analyses were then carried out in Abaqus, the outcomes of which showed good agreement with the laboratory results. Both numerical and laboratory results have shown that the thickness of the adhesive layer had a minor effect on the reduction in deformation and displacement of the tested beams. At the same time, with the increase in the thickness of the adhesive layer, the value of the load at which the CFRP tapes detached from the beam surface significantly decreased.

Keywords: thin-walled steel beam; CFRP tapes; reinforcement method; adhesive connection

check for updates

Citation: Szewczak, I.; Snela, M.; Rozylo, P. Influence of Adhesive Layer Thickness on the Effectiveness of Reinforcing Thin-Walled Steel Beams with CFRP Tapes—A Pilot Study. *Materials* **2022**, *15*, 8365. https://doi.org/10.3390/ma15238365

Academic Editor: Ayman El-Zohairy

Received: 7 November 2022
Accepted: 22 November 2022
Published: 24 November 2022

Publisher's Note: MDPI stays neutral with regard to jurisdictional claims in published maps and institutional affiliations.

1. Introduction

The intensive development of structures with thin walled cross sections is mainly related to technical advances in the fabrication and assembly of steel components and the desire to minimize material consumption. According to [1], the development of thin-walled structures has reduced the consumption of steel by up to 50% compared with conventional hot-rolled cross-sectional structures, reduced assembly time by up to 60% and reduced construction costs by up to 25%. Thanks to these advantages, thin-walled elements are increasingly being used not only as building envelope elements, but also as the main elements of the load-bearing structure. It should be noted that the increasing use of thin-walled steel components is also linked to the possible need to reinforce them, for example, in the event of design errors or additional external loads. Unfortunately, thin-walled steel structures are characterized by limited reinforcement possibilities. The very thin cross-sectional walls (1 mm to 3 mm) limit the possibility of welded connections, and the implementation of connections using mechanical fasteners is significantly restricted. It is, therefore, necessary to find an efficient and easy method for strengthening this type of structure. One such method may be the use of carbon fiber strips embedded in an epoxy matrix (carbon fiber-reinforced polymer (CFRP)) bonded to the beam with adhesive.

Based on the available number of publications, the use of composite materials to reinforce hot-rolled steel structures is fairly well recognized [2,3]. Unfortunately, the situation is somewhat different for cold-formed steel components.

In [4–8], experiments on axially pressed thin-walled steel elements with square, circular and channel sections reinforced with CFRP were described. All studies confirmed that the use of CFRP tapes allows a significant increase in the maximum load capacity of the tested elements (from 33% to more than 200%, depending on the method of CFRP sheet placement). In addition, studies on the reinforcement of flexural steel beams with sigma sections [9,10] have shown a reduction in the horizontal displacement of reinforced beams of up to 49% relative to unreinforced beams. The cited research results give cause for optimism that CFRP tapes may find wide application to strengthen thin-walled steel structures.

According to the authors of the study, to develop the optimum method for reinforcing thin-walled steel members with CFRP tapes, attention must be paid to three issues: choosing the correct anchorage length for the CFRP tapes, determining the proper adhesive layer thickness for fixing the CFRP tapes and making the correct end of the joint. In the aforementioned studies described in [9,10], the anchorage length and adhesive thickness were based on work [3] on the reinforcement of hot-rolled steel I-section beams. In [11], the effect of the shape of the adhesive joint end on the performance of the steel composite joint is discussed. No publications describing the selection of the correct anchorage length of CFRP tapes when reinforcing thin-walled steel beams were encountered during the literature study. In addition, studies on thin-walled steel beams did not address the thickness of the adhesive layer (which is important, for example, for the failure mode of the thin-walled steel beams).

Therefore, the authors of this paper decided to address this topic and conduct a pilot laboratory study to determine the optimum thickness of the adhesive layer concerning the strengthening of steel beams with a thin-walled rectangular cross-section.

2. Laboratory Tests

Eight thin-walled steel beams with a rectangular cross-section of 120 × 60 × 3, 3 m long, made of S235 steel, were used for the study. Two reference beams (B1R and B2R), and six beams reinforced with Sika CarboDur S512 CFRP tape, 50 mm wide, 1.2 mm thick and 171 cm long, were subjected to bending in a four-point scheme. Based on material tests, the fundamental strength parameters of the CFRP tape were determined: Young's modulus $E = 165$ GPa and Poisson's ratio $\nu = 0.308$. The CFRP tape was bonded to the beams with SikaDur-30 adhesive, using three adhesive layer thicknesses: in two beams, the adhesive layer thickness was 0.65 mm (beams marked B1_0.65 and B2_0.65); in two beams, 1.3 mm (B1_1.3, B2_1.3); and in two beams, 1.75 mm (B1_1.75, B2_1.75). The essential strength characteristics of the adhesive, as stated in the manufacturer's material sheet, are a minimum compressive strength after 7 days of 75 MPa, a compressive modulus of 9600 MPa and a minimum tensile strength after 7 days equal to 26 MPa. Basic strength characteristics of the adhesive included also a minimum shear strength of 16 MPa, a minimum peel strength after 7 days equal to 21 MPa and a shrinkage of 0.04%. The adhesive was prepared according to the manufacturer's instructions. The strength parameters of the steel were determined on the basis of laboratory tests. The steel tensile test was carried out on five samples cut from the tested beams. The shape of the test samples and the course of the test were developed in accordance with the ISO 6892-1:2009 standard. The results obtained for individual samples were repeatable. In order to determine Young's modulus and Poisson's ratio, the results of tensile coupon tests were statistically processed. The Young's modulus was 207.87 GPa and the Poisson's ratio was 0.307. The stress–strain diagram for the samples cut from the tested beams is shown below (Figure 1).

To determine the full strength of the bonded joint, the beams were tested in a Zwick&Roel testing machine (ZwickRoell GmbH & Co. KG, Ulm, Germany) 28 days after the CFRP tape was glued. The load increment was carried out with the press piston assuming a piston extension speed of 1 mm/min and recording the force every 0.01 s. Figure 2 shows a photograph of the test stand.

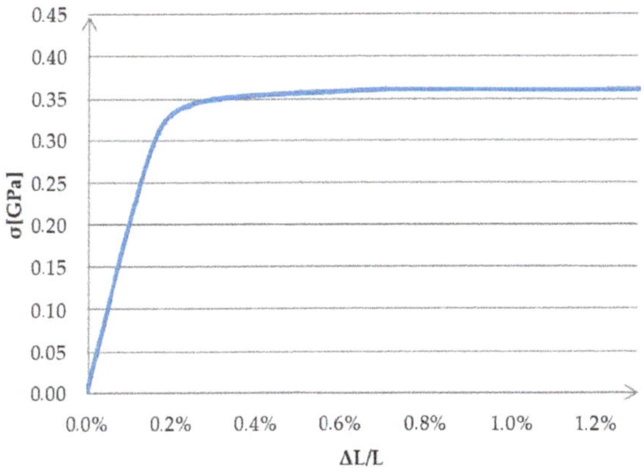

Figure 1. Material characteristics of steel.

Figure 2. Photograph of the test stand and one of the supports.

The load application point and the support points of the beams were dimensioned in the diagram (Figure 3). During the test, the deflection of the beams was measured using an inductive sensor, and the deformation of the beams was measured using TENMEX TFs-10 120 Ω ± 0.2% electrofusion strain gauges. The inductive sensor (LVDT2) and strain gauges (T2) were placed at the center of the beam span, as marked in Figure 3. The dimensions shown in Figure 3 are expressed in cm.

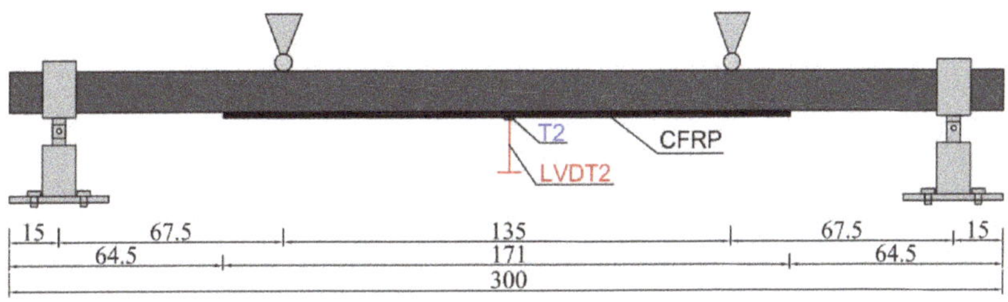

Figure 3. The scheme of the laboratory stand.

3. Laboratory Test Results

Figure 4a,b show load–displacement and load–deformation diagrams. To compare the results obtained, it was decided to limit the diagrams to the load value of 26.5 kN, at which the tape was detached from one of the beams.

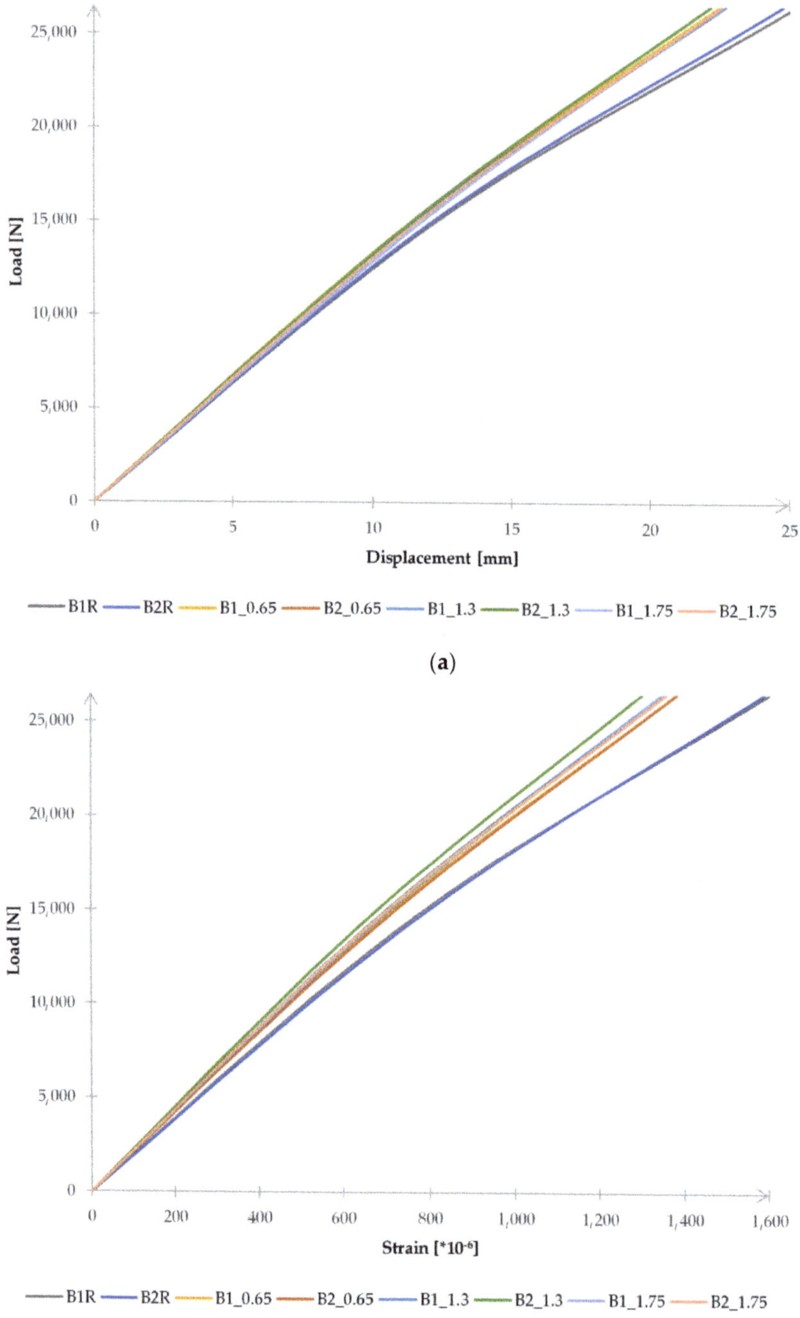

(b)

Figure 4. Diagram of the correlation of (**a**) vertical displacement–load at the location of the LVDT2 sensor and (**b**) deformation–load at the location of strain gauge T2.

Since the measured values obtained were highly reproducible and the graphs presented were therefore unreadable, Table 1 was prepared to analyze the effect of the adhesive layer thickness on the displacement and strain results obtained from tests on individual beams.

Table 1. Laboratory test results.

Load (kN)	Starin measured with a strain gauge T2 ($\times 10^{-6}$)							
	B1R	B2R	B1_0.65	B2_0.65	B1_1.3	B2_1.3	B1_1.75	B2_1.75
26.5	1598	1587	1382	1382	1347	1300	1351	1358
Load (kN)	Vertical deflection (mm)							
	B1R	B2R	B1_0.65	B2_0.65	B1_1.3	B2_1.3	B1_1.75	B2_1.75
26.5	25.26	24.77	22.5	22.63	22.71	22.17	22.59	22.58
	Load at the CFRP tape detachment moment (kN)							
	B1R	B2R	B1_0.65	B2_0.65	B1_1.3	B2_1.3	B1_1.75	B2_1.75
	-	-	37.4	40.9	31.4	31.1	29.9	26.5

By averaging the results obtained for each group of beams and comparing the results of the reference and reinforced beams, it was found that strengthening the beams with CFRP tapes allows a significant reduction in displacement and deformation of the beams tested in each case. For steel members with a 0.65 mm adhesive layer thickness, vertical displacements were reduced by 9.7% and strains measured by the T2 strain gauge by 13.2%; with a 1.3 mm adhesive layer thickness, displacements were decreased by 10.3% and strains by 16.8%; and for beams with a 1.75 mm adhesive layer thickness, reductions of 9.7% and 14.9% were achieved, respectively. As can be seen, the discrepancy between the results obtained by the different groups of reinforced beams is a maximum of 3.6% when measuring strains in the tension zone and 0.6% when measuring vertical displacements. The most favorable results were obtained in beams with an adhesive layer thickness of 1.3 mm. By far, the more critical factor appears to be the moment of tape detachment from the beam. Both reference beams failed under a load of 40.7 kN. The beams with the lowest adhesive layer thickness had the highest load values at the moment of tape release.

4. Numerical Simulations

Numerical simulations were conducted using the Abaqus software (2022, Dassault Systemes Simulia Corporation, Velizy Villacoublay, France). As part of the numerical studies, four original numerical models were developed using the finite element method to represent parallel experimental studies. Within the framework of the first numerical model, a steel beam profile (with geometric parameters presented at the stage of describing the subject of the study), the elements supporting the structure and the element loading the structure were modeled. The other three numerical models additionally modeled a composite tape made of carbon-epoxy composite CFRP, as well as an adhesive layer (with three different thicknesses of 0.65, 1.3 and 1.75 mm, respectively, for each of the three numerical models) which directly connected the composite tape to the bottom surface of the previously modeled steel beam profile. The above approach made it possible to analyze both the behavior of a bent steel beam without reinforcement and the behavior of bent steel beams reinforced with composite tape where three different adhesive layer thicknesses were used.

The FEA models developed allowed a faithful representation of the actual test conditions. Between the supports and the steel beam, as well as the loading element and the steel beam, contact relations were used in the tangential and normal directions, with a friction coefficient of 0.2. For numerical models containing different adhesive layer thicknesses and composite tape, an additional "tie" connection was used between the composite tape and the adhesive layer and the adhesive layer and the adjacent bottom surface of the steel beam. The steel beam was modeled as a shell model using S4R-type finite elements (shell finite elements, with a linear shape function, having four nodes for each finite element, where all translational and rotational degrees of freedom were at each node). Non-deformable finite elements of type R3D4 were used for the elements that constitute the supports and the loading element of the steel structure. Special reference points were also defined in these components with the necessary boundary conditions shown in Figure 5.

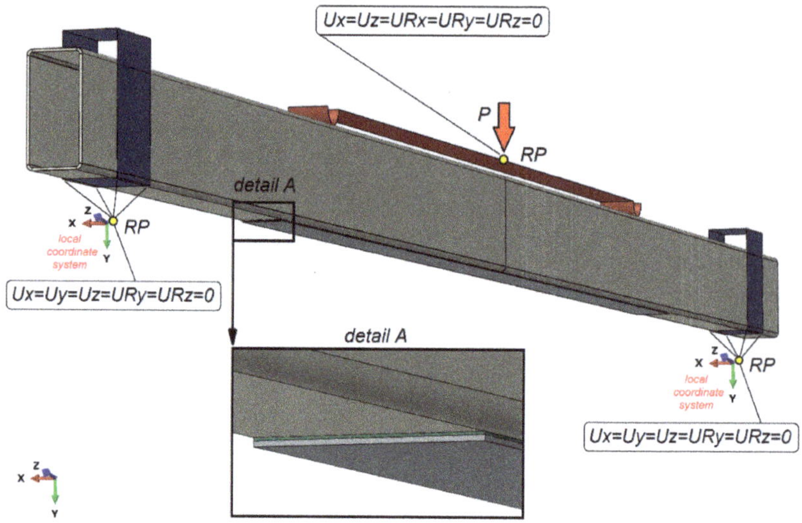

Figure 5. Numerical model.

The composite tape was also modeled as a shell element with S4R-type elements, while the adhesive layer was described with COH3D8-type finite elements (having eight nodes in each finite element). The discrete model in the basic form, i.e., without the adhesive layer and composite tapes, consisted of 16,900 finite elements and 17,325 computational nodes, while each model containing a composite tape and adhesive layer was described by 21,175 finite elements and 25,903 nodes. The size of the finite elements describing the adhesive joint was 5×5 mm, and the composite tape and steel beam were described by elements of 10×10 mm. The size of the elements in the case of the modeled adhesive layer was therefore half the size of the adjacent components. The effect of mesh density was tested, and no significant deviation from the adopted solution was observed. More significant information regarding the numerical studies, using a similar solution to the present paper, is presented in [12]. An example of a discrete model is presented in Figure 6.

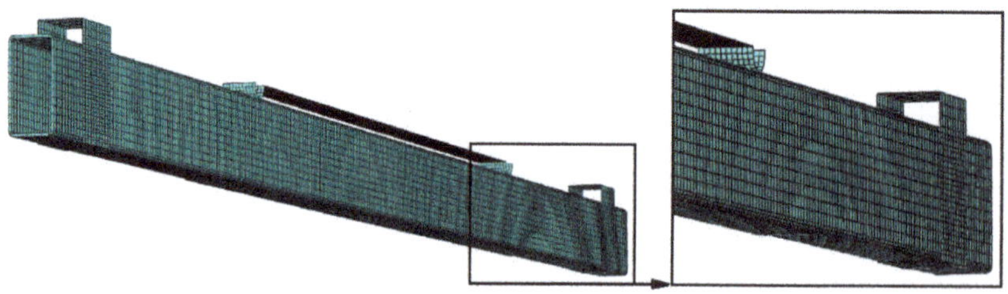

Figure 6. Discrete model.

The material parameters for the main test object, which was a steel beam with a rectangular cross-section, were described by an elastic–plastic material model, the data of which were presented in the test subject description stage. The adhesive layer and the composite tape had data adequate in comparison to the information presented in a thematically similar scientific publication [13]. Material data for the composite tape and adhesive were taken mainly from [14].

5. Results of Laboratory Tests and Numerical Analyses

The developed numerical model showed high compliance of the results obtained numerically with those from laboratory tests. The beam failure mode was the same in the case of laboratory tests and numerical analyses, as shown in Figures 7 and 8.

Figure 7. The damage of the beams obtained in numerical analyses and laboratory tests.

Figure 8. The beam failure modes obtained in numerical analyses and laboratory tests.

The high compliance of the results was also obtained when comparing the values of the vertical deflection and the deformation in the middle of the beam span, in the tension flange and in the place where the T2 strain gauge was placed. Examples of diagrams are presented in Figure 9. Figure 9 is limited to presenting the results obtained for two beams due to the legibility of the diagrams. On the graphs, the beams are marked as in the description of laboratory tests (B1R, B1_0.65, etc.), while the numerical models corresponding to the individual samples are marked by adding the letter "a" (B1Ra, B1_0.65a) at the end. The results obtained indicate that the model can be used for further numerical analyses. As in the case of laboratory tests, it was found that the reinforcement of the beams with the CFRP tapes each time allowed for a significant reduction in displacement and strain of the tested beams. However, a change in the thickness of the adhesive layer has a minimal effect on the value of vertical displacements or strain, which is presented in Figure 10.

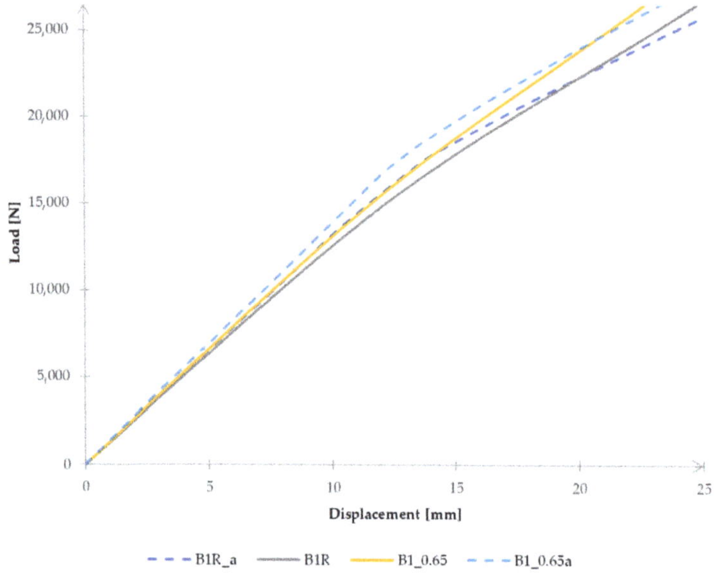

(a)

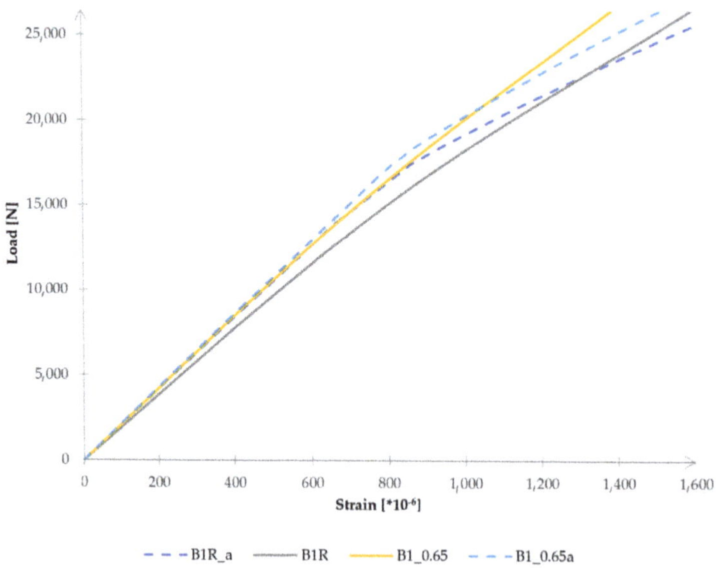

(b)

Figure 9. Comparison of the results of laboratory tests and numerical analyses: (**a**) vertical deflections at the location of LVDT2 sensor, (**b**) strain at the T2 strain gauge site.

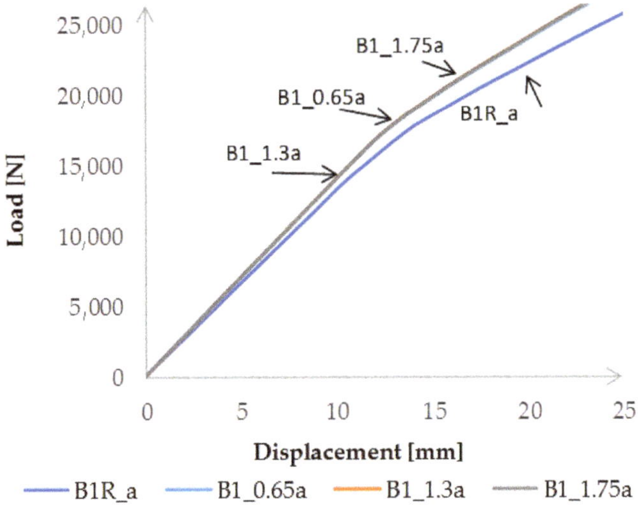

(a)

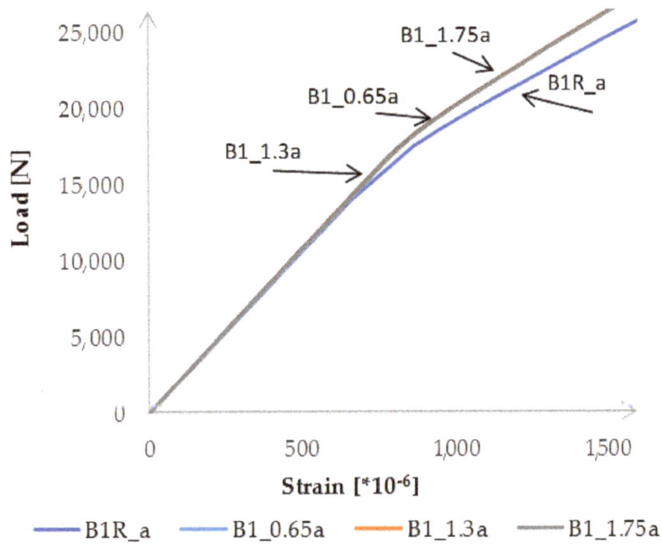

(b)

Figure 10. The results of numerical analyses: (**a**) vertical deflections at the LVDT2 sensor site, (**b**) strain at the T2 strain gauge site.

However, in the case of the numerical analyses, the key influence of the thickness of the adhesive layer on the moment of the CFRP tape detaching from the beam, and thus the loss of reinforcement, was confirmed again. In FEM, defects in the adhesive layer appeared at a load of 37.8 kN for the 0.65 mm adhesive layer, 31.35 kN for the 1.3 mm layer, and 28.04 kN for the 1.75 mm layer, which is in line with the average results obtained by individual groups of beams in laboratory tests.

In numerical simulations, the damage to the adhesive layer was observed with the use of the SDEG parameter. The scalar stiffness degradation (SDEG) represents the damage status of interface elements adopted in a cohesive element. Moreover, damage evolution in cohesive elements is tracked via the above-mentioned parameter. The SDEG variable tells about the state of damage in the element. Obviously, initially, damage initiation was based on the maximum stress criterion (MAXS), while damage evolution was based on the energy criterion. When SDEG parameter achieves a value of 1.0, it means that the bonded joint in a given area is damaged. Figure 11 shows the damage of the glued joint.

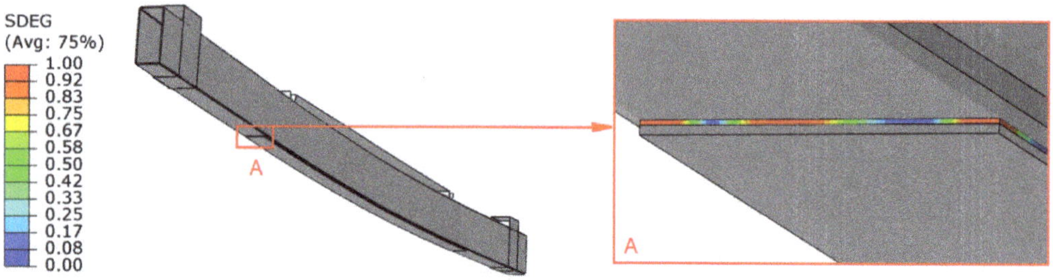

Figure 11. The damage of the glued (adhesive) joint.

6. Conclusions

The conducted laboratory tests allowed the following conclusions:

- Reinforcement of the beams with CFRP tapes each time reduced the value of vertical displacements and strain in the tension zone, which at the load level of 26.5 kN were reduced by up to 10.3% and 16.8%, respectively;
- The most favorable results in terms of reduction in strain and displacement values were obtained for beams with an adhesive layer thickness of 1.3 mm, which is in line with the results obtained in [3]; however, the results obtained for beams with different adhesive layer thicknesses were very similar;
- As the thickness of the adhesive layer increased, the load value at which the bonded joint was damaged decreased. The difference in the load values obtained for the smallest and largest adhesive layer thicknesses was 14.4 kN, and in the case of averaged results, this value was 10.95 kN. It should be noted that this difference is particularly significant when related to the maximum value of the load, which was 40.9 kN.

The outcomes of conducted numerical analyses are consistent with the laboratory test results. They showed that the use of CFRP tapes significantly reduced the level of deflection and strain in the analyzed points of individual numerical models. In turn, the change in the adhesive layer thickness had a negligible impact on the change in the level of deflections of the beam and strain in the tension beam flange. Performed numerical analyses also confirmed the influence of the thickness of the glued joint layer on the moment of failure of the joint, which was also demonstrated in numerical analyses for a different beam cross-section described in [13]. For the beams analyzed in this study, numerical analyses showed that the increase in the thickness of the adhesive layer significantly affected the decrease in the load value at which the tape detached; the difference in the load values obtained for the extreme adhesive layer thickness was 9.76 kN.

Based on preliminary laboratory tests and numerical analyses, it can be concluded that when thin-walled steel beams are reinforced with CFRP tapes, a 0.65 mm adhesive layer thickness is the most advantageous.

Author Contributions: Conceptualization, I.S.; methodology, I.S.; software, P.R.; validation, P.R.; formal analysis, I.S.; investigation, I.S.; resources, P.R. and I.S.; data curation, I.S. and P.R.; writing—original draft preparation, I.S.; writing—review and editing, I.S., M.S. and P.R.; visualization, I.S. and M.S.; supervision, I.S.; project administration, I.S. and M.S.; funding acquisition, P.R., I.S. and M.S. All authors have read and agreed to the published version of the manuscript.

Funding: The research was financed by Lublin University of Technology, Faculty of Civil Engineering and Architecture, grants number FD-20/IL-4/049 and FD-20/IL-4/054. The APC was funded by: Discount Voucher 1463d06cb8288147 (100.00%) (p.rozylo@pollub.pl).

Institutional Review Board Statement: Not applicable.

Informed Consent Statement: Not applicable.

Data Availability Statement: Data are contained within the article.

Acknowledgments: The authors acknowledge Lublin University of Technology, Faculty of Civil Engineering and Architecture.

Conflicts of Interest: The authors declare no conflict of interest.

References

1. Bródka, J.; Łubiński, M. *Lekkie Konstrukcje Stalowe*; Arkady: Warszawa, Poland, 1978.
2. Machelski, C.; Wrzesiński, M. Wzmacnianie elementów konstrukcji stalowych z wykorzystaniem taśm CFRP. *Mater. Bud.* **2015**, *10*, 51–54. [CrossRef]
3. Kowal, M. Wzmacnianie Elementów Stalowych Kompozytami FRP. Politechnika Lubelska 2015. Lublin. Available online: http://bc.pollub.pl/dlibra/docmetadata?id=12764 (accessed on 22 November 2022).
4. Han, X.; Hou, S.; Ying, L.; Hou, W.; Aliyev, H. On the fracture behaviour of adhesively bonded CFRP hat-shaped thin-walled beam under axial crushing load: An experimental and modelling study. *Compos. Struct.* **2019**, *215*, 258–265. [CrossRef]
5. Subramanian, S. Study on the performance of CFRP strengthened circular hollow steel sections. *Int. J. Res. Eng. Technol.* **2015**, *4*, 118–121.
6. Tang, H.; Wang, C.; Wang, R. Enhancing Stability of Thin-Walled Short Steel Channel Using CFRP under Eccentric Compression. *Int. J. Polym. Sci.* **2016**, *2016*, 2790385. [CrossRef]
7. Amous, E.A.; Ghanem, M.A. Axially loaded capacity of alternatively strengthened cold-formed channel columns with CFRP. *J. Constr. Steel Res.* **2018**, *140*, 139–152. [CrossRef]
8. Kalavagunta, S.; Naganathan, S.; Nasharuddin, K.; Mustapha, K.N.B. Proposal for design rules of axially loaded CFRP strengthened cold formed lipped channel steel sections. *Thin-Walled Struct.* **2013**, *72*, 14–19. [CrossRef]
9. Szewczak, I.; Rzeszut, K.; Rozylo, P.; Samborski, S. Laboratory and numerical analysis of steel cold-formed sigma beams retrofitted by bonded CFRP tapes. *Materials* **2020**, *13*, 4339. [CrossRef] [PubMed]
10. Szewczak, I.; Rzeszut, K.; Rozylo, P.; Snela, M. Laboratory and numerical analysis of steel cold-formed sigma beams retrofitted by bonded CFRP tapes—Extended research. *Materials* **2020**, *13*, 4960. [CrossRef] [PubMed]
11. Kowal, M.; Rozylo, P. Effect od bond end shape on CFRP/steel joint strength. *Compos. Struct.* **2022**, *284*, 115186. [CrossRef]
12. Castagnetti, D.; Dragoni, E.; Spaggiari, A. Failure analysis of complex bonded structures: Experimental tests and efficient finite element modelling by tied mesh method. *Int. J. Adhes. Adhes.* **2011**, *31*, 338–346. [CrossRef]
13. Szewczak, I.; Rozylo, P.; Snela, M.; Rzeszut, K. Impact of adhesive layer thickness on the behavior of reinforcing thin-walled sigma-type steel beams with CFRP tapes. *Materials* **2022**, *15*, 1250. [CrossRef] [PubMed]
14. Szewczak, I.; Rozylo, P.; Rzeszut, K. Influence of Mechanical Properties of Steel and CFRP Tapes on the Effectiveness of Strengthening Thin-Walled Beams. *Materials* **2021**, *14*, 2388. [CrossRef] [PubMed]

materials

Article

Improvements in the Engineering Properties of Cementitious Composites Using Nano-Sized Cement and Nano-Sized Additives

Ibadur Rahman [1], Priyanka Singh [2], Nirendra Dev [3], Mohammed Arif [4], Faiz Noor Khan Yusufi [5], Ameer Azam [6], M. Masroor Alam [4], Sandeep Singh [7], Jasgurpreet Singh Chohan [8], Raman Kumar [8], Lovneesh Sharma [9], Elsayed Tag-Eldin [10,*], Shubham Sharma [8,11,*] and Muhammad Rizal Muhammad Asyraf [12,13]

1 Department of Civil Engineering, Jamia Millia Islamia, Jamia Nagar, New Delhi 110025, India
2 Department of Civil Engineering, Amity School of Engineering & Technology, Amity University Uttar Pradesh, Noida 201313, India
3 Department of Civil Engineering, Delhi Technological University, Shahbad, Daulatpur, Bawana Road, New Delhi 110042, India
4 Department of Civil Engineering, Aligarh Muslim University, Aligarh 202002, India
5 Department of Statistics & Operations Research, Aligarh Muslim University, Aligarh 202002, India
6 Department of Applied Physics, Aligarh Muslim University, Aligarh 202002, India
7 Department of Civil Engineering, University Center for Research and Development, Chandigarh University, Mohali 140413, India
8 Mechanical Engineering Department, University Center for Research & Development, Chandigarh University, Mohali 140413, India
9 Department of Civil Engineering, Universal Institute of Engineering & Technology, Mohali 140413, India
10 Faculty of Engineering and Technology, Future University in Egypt, New Cairo 11835, Egypt
11 School of Mechanical and Automotive Engineering, Qingdao University of Technology, Qingdao 266520, China
12 Engineering Design Research Group (EDRG), Faculty of Mechanical Engineering, Universiti Teknologi Malaysia, Johor Bahru 81310, Malaysia
13 Centre for Advanced Composite Materials (CACM), Universiti Teknologi Malaysia, Johor Bahru 81310, Malaysia
* Correspondence: elsayed.tageldin@fue.edu.eg (E.T.-E.); shubham543sharma@gmail.com or shubhamsharmacsirclri@gmail.com (S.S.)

Abstract: The findings of an extensive experimental research study on the usage of nano-sized cement powder and other additives combined to form cement–fine-aggregate matrices are discussed in this work. In the laboratory, dry and wet methods were used to create nano-sized cements. The influence of these nano-sized cements, nano-silica fumes, and nano-fly ash in different proportions was studied to the evaluate the engineering properties of the cement–fine-aggregate matrices concerning normal-sized, commercially available cement. The composites produced with modified cement–fine-aggregate matrices were subjected to microscopic-scale analyses using a petrographic microscope, a Scanning Electron Microscope (SEM), and a Transmission Electron Microscope (TEM). These studies unravelled the placement and behaviour of additives in controlling the engineering properties of the mix. The test results indicated that nano-cement and nano-sized particles improved the engineering properties of the hardened cement matrix. The wet-ground nano-cement showed the best result, 40 MPa 28th-day compressive strength, without mixing any additive compared with ordinary and dry-ground cements. The mix containing 50:50 normal and wet-ground cement exhibited 37.20 MPa 28th-day compressive strength. All other mixes with nano-sized dry cement, silica fume, and fly ash with different permutations and combinations gave better results than the normal-cement–fine-aggregate mix. The petrographic studies and the Scanning Electron Microscope (SEM) and Transmission Electron Microscope (TEM) analyses further validated the above findings. Statistical analyses and techniques such as correlation and stepwise multiple regression analysis were conducted to compose a predictive equation to calculate the 28th-day compressive strength. In addition to these methods, a repeated measures Analysis of Variance (ANOVA) was also implemented to analyse the statistically significant differences among three differently timed strength readings.

check for updates

Citation: Rahman, I.; Singh, P.; Dev, N.; Arif, M.; Yusufi, F.N.K.; Azam, A.; Alam, M.M.; Singh, S.; Chohan, J.S.; Kumar, R.; et al. Improvements in the Engineering Properties of Cementitious Composites Using Nano-Sized Cement and Nano-Sized Additives. *Materials* **2022**, *15*, 8066. https://doi.org/10.3390/ma15228066

Academic Editor: Ayman El-Zohairy

Received: 6 September 2022
Accepted: 3 November 2022
Published: 15 November 2022

Keywords: cementitious composites; dry- and wet-grinding; nano-sized cement; nano-sized additives; morphological studies; stepwise multiple regression; repeated measures ANOVA

1. Introduction

In the past decade, construction technology has undergone some profound changes. On the one hand, the emphasis has been on the construction techniques, and on the other, on using a variety of construction materials, specifically alternate materials. Many research studies from the viewpoint of both the short- and long-term mechanical characteristics of cementitious composites have been undertaken. This has resulted in the establishment of different additives to improve the physicochemical and mechanical properties of conventional concrete. The neo-improved concrete now shows a much better crack arrest mechanism, better creep–shrinkage performance, high durability, etc. [1–4]. There are few critical studies on the application of nanotechnology in concrete production. Nano-sized particles, having greater surface area than volume, have a high potential for tremendous chemical activity [5–7]. Nano-engineering and nano-modification of cement represent an important arena, though in the embryonic stage as far as mass production and utilization are concerned [7–10]. A limited number of investigations are available, dealing with the modification of cement in nano-sized particles and their noteworthy effect as nano-binders for concrete [10].

Nanotechnology is another promising field of study (e.g., see Feynman [6]) that is bound to revolutionize material and construction technology when applied to the construction industry [9–12]. Furthermore, with the increasing demand for optimal material resource realization and sustainable development approaches, research on nanotechnology and its application for improving the performance of cementitious composites are very much warranted [1]. A number of recent studies have demonstrated that the use of cement hydration products on the nano-scale enhances the strength of cementitious composites, and nano-fly ash has been combined with cement matrix in varied quantities in order to collate and compare conventional cement matrix's engineering features with those of nano-additives [13–15]. A significant improvement in the properties of the cementitious matrix was discerned in this study, showing agreement with other studies [16–18].

Combining the nano-scale of different materials offers avenues for developing new additives for cement, such as superplasticizers, nano-particles, or nano-reinforcements [16–18]. Much of the work to date with nano-particles has been performed with nano-silica (nano-SiO_2 and nano-titanium oxide (nano-TiO_2). There are a few studies on incorporating nano-iron (nano-Fe_2O_3), nano-alumina (nano-Al_2O_3), and nano-clay particles [12,19–22].

Nano-SiO_2 is more efficient in enhancing strength and other engineering properties than silica fume. The addition of 10% nano-SiO_2 with dispersing agents increases the compressive strength of cement mortars by 26% in 28 days compared with only a 10% increase with the addition of 15% silica fume [11,23–25]. In this research, nano-cement and nano-silica fumes were developed in the laboratory, and their combined effects were focused on to determine the strength parameters. To the best of our knowledge, this type of work has not been performed previously [24–26].

Nano-Al_2O_3 shows a significant increase in the modulus of elasticity, i.e., up to 143% at a dosage of 5%. However, it has a limited effect on the compressive strength, and no other significant changes have been reported [21,26,27].

The ability to insert a variety of organic molecules, preferably nano-sized, into the basic C-S-H structure can provide the potential for creative manipulation. This is in addition to the layered structure and the tendency to have structural defects in the silicate chains, such as the missing bridging tetrahedron of C-S-H [28–30]. Three schemes for hybridizing, or incorporating, "guest molecules" into C–S–H were proposed by Minet et al. [18] who demonstrated that it could accommodate small-sized organic groups directly linked to the silicate chains in the interlayer space of C–S–H [11,26,31,32].

A recent study by Reza found that "nano-MgO" influences the "microstructure" and "strength" of "cement composites" [31]. By comparing the "treated composites" with the "plain composite", the findings indicate that, in 7 and 28 days, 1% of "nano-MgO binder" by weight increases the "compressive strength" by 103 and 80%, respectively, while 95 and 70% of the "flexural strength" are the increases for the plain composites. The "microstructure" of the "cementitious composites" with "nano-MgO" is more "compact" and "homogeneous" than that of "plain composites" due to the "expansive properties" of "nano-MgO" [31].

From the existing studies, it is possible to develop novel "multifunctional", efficient, performance-enhancing "cementitious composites" that can be suited for a wide range of applications by manipulating the "material's structure" on the "nanoscale", such as with "nanoparticles" and "nanoscale fibers" [23]. "Conventional concrete" has been shown to be improved in its "pore structure" to form a "C–S–H gel" faster and to be stronger, more "flexible", and more "durable" when reinforced with "nanoparticles". The reinforcement characteristics of "nano-additives" and "nano-sized cement" have been demonstrated to enhance the "strength", "fracture characteristics", and "durability" of "cementitious composites" [23].

Hence, nanotechnology researchers have begun to develop novel materials in recent years due to the need to use them for increasing the characteristics of various materials in general and materials utilized in the construction sector in particular [33–35]. Previous studies have investigated normal cement composites with nano-additives [36,37], whereas this research study focused on converting cement into nano-cement, followed by converting other additives into nano-additives; then, their individual and combined effects were studied.

Hence, nano-sized cement as well as nano-additives have nano-sized characteristics that change "cement hydration", the "compaction degree", and the "thixotropy behavior", contributing to the "modification of cement hydration". It is possible to explore all of these effects to make "concrete" that is "stronger", "greener", and "workable". In addition to raising the "mechanical properties" of "cement-based composite materials", the "chemical reactivity" of "nanomaterials" can be used to modify the "morphology", thus optimizing the "C–S–H gel structure" and improving the "nanomechanical properties", which have been proved to have high "durability enhancement capabilities".

According to literary sources, "nano-additives" and "nanosized cement" increase the "packing density" both in the "cementitious matrix" as well as in the "interfacial transition zone". In the "interfacial transition zone" between the "nano-reinforcements" or "nano-additives" and the "cementitious matrix," the "friction at the interface" is dependent on "packing density" and "stiffness" [38]. Therefore, a "cementitious composite" that is frictionally bound has superior "mechanical characteristics." Several studies have found that "nano-additives" and "nanosized cement" accelerate "hydration," thereby contributing to the development of "flexural strength" in the early stages. It is also worth noting that lower "CH content" and higher "C–S–H content," along with the "seeding effect" of "nano-sized particles," extensively contribute to strengthening the "interfacial-frictional bond."

It was found through a literature search that the utilisation of nanotechnology in cement paste, mortar, and concrete is on the rise. This growth can be attributed to the availability of freshly discovered nanomaterials, as well as the rising popularity of nanomaterial-modified cement composites. Nanotechnology in cements focuses on two key areas: the creation of novel nano-scale-engineered products for the concrete industry and the characterization and understanding of materials on the nano- and micro-scales using cutting-edge characterization techniques. Hitherto, limited studies can be found where comparisons are made between the dry- and wet-grinding of the constituents. Moreover, the literature survey reported a lack of investigations into the mechanical influence of different permutations and combinations of nano-sized dry cement, silica fume, and fly ash. Thus, the influence of these nano-sized cements, nano-silica fumes, and nano-fly ash in different proportions was studied to evaluate the engineering properties of cement–fine-aggregate matrices using normal-sized, commercially available cement.

2. Materials and Experimental Programme

To assess the effect of nano-modification on cement matrices, nano-sized dry- and wet-ground cements were produced and were used as additives in cement matrices to test their strength. Other construction materials used in different proportions in the mix were cement, fine aggregates, coarse aggregates, silica fumes, fly ash, and acetone.

Cement, fine aggregates, silica fume, and fly ash were studied using a standard petrographic microscope and a Scanning Electron Microscope (SEM). Nano-cement was also studied under a Transmission Electron Microscope (TEM), as well as a SEM. Finally, cement concrete blocks were cast, and their compressive strength was measured as per the recommendation of the relevant Indian Standard Code of Practices. It is necessary to evaluate the compressive strength of cement mortar cubes in order to determine whether or not cement satisfies the requirements of the Indian standard specifications and whether or not it is capable of producing concrete with the necessary compressive strength.

Concrete is utilised in construction because of its ability to withstand compressive stresses. However, in situations where the tensile strength or shear strength is of major concern, the compressive strength is estimated to be the most important property of a cement mortar cube.

2.1. Materials Used for Experimentation

Shree Ultra 43 grade ordinary Portland cements were utilized for the research work. In the laboratory, the physical properties of cement were determined, as indicated in Table 1.

Table 1. Properties of cements.

S. No.	Type of Cement	Consistency	Initial Setting Time (min)	Final Setting Time (min)
1.	Normal cement	28.50%	61	294
2.	Dry-ground nano-cement	30.00%	58	291
3.	Wet-ground nano-cement	32.00%	54	290

Badarpur sand was selected since it was readily accessible. The sand was separated and screened to remove lumps of clay and other foreign and detrimental materials, after which it was washed with water and air-dried. Table 2 shows the grading and fineness modulus of sand.

$$Fineness\ Modulus = \frac{250.35}{100} = 2.5035$$

Table 2. Sieve analysis results of fine aggregates (sand).

S. No.	IS: Sieve Designation	Weight Retained (g)	Cumulative Weight Retained (g)	% Cumulative Weight Retained
1.	4.75 mm	Nil	0	0
2.	2.36 mm	20	20	1
3.	1.18 mm	235	255	12.75
4.	600 micron	560	815	40.75
5	300 micron	1102	1917	95.85
6.	150 micron	83	2000	100.00

Fly ash was collected from the Panipat (Haryana) thermal power plant. The physical properties of fly ash are given in Table 3.

Table 3. Physical properties of fly ash.

S. No.	Constituent/Property	Value
1.	Colour	Grey
2.	Percent passing 45-micron sieve	90%
3.	Size of the particle	4.70×10^{-7} m
4.	Maximum dry density (MDD)	9.30 kN/m^3
5.	Optimum moisture content (OMC)	27.5%
6.	Specific gravity	2.02 at 27 °C
7.	Surface area	3060 cm^2/g

Coarse aggregates of locally available quartzite were used in the mix. The properties of the coarse aggregates were determined via sampling and testing in accordance with the requirements of BS 812: Part 103:1985. The specific gravity of the aggregates was 2.76, and the water absorption was 0.40 percent. The particle size distribution of the coarse aggregates used was as presented in Table 4 and illustrated in Figure 1.

Table 4. Gradation and properties of locally available quartzite coarse aggregates.

Sieve No.	Mass Retained (g)	Percentage Retained (%)	Cumulative Percentage Retained (%)	Cumulative Percentage Passing (%)
40.00 mm	0	0	0	100
20.00 mm	330	7.2	7.2	92.8
12.50 mm	3060	66.8	74.0	26.0
10.00 mm	1100	24.0	98.0	2
4.75 mm	92	2.0	100.0	0

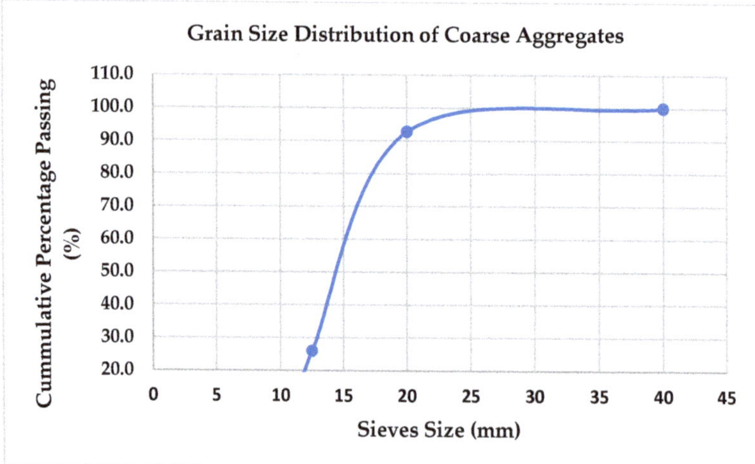

Figure 1. Grain size distribution of coarse aggregates.

For the experiments, we obtained silica fume from Elkem Company. The properties of silica fume are given in Table 5.

Table 5. Physical properties of silica fume.

S. No.	Description	Result
1.	Particle size	6.09×10^{-8} m
2.	Surface area	$14\ m^2/gm$
3.	Density	$2.2\ g/cm^3$

2.2. Conversion of the Constituent Material to Nano-Size

The conversion of the constituent particles to the nano-scale was carried out with the help of Ball Mill Machine (FRITSCH, Germany) in two ways, i.e., via dry-grinding (Figure 2) and wet-grinding (Figure 3). Dry-grinding was carried out for all the constituents, while wet-grinding was only carried out for cement. Dry-grinding reduces the size via particle-on-particle impacts, while wet-grinding smashes particles against solid grinding media in a liquid, dispersing the raw material in a slurry and then circulating it. In areas with ample water supply, contractors are naturally inclined to use concrete wet-grinding and -polishing, since they were developed long before dry-grinding. Throughout this process, water is primarily used to prevent the diamond tool from overheating and to provide lubrication to reduce friction. It is generally more efficient to grind wetly. This is because thorough mixing takes place when the material is mixed, enabling more balanced feed to be sent directly to the grinding mill. This hazard is eliminated, since dust is not prevalent. This results in a cleaner plant. It is possible to perform a more efficient classification for kiln feed, although thickeners are required due to high dilutions. It is less expensive to resort to less efficient dry-grinding methods where low-cost fuel is available, because the extra heat needed during calcining drives off the water.

Due to the development of better air separators and dust collectors, some of these problems have been minimized to the point where present-day costs are virtually identical.

Acetone was utilized for the wet-grinding of cement. All the converted materials were studied under a Scanning Electron Microscope (SEM) before and after conversion to ascertain the morphological properties of the particles of cement, fly ash, and silica fumes, and the results are presented in Figure 4, respectively. Transmitted electron microscopy was also used for ordinary cement, dry-ground cement, and wet-ground cement, and the results are shown in Figure 5, respectively. Numerous analytical attachments, including multiple EDS, EBSD (co-planar with EDS), WDS, CL, STEM, heating/cooling sub-stages, etc., may be easily mounted onto the SEM (Jeol 700SM) large specimen chamber through ports carefully located on the instrument. Similarly to the SEM, the TEM is an instrument used for electronic spectroscopic imaging. The TEM has the capacity to perform analytical measurements and has a higher spatial resolution than the SEM.

Figure 2. Inside view of ball mill machine.

Figure 3. Inside view of wet-grinding of cement.

Figure 4. SEM images of (**a**) normal cement, (**b**) converted nano-cement, (**c**) fly ash, (**d**) nano-fly ash, (**e**) silica fume, and (**f**) nano-silica fume.

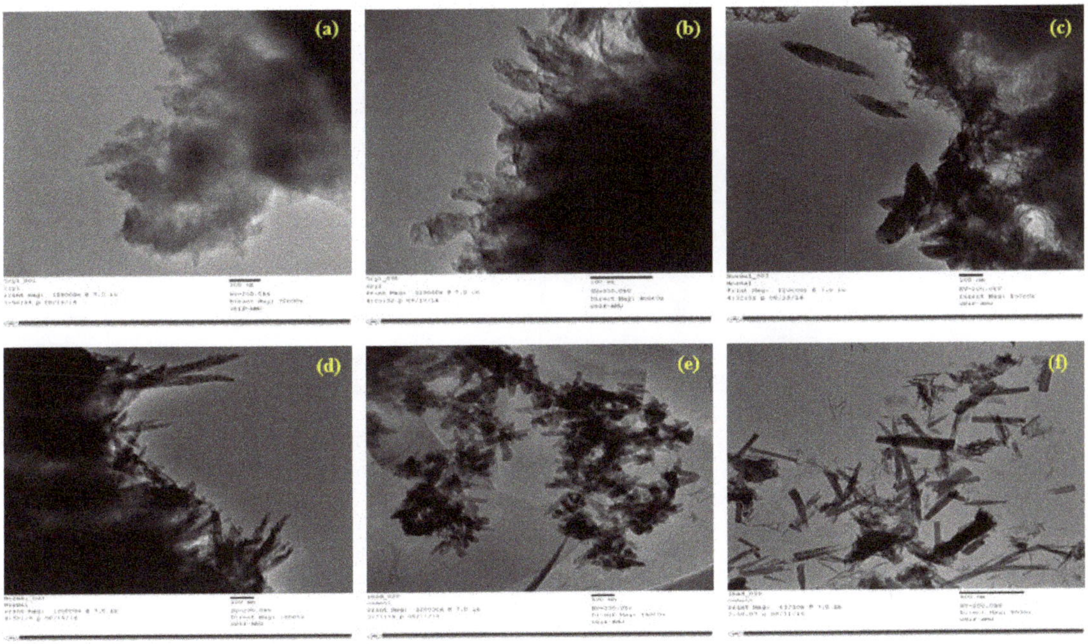

Figure 5. TEM images of (**a**) normal cement, (**b**) normal cement (another view), (**c**) dry-ground nano-cement, (**d**) dry-ground nano-cement (another view), (**e**) wet-ground nano-cement, and (**f**) wet-ground nano-cement (another view).

2.3. Compressive Strength Test of Cement Matrix

To study the materials' performance and mix, 70.06 mm size cubes were cast in the laboratory (Figure 6). In total, 176 cement cube samples were cast, and specific designations were assigned to them. The details of the work plan are given in Table 6. The constituent materials of the cubes were cement, fine aggregates, and potable water. The ratio of cement to fine aggregates was taken as 1:3 by weight. The water–cement ratio was kept as 0.45. The proportions of the material matrices included a fixed sand proportion, i.e., 3, and variations in the cement content, i.e., from 0.45 to 1, in the five materials (i.e., nano-cement, silica fume, nano-silica fume, nano-fly ash, and fly ash), forming binary, tertiary, and quaternary blends of the cementitious mixture. The fractional variations of cement and nano-cement were 0.1, 0.45, 0.5, and 1, respectively, whereas the shares of silica fume, nano-silica fume, fly ash, and nano-fly ash were kept fixed, i.e., 0.1 only. After casting the specimens, the cubes were cured for the specified durations of 7, 14, and 28 days. To determine their crushing strength, the cubes were put through compression testing equipment until they failed.

The compressive strength of the cement-paste cubes is given in Table 7. The cement matrix cubes including the nano-particles demonstrated outstanding strength after 7, 14, and 28 days, as can be observed. As a result, the material system could be trusted for its intended use.

Figure 6. Concrete specimen in the mould on vibrating table.

Table 6. Work plan for compressive strength test on cement matrix.

S. No.	Type	Composition	Abbreviation	Ratio
1.	I	Cement:sand	C:S	1:3
2.	II	Nano-cement:sand	NC:S	1:3
3.	II	Nano-cement (wet):sand	NC:S	1:3
4.	III	Cement:nano-cement:sand	C:NC:S	0.5:0.5:3
5.	IV	Cement:silica fume:sand	C:SF:S	0.9:0.1:3
6.	V	Nano-cement:silica fume:sand	NC:SF:S	0.9:0.1:3
7.	VI	Cement:nano-cement:silica fume:sand	C:NC:SF:S	0.45:0.45:0.1:3
8.	VII	Cement:nano-silica fume:sand	C:NSF:S	0.9:0.1:3
9.	VIII	Nano-cement:nano-silica fume:sand	NC:NSF:S	0.9:0.1:3
10.	IX	Cement:nano-cement:nano-silica fume:sand	C:NC:NSF:S	0.45:0.45:0.1:3
11.	X	Cement:fly ash:sand	C:FA:S	0.9:0.1:3
12.	XI	Nano-cement:fly ash:sand	NC:FA:S	0.9:0.1:3
13.	XII	Cement:nano-cement:fly ash:sand	C:NC:FA:S	0.45:0.45:0.1:3
14.	XIII	Cement:nano-fly ash:sand	C:NFA:S	0.9:0.1:3
15.	XIV	Nano-cement:nano-fly ash:sand	NC:NFA:S	0.9:0.1:3
16.	XV	Cement:nano-cement:nano-fly ash:sand	C:NC:NFA:S	0.45:0.45:0.1:3

Table 7. Comparison of test results of cubes with cement (C) and nano-cement (NC) with test results of cubes with cement, nano-cement, and silica fume (SF); test results of cubes with cement, nano-cement, and nano-silica fume (NSF); test results of cubes with cement, nano-cement, and fly ash (FA); test results of cubes with cement, nano-cement, and nano-fly ash (NFA); and test results of cubes with cement.

Type	Composition	Ratio	W/C Ratio	Grinding Type (Cement)	Compressive Strength, 7 Days (MPa)	Compressive Strength, 14 Days (MPa)	Compressive Strength, 28 Days (MPa)
I	C:S	1:3	0.45	—	15.33	17.33	20.88
II	NC:S	1:3	0.45	Dry	24.33	26.73	34.33
II	NC:S	1:3	0.45	Wet	27.00	30.33	40.00
III	C:NC:S	0.5:0.5:3	0.45	Wet	24.45	27.30	37.20

Table 7. *Cont.*

Type	Composition	Ratio	W/C Ratio	Grinding Type (Cement)	Compressive Strength, 7 Days (MPa)	Compressive Strength, 14 Days (MPa)	Compressive Strength, 28 Days (MPa)
IV	C:SF:S	0.9:0.1:3	0.45	—	20.50	22.60	26.75
V	NC:SF:S	0.9:0.1:3	0.45	Dry	22.30	24.40	27.00
VI	C:NC:SF:S	0.45:0.45:0.1:3	0.45	Dry	21.70	23. 05	26.80
VII	C:NSF:S	0.9:0.1:3	0.45	—	21.00	23.10	26.90
VIII	NC:NSF:S	0.9:0.1:3	0.45	Dry	23.46	25.50	29.20
IX	C:NC:NSF:S	0.45:0.45:0.1:3	0.45	Dry	22.24	24.63	28.86
X	C:FA:S	0.9:0.1:3	0.45	—	22.70	24.35	27.03
XI	NC:FA:S	0.9:0.1:3	0.45	Dry	26.26	27.45	31.00
XII	C:NC:FA:S	0.45:0.45:0.1:3	0.45	Dry	23.30	24.50	29.00
XIII	C:NFA:S	0.9:0.1:3	0.45	—	24.13	26.75	29.03
XIV	NC:NFA:S	0.9:0.1:3	0.45	Dry	29.20	31.60	35.20
XV	C:NC:NFA:S	0.45:0.45:0.1:3	0.45	Dry	26.33	28.33	31.33

2.4. Compressive Strength Test on Concrete

To study the performance of concrete, cubes of the size of 150 mm × 150 mm × 150 mm × 150 mm were cast in the laboratory. A total of six concrete cube samples were cast with the specific designation assigned to these. The detailed work plan is given in Table 8. The constituent materials used for the concrete cubes were cement, fine aggregates, coarse aggregates, and water. M20 grade nominal mix was used. In this case, the water-to-cement ratio was maintained as 0.50. These cubes were demoulded after 24 h and were allowed to cure in the curing tank for a total of seven days. The cubes were subjected to compressive testing in the laboratory under compression testing equipment until they failed in order to determine their crushing strength. According to the results in Table 9, the compressive strength of concrete cubes was good after seven days, indicating that the concrete cubes including the nanomaterials had excellent strength after seven days (Figure 6).

Table 8. Work plan of compressive strength test on concrete.

S. No.	Type	Mix Design	Composition	Ratio
1.	XVI	M20	Cement: sand: coarse aggregates	1:1.5:3
2.	XVII	M20	Cement: nano-cement: sand: coarse aggregates	0.75:0.25:1.5:3

Table 9. Test results of concrete cubes with cement and 25% nano-cement.

Type	Composition	Ratio	Grinding Type (Cement)	Compressive Strength, 7 Days (MPa)
VI	Cement:sand:coarse aggregates	1:1.5:3	—	23.33
XVII	Cement:nano-cement:sand:coarse aggregates	0.75:0.25: 1.5:3	Wet	26.57

2.5. Microscopic Examination of Cement Matrix

The cement cubes, cast and tested using a universal testing machine, were subjected to petrographic studies under a microscope [22–25]. The pieces of the cube after testing were converted into thin sections for studying under a petrological microscope [26–29]. The slides of different types of cement cubes were prepared and studied under normal and polarized light. The petrographic studies of cements began with transferring a small amount with a needle to refractive index oil on a clean glass slide for the inspection of individual grains (oil immersion mounts). After that, grains were sprayed on an epoxy-coated frosted glass slide for petrographic thin-section investigation. The representative sample was then encased with epoxy in a castable mould for thin sectioning and petrographic analyses. Then, polished sections were obtained for scanning electron microscopy, and a small bulk sample (10 grammes) was quartered for X-ray diffraction or chemical analyses. It was found that concrete chips were denser when they included nano-sized cement [33–35]. The effect of dry- and wet-ground cements on the matrices could not be found out on this scale of magnification. The following were analysed:

i. Normal cement cubes (Figures 7 and 8);
ii. Dry-ground cement cubes (Figures 9 and 10);
iii. Wet-ground cement cubes (Figures 11 and 12).

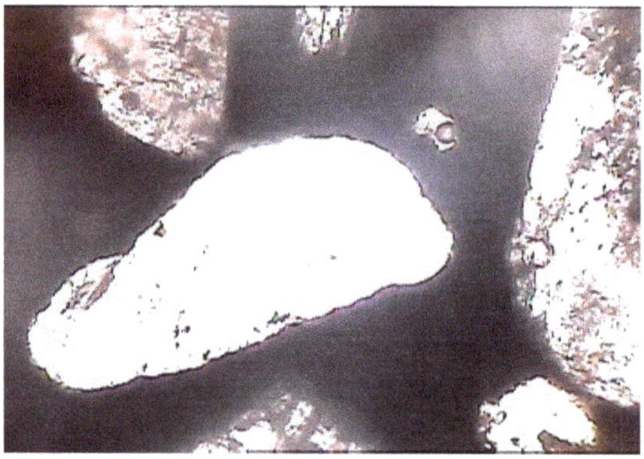

Figure 7. Photomicrograph of normal cement–fine-aggregate mix showing micropores within cement and loose contact between cement paste and aggregate grains (magnification of 60× under non-polarized light).

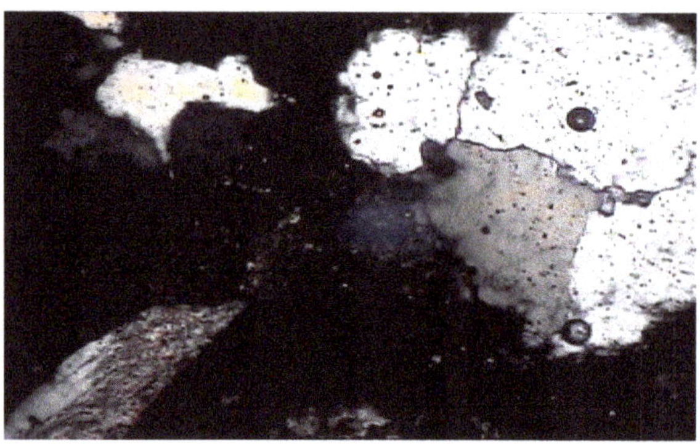

Figure 8. Photomicrograph of mix of nano-sized dry-ground cement and fine aggregates showing dense cement phase and good contact between cement paste and aggregate grains (magnification of 60× under non-polarized light).

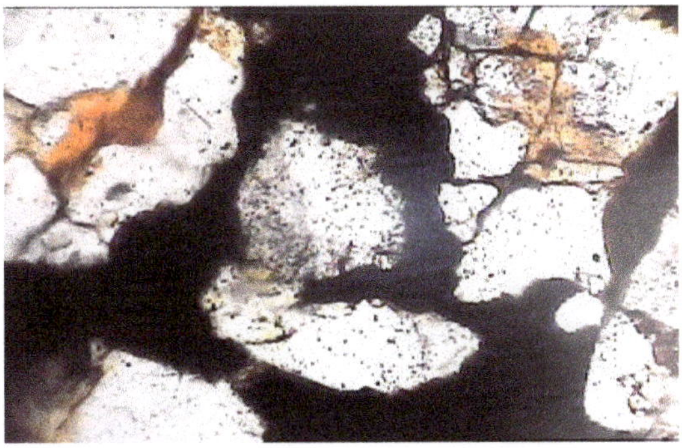

Figure 9. Photomicrograph of mix of nano-sized wet-ground cement and fine aggregates showing dense cement phase and good contact between cement paste and aggregate grains (magnification of 60× under non-polarized light).

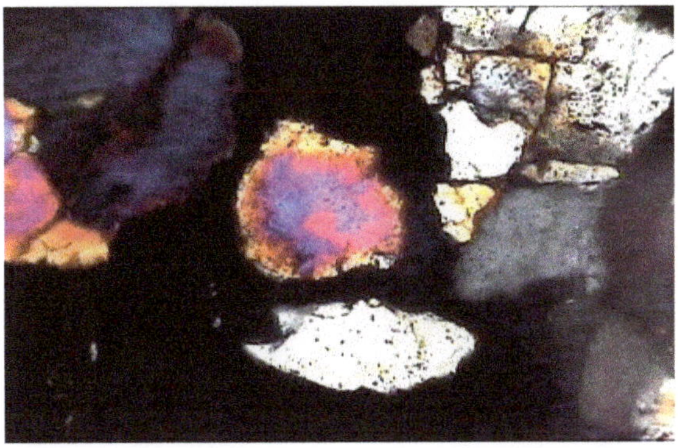

Figure 10. Photomicrograph of normal cement–fine-aggregate mix showing micropores within cement and loose contact between cement paste and aggregate grains (magnification of 60× under polarized light).

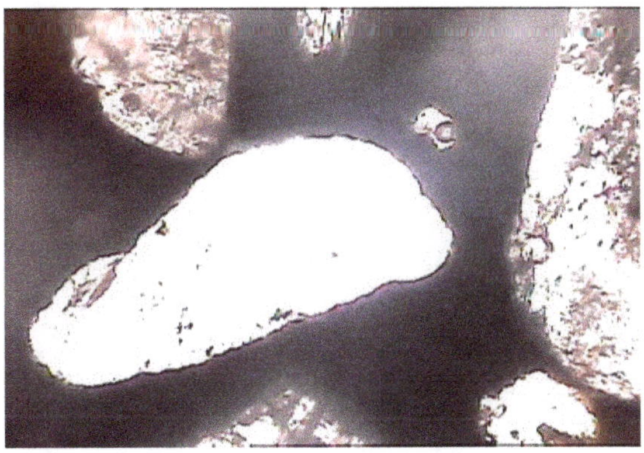

Figure 11. Photomicrograph of mix of nano-sized dry-ground cement and fine aggregates showing dense cement phase and good contact between cement paste and aggregate grains (magnification of 60× under polarized light).

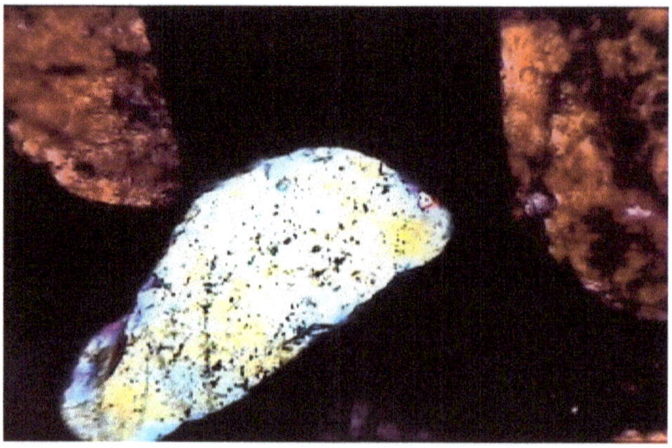

Figure 12. Photomicrograph of mix nano-sized wet-ground cement and fine aggregates showing dense cement phase and good contact between cement paste and aggregate grains (magnification of 60× under polarized light).

2.6. Statistical Analysis

Several statistical techniques were applied on the compressive strength values of different material compositions to statistically justify our study findings and create a predictive model using the variables [36–38].

A repeated measures ANOVA was implemented to test the difference between the compressive strength readings collected after 7, 14, and 28 days. A p-value below 0.05 defined significance among the groups [39–41]. Pearson correlation and stepwise multiple regression analyses were also executed to develop a predictive equation, with strength on the 28th day as the dependent variable and strengths on the 7th and 14th days as two categorical variables, while the type of grinding and a grouping variable were taken as the independent variables. Grouping was defined as the type of mixing, and the grinding variable was categorized as dry, wet, or standard [42–44]. Group 1 denotes the composition of cement and sand. Group 2 is the mixture of cement, sand, and silica fume. Lastly, Group 3 is the composition of cement, sand, and fly ash.

3. Results and Discussion

This paper presents a discussion of the test findings in order to derive both qualitative and quantitative inferences. Nanotechnology has begun to develop novel materials in recent years due to the need to use them for increasing the characteristics of various materials in general and materials utilized in the construction sector in particular [45–48]. The results obtained from compressive tests on cement samples are given in Table 6. Previous studies have investigated normal cement composites with nano-additives, whereas this research study focused on converting cement into nano-cement, followed by converting other additives into nano-additives; then, their individual and combined effects were studied [48–50]. Previous literature works did not study the nano-cement effects on the composites [51–53].

3.1. Compressive Strength of Cement Matrix

The results of 176 cement cube specimens of size 70.7 mm are given in Table 6. The graphs show variations mainly where the nano-cement and nano-construction materials with aggregates were used. The variations are discussed in detail below.

3.1.1. Test Results of Cubes with Cement and Nano-Cement

As shown in Figure 13, the compressive strength of nano-cement (converted cement) after 7, 14, and 28 days was greater than the compressive strength of normal cement. Nano-cement (wet-ground) showed greater strength than nano-cement (dry-ground) and normal cement [54–56]. The percentage increases in the strength of wet-ground cement compared with normal cement after 7, 14, and 28 days were 76, 75, and 90%, respectively.

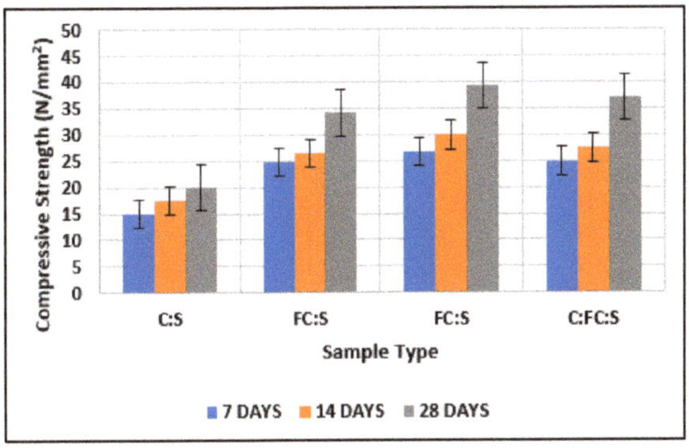

Figure 13. Comparison of test results of cubes with cement and fine cement.

In addition, the percentage increases in the strength of dry-ground cement compared with normal cement after 7, 14, and 28 days were 59, 55, and 65%, respectively. The strength of cubes made of cement and fine aggregate lies in the strength of cubes made of nano-cement matrix and normal cement matrix. The fineness of cement, which causes a greater surface area of cement to come into contact with water at successive phases, is responsible for the increase in strength in each step [57,58]. Furthermore, because there are fewer voids, the strength of the structure increases because cracks require space or gaps to propagate through [58,59].

3.1.2. Comparison of Test Results of Cubes with Cement, Nano-Cement, and Silica Fume with Those of Cubes with Cement, Nano-Cement, and Nano-Silica Fume

Figure 14 shows the compressive strength comparison of samples with the addition of 10% nano-silica fume and 10% normal silica fume. The graph shows that nano-cement having 10% nano-silica fume showed greater compressive strength than nano-cement having 10% normal silica fume. In addition, nano-cement with 10% nano-silica fume had greater compressive strength than normal cement with 10% normal silica fumes, and the findings depict it to be superior to the results of existing studies [60–62].

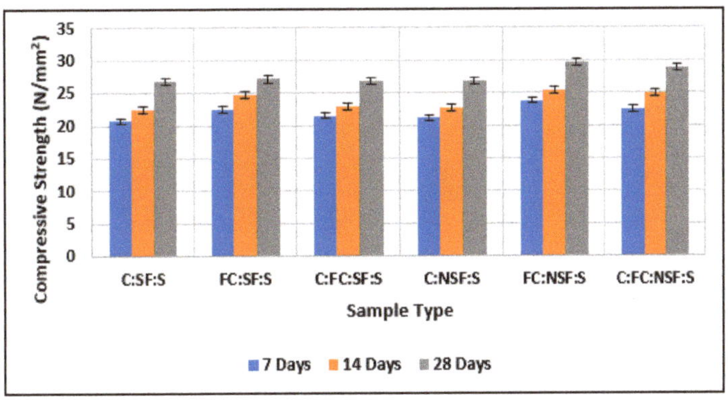

Figure 14. Comparison of test results of cubes with cement, fine cement, and silica fume with test results of cubes with cement, fine cement, and nano-silica fume.

3.1.3. Comparison of Test Results of Cubes with Cement and Nano-Cement with Test Results of Cubes with Cement, Nano-Cement, and Silica Fume and of Cubes with Cement, Nano-Cement, and Nano-Silica Fume

In Figure 15, it can be seen that the maximum strength was shown by the nano-cement matrix and the cement matrix that had nano-cement and nanomaterials, and the findings are superior to those of existing studies [63–65].

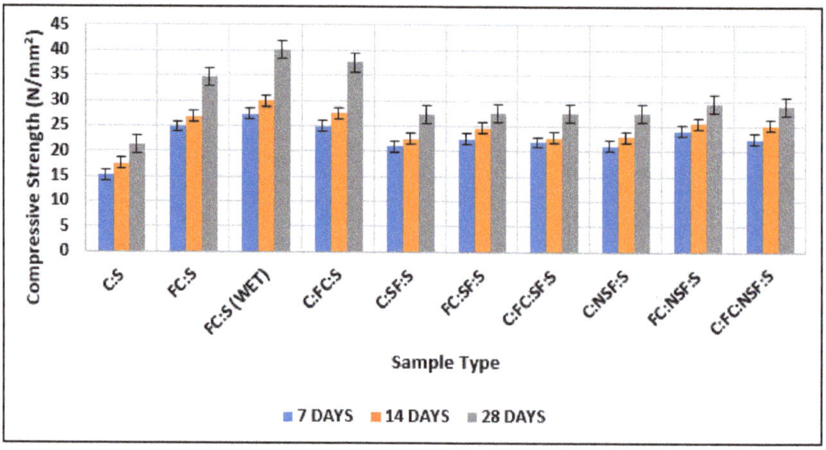

Figure 15. Comparison of test results of cubes with cement and fine cement with test results of cubes with cement, fine cement, and silica fume and with test results of cubes with cement, fine cement, and nano-silica fume.

3.1.4. Comparison of Test Results of Cubes with Cement, Nano-Cement, and Fly Ash with Those of Cubes with Cement, Nano-Cement, and Nano-Fly Ash

Table 7 and Figure 16 compare the compressive strength obtained by adding 10% nano-fly ash and 10% normal fly ash. This showed that nano-cement containing 10% nano-fly ash had greater compressive strength than nano-cement containing 10% regular fly ash. In addition, nano-cement with 10% nano-fly ash showed greater compressive strength than normal cement with 10% fly ash, and the related findings have been proved in numerous literature studies [66–68].

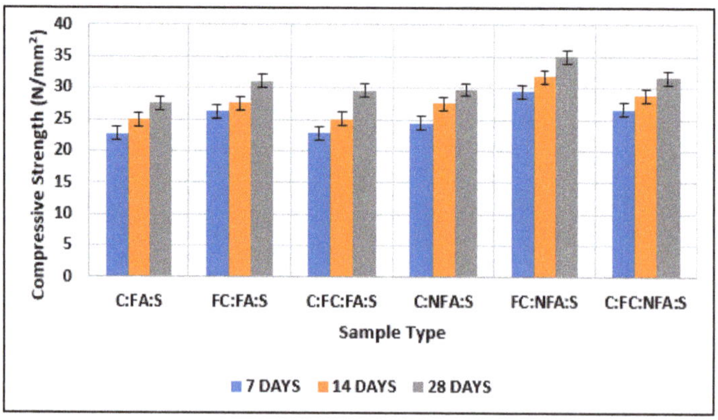

Figure 16. Comparison of test results of cubes with cement, fine cement, and fly ash with test results of cubes with cement, fine cement, and nano-fly ash.

3.1.5. Comparison of Test Results of Cubes with Cement and Nano-Cement with Those of Cubes with Cement, Nano-Cement, and Fly Ash and of Cubes with Cement, Nano-Cement, and Nano-Fly Ash

The maximum strength was shown by the nano-cement matrix and the cement matrix with nano-cement and nanomaterials, as shown in Figure 17. Wet-ground cement attained the maximum value, and the normal cement matrix attained the minimum, which is somewhat related to the results of existing studies [69–72].

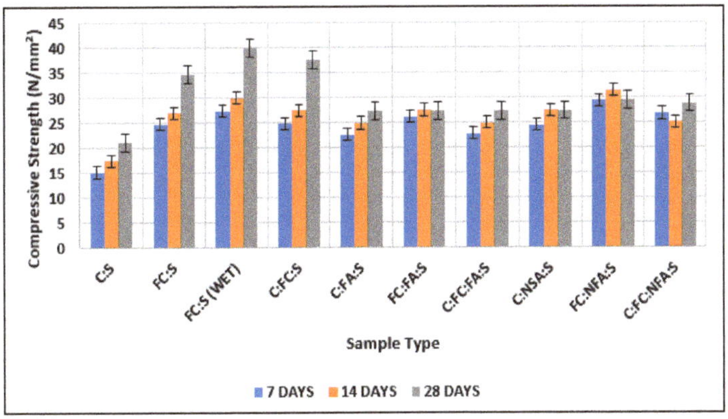

Figure 17. Comparison of test results of cubes with cement and fine cement with test results of cubes with cement, fine cement, and fly ash and with test results of cubes with cement, fine cement, and nano-fly ash.

3.1.6. Overall Comparison of Test Results of Cubes with Cement, Nano-Cement, Silica Fume, Nano-Silica Fume, Fly Ash, and Nano-Fly Ash

Figure 18 shows the summary of all the tables. We can easily conclude that the matrix gained the maximum strength with nano-cement, nano-silica fume, and nano-fly ash. In this research work, the focus was on the comparison of the strength parameters of normal cement composites with those of nano-cement composites. This work can be extended in the future to study the effect of nano-additives, including particle shape, impurity, defects, etc., within cement matrices.

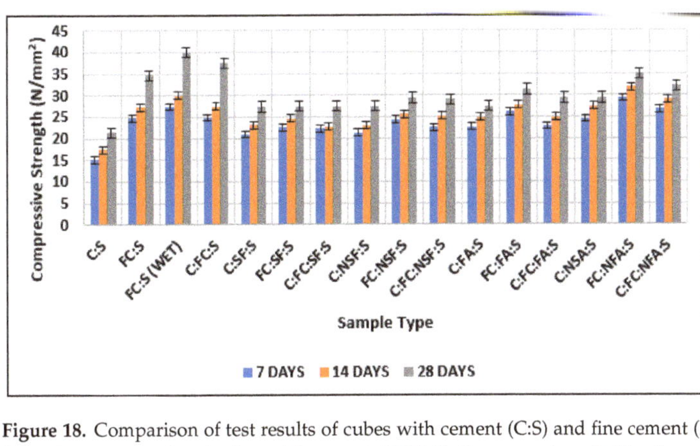

Figure 18. Comparison of test results of cubes with cement (C:S) and fine cement (FC:S) with the following: test results of cubes with cement (C:FC:S), fine cement, and silica fume; test results of cubes with cement, fine cement, and nano-silica fume; test results of cubes with cement, fine cement, and fly ash; test results of cubes with cement, fine cement, and nano-fly ash; and test results of cubes with cement.

3.2. Compressive Strength of Concrete Matrix

From Figure 19, it is clearly visible that the concrete cubes that contained only 25% wet-ground cement had greater compressive strength than the concrete cubes with normal cement for the M20 mix design.

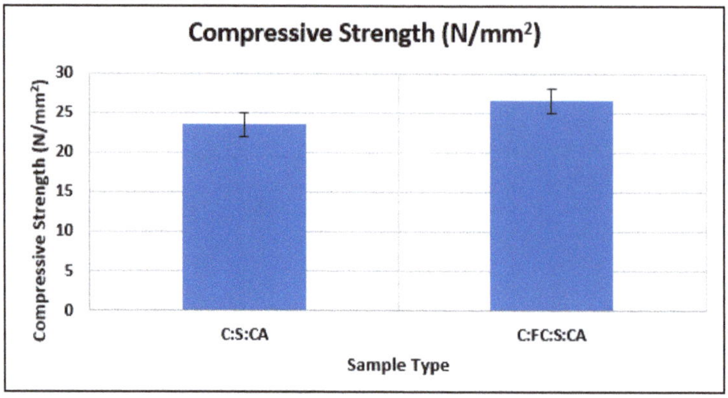

Figure 19. Comparison of test results of concrete cubes with cement and 25% fine cement.

Since the nano-particles fill the gaps among cement grains and consume a portion of the calcium hydroxide present, the production of extra calcium silicate hydrate (C-S-H) results in an improvement in the interface structure over and beyond what has been previously achieved. The increase in strength can be attributed to the fineness of cement, which allows a higher surface area of cement to come into contact with water and denser micro-fabric to be formed, limiting the number and size of voids, thus improving strength [72–74]. The properties of nano-sized wet-ground cement and fine aggregates were comparable to those of nano-sized dry-ground cement. As a result, no distinction could be detected at the petrographic microscope magnification level. Though transmission electron microscopy, we could distinguish the two in terms of the growth of interfingering crystal mesh [75–77].

The matrix with nano-cement, nano-silica fume, and nano-fly ash gained the most strength and was shown to have substantial improvement in terms of the compressive strength of cementitious composites after 7–28 days of curing.

3.3. Microscopic Examination of Cement Matrix

From the general photomicrographic studies, it seemed that micro-pores were evident within the cement paste in the case of normal cement. The contacts with the aggregate grains were also not very dense (Figures 7 and 8).

In nano-sized dry-ground cement, the mix showed a dense cement phase and good cement–grain contact (Figures 9 and 10). Nano-sized dry-ground cement, in some places, showed that rounded pores had formed due to air tramping in the mix.

The mix made of nano-sized wet-ground cement and aggregates showed characteristics similar to those of nano-sized dry-ground cement [76–78]. As such, no differentiation could be made at the magnification level of petrographic microscopy (Figures 11 and 12). Detailed scanning electron microscopy may help to unravel the difference between the two. However, the TEM studies on cement showed the development of nanorods with increasing length and slenderness in dry- and wet-ground nano-cement compared with normal cement, which may have played an essential role in the hydration and solidification process involved in the hardening of cement [77,78]. The densifying properties of nano-composites resulted in the reduction in pores in the cement matrix, which enhanced the overall strength. This was evident in the photomicrographic studies. These findings explain the relationship between the strength and the microscopic structure of the cementitious composites.

However, Y.F. Ling et al., Y.R. Jeng et al., and L. Wei et al. have provided comprehensive reviews on the effect of grain size and orientation on the mechanical strength of crystalline nanomaterials [76–79].

3.4. Statistical Findings

Since the compressive strength readings were measured repeatedly in three different periods, the three readings were dependent on each other; hence, there was a strong correlation among them [78,79]. The correlation coefficient between the 7th- and 14th-day strength was 0.986 (p-value < 0.05); that between the 7th- and 28th-day strength was 0.828 (p-value < 0.05); that between the 14th- and 28th-day strength was 0.882 (p-value < 0.05).

The final model consisted of two independent variables, the 14th-day compressive strength and the grouping variable, for the implementation of stepwise multiple regression analyses. These two factors were the only significant variables that the 28th-day strength depended on (p-value <0.05). The 7th-day strength and grinding factors were not significant. The regression equation is given below, and this is justified by a very strong, adjusted R-squared value of 93.8%.

$$\text{28th day compressive strength} = 1.391 * \text{14th day compressive strength} — 2.483 * \text{Group} — 0.158$$

With the help of the above equation, it is possible to calculate the 28th-day compressive strength, and only the 14th-day strength and the type of composition are required.

3.5. Repeated Measures ANOVA

The assumptions of equal variance and equal covariance for applying the repeated measures ANOVA test were satisfied. Box's test of equality of covariance matrices, Levene's test of equality of variance, and Mauchly's test of sphericity were all non-significant (p-value > 0.05) [77–79]. The repeated measures ANOVA method resulted in a significant difference between the three compressive strength readings (p-value <0.05) for the group and the grinding categories. The exact difference was identified using Scheffe's multiple comparison post hoc test. There was a significant difference among all three types of methods in the case of the grinding factor: standard, wet, and dry (p-value < 0.05). However, the difference was only found between Group 1 and Group 2 (p-value < 0.05) for the group variable. Within-group differences were also tested, and a significant difference was found among the three groups and in the grinding process.

4. Conclusions

The following findings were reached as a result of experimental research: The use of nano-additions improved the characteristics of cement matrix and concrete. Because nano-particles filled the gaps between cement grains and consumed a portion of the calcium hydroxide present, the production of extra calcium silicate hydrate (C-S-H) resulted in an improvement in the interface structure over and beyond what has been achieved previously. The increase in strength can be attributed to the fineness of cement, which allows a higher surface area of the cement to come into contact with water and denser micro-fabric to be formed, limiting the number and size of voids, thus improving strength. The properties of nano-sized wet-ground cement and fine aggregates were comparable to those of nano-sized dry-ground cement. As a result, no distinction could be detected at the petrographic microscope magnification level. Though transmission electron microscopy, the two could be distinguished in terms of the growth of interfingering crystal mesh.

The matrix with nano-cement, nano-silica fume, and nano-fly ash gained the greatest strength and was shown to have substantial improvement in terms of the compressive strength of cementitious composites after 7–28 days of curing.

All of the varied mixes using nano-sized dry cement, silica fume, and fly ash in various permutations and combinations provided strength values ranging from 26.75 to 35.20 MPa, which were greater than that of the typical cement–fine-aggregate mix (20.88 MPa). The 50:50 combination of normal and wet-ground cement produced a compressive strength of 37.20 MPa after 28 days. When comparing cube specimens utilizing wet-ground nano-cement (40 MPa) with cube specimens using conventional cement, an increase of up to 92 percent was found (20.88 MPa).

A strong correlation was found among the three compressive strength readings, and a linear equation was composed to calculate the 28th-day compressive strength. A significant difference was identified among the three strength values for the group and the grinding categories.

Author Contributions: Conceptualization, I.R., P.S., N.D., M.A., F.N.K.Y., A.A., M.M.A., S.S. (Sandeep Singh), J.S.C., R.K. and S.S. (Shubham Sharma); methodology, I.R., P.S., N.D., M.A., F.N.K.Y., A.A., M.M.A., S.S. (Sandeep Singh), J.S.C., R.K. and S.S. (Shubham Sharma); formal analysis, S.S. (Sandeep Singh), J.S.C., R.K., L.S. and S.S. (Shubham Sharma); investigation, I.R., P.S., N.D., M.A., F.N.K.Y., A.A., M.M.A., S.S. (Sandeep Singh), J.S.C., R.K., L.S. and S.S. (Shubham Sharma); writing—original draft preparation, I.R., P.S., N.D., M.A., F.N.K.Y., A.A., M.M.A., S.S. (Sandeep Singh), J.S.C., R.K. and S.S. (Shubham Sharma); writing—review and editing, L.S., S.S. (Shubham Sharma), M.R.M.A. and E.T.-E.; supervision, S.S. (Sandeep Singh), S.S. (Shubham Sharma), M.R.M.A. and E.T.-E.; project administration, S.S. (Shubham Sharma) and E.T.-E.; funding acquisition, S.S. (Shubham Sharma) and E.T.-E. All authors have read and agreed to the published version of the manuscript.

Funding: This research received no external funding.

Institutional Review Board Statement: Not applicable.

Informed Consent Statement: Not applicable.

Data Availability Statement: No data were used to support this study.

Acknowledgments: The authors would like to acknowledge the support of Future University, Egypt, for paying the Article Processing Charges (APC) of this publication.

Conflicts of Interest: The authors declare no conflict of interest.

References

1. Bartos, P.J. Nanotechnology in construction: A roadmap for development. In *Nanotechnology in Construction 3*; Springer: Berlin/Heidelberg, Germany, 2009; pp. 15–26.
2. Björnström, J.; Martinelli, A.; Matic, A.; Börjesson, L.; Panas, I. Accelerating effects of colloidal nano-silica for beneficial calcium–silicate–hydrate formation in cement. *Chem. Phys. Lett.* **2004**, *392*, 242–248. [CrossRef]
3. Chen, J.J.; Thomas, J.J.; Taylor, H.F.; Jennings, H.M. Solubility and structure of calcium silicate hydrate. *Cem. Concr. Res.* **2004**, *34*, 1499–1519. [CrossRef]
4. Chang, T.P.; Shih, J.Y.; Yang, K.M.; Hsiao, T.C. Material properties of Portland cement paste with nano-montmorillonite. *J. Mater. Sci.* **2007**, *42*, 7478–7487. [CrossRef]
5. Cong, X.; Kirkpatrick, R.J. 29Si and 17O NMR investigation of the structure of some crystalline calcium silicate hydrates. *Adv. Cem. Based Mater.* **1996**, *3*, 133–143. [CrossRef]
6. Feynman, R. There's plenty of room at the bottom. Presented at the Annual Meeting of the American Physical Society, Pasadena, CA, USA, 29 December 1959.
7. Ji, T. Preliminary study on the water permeability and microstructure of concrete incorporating nano-SiO_2. *Cem. Concr. Res.* **2005**, *35*, 1943–1947. [CrossRef]
8. Jo, B.W.; Kim, C.H.; Tae, G.H.; Park, J.B. Characteristics of cement mortar with nano-SiO_2 particles. *Constr. Build. Mater.* **2007**, *21*, 1351–1355. [CrossRef]
9. Kuo, W.Y.; Huang, J.S.; Lin, C.H. Effects of organo-modified montmorillonite on strengths and permeability of cement mortars. *Cem. Concr. Res.* **2006**, *36*, 886–895. [CrossRef]
10. Lee, S.J.; Kriven, W.M. Synthesis and hydration study of Portland cement components prepared by the organic steric entrapment method. *Mater. Struct.* **2005**, *38*, 87–92. [CrossRef]
11. Li, H.; Zhang, M.H.; Ou, J.P. Abrasion resistance of concrete containing nano-particles for pavement. *Wear* **2006**, *260*, 1262–1266. [CrossRef]
12. Li, H.; Xiao, H.G.; Ou, J.P. A study on mechanical and pressure-sensitive properties of cement mortar with nanophase materials. *Cem. Concr. Res.* **2004**, *34*, 435–438. [CrossRef]
13. Lin, D.F.; Lin, K.L.; Chang, W.C.; Luo, H.L.; Cai, M.Q. Improvements of nano-SiO_2 on sludge/fly ash mortar. *Waste Manag.* **2008**, *28*, 1081–1087. [CrossRef] [PubMed]

14. Lin, K.L.; Chang, W.C.; Lin, D.F.; Luo, H.L.; Tsai, M.C. Effects of nano-SiO_2 and different ash particle sizes on sludge ash–cement mortar. *J. Environ. Manag.* **2008**, *88*, 708–714. [CrossRef] [PubMed]
15. Li, H.; Xiao, H.G.; Yuan, J.; Ou, J. Microstructure of cement mortar with nano-particles. *Compos. Part B Eng.* **2004**, *35*, 185–189. [CrossRef]
16. Li, H.; Zhang, M.H.; Ou, J.P. Flexural fatigue performance of concrete containing nano-particles for pavement. *Int. J. Fatigue* **2007**, *29*, 1292–1301. [CrossRef]
17. Li, Z.; Wang, H.; He, S.; Lu, Y.; Wang, M. Investigations on the preparation and mechanical properties of the nano-alumina reinforced cement composite. *Mater. Lett.* **2006**, *60*, 356–359. [CrossRef]
18. Minet, J.; Abramson, S.; Bresson, B.; Franceschini, A.; Van Damme, H.; Lequeux, N. Organic calcium silicate hydrate hybrids: A new approach to cement based nanocomposites. *J. Mater. Chem.* **2006**, *16*, 1379–1383. [CrossRef]
19. Sanchez, F.; Sobolev, K. Nanotechnology in concrete—A review. *Constr. Build. Mater.* **2010**, *24*, 2060–2071. [CrossRef]
20. Scrivener, K.L.; Kirkpatrick, R.J. Innovation in use and research on cementitious material. *Cem. Concr. Res.* **2008**, *38*, 128–136. [CrossRef]
21. Sobolev, K.; Gutiérrez, M.F. How nanotechnology can change the concrete world: Part two of a two-part series. *Am. Ceram. Soc. Bull.* **2005**, *84*, 16–19.
22. Sobolev, K.; Flores, I.; Torres-Martinez, L.M.; Valdez, P.L.; Zarazua, E.; Cuellar, E.L. Engineering of SiO_2 nanoparticles for optimal performance in nano cement-based materials. In *Nanotechnology in Construction 3*; Springer: Berlin/Heidelberg, Germany, 2009; pp. 139–148.
23. Shah, S.P.; Hou, P.; Konsta-Gdoutos, M.S. Nanomodification of cementitious material: Toward a stronger and durable concrete. *J. Sustain. Cem.-Based Mater.* **2015**, *5*, 1–22. [CrossRef]
24. Qing, Y.; Zenan, Z.; Deyu, K.; Rongshen, C. Influence of nano-SiO_2 addition on properties of hardened cement paste as compared with silica fume. *Constr. Build. Mater.* **2007**, *21*, 539–545. [CrossRef]
25. Qing, Y.E.; Zenan, Z.; Li, S.; Rongshen, C. A comparative study on the pozzolanic activity between nano-SiO_2 and silica fume. *J. Wuhan Univ. Technol.-Mater. Sci. Ed.* **2006**, *21*, 153–157. [CrossRef]
26. Zhang, R.; Cheng, X.; Hou, P.; Ye, Z. Influences of nano-TiO_2 on the properties of cement-based materials: Hydration and drying shrinkage. *Constr. Build. Mater.* **2015**, *81*, 35–41. [CrossRef]
27. Roslan, Z.B.; Ramli, Z.; Razman, M.R.; Asyraf, M.R.M.; Ishak, M.R.; Ilyas, R.A.; Nurazzi, N.M. Reflections on Local Community Identity by Evaluating Heritage Sustainability Protection in Jugra, Selangor, Malaysia. *Sustainability* **2021**, *13*, 8705. [CrossRef]
28. Izadifar, M.; Natzeck, C.; Emmerich, K.; Weidler, P.G.; Gohari, S.; Burvill, C.; Thissen, P. Unexpected Chemical Activity of a Mineral Surface: The Role of Crystal Water in Tobermorite. *J. Phys. Chem. C* **2022**, *126*, 12405–12412. [CrossRef]
29. Izadifar, M.; Koöniger, F.; Gerdes, A.; Woöll, C.; Thissen, P. Correlation between composition and mechanical properties of calcium silicate hydrates identified by infrared spectroscopy and density functional theory. *J. Phys. Chem. C* **2019**, *123*, 10868–10873. [CrossRef]
30. Izadifar, M.; Dolado, J.S.; Thissen, P.; Ayuela, A. Interactions between reduced graphene oxide with monomers of (calcium) silicate hydrates: A first-principles study. *Nanomaterials* **2021**, *11*, 2248. [CrossRef]
31. Moradpour, R.; Taheri-Nassaj, E.; Parhizkar, T.; Ghodsian, M. The effects of nanoscale expansive agents on the mechanical properties of non-shrink cement-based composites: The influence of nano-MgO addition. *Compos. Part B Eng.* **2013**, *55*, 193–202. [CrossRef]
32. Tiwary, A.K.; Singh, S.; Kumar, R.; Chohan, J.S.; Sharma, S.; Singh, J.; Li, C.; Ilyas, R.A.; Asyraf, M.R.M.; Malik, M.A. Effects of Elevated Temperature on the Residual Behavior of Concrete Containing Marble Dust and Foundry Sand. *Materials* **2022**, *15*, 3632. [CrossRef]
33. Tiwary, A.K.; Singh, S.; Kumar, R.; Sharma, K.; Chohan, J.S.; Sharma, S.; Singh, J.; Kumar, J.; Deifalla, A.F. Comparative Study on the Behavior of Reinforced Concrete Beam Retrofitted with CFRP Strengthening Techniques. *Polymers* **2022**, *14*, 4024. [CrossRef]
34. Singh, S.; Khairandish, M.I.; Razahi, M.M.; Kumar, R.; Chohan, J.S.; Tiwary, A.; Sharma, S.; Li, C.; Ilyas, R.A.; Asyraf, M.R.M.; et al. Preference Index of Sustainable Natural Fibers in Stone Matrix Asphalt Mixture Using Waste Marble. *Materials* **2022**, *15*, 2729. [CrossRef] [PubMed]
35. Kumar, M.; Anand, A.; Chatterjee, R.; Sharma, S.; Maiti, T.K.; Dwivedi, S.P.; Saxena, A.; Li, C.; Eldin, E.M.T. Investigation on Carbonation and Permeability of Concrete with Rice Hush Ash and Shop Solution Addition. *Materials* **2022**, *15*, 6149. [CrossRef]
36. Ilyas, R.A.; Zuhri, M.Y.M.; Norrrahim, M.N.F.; Misenan, M.S.M.; Jenol, M.A.; Samsudin, S.A.; Nurazzi, N.M.; Asyraf, M.R.M.; Supian, A.B.M.; Bangar, S.P.; et al. Natural Fiber-Reinforced Polycaprolactone Green and Hybrid Biocomposites for Various Advanced Applications. *Polymers* **2022**, *14*, 182. [CrossRef] [PubMed]
37. Ilyas, R.A.; Zuhri, M.Y.M.; Aisyah, H.A.; Asyraf, M.R.M.; Hassan, S.A.; Zainudin, E.S.; Sapuan, S.M.; Sharma, S.; Bangar, S.P.; Jumaidin, R.; et al. Natural Fiber-Reinforced Polylactic Acid, Polylactic Acid Blends and Their Composites for Advanced Applications. *Polymers* **2022**, *14*, 202. [CrossRef] [PubMed]
38. Wang, S.; Li, V.C. Engineered cementitious composites with high-volume fly ash. *ACI Mater. J.* **2007**, *104*, 233–241.
39. Rajawat, A.S.; Singh, S.; Gangil, B.; Ranakoti, L.; Sharma, S.; Asyraf, M.R.M.; Razman, M.R. Effect of Marble Dust on the Mechanical, Morphological, and Wear Performance of Basalt Fibre-Reinforced Epoxy Composites for Structural Applications. *Polymers* **2022**, *14*, 1325. [CrossRef]

40. Asyraf, M.R.M.; Ishak, M.R.; Sapuan, S.M.; Yidris, N.; Ilyas, R.A.; Rafidah, M.; Razman, M.R. Potential Application of Green Composites for Cross Arm Component in Transmission Tower: A Brief Review. *Int. J. Polym. Sci.* **2020**, *2020*, 8878300. [CrossRef]
41. Alsubari, S.; Zuhri, M.Y.M.; Sapuan, S.M.; Ishak, M.R.; Ilyas, R.A.; Asyraf, M.R.M. Potential of Natural Fiber Reinforced Polymer Composites in Sandwich Structures: A Review on Its Mechanical Properties. *Polymers* **2021**, *13*, 423. [CrossRef]
42. Asyraf, M.R.M.; Ishak, M.R.; Syamsir, A.; Amir, A.L.; Nurazzi, N.M.; Norrrahim, M.N.F.; Asrofi, M.; Rafidah, M.; Ilyas, R.A.; Rashid, M.Z.A.; et al. Filament-wound glass-fibre reinforced polymer composites: Potential applications for cross arm structure in transmission towers. *Polym. Bull.* **2022**, 1–26. [CrossRef]
43. Sharma, S.; Patyal, V.; Sudhakara, P.; Singh, J.; Petru, M.; Ilyas, R.A. Mechanical, Morphological, and Fracture-deformation behavior of MWCNTs reinforced (Al-Cu-Mg-T351) alloy cast nanocomposites fabricated by optimized Mechanical milling and Powder metallurgy techniques. *Nanotechnol. Rev.* **2022**, *11*, 65–85. [CrossRef]
44. Chohan, J.S.; Mittal, N.; Kumar, R.; Singh, S.; Sharma, S.; Dwivedi, S.; Saxena, A.; Chattopadhyaya, S.; Ilyas, R.; Le, C.; et al. Optimization of FFF process parameters by naked mole-rat algorithms with enhanced exploration and exploitation capabilities. *Polymers* **2021**, *13*, 1702. [CrossRef] [PubMed]
45. Ilyas, R.A.; Sapuan, S.M.; Asyraf, M.R.M.; Dayana, D.A.Z.N.; Amelia, J.J.N.; Rani, M.S.A.; Norrrahim, M.N.F.; Nurazzi, N.M.; Aisyah, H.A.; Sharma, S.; et al. Polymer Composites Filled with Metal Derivatives: A Review of Flame Retardants. *Polymers* **2021**, *13*, 1701. [CrossRef] [PubMed]
46. Chohan, J.S.; Mittal, N.; Kumar, R.; Singh, S.; Sharma, S.; Singh, J.; Rao, K.V.; Mia, M.; Pimenov, D.Y.; Dwivedi, S.P. Mechanical Strength Enhancement of 3D Printed Acrylonitrile Butadiene Styrene Polymer Components Using Neural Network Optimization Algorithm. *Polymers* **2020**, *12*, 2250. [CrossRef] [PubMed]
47. Zhang, C.; Ali, A.; Sun, L. Investigation on low-cost friction-based isolation systems for masonry building structures: Experimental and numerical studies. *Eng. Struct.* **2021**, *243*, 112645. [CrossRef]
48. Sharma, S.; Singh, J.; Kumar, H.; Sharma, A.; Aggarwal, V.; Gill, A.S.; Jayarambabu, N.; Kailasa, S.; Rao, K.V. Utilization of rapid prototyping technology for the fabrication of an orthopedic shoe inserts for foot pain reprieve using thermo-softening viscoelastic polymers: A novel experimental approach. *Meas. Control* **2020**, *53*, 519–530. [CrossRef]
49. Chohan, J.S.; Kumar, R.; Singh, T.H.B.; Singh, S.; Sharma, S.; Singh, J.; Mia, M.; Pimenov, D.Y.; Chattopadhyaya, S.; Dwivedi, S.P.; et al. Taguchi S/N and TOPSIS based optimization of fused deposition modelling and vapor finishing process for manufacturing of ABS plastic parts. *Materials* **2020**, *13*, 5176. [CrossRef]
50. Huang, H.; Huang, M.; Zhang, W.; Pospisil, S.; Wu, T. Experimental Investigation on Rehabilitation of Corroded RC Columns with BSP and HPFL under Combined Loadings. *J. Struct. Eng.* **2020**, *146*, 04020157. [CrossRef]
51. Sharma, S.; Sudhakara, P.; Singh, J.; Ilyas, R.A.; Asyraf, M.R.M.; Razman, M.R. Critical review of biodegradable and bioactive polymer composites for bone tissue engineering and drug delivery applications. *Polymers* **2021**, *13*, 2623. [CrossRef]
52. Sharma, S.; Sudhakara, P.; Omran, A.A.B.; Singh, J.; Ilyas, R.A. Recent Trends and Developments in Conducting Polymer Nanocomposites for Multifunctional Applications. *Polymers* **2021**, *13*, 2898. [CrossRef]
53. Jha, K.; Tyagi, Y.K.; Kumar, R.; Sharma, S.; Huzaifah, M.R.M.; Li, C.; Ilyas, R.A.; Dwivedi, S.P.; Saxena, A.; Pramanik, A. Assessment of dimensional stability, biodegradability, and fracture energy of bio-composites reinforced with novel pine cone. *Polymers* **2021**, *13*, 3260. [CrossRef]
54. Kadier, A.; Ilyas, R.A.; Huzaifah, M.R.M.; Harihastuti, N.; Sapuan, S.M.; Harussani, M.M.; Azlin, M.N.M.; Yuliasni, R.; Ibrahim, R.; Atikah, M.S.N.; et al. Use of Industrial Wastes as Sustainable Nutrient Sources for Bacterial Cellulose (BC) Production: Mechanism, Advances, and Future Perspectives. *Polymers* **2021**, *13*, 3365. [CrossRef] [PubMed]
55. Singh, Y.; Singh, J.; Sharma, S.; Lam, T.-D.; Nguyen, D.-N. Fabrication and characterization of coir/carbon-fiber reinforced epoxy based hybrid composite for helmet shells and sports-good applications: Influence of fiber surface modifications on the mechanical, thermal and morphological properties. *J. Mater. Res. Technol.* **2020**, *9*, 15593–15603. [CrossRef]
56. MSuriani, M.J.; Ilyas, R.A.; Zuhri, M.Y.M.; Khalina, A.; Sultan, M.T.H.; Sapuan, S.M.; Ruzaidi, C.M.; Wan, F.N.; Zulkifli, F.; Harussani, M.M.; et al. Critical Review of Natural Fiber Reinforced Hybrid Composites: Processing, Properties, Applications and Cost. *Polymers* **2021**, *13*, 3514. [CrossRef]
57. Kumar, R.; Ranjan, N.; Kumar, V.; Kumar, R.; Chohan, J.S.; Yadav, A.; Piyush; Sharma, S.; Prakash, C.; Singh, S.; et al. Characterization of Friction Stir-Welded Polylactic Acid/Aluminum Composite Primed through Fused Filament Fabrication. *J. Mater. Eng. Perform.* **2021**, *31*, 2391–2409. [CrossRef]
58. Azlin, M.N.M.; Ilyas, R.A.; Zuhri, M.Y.M.; Sapuan, S.M.; Harussani, M.M.; Sharma, S.; Nordin, A.H.; Nurazzi, N.M.; Afiqah, A.N. 3D Printing and Shaping Polymers, Composites, and Nanocomposites: A Review. *Polymers* **2022**, *14*, 180. [CrossRef] [PubMed]
59. Huang, H.; Huang, M.; Zhang, W.; Yang, S. Experimental study of predamaged columns strengthened by HPFL and BSP under combined load cases. *Struct. Infrastruct. Eng.* **2020**, *17*, 1210–1227. [CrossRef]
60. Wei, J.; Xie, Z.; Zhang, W.; Luo, X.; Yang, Y.; Chen, B. Experimental study on circular steel tube-confined reinforced UHPC columns under axial loading. *Eng. Struct.* **2021**, *230*, 111599. [CrossRef]
61. Zhang, W.; Liu, X.; Huang, Y.; Tong, M.-N. Reliability-based analysis of the flexural strength of concrete beams reinforced with hybrid BFRP and steel rebars. *Arch. Civ. Mech. Eng.* **2022**, *22*, 171. [CrossRef]
62. Wang, X.; Yang, Y.; Yang, R.; Liu, P. Experimental Analysis of Bearing Capacity of Basalt Fiber Reinforced Concrete Short Columns under Axial Compression. *Coatings* **2022**, *12*, 654. [CrossRef]

63. Yuan, J.; Lei, D.; Shan, Y.; Tong, H.; Fang, X.; Zhao, J. Direct Shear Creep Characteristics of Sand Treated with Microbial-Induced Calcite Precipitation. *Int. J. Civ. Eng.* **2022**, *20*, 763–777. [CrossRef]
64. Shan, Y.; Zhao, J.; Tong, H.; Yuan, J.; Lei, D.; Li, Y. Effects of activated carbon on liquefaction resistance of calcareous sand treated with microbially induced calcium carbonate precipitation. *Soil Dyn. Earthq. Eng.* **2022**, *161*, 107419. [CrossRef]
65. Hu, Z.; Shi, T.; Cen, M.; Wang, J.; Zhao, X.; Zeng, C.; Zhou, Y.; Fan, Y.; Liu, Y.; Zhao, Z. Research progress on lunar and Martian concrete. *Constr. Build. Mater.* **2022**, *343*, 128117. [CrossRef]
66. Cheng, H.; Sun, L.; Wang, Y.; Chen, X. Effects of actual loading waveforms on the fatigue behaviours of asphalt mixtures. *Int. J. Fatigue* **2021**, *151*, 106386. [CrossRef]
67. Cheng, H.; Liu, L.; Sun, L. Bridging the Gap between Laboratory and Field Moduli of Asphalt Layer for Pavement Infrastructure Design and Assessment: A Comprehensive Loading Frequency-Based Approach. *Front. Struct. Civ. Eng.* **2022**, *16*, 267–280. [CrossRef]
68. Guo, Y.; Yang, Y.; Kong, Z.; He, J.; Wu, H. Development of Similar Materials for Liquid-Solid Coupling and Its Application in Water Outburst and Mud Outburst Model Test of Deep Tunnel. *Geofluids* **2022**, *2022*, 8784398. [CrossRef]
69. Zhu, Z.; Wu, Y.; Liang, Z. Mining-Induced Stress and Ground Pressure Behavior Characteristics in Mining a Thick Coal Seam with Hard Roofs. *Front. Earth Sci.* **2022**, *10*, 843191. [CrossRef]
70. Shi, L.; Xiao, X.; Wang, X.; Liang, H.; Wang, D. Mesostructural characteristics and evaluation of asphalt mixture contact chain complex networks. *Constr. Build. Mater.* **2022**, *340*, 127753. [CrossRef]
71. Xu, H.; Wang, X.; Liu, C.; Chen, J.; Zhang, C. A 3D root system morphological and mechanical model based on L-Systems and its application to estimate the shear strength of root-soil composites. *Soil Tillage Res.* **2021**, *212*, 105074. [CrossRef]
72. Islam, S.; Dwivedi, S.P.; Dwivedi, V.K.; Sharma, S.; Kozak, D. Development of Marble Dust/Waste PET Based Polymer Composite Material for Environmental Sustainability: Fabrication and Characterizations. *J. Mater. Perform. Charact.* **2021**, *10*, 538–552. [CrossRef]
73. Lan, M.Y.; Zheng, M.B.; Shi, T.; Ma, C.; Liu, Y.; Zhao, Z. Crack resistance property of carbon nanotubes-modified concrete. *Mag. Concr. Res.* **2022**, *74*, 1165–1175. [CrossRef]
74. Wang, J.; Meng, Q.; Zou, Y.; Qi, Q.; Tan, K.; Santamouris, M.; He, B. Performance synergism of pervious pavement on stormwater management and urban heat island mitigation: A review of its benefits, key parameters, and co-benefits approach. *Water Res.* **2022**, *221*, 118755. [CrossRef]
75. Khare, J.M.; Dahiya, S.; Gangil, B.; Ranakoti, L.; Sharma, S.; Huzaifah, M.R.M.; Ilyas, R.A.; Dwivedi, S.P.; Chattopadhyaya, S.; Kilinc, H.C.; et al. Comparative Analysis of Erosive Wear Behaviour of Epoxy, Polyester and Vinyl Esters Based Thermosetting Polymer Composites for Human Prosthetic Applications using Taguchi Design. *Polymers* **2021**, *13*, 3607. [CrossRef] [PubMed]
76. Asyraf, M.R.M.; Syamsir, A.; Zahari, N.M.; Supian, A.B.M.; Ishak, M.R.; Sapuan, S.M.; Sharma, S.; Rashedi, A.; Razman, M.R.; Zakaria, S.Z.S.; et al. Product Development of Natural Fibre-Composites for Various Applications: Design for Sustainability. *Polymers* **2022**, *14*, 920. [CrossRef]
77. Ling, Y.-F.; Zhang, P.; Wang, J.; Shi, Y. Effect of Sand Size on Mechanical Performance of Cement-Based Composite Containing PVA Fibers and Nano-SiO_2. *Materials* **2020**, *13*, 325. [CrossRef] [PubMed]
78. Jeng, Y.R.; Tsai, P.C.; Chiang, S.H. Effects of grain size and orientation on mechanical and tribological characterizations of nanocrystalline nickel films. *Wear* **2013**, *303*, 262–268. [CrossRef]
79. Wei, L.; Zuo, W.; Pan, H.; Lyu, K.; Zhang, W.; She, W. Rational design of lightweight cementitious composites with reinforced mechanical property and thermal insulation: Particle packing, hot pressing method, and microstructural mechanisms. *Compos. Part B Eng.* **2021**, *226*, 109333. [CrossRef]

materials

Article

Prediction of Axial Capacity of Concrete Filled Steel Tubes Using Gene Expression Programming

Kaffayatullah Khan [1,*], Mudassir Iqbal [2], Muhammad Raheel [3], Muhammad Nasir Amin [1], Anas Abdulalim Alabdullah [1], Abdullah M. Abu-Arab [1] and Fazal E. Jalal [4]

1 Department of Civil and Environmental Engineering, College of Engineering, King Faisal University, Al-Ahsa 31982, Saudi Arabia
2 Department of Civil Engineering, University of Engineering and Technology Peshawar, Peshawar 25120, Pakistan
3 Department of Civil Engineering, University of Engineering and Technology Mardan, Mardan 23200, Pakistan
4 Department of Civil Engineering, Shanghai Jiao Tong University, Shanghai 200240, China
* Correspondence: kkhan@kfu.edu.sa; Tel.: +966-13-589-5408; Fax: +966-13-581-7068

Abstract: The safety and economy of an infrastructure project depends on the material and design equations used to simulate the performance of a particular member. A variety of materials can be used in conjunction to achieve a composite action, such as a hollow steel section filled with concrete, which can be successfully utilized in the form of an axially loaded member. This study aims to model the ultimate compressive strength (P_u) of concrete-filled hollow steel sections (CFSS) by formulating a mathematical expression using gene expression programming (GEP). A total of 149 datapoints were obtained from the literature, considering ten input parameters, including the outer diameter of steel tube (D), wall thickness of steel tube, compressive strength of concrete (f_c'), elastic modulus of concrete (E_c), yield strength of steel (f_y), elastic modulus of steel (E_s), length of the column (L), confinement factor (ζ), ratio of D to thickness of column, and the ratio of length to D of column. The performance of the developed models was assessed using coefficient of regression R^2, root mean squared error RMSE, mean absolute error MAE and comparison of regression slopes. It was found that the optimal GEP Model T3, having number of chromosomes N_c = 100, head size H_s = 8 and number of genes N_g = 3, outperformed all the other models. For this particular model, $R^2_{overall}$ equaled 0.99, RMSE values were 133.4 and 162.2, and MAE = 92.4 and 108.7, for training (TR) and testing (TS) phases, respectively. Similarly, the comparison of regression slopes analysis revealed that the Model T3 exhibited the highest R^2 of 0.99 with m = 1, in both the TR and TS stages, respectively. Finally, parametric analysis showed that the P_u of composite steel columns increased linearly with the value of D, t and f_y.

Keywords: concrete filled steel tubes; compression strength; GEP modelling; hyperparameters tuning; parametric and sensitivity analysis

Citation: Khan, K.; Iqbal, M.; Raheel, M.; Amin, M.N.; Alabdullah, A.A.; Abu-Arab, A.M.; Jalal, F.E. Prediction of Axial Capacity of Concrete Filled Steel Tubes Using Gene Expression Programming. *Materials* **2022**, *15*, 6969. https://doi.org/10.3390/ma15196969

Academic Editor: Ayman El-Zohairy

Received: 10 August 2022
Accepted: 1 September 2022
Published: 7 October 2022

1. Introduction

Columns are a desideratum in designing the structural elements of a building. Depending upon the locality, application and resource availability, different materials, for instance, concrete, steel and their combination can be utilized for column construction [1–5]. Reinforced concrete (RC) columns are vastly used around the globe due to a number of reasons, i.e., easy availability of their constituent materials, known behavior under different loading conditions, and development of design codes, such as the ACI 318 [6] design manual and Euro code [7], among others. However, hollow steel sections filled with concrete (HSSFC) represent an improved version in contrast to the other traditional RC columns, both in terms of performance as well as construction costs, alongside possessing the merit of no longitudinal and transverse reinforcement requirements [8]. For example, the hollow steel section (from hereon, referred to as HSS) confines the concrete thus improving its mechanical properties while the buckling resistance of the steel section is enhanced due to the presence of concrete. The composite action of the two materials increases the overall strength, stiffness, ductility, buckling resistance of the element, and provides an improved fire resistance. Since concrete is poured inside the HSS, the need for formwork is eliminated and labor cost is also reduced [9–13].

The cross-sections of the HSS used in concrete-filled hollow steel sections (CFSS) exhibit numerous shapes, i.e., circular, rectangular and square, wherein the most common one is circular because of its high confinement performance [14–16]. On a global scale, a number of design codes encompasses the provisions for axial and flexure capacity design of the CFSS [17]. Examples include the Eurocode 4: Design of composite steel and concrete structures (EN1994) [7], AISC-360: Specification for structural steel buildings [18], ACI-318 [6], etc. In addition, the behavior of CFSS has been studied by a number of researchers by considering a number of design parameters such as the slenderness ratio, eccentricity ratio, end-moment ratio, thickness of internal steel tube and compressive strength of concrete, in addition to others [19–21], using laboratory experiments and numerical modelling. Although the current codes provide enough guidelines about the axial capacity of CFSS columns, however, uncertainties in the nominal and actual mechanical properties of steel and concrete make the design conservative. In this situation, artificial intelligence (AI) models provide an accurate solution based on mechanical properties of steel and concrete.

Besides the experimental studies for developing knowhow about the influence of various input parameters on the structural performance, soft computing (SC)/machine learning (ML)/(AI)/techniques are gaining popularity nowadays, because of their ability to learn from training data so as to formulate a trained algorithm which can be used for accurate prediction of the output(s) [22–25]. The accuracy of a typical AI model depends on the number of data points used during the training process and the selection of influential input variables (i.e., high Pearson correlation value). Therefore, several studies have been undertaken recently for evaluating the behavior of CFSS columns under different loading conditions. For example, Albero et al. [26] studied the ultimate resisting load of CFSS under unequal eccentricities at both ends. It was found that the application of unequal load eccentricities enhanced the ultimate resisting load with columns having higher slenderness ratio of 27.78. Similarly, neural networks were successfully used for modelling the strength of columns considering f_y, f_c' and the diameter and wall thickness of the HSS and they concluded that the ANN model successfully predicted the columns' performance (coefficient of regression $R^2 > 0.98$ in the training, test and validation phases) and recommended it to be a reliable tool for assessing the columns' performance based on the ANN-based Monte Carlo method [27]. Likewise, the axial load capacity of a CFSS having a circular cross-section was modelled using Artificial Neural Networks (ANN) and Gene Expression Programming (GEP). For example, Naderpour et al. [28] deployed GEP, ANN and group methods of data handling to predict the compressive strength of columns confined with fiber reinforced polymers. It was observed that the ANN model had the highest accuracy among its competitors, with $R^2 > 0.98$. Similarly, the error percentage ($\pm$ 20% error range) of the forecasted output by ANN, GEP and GMDH was recorded as 94.7, 84.2 and 88.4%, respectively. Similarly, Azim et al. [29] utilized the GEP approach to develop a prediction model for the compressive arch action capacity of RC beam-column substructures. Again, GEP was successfully used for studying the behavior of H-section steel columns ($R^2 > 0.94$) under blast loading by Momeni et al. [30]. In addition, an empirical equation was also generated from the GEP model for relating the damage index to the displacement/rotational index. Wang et al. [31] studied the blast resistance and residual strength of CFSS considering the thickness of steel tube and cross-section geometry. It was observed that the CFSS retained up to 60% of its ultimate axial loading capacity even after close-range blast loading and that their axial load capacity retention was enhanced with the thickness of steel tubes. Similarly, Zhang et al. [32] also concluded that concrete filled steel tubular columns showed excellent performance against flexural loads under both static and dynamic loads.

A literature survey reveals successful application of different AI techniques, such as ANN, GEP and ANFIS alongside the combination of meta-heuristic optimization algorithms and ML algorithms for predicting the mechanical performance of concrete as well as soils [24,33–39]. However, the ANN and ANFIS provide lesser insight about the models pertaining to their practical implications, e.g., to derive an empirical relation between the input parameters and output(s), which can further be used for predicting the output(s) and performing parametric and sensitivity analysis [40,41]. In addition, both these models are termed as 'black-box models' in the literature [40,41], due to the complex interaction of neurons present in different hidden layers. As a result, a useful empirical relation between the input and output parameters is difficult to develop and has reliability issues [42]. In contrast, the white-box models, such as GEP, can provide a simple and easy-to-use mathematical expression to forecast the output for a specific range of input parameters. Using that expression sensitivity and parametric analyses could enable validation of the developed GEP model [43]. Shahmansouri et al. [44] developed an empirical equation for predicting the f_c' of geopolymer concrete by considering the specimen's age, concentration of alkaline activator (i.e., NaOH), natural zeolite, silica fume and blast furnace slag content to be the most influential input parameters. Similarly, Naser et al. [45] developed design equations for the structural response of the CFSS by utilizing the Genetic Algorithm (GA) and GEP technique. They concluded that both GA and GEP models outperformed the current design codes, such as Eurocode 4 [7], AISC 360-16 [18] and New Zealand code (NZS 2327) [46]. Both the models had high predictive capability as a majority of the data points ranged within a 10% bounding error. In an effort to develop an empirical relation between the mix design parameters of lightweight foamed concrete, Sami et al. [47] employed GEP by using 191 data points to develop a model which could simulate the influence of input parameters (i.e., amount of cement, fine aggregate, water to cement ratio and foam volume) on two outputs (i.e., dry density and f_c'). They revealed that GEP accurately modelled the dry density and f_c' of foamed concrete, as evident from their higher values of the coefficient of determination (R^2) (i.e., 0.79 and 0.94, respectively).

In connection to the use of AI models related to predicting the P_u of CFSS columns, Sarir et al. [48] used a dataset of 303 points to compare the performance of a GEP model with ANN model optimized by particle swarm optimization (PSO) algorithm, as listed in Table 1. The concrete compressive strength (f_c'), the column length (L), outer diameter (D), tensile yield stress of the steel column (f_y), and steel cover thickness (t) were considered as input variables. The best performance was obtained for the GEP model interpreting R^2 equaling 0.939. Javed et al. [49] included eccentricities at end supports (e_t, e_b) alongside the previously mentioned variables by employing 227 sample specimens. Khan et al. used an extensive database yielding R^2 of 0.9812 for test data; however, the value of MAE recorded was comparatively high. Several other researchers (Ngo et al. [50], Jiang et al. [51], Jayalekshmi et al. [52], Ahmadi et al. [53]) also developed Support Vector Regression (SVR) optimized by Grey Wolf Optimization (GWO), GEP, and ANN models, and evaluated the accuracy in terms of R^2. The capabilities of the GEP model were not fully explored by changing its genetic parameters, i.e., number of chromosomes (N_c), genes (N_g) and head sizes (h_s) to obtain the best hyperparameters. Nevertheless, this study utilized 10 variables including the confinement factor (ζ) which was not used in the reported literature.

The ANN and various other such hybrid algorithms (e.g., ANFIS) are unable to provide an empirical equation between the inputs and output(s). As a result, GEP is deployed to robustly simulate the ultimate compressive strength (P_u) of the CFSS considering different inputs parameters, i.e., (i) outer diameter of steel tube (D), (ii) wall thickness of steel tube (t), (iii) compressive strength of concrete (f_c'), (iv) elastic modulus of concrete (E_c), (v) yield strength of steel (f_y), (vi) elastic modulus of steel (E_s), (vii) length of the column (L), (viii) confinement factor (ζ), (ix) D/t ratio of column and (x) the L/D ratio of column. The performance of the developed model was assessed using different statistical indices, i.e., R^2, root mean square error (RMSE), mean absolute error (MAE), comparison of regression slopes, predicted to experimental (P/E) ratios. In addition, parametric and sensitivity analyses were also carried out to assess the effect and contribution of input parameters on the P_u of CFSS.

Table 1. Summary of the previous studies in the literature pertaining to prediction of axial capacity of CFSS.

Reference	Model	Number of Data Points	Input Variables	Testing Data Performance	
				R^2	*MAE*
Sarir et al. [48]	GEP	303	f_c', L, D, f_y, t	0.939	–
Javed et al. [49]	GEP	227	f_c', L, D, f_y, t, L/D	0.980	153.9
Khan et al. [54]	GEP	702	f_c', L, D, f_y, t, e_t, e_b	0.981	290.36
Ngo et al. [50]	SVR-GWO	802	f_c', L, D, f_y, t, D/t	0.996	–
Jiang et al. [51]	GEP	32	D, L, t, L/D, D/t, f_y, E_s, f′c, E_c, υ	–	–
Jayalekshmi et al. [52]	ANN	633	D, t, f_y, f_c', L	0.962	–
Ahmadi et al. [53]	ANN	272	D, t, f_y, f_c', L	0.801	–
This study	GEP	149	D, t, f_c', E_c, f_y, E_s, L, ζ, D/t, L/D	0.99	108.7

The flow of the paper is that the database is compiled, and the GEP modelling is performed. The formulated GEP model is obtained after undertaking several trials and the effect of genetic parameters on the model performance is studied. After selecting the most optimal trial, the mathematical expression is determined. A variety of performance measures (error indices) were calculated to check the performance of the optimal model. Parametric and sensitivity analyses were also performed for the input parameters considered in the formulated GEP model.

2. Methodology

2.1. Database Compilation

The database was compiled from the work published by Bardhan et al. [55], which comprises a total of 149 datapoints of which 104 datapoints were used for the training (TR) phase whereas the remaining 49 points were utilized for the testing (TS) phase. Ten input parameters, i.e., D, t, f_c', E_c, f_y, E_s, L, ζ, ratio of D and t (i.e., D/t) and the ratio of L and D (i.e., L/D) were considered in order to model their influence on the P_u of CFSS using a genetic programming approach. The descriptive statistics of the dataset used for developing the GEP model are listed in Table 2. In order to indicate the frequency distribution of input parameters, the departure of data from the horizontal symmetry (skewness) and the sharpness of the central peak, relative to a standard normal distribution curve [56], the frequency histograms of the input parameters along with their respective normal distribution fit are shown in Figure 1. The values of kurtosis and skewness in Table 2, are in accordance with the aforementioned normal distribution curves. For example, the values of kurtosis and skewness are positive for all the input parameters except for E_c and E_s, respectively. Referring to Figure 1a–c,e–j), the distribution is peaked with a thick tail while Figure 1d shows that the distribution of E_c is flatter, which justifies the negative kurtosis value. Similarly, the negative value of skewness for E_s can be corroborated from Figure 1a, since the distribution has a fatter tail on its left side.

Table 2. Descriptive statistics of the input variables.

Descriptive Statistics	D (mm)	T (mm)	f_c' (MPa)	E_c (MPa)	f_y (MPa)	E_s (MPa)	L (mm)	ζ	D/t (mm/mm)	L/D (mm/mm)
Average	164.38	3.71	65.60	3.5×10^4	339.85	201,767	485.07	0.86	58.07	2.94
Standard Error	5.17	0.17	3.82	1.1×10^3	8.16	575.81	18.08	0.06	3.38	0.04
Standard Deviation	63.09	2.08	46.58	1.3×10^4	99.57	7029	220.73	0.73	41.21	0.53
Sample Variance	3980	4.34	2170	1.66×10^8	9914	49,401,978	48,721	1	1698	0
Kurtosis	5.18	1.90	0.80	−0.18	11.50	2.21	9.12	0.94	3.03	0.84
Skewness	1.79	1.40	1.24	0.83	2.75	−1.01	2.32	1.19	1.78	0.08
Minimum	60	0.86	18.03	1.8×10^4	186	177,000	180	0.05	17	1.8
Maximum	450	10.36	193.30	6.6×10^4	853	213,000	1760	3.22	221	4.90

2.2. GEP Modelling

GeneXprotools v5.0, developed by Candida Ferreira (Portugal), was employed for developing the desired GEP models. For this purpose, the data was fed into the GEP interface. Afterwards, the dataset was divided into two subsets namely; the TR dataset (70%) and the TS dataset (30%). As a result, 104 datapoints were used for the TR phase whereas the remaining 45 datapoints were used during the TS phase. The option of normalization is readily available in the data tab of this tool; however, the authors did not utilize this option in the current study and the models were trained using the actual values. The authors opine that the statistical models are based upon the descriptive statistics of the given dataset. Similar studies have been widely reported in the past literature [57,58]. In order to achieve the best optimal model, the hyperparameter settings of GEP parameters were adjusted accordingly. For example; the number of chromosomes (N_c) were varied from 30 to 200, the number of genes (N_g) from 3 to 5 and, the head size (H_s) from 8 to 12. Similarly, it was observed that the addition function provides the optimal performance. This was achieved by exploring different linking functions (+, −, ×, /). The details of the genetic parameters settings, i.e., mutation, transposition, and recombination rates are given in Table 3. The details of 11 number of trials employing different values of hyperparameters and the resulting model performance in terms of R^2, *RMSE* and *MAE* for both the TR and TS phases, respectively, are shown in Table 4. Moreover, Figure 2 shows the flowchart of GEP modelling. The process starts with feeding the input parameters data followed by the random partitioning of datasets. The process is continued by selecting the fitness function, N_c, H_s, N_g and assigning suitable genetic operators. After assignment of linking functions and terminals setting, the model is run and its performance can be checked using different statistical tools such as R^2, *RMSE* and *MAE*, as shown in Equations (1)–(3), respectively.

$$R^2 = \left(\frac{\sum_{i=1}^{n}(e_i - \bar{e}_i)(p_i - \bar{p}_i)}{\sum_{i=1}^{n}(e_i - \bar{e}_i)^2 \sum_{i=1}^{n}(p_i - \bar{p}_i)^2}\right)^2 \tag{1}$$

$$RRMSE = \frac{1}{|\bar{e}|}\sqrt{\frac{\sum_{i=1}^{n}(e_i - p_i)^2}{n}} \tag{2}$$

$$MAE = \frac{\sum_{i=1}^{n}|e_i - p_i|}{n} \tag{3}$$

where e_i and p_i are the ith experimental and predicted output values, respectively; $\bar{e}_i$ and $\bar{p}_i$ are the average values of the experimental and predicted output values, respectively, and n are the total samples.

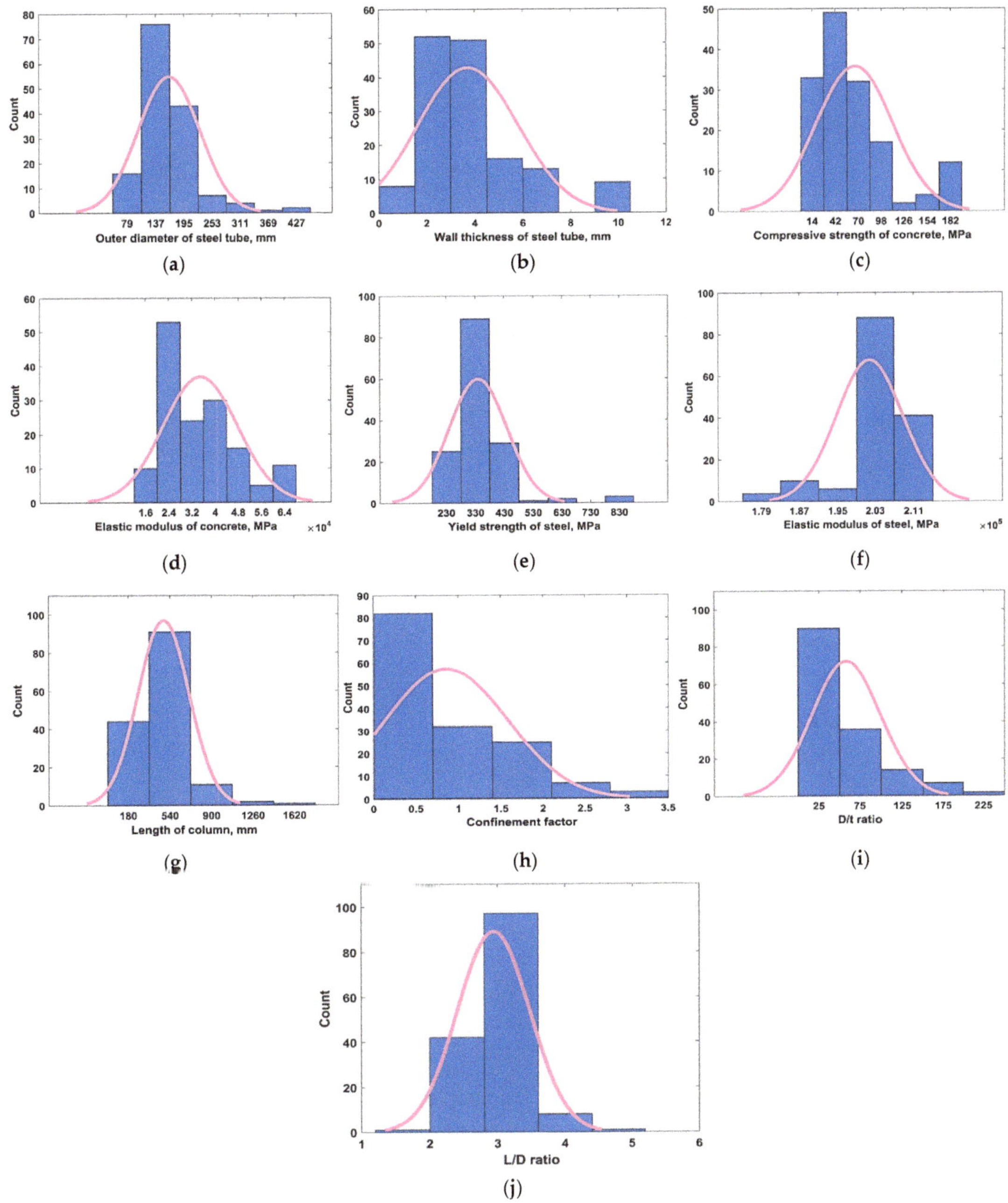

Figure 1. Frequency histograms of input variables. (**a**) Elastic modulus of concrete, (**b**) wall thickness of steel tube, (**c**) compressive strength of concrete, (**d**) Elastic modulus of concrete, (**e**) Yield strength of steel, (**f**) Elastic modulus of steel, (**g**) Length of column, (**h**) Confinement factor, (**i**) D/t, (**j**) L/D.

Table 3. Parameters setting for GEP algorithms.

Parameters	Settings
	Axial Capacity (P_u) of CFSS
Numerical constants	
Constant per gene	10
Type of data	Floating number
Maximum complexity	10
Ephemeral random constant	[−10, 10]
Genetic operators	
Rate of mutation	0.00138
Inversion rate	
IS transposition rate	0.00546
RIS transposition rate	
1-point recombination rate	
2-point recombination rate	
Gene recombination rate	0.00277
Gene transposition rate	

Table 4. Details of the different trials/models conducted to obtain optimal model.

Trial/Model	No. of Variables	No. of Chromosomes	Head Size	No. of Genes	TR Phase			TS Phase		
					R^2	*RMSE*	*MAE*	R^2	*RMSE*	*MAE*
T1	5	30	8	3	0.92	644.4	443.6	0.94	432.9	323.9
T2	7	50	8	3	0.98	236.5	173.8	0.98	205.3	159.6
T3	**7**	**100**	**8**	**3**	**0.99**	**133.4**	**92.4**	**0.99**	**162.2**	**108.7**
T4	8	150	8	3	0.98	287.8	198.1	0.97	325.5	221.7
T5	9	200	8	3	0.98	255.1	154.9	0.99	162.3	126.0
T6	7	100	9	3	0.98	298.5	202.1	0.96	363.1	243.8
T7	8	100	10	3	0.97	389.0	257.9	0.96	358.3	242.7
T8	7	100	11	3	0.93	605.8	373.1	0.84	828.0	501.3
T9	8	100	12	3	0.97	385.1	276.0	0.96	386.0	279.8
T10	8	100	8	4	0.98	313.6	209.9	0.97	298.2	222.7
T11	8	100	8	5	0.96	443.8	286.2	0.98	288.9	218.5

A trial and error approach was used for setting the GEP parameters, such that an optimal performing model with the best hyperparameter settings could be obtained. This practice would help avoiding overfitting the data during the TR phase, and, subsequently, improve their performance in the TS phase. In the past, researchers have addressed the problem of overfitting [59]. For example, Gandomi and Roke [60] selected the model with minimum objective function (OF) value which varies from 0 to maximum value, with the model having OF ≈ 0 is considered to have the best performance. Here, statistical evaluation was used to select a non-overfitted model. Table 5 shows the ideal values of the three performance indices, i.e., R^2, *RMSE* and *MAE*, used to assess the models' performance.

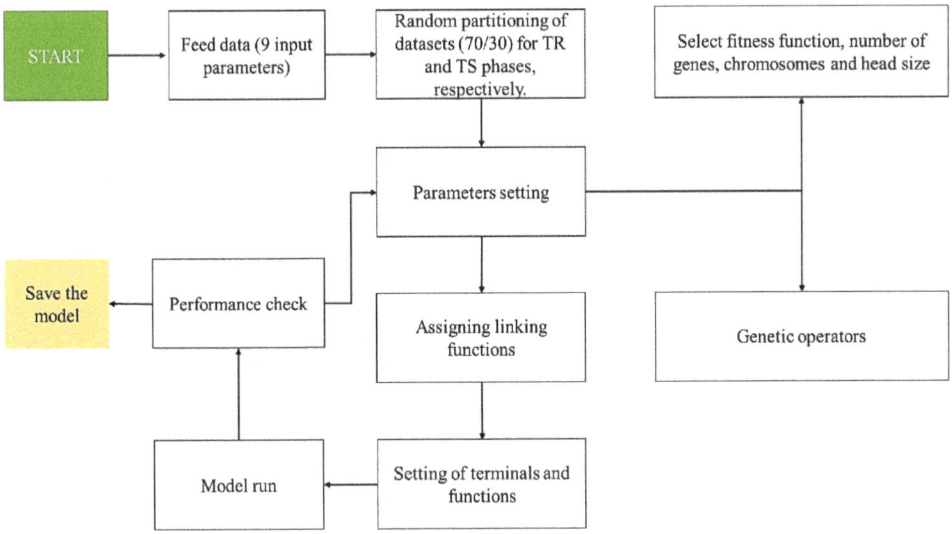

Figure 2. Flowchart of GEP modelling.

Table 5. Ideal values of performance indices.

Index	Range/Ideal Value
R^2	(0–1)/1
RMSE	(0–∞)/0
MAE	(0–∞)/0

3. Results & Discussion

3.1. Variation of Genetic Parameters

Table 4 depicts that a total of 11 trials (Model T1 to T11) were conducted by varying the different parameters of the GEP model (N_c, H_s and N_g). As discussed earlier, the hyperparameters of GEP models were varied to achieve an optimal GEP model. A total of 11 trials (i.e., Model T1 to T11) were employed with varying values of N_c, N_g and H_s (Table 4). Firstly, the values of N_c were varied from 30 to 200, while keeping the H_s and N_g constant (i.e., 8 and 3, respectively). In the second stage, the H_s was changed from 8 to 12, while keeping both N_c and N_g constant. Similar practice was followed for N_g as well. The values of N_c, H_s and N_g for the best performing model (Model T3) came out to be 100, 8 and 3, respectively. For all the trials, the performance of the models was assessed using R^2, *RMSE* and *MAE*. It is evident from Table 4 that the model T3 has the highest R^2 and, lowest *RMSE* and *MAE*, in both the TR and TS phases, respectively.

Figure 3 shows the influence of N_c on the R^2, *RMSE* and *MAE* of the models, in both the TR and TS phases, respectively. It can be seen from Figure 3a that, the value of R^2 increases with the N_c till it drops for $N_c = 150$. A further increase in N_c (i.e., from 150 to 200) enhances the R^2 again. Similar improvement in the performance of the models with the increase in N_c up to 100 can be observed in Figure 3b,c. It is evident from these Figures that the value of *RMSE* and *MAE* plummets with the increase in N_c; however, the model performs poorly for $N_c = 150$ as evident from its lower $R^2 = 0.98, 0.97$, higher *RMSE* = 287.8, 325.5, and MAE = 198.1, 221.7, in the TR and TS phases, respectively.

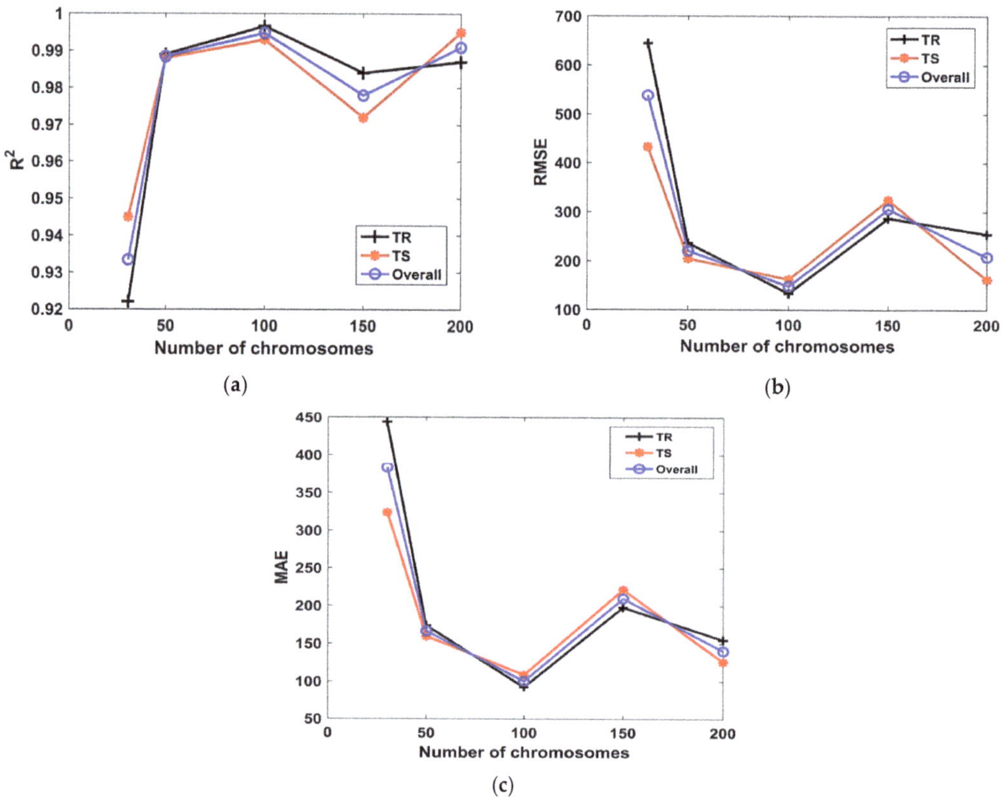

Figure 3. Effect of N_c on the performance of models. (**a**) R^2, (**b**) *RMSE*, (**c**) *MAE*.

The influence of H_s on the performance of models (both TR and TS phases) in terms of R^2, *RMSE* and *MAE* can be observed in Figure 4a–c, respectively. It is evident from Figure 4a that, R^2 decreases with increasing H_s. The model depicts immensely poor performance for $H_s = 11$; however, upon further increase in the $H_s = 12$, the value of R^2 improves. A similar trend can be observed in the case of Figure 4b,c, wherein the *RMSE* and *MAE* increase with the H_s. The model performs poorly as is evident from its high *RMSE* and *MAE* values for $H_s = 11$. The influence of N_g on the performance of models (TR and TS phases) in terms of R^2, *RMSE* and *MAE* can be observed in Figure 5a–c, respectively. Considering the TR phase, it is evident from Figure 5a, that the value of R^2 increases with the increase in the N_g up to 4. Further increase in value of N_g (i.e., 4 to 5) lowers the value of R^2. Similar trend can be observed from Figure 5b,c, wherein the value of *RMSE* and MAE decrease with the value of N_g (i.e., 3 to 4). When the N_g is further increased from 4 to 5, the accuracy of the model decreased as is evident from its high *RMSE* and *MAE* values for $N_g = 5$. However, the performance of the model improves at higher N_g in the TS phase, as evident from its higher R^2 value = 0.98, and lower *RMSE* and *MAE* values of 288.9 and 218.5, respectively.

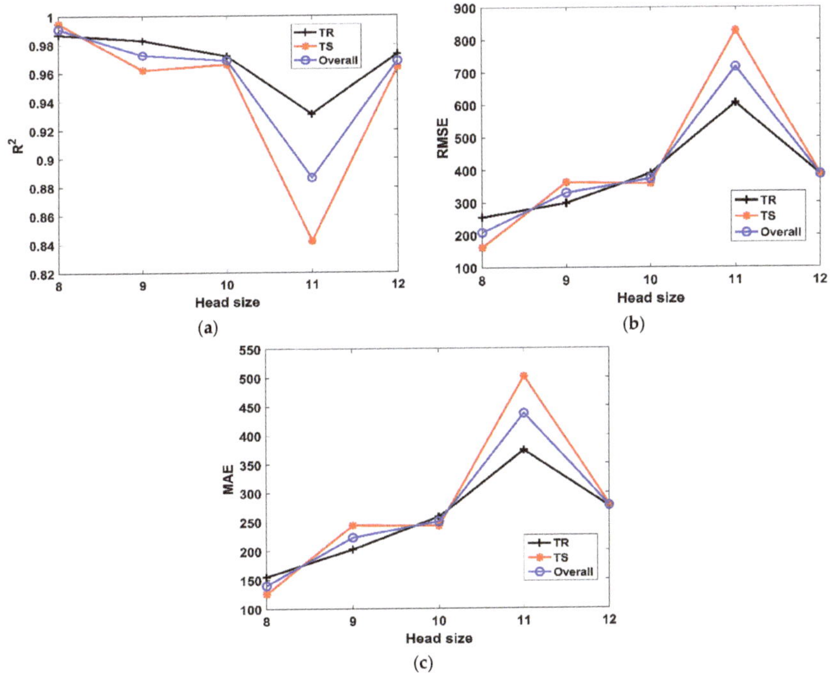

Figure 4. Effect of H_s on the performance of models. (**a**) R^2, (**b**) *RMSE*, (**c**) *MAE*.

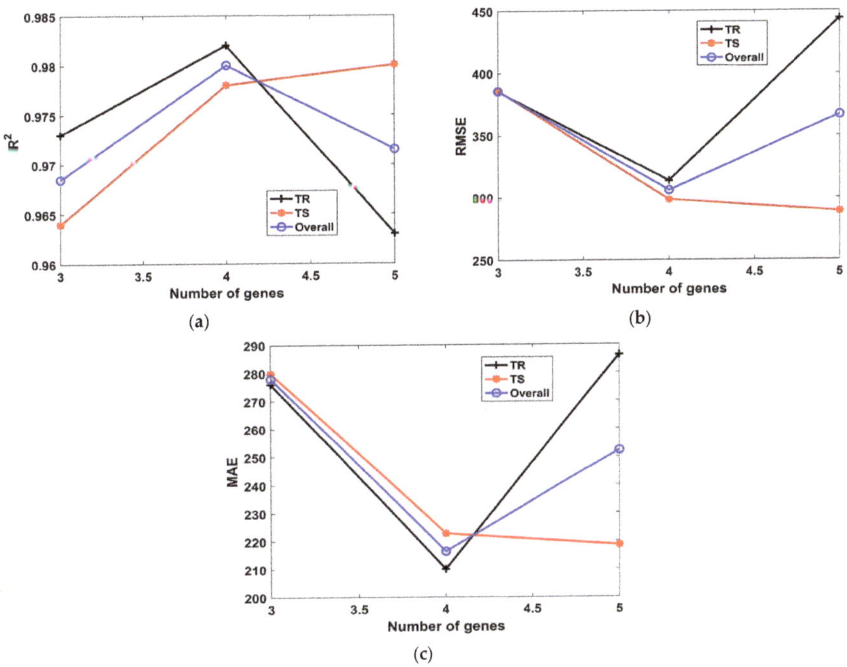

Figure 5. Effect of N_g on the performance of models. (**a**) R^2, (**b**) *RMSE*, (**c**) *MAE*.

3.2. Models' Performance

3.2.1. Statistical Evaluation

It can be seen from Table 4 that the performance of the Model T3 is better (R^2 = 0.99 for both the TR and TS phase) followed by Model T5 (R^2 = 0.98 in TR phase, and 0.99 for TS phase). The observed values of R^2 shows a good agreement between the predicted and experimental values. However, a model cannot be declared as the "best performing model" solely on the basis of R^2. Other statistical error indices must also be considered, such as *RMSE* and *MAE*, in addition to others. In this regard, the values of *RMSE* and MAE were also studied in the current study while assessing the performance of different models. It is evident from Table 4 that, in addition to a higher R^2 value, Model T3 exhibited the lowest *RMSE* (133.4 in TR phase, and 162.2 in TS phase) and MAE (92.4 in TR phase and 108.7 in TS phase). Similarly, the Model T5 performed as second-best model having *RMSE* = 255.1 in the TR phase, and, *RMSE* = 478.2, and MAE = 386.6 in the TS phase, respectively. Model T2 also performed better as it possessed the second highest R^2 and second lowest value of *RMSE* in the TR phase, respectively. The ranking of the models based on the different statistical indices has also been shown in Table 6.

Table 6. Ranking of models based on R^2 and *RMSE*.

Statistic	R^2		*RMSE*		*MAE*	
Rank	**1st**	**2nd**	**1st**	**2nd**	**1st**	**2nd**
TR Phase	T3	T2, T5	T3	T2	T3	T5
TS Phase	T3	T5	T3	T5	T3	T5

3.2.2. Comparison of Regression Slopes

Prediction models can be evaluated by plotting a trend line between the experimental and predicted values. This assessment method has also been used in this study and as a result, regression slopes have been plotted for all the 11 models, in both the TR and TS phases, respectively. It is noteworthy to mention that an ideal trend line has a slope "m" value of unity (=1.0) and, its angle of inclination with both the *X*- and *Y*-axis in the cartesian coordinate system equals 45°. The performance of the model is considered to be reliable and accurate provided the plot between the forecasted and experimental values follow the ideal trend line (i.e., inclined at an angle of 45° with the *X*-axis). A regression line whose m value approaches one, and, its correlation value i.e., R $\geq$ 0.8 are considered reliable in forecasting new data [61].

Figure 6 shows the values of R^2 and m for both the TR and TS phases, respectively. It is evident from the Figure 6 that R^2 exceeds 0.90 for all of the models. In the TR phase, Model T3 has the best fit with an R^2 value of 0.99 and m = 1 whereas, Model T1 has a lower value of R^2 = 0.92 and m = 0.91. Similarly, the performance of the models improved in the TS phase, as indicated from the higher R^2 values and the m values becoming closer to one. All the models have $R^2 \geq 0.90$. It can also be observed from Figure 6 that, the highest value of m = 1 has been obtained for the optimal model T3 in both the TR and TS phases, respectively. However, Figure 7 shows that in addition to model T3, other models (except T2 and T6) had m values equaling one. It is important to mention here that for m = 1, the slope of the regression line will be exactly 45°. In comparison to other models, the values of 'm' and R^2 observed for the Model T3 are closer to one. Therefore, it can be concluded that the Model T3 is the best performing model, compared to the others.

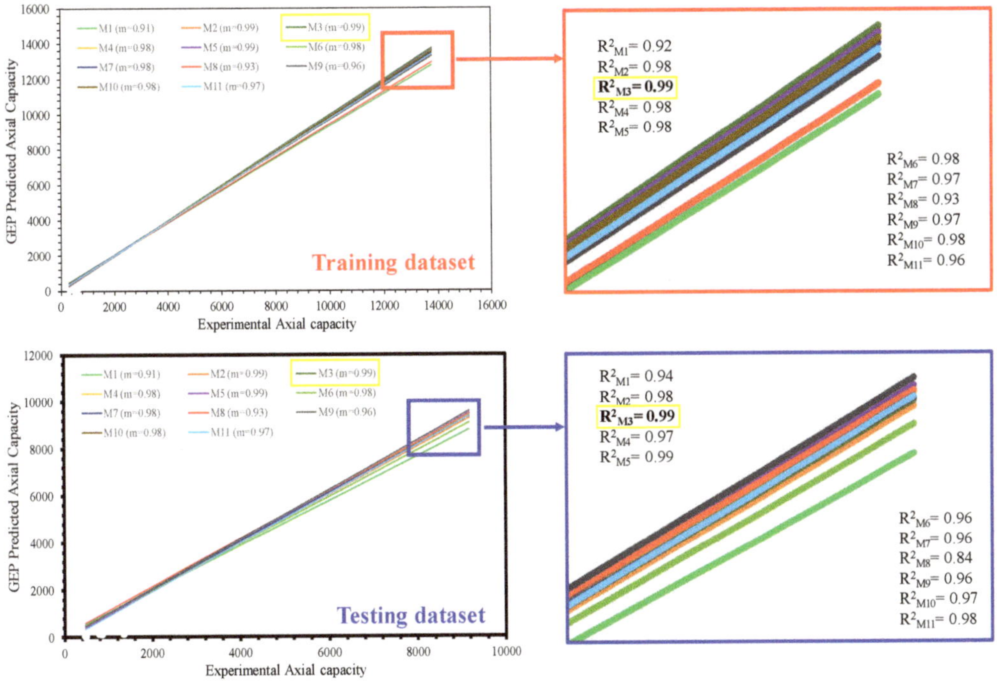

Figure 6. Comparison of regression slopes in the TR phase.

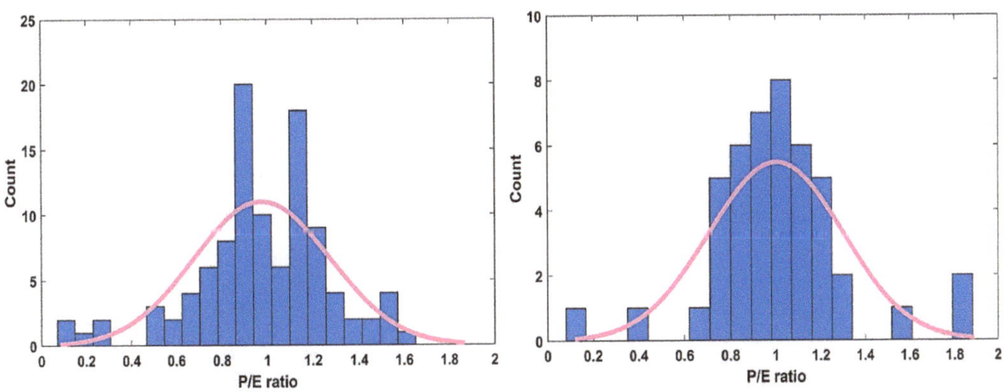

(a) (b)

Figure 7. P/E ratio distribution of the best performing model T3, (**a**) TR phase, (**b**) TS phase.

3.2.3. Model Predicted to Experimental Ratio (P/E)

The performance of the generated models as a result of different trials (by varying N_c, H_s and N_g) was further studied using the P/E ratio. Figure 7a,b depict the distribution of the P/E ratios for the optimal performing Model T3 in the TR and TS phases, respectively. The bin range has been varied between 0 and 2 using a uniform interval of 0.2. It can be observed from both Figure 7a,b that most of the P/E values (higher frequencies) concentrated in the bin range proximal to one. This serves another statistical check in evaluating the performance of the formulated model and acts as visual justification in case of the optimal performance Model T3.

3.3. GEP Formulations

In addition to successful simulation of the P_u of CFSS considering a number of input parameters, another novel achievement of this research work is to obtain an empirical equation which can be used for predicting the P_u of the CFSS using the different input variables. For this purpose, the optimal performing Model T3 was used to develop the empirical equation. The expression tree for the Model T3 (Figure 8) and the MATLAB model were utilized to obtain the mathematical expression, which can be further used for forecasting the P_u of the CFSS, and, sensitivity and parametric analysis can be performed as well. As a result, Equation (4) was obtained, which is able to predict the P_u of CFSS using various input variables (i.e., D, t, f_c', E_c, f_y, E_s, L, ζ, ratio of D to thickness of column and, the ratio of length to D of column). It is highly recommended to use the prediction equation for input variables whose range and other details have already been discussed in Section 2.1 [62,63].

$$P_u = \frac{(-36.9 - f_y) * (D - 25.2) * t}{17.2 * (\zeta - 17.2)} + \frac{((2.3 * f_y) - L) * 21.2}{L/D * \zeta * fc'} + \frac{((f_y - 17.2) * t) * (D + 2.2)}{\zeta * 34.4} \quad (4)$$

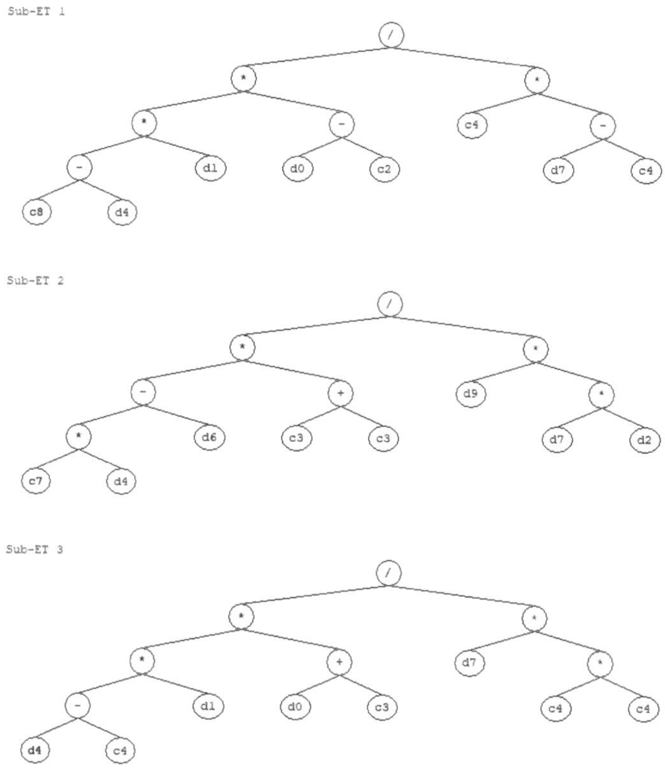

Figure 8. Expression tree for the model T3 depicting the three sub expression trees.

The values of the constants (c), as present in Figure 8 are as follows:

Sub-ET1: c8 = −36.9; c2 = 25.2; c4 = 17.2; Sub-ET2: c7 = 2.3; c3 = 10.6; Sub-ET3: c4 = 17.2; c3 = 2.2;

Whereas, d0, d1, d2, d3, d4, d5, d6, d7, d8, d9 represents the input parameters such as D, t, f_c', E_c, f_y, E_s, L, ζ, D/t and L/D, respectively.

This Equation (4) is a simple-to-use mathematical expression for designers and practitioners, who can robustly determine the Pu of the CFSS when the easily determinable parameters are available.

3.4. Parametric and Sensitivity Analysis

The reliability of different AI models can be verified by conducting parametric analysis of the input features. In this study, the parametric analysis was also carried out for all the input parameters viz., D, t, f_c', E_c, f_y, E_s, L, ζ, ratio of D to column, and the ratio of L to D, in order to evaluate their effect on the resulting P_u of the CFSS columns. Table 7 shows the possible combination of the study input parameters, considered for the parametric analysis. Hence, a dataset was generated such that one of the input variables (first variable, D) was varied between its extreme values (lower to maximum value in the dataset) in equal increments, while, keeping the remaining input parameters at their average values. In the next step, a second input parameter (i.e., t) was altered in a similar manner, while, keeping the other input features at their mean values. This practice was repeated for all the input parameters. The prediction equation (as given by Equation (1)) was utilized to obtain the corresponding change in the target variable. The range of input variables and their corresponding influence on the P_u of CFSS columns was plotted to study their relationship. The net change in the output (P_u of CFSS columns) due to changing a particular input feature, was calculated in terms of weighted percentage, in order to find the sensitivity of each input attribute, as well. Note that the sensitivity analysis shows the response of the prediction model by varying the input parameters [64]. The relative contribution of the input parameters can be studied using this analysis as shown in Equation (5) and (6) below.

$$P_i = f_{\max}(s_i) - f_{\min}(s_i) \times N \quad (5)$$

$$Sensitivity(\%) = \frac{P_i}{\sum_{i=1}^{n} P_j} \times 100 \quad (6)$$

Table 7. Dataset used for parametric analysis.

Input Variables		Constant Input Parameters	No. of DataPoints
Parameter	Range		
D	60–450	t = 3.71, f_c' = 65.60, f_y = 339.85, L = 485.07, ζ = 0.86, L/D = 2.94	10
t	0.86–10.36	D = 164.34, f_c' = 65.60, f_y = 339.85, L = 485.07, ζ = 0.86, L/D = 2.94	
f_c'	18.03–193.30	D = 164.34, t = 3.71, f_y = 339.85, L = 485.07, ζ = 0.86, L/D = 2.94	
f_y	186–853	D = 164.34, t = 3.71, f_c' = 65.60, L = 485.07, ζ = 0.86, L/D = 2.94	
L	180–1760	D = 164.34, t = 3.71, f_c' = 65.60, f_y = 339.85, ζ = 0.86, L/D = 2.94	
ζ	0.045–3.221	D = 164.34, t = 3.71, f_c' = 65.60, f_y = 339.85, L = 485.07, L/D = 2.94	
L/D	1.8–4.9	D = 164.34, t = 3.71, f_c' = 65.60, f_y = 339.85, L = 485.07, ζ = 0.86,	

Here, $f_{max}(s_i)$ and $f_{min}(s_i)$ refer to the maximum and minimum GEP estimated values for the i'th input domains, where the remaining input factors = 1. The value of sensitivity analysis is between 0 and 1 which depicts the relative contribution of input parameters among each input attribute as well as the predicted output variable.

Figure 9 shows the variation of the P_u of CFSS columns in response to the change in each input variable, as explained above. It is apparent from Figure 9a,b,d that the P_u of CFSS columns increases linearly with the values of D, t and f_y while it decreases with the amount of f_c', L, ζ and L/D ratio, as is evident from Figure 9c,e–g), respectively. The parametric analysis also revealed that the input parameters, such as E_c, E_s and the ratio of D/t had least influence on the output parameter, i.e., P_u. Moreover, Figure 9c,f reveal that the value of P_u plummets with the increasing values of f_c' and ζ. Similarly, different equations such as linear, 3-, 4- and 6-degree polynomial equations were fitted to the resulting parametric analysis, which illustrate good agreement with the datapoints, i.e., all the parameters exhibit $R^2 \geq 0.99$. The sensitivity analysis of the input variables (as shown in Figure 10) reveals that ζ has the highest value of sensitivity (59%) followed by D and t (14.5%, each). It is also evident that L/D has the least sensitivity value = 0.2. This can be explained from the fact that the confinement of concrete enhances its strength and ductility, which enhances the P_u. Similarly, increasing value of D and t reduces the slenderness ratio of a column and enables the steel casing to withstand higher tensile stresses imparted by the confined concrete, respectively [21].

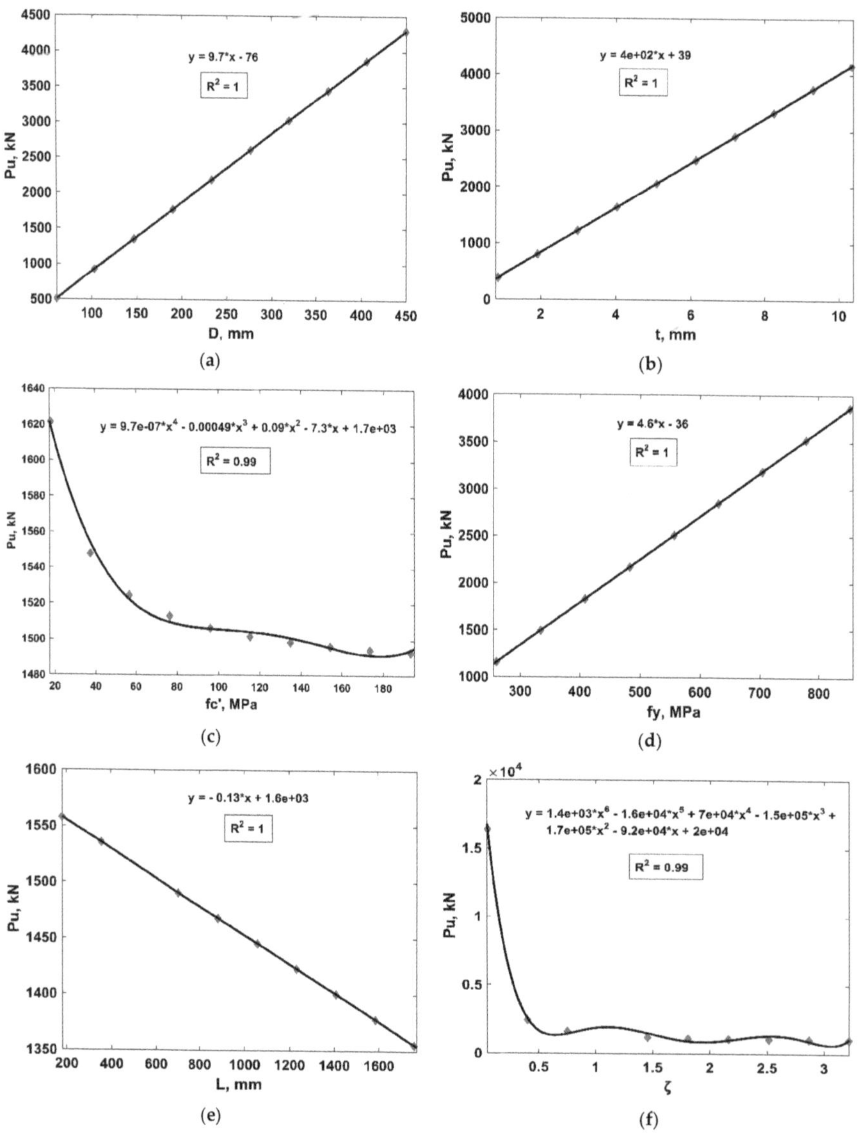

Figure 9. *Cont.*

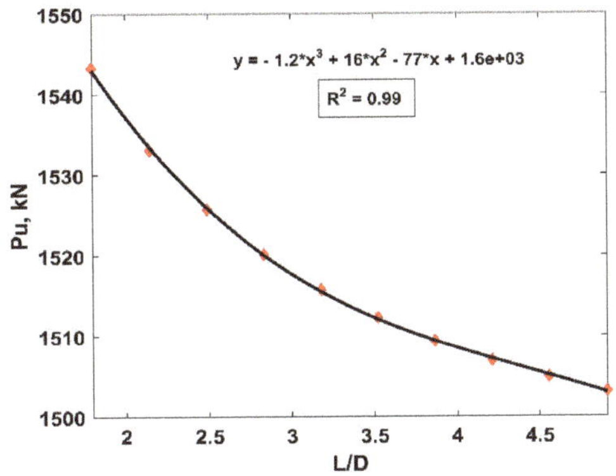

(g)

Figure 9. Parametric analysis of input variables. (**a**) Outer diameter of steel tube (D), (**b**) wall thickness of steel tube (t), (**c**) compressive strength of concrete (f_c'), (**d**) yield strength of steel (f_y), (**e**) length of column, (L), (**f**) confinement factor (ζ), (**g**) ratio of L/D.

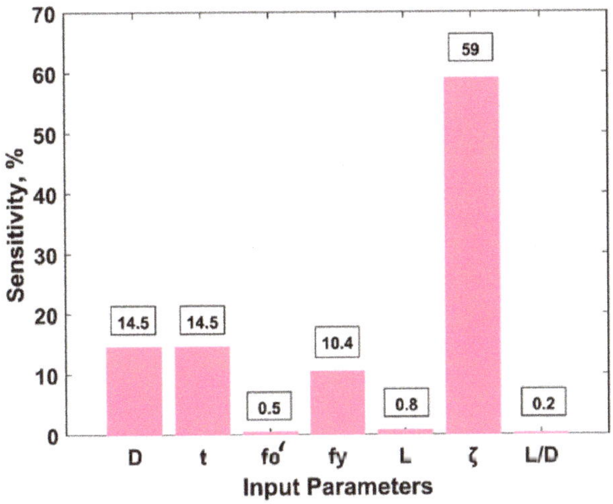

Figure 10. Sensitivity analysis of input variables.

4. Conclusions

This research study aims to model the ultimate compressive strength (P_u) of concrete-filled hollow steel sections (CFSS) columns by formulating an empirical relation between the output and input parameters. A total of 149 datapoints were taken from the literature, considering a number of input parameters such as outer diameter of steel tube (D), wall thickness of steel tube, compressive strength of concrete, elastic modulus of concrete (E_c), yield strength of steel (f_y), elastic modulus of steel (E_s), length of the column (L), confinement factor (ζ), ratio of D to thickness of column and the ratio of length to D of column. Different trials were undertaken to develop GEP models using various settings of the hyperparameters.

1. The performance of the developed models was assessed using variety of performance indices, i.e., R^2, *RMSE, MAE* and comparison of regression slopes. It was found that Model T3 having N_c = 100, H_s = 8 and N_g = 3 was the optimally performing model among all others. The model exhibits the highest R^2 value of 0.99, and the

lowest RMSE = 133.4 and 162.2 and MAE = 92.4 and 108.7, in the training and testing phases, respectively.
2. Similarly, the comparison of regression slopes analysis reveals that the Model T3 possess the highest value of R^2 equaling 0.99 and m = 1, which represents its high performance and robustness.
3. Finally, the parametric analysis depicts that the P_u of CFSS columns increases linearly with the value of D, t and f_y while, E_c, E_s and the ratio of D/t had the least influence on the output parameter. The sensitivity analysis of input variables reveals that the ζ is recorded to have the highest value of sensitivity (59%) whereas L/D has the least effect (i.e., 0.2%) in governing the P_u of the CFSS.
4. It is highly recommended to use the prediction equation (Equation (4)) for input variables whose range and other details are considered in the descriptive statistics of the current study. The following simple-to-use mathematical expression can be used to predict the P_u of the CFSS with higher accuracy.

Author Contributions: Conceptualization, K.K.; Data curation, M.I. and F.E.J.; Formal analysis, F.E.J.; Funding acquisition, K.K.; Investigation, M.I., M.R., A.A.A., A.M.A.-A. and F.E.J.; Methodology, M.R., M.N.A., A.A.A., A.M.A.-A. and F.E.J.; Project administration, K.K.; Resources, K.K.; Software, M.I. and F.E.J.; Validation, M.N.A.; Visualization, M.N.A.; Writing – original draft, K.K., M.I., M.R. and F.E.J. All authors have read and agreed to the published version of the manuscript.

Funding: This work was supported by the Deanship of Scientific Research, Vice Presidency for Graduate Studies and Scientific Research, King Faisal University, Saudi Arabia [Project No. GRANT762]. The APC was funded by the same "Project No. GRANT762".

Institutional Review Board Statement: Not applicable.

Informed Consent Statement: Not applicable.

Data Availability Statement: The data used in this research have been properly cited and reported in the main text.

Acknowledgments: The authors acknowledge the Deanship of Scientific Research, Vice Presidency for Graduate Studies and Scientific Research, King Faisal University, Saudi Arabia [Project No. GRANT762]. The authors extend their appreciation for the financial support that has made this study possible.

Conflicts of Interest: The authors declare no conflict of interest.

References

1. Sun, Z.; Chen, Y.; Xu, S.Y.; Sun, Y.; Wu, G. Shaking table test of concrete columns hybrid reinforced by steel/FRP bars. *J. Build. Eng.* **2022**, *48*, 103938. [CrossRef]
2. Melo, J.; Afshan, S.; Rossetto, T.; Varum, H.; Arêde, A. Experimental and numerical investigation of the cyclic response of stainless steel reinforced concrete columns. *Eng. Struct.* **2022**, *252*, 113607. [CrossRef]
3. Zhao, X.; Wen, F.; Yan, S. Behavior of multi-steel reinforced concrete columns with encased steel H-sections. *Eng. Struct.* **2022**, *264*, 114463. [CrossRef]
4. Zhuang, M.-L.; Sun, C.; Dong, B. Experimental and numerical investigations on seismic performance of HTRB630 high-strength steel bars reinforced concrete columns. *Case Stud. Constr. Mater.* **2022**, *17*, e01185. [CrossRef]
5. Golewski, G.L. A novel specific requirements for materials used in reinforced concrete composites subjected to dynamic loads. *Compos. Struct.* **2019**, *223*, 110939. [CrossRef]
6. *ACI 318-14*; ACI Committee 318, Building Code Requirements for Structural Concrete (ACI 318-14): Commentary on Building Code Requirements for Structural Concrete (ACI 318R-14)—An ACI Standard and Report. American Concrete Institute: Farmington Hills, MI, USA, 2014.
7. *EN1994-1*; Design of Composite Steel and Concrete Structures—Part 1-1: General Rules and Rules for Buildings. European Committee for Standardization: Brussels, Belgium, 1994.
8. Radovanovic, M.M.L.; Nikolic, J.Z.; Radovanovic, J.R.; Kostic, S.M. Structural Behaviour of Axially Loaded Concrete-Filled Steel Tube Columns during the Top-Down Construction Method. *Appl. Sci.* **2022**, *12*, 3771. [CrossRef]
9. Zhang, J.; Liu, Y.; Chen, Z.; Cai, R.; Li, X. Seismic behavior of high-strength concrete-filled square steel tube columns reinforced with ultrahigh-strength reinforcing bars. *Structures* **2021**, *34*, 3125–3140. [CrossRef]
10. Shao, Z.; Zha, X.; Wan, C. Design method of fire-resistance capacity of reinforced-concrete-filled steel tube column under axial compression. *Fire Saf. J.* **2022**, *129*, 103572. [CrossRef]

11. Zhang, T.; Ding, F.X.; Liu, X.M.; Yu, Z.W. Seismic behavior of terminal stirrup-confined concrete-filled elliptical steel tube columns: Experimental investigation. *Thin-Walled Struct.* **2021**, *167*, 108251. [CrossRef]
12. Huang, Y.; Zhao, P.; Lu, Y.; Zhang, H. Push-out tests of CFST columns strengthened with self-compacting and self-stressing concrete filled square steel tube. *J. Constr. Steel Res.* **2022**, *193*, 107263. [CrossRef]
13. Feng, W.; Wang, Y.; Sun, J.; Tang, Y.; Wu, D.; Jiang, Z.; Wang, J.; Wang, X. Prediction of thermo-mechanical properties of rubber-modified recycled aggregate concrete. *Constr. Build. Mater.* **2022**, *318*, 125970. [CrossRef]
14. Zhao, P.; Huang, Y.; Liu, Z.; Wang, H.; Lu, Y. Experimental research on seismic performance of steel fiber-reinforced recycled concrete-filled circular steel tube columns. *J. Build. Eng.* **2022**, *54*, 104683. [CrossRef]
15. Wang, X.; Fan, F.; Lai, J. Strength behavior of circular concrete-filled steel tube stub columns under axial compression: A review. *Constr. Build. Mater.* **2022**, *322*, 126144. [CrossRef]
16. Jahangir, H.; Soleymani, A.; Esfahani, M.R. Investigating the Confining Effect of Steel Reinforced Polymer and Grout Composites on Compressive Behavior of Square Concrete Columns. *Iran. J. Sci. Technol. Trans. Civ. Eng.* **2022**, 1–17. [CrossRef]
17. Tang, Y.; Zhu, M.; Chen, Z.; Wu, C.; Chen, B.; Li, C.; Li, L. Seismic performance evaluation of recycled aggregate concrete-filled steel tubular columns with field strain detected via a novel mark-free vision method. *Structures* **2022**, *37*, 426–441. [CrossRef]
18. *ANSI/AISC 360-16*; Specification for Structural Steel Buildings. American Institute of Steel Construction (AISC): Chicago, IL, USA, 2016.
19. Won, D.; Kim, S.; Seo, J.; Kang, Y.-J. Experimental Study of Composite Hollow RC Column under Uniaxial Compressive Load. *Appl. Sci.* **2019**, *9*, 373. [CrossRef]
20. Lin, S.; Li, Z.; Lu, Z.-H.; Zhao, Y.-G. Experimental study on the behavior of circular ultra-high strength concrete-filled steel tube columns subjected to unequal end moments. *Eng. Struct.* **2022**, *267*, 114709. [CrossRef]
21. Ci, J.; Ahmed, M.; Liang, Q.Q.; Chen, S.; Chen, W.; Sennah, K.; Hamoda, A. Experimental and numerical investigations into the behavior of circular concrete-filled double steel tubular slender columns. *Eng. Struct.* **2022**, *267*, 114644. [CrossRef]
22. Mangalathu, S.; Jeon, J.S. Classification of failure mode and prediction of shear strength for reinforced concrete beam-column joints using machine learning techniques. *Eng. Struct.* **2018**, *160*, 85–94. [CrossRef]
23. Munir, M.J.; Kazmi, S.M.S.; Wu, Y.F.; Lin, X.; Ahmad, M.R. Development of novel design strength model for sustainable concrete columns: A new machine learning-based approach. *J. Clean. Prod.* **2022**, *357*, 131988. [CrossRef]
24. Santarsiero, G.; Mishra, M.; Singh, M.K.; Masi, A. Structural health monitoring of exterior beam–column subassemblies through detailed numerical modelling and using various machine learning techniques. *Mach. Learn. Appl.* **2021**, *6*, 100190. [CrossRef]
25. Naderpour, H.; Mirrashid, M.; Parsa, P. Failure mode prediction of reinforced concrete columns using machine learning methods. *Eng. Struct.* **2021**, *248*, 113263. [CrossRef]
26. Albero, V.; Ibañez, C.; Hernández-Figueirido, D.; Piquer, A. Experimental analysis on circular concrete-filled steel tubular beam-columns under unequal load eccentricities. *Eng. Struct.* **2022**, *259*, 114206. [CrossRef]
27. Irandegani, M.A.; Zhang, D.; Shadabfar, M. Probabilistic assessment of axial load-carrying capacity of FRCM-strengthened concrete columns using artificial neural network and Monte Carlo simulation. *Case Stud. Constr. Mater.* **2022**, *17*, e01248. [CrossRef]
28. Naderpour, H.; Nagai, K.; Fakharian, P.; Haji, M. Innovative models for prediction of compressive strength of FRP-confined circular reinforced concrete columns using soft computing methods. *Compos. Struct.* **2019**, *215*, 69–84. [CrossRef]
29. Azim, I.; Yang, J.; Javed, M.F.; Iqbal, M.F.; Mahmood, Z.; Wang, F.; Liu, Q.-F. Prediction model for compressive arch action capacity of RC frame structures under column removal scenario using gene expression programming. *Structures* **2020**, *25*, 212–228. [CrossRef]
30. Momeni, M.; Hadianfard, M.A.; Bedon, C.; Baghlani, A. Damage evaluation of H-section steel columns under impulsive blast loads via gene expression programming. *Eng. Struct.* **2020**, *219*, 110909. [CrossRef]
31. Wang, H.; Wu, C.; Zhang, F.; Fang, Q.; Xiang, H.; Li, P.; Li, Z.; Zhou, Y.; Zhang, Y.; Li, J. Experimental study of large-sized concrete filled steel tube columns under blast load. *Constr. Build. Mater.* **2017**, *134*, 131–141. [CrossRef]
32. Zhang, F.; Wu, C.; Wang, H.; Zhou, Y. Numerical simulation of concrete filled steel tube columns against BLAST loads. *Thin-Walled Struct.* **2015**, *92*, 82–92. [CrossRef]
33. Koopialipoor, M.; Fallah, A.; Armaghani, D.J.; Azizi, A.; Mohamad, E.T. Three hybrid intelligent models in estimating flyrock distance resulting from blasting. *Eng. Comput.* **2019**, *35*, 243–256. [CrossRef]
34. Iftikhar, B.; Alih, S.C.; Vafaei, M.; Elkotb, M.A.; Shutaywi, M.; Javed, M.F.; Deebani, W.; Khan, M.I.; Aslam, F. Predictive modeling of compressive strength of sustainable rice husk ash concrete: Ensemble learner optimization and comparison. *J. Clean. Prod.* **2022**, *348*, 131285. [CrossRef]
35. Kaloop, M.R.; Bardhan, A.; Kardani, N.; Samui, P.; Hu, J.W.; Ramzy, A. Novel application of adaptive swarm intelligence techniques coupled with adaptive network-based fuzzy inference system in predicting photovoltaic power. *Renew. Sustain. Energy Rev.* **2021**, *148*, 111315. [CrossRef]
36. Bui, D.T.; Nhu, V.H.; Hoang, N.D. Prediction of soil compression coefficient for urban housing project using novel integration machine learning approach of swarm intelligence and Multi-layer Perceptron Neural Network. *Adv. Eng. Inform.* **2018**, *38*, 593–604. [CrossRef]
37. Abukhettala, M.; Fall, M. Geotechnical characterization of plastic waste materials in pavement subgrade applications. *Transp. Geotech.* **2021**, *27*, 100472. [CrossRef]

38. Boğa, A.R.; Öztürk, M.; Topçu, I.B. Using ANN and ANFIS to predict the mechanical and chloride permeability properties of concrete containing GGBFS and CNI. *Compos. Part B: Eng.* **2013**, *45*, 688–696. [CrossRef]
39. Lemonis, M.E.; Daramara, A.G.; Georgiadou, A.G.; Siorikis, V.G.; Tsavdaridis, K.D.; Asteris, P.G. Ultimate axial load of rectangular concrete-filled steel tubes using multiple ANN activation functions. *Steel Compos. Struct.* **2022**, *42*, 459. [CrossRef]
40. Mjalli, F.S.; Al-Asheh, S.; Alfadala, H.E. Use of artificial neural network black-box modeling for the prediction of wastewater treatment plants performance. *J. Environ. Manag.* **2007**, *83*, 329–338. [CrossRef]
41. Sarker, I.H. Deep Learning: A Comprehensive Overview on Techniques, Taxonomy, Applications and Research Directions. *SN Comput. Sci.* **2021**, *2*, 420. [CrossRef]
42. Khan, M.A.; Aslam, F.; Javed, M.F.; Alabduljabbar, H.; Deifalla, A.F. New prediction models for the compressive strength and dry-thermal conductivity of bio-composites using novel machine learning algorithms. *J. Clean. Prod.* **2022**, *350*, 131364. [CrossRef]
43. Amar, M.N.; Ghahfarokhi, A.J. Prediction of CO2 diffusivity in brine using white-box machine learning. *J. Pet. Sci. Eng.* **2020**, *190*, 107037. [CrossRef]
44. Shahmansouri, A.A.; Akbarzadeh Bengar, H.; Ghanbari, S. Compressive strength prediction of eco-efficient GGBS-based geopolymer concrete using GEP method. *J. Build. Eng.* **2020**, *31*, 101326. [CrossRef]
45. Naser, M.Z.; Thai, S.; Thai, H.T. Evaluating structural response of concrete-filled steel tubular columns through machine learning. *J. Build. Eng.* **2021**, *34*, 101888. [CrossRef]
46. *AS/NZS2327*; Composite Steel-Concrete Construction for Buildings. Standards Austalia: Sydney, Australia; Standards New Zealand: Wellington, New Zealand, 2017.
47. Sami Ullah, H.; Khushnood, R.A.; Ahmad, J.; Farooq, F. Predictive modelling of sustainable lightweight foamed concrete using machine learning novel approach. *J. Build. Eng.* **2022**, *56*, 104746. [CrossRef]
48. Sarir, P.; Chen, J.; Asteris, P.G.; Armaghani, D.J.; Tahir, M.M. Developing GEP tree-based, neuro-swarm, and whale optimization models for evaluation of bearing capacity of concrete-filled steel tube columns. *Eng. Comput.* **2019**, *37*, 1–19. [CrossRef]
49. Javed, M.F.; Farooq, F.; Memon, S.A.; Akbar, A.; Khan, M.A.; Aslam, F.; Alyousef, R.; Alabduljabbar, H.; Rehman, S.K.U.; Rehman, S.K.U.; et al. New Prediction Model for the Ultimate Axial Capacity of Concrete-Filled Steel Tubes: An Evolutionary Approach. *Crystals* **2020**, *10*, 741. [CrossRef]
50. Ngo, N.T.; Le, H.A.; Pham, T.P. Integration of support vector regression and grey wolf optimization for estimating the ultimate bearing capacity in concrete-filled steel tube columns. *Neural Comput. Appl.* **2021**, *33*, 8525–8542. [CrossRef]
51. Jiang, H.; Mohammed, A.S.; Kazeroon, R.A.; Sarir, P. Use of the Gene-Expression Programming Equation and FEM for the High-Strength CFST Columns. *Appl. Sci.* **2021**, *11*, 10468. [CrossRef]
52. Jayalekshmi, S.; Jegadesh, J.S.S.; Goel, A. Empirical Approach for Determining Axial Strength of Circular Concrete Filled Steel Tubular Columns. *J. Inst. Eng. Ser. A* **2018**, *99*, 257–268. [CrossRef]
53. Ahmadi, M.; Naderpour, H.; Kheyroddin, A. Utilization of artificial neural networks to prediction of the capacity of CCFT short columns subject to short term axial load. *Arch. Civ. Mech. Eng.* **2014**, *14*, 510–517. [CrossRef]
54. Khan, S.; Khan, M.A.; Zafar, A.; Javed, M.F.; Aslam, F.; Musarat, M.A.; Vatin, N.I. Predicting the Ultimate Axial Capacity of Uniaxially Loaded CFST Columns Using Multiphysics Artificial Intelligence. *Materials* **2022**, *15*, 39. [CrossRef]
55. Bardhan, A.; Biswas, R.; Kardani, N.; Iqbal, M.; Samui, P.; Singh, M.; Asteris, P.G. A novel integrated approach of augmented grey wolf optimizer and ANN for estimating axial load carrying-capacity of concrete-filled steel tube columns. *Constr. Build. Mater.* **2022**, *337*, 127454. [CrossRef]
56. Baykasoğlu, A.; Öztaş, A.; Özbay, E. Prediction and multi-objective optimization of high-strength concrete parameters via soft computing approaches. *Expert Syst. Appl.* **2009**, *36*, 6145–6155. [CrossRef]
57. Azimi-Pour, M.; Eskandari-Naddaf, H. ANN and GEP prediction for simultaneous effect of nano and micro silica on the compressive and flexural strength of cement mortar. *Constr. Build. Mater.* **2018**, *189*, 978–992. [CrossRef]
58. Yang, L.; Xu, Z.; Xu, R.; Lu, J.; Xu, Z.; Li, K. A New Dual-Mode GEP Prediction Algorithm Based on Irregularity and Similar Period. *Discret. Dyn. Nat. Soc.* **2021**, *2021*, 3594271. [CrossRef]
59. Iqbal, M.F.; Liu, Q.-F.; Azim, I.; Zhu, X.; Yang, J.; Javed, M.F.; Rauf, M. Prediction of mechanical properties of green concrete incorporating waste foundry sand based on gene expression programming. *J. Hazard. Mater.* **2020**, *384*, 121322. [CrossRef]
60. Gandomi, A.H.; Roke, D.A. Assessment of artificial neural network and genetic programming as predictive tools. *Adv. Eng. Softw.* **2015**, *88*, 63–72. [CrossRef]
61. Onyelowe, K.C.; Iqbal, M.; Jalal, F.E.; Onyia, M.E.; Onuoha, I.C. Application of 3-algorithm ANN programming to predict the strength performance of hydrated-lime activated rice husk ash treated soil. *Multiscale Multidiscip. Model. Exp. Des.* **2021**, *4*, 259–274. [CrossRef]
62. Wang, Z.; Zhao, X.-L.; Xian, G.; Wu, G.; Raman, R.S.; Al-Saadi, S.; Haque, A. Long-term durability of basalt- and glass-fibre reinforced polymer (BFRP/GFRP) bars in seawater and sea sand concrete environment. *Constr. Build. Mater.* **2017**, *139*, 467–489. [CrossRef]
63. Kafodya, I.; Xian, G.; Li, H. Durability study of pultruded CFRP plates immersed in water and seawater under sustained bending: Water uptake and effects on the mechanical properties. *Compos. Part B Eng.* **2015**, *70*, 138–148. [CrossRef]
64. Hanandeh, S.; Ardah, A.; Abu-Farsakh, M. Using artificial neural network and genetics algorithm to estimate the resilient modulus for stabilized subgrade and propose new empirical formula. *Transp. Geotech.* **2020**, *24*, 100358. [CrossRef]

materials

MDPI

Article

Optimization and Predictive Modeling of Reinforced Concrete Circular Columns

Gebrail Bekdaş [1,*], Celal Cakiroglu [2], Sanghun Kim [3] and Zong Woo Geem [4,*]

[1] Department of Civil Engineering, Istanbul University-Cerrahpasa, 34320 Istanbul, Turkey
[2] Department of Civil Engineering, Turkish-German University, 34820 Istanbul, Turkey
[3] Department of Civil and Environmental Engineering, Temple University, Philadelphia, PA 19122, USA
[4] Department of Smart City & Energy, Gachon University, Seongnam 13120, Korea
* Correspondence: bekdas@iuc.edu.tr (G.B.); geem@gachon.ac.kr (Z.W.G.)

Abstract: Metaheuristic optimization techniques are widely applied in the optimal design of structural members. This paper presents the application of the harmony search algorithm to the optimal dimensioning of reinforced concrete circular columns. For the objective of optimization, the total cost of steel and concrete associated with the construction process were selected. The selected variables of optimization include the diameter of the column, the total cross-sectional area of steel, the unit costs of steel and concrete used in the construction, the total length of the column, and applied axial force and the bending moment acting on the column. By using the minimum allowable dimensions as the constraints of optimization, 3125 different data samples were generated where each data sample is an optimal design configuration. Based on the generated dataset, the SHapley Additive exPlanations (SHAP) algorithm was applied in combination with ensemble learning predictive models to determine the impact of each design variable on the model predictions. The relationships between the design variables and the objective function were visualized using the design of experiments methodology. Applying state-of-the-art statistical accuracy measures such as the coefficient of determination, the predictive models were demonstrated to be highly accurate. The current study demonstrates a novel technique for generating large datasets for the development of data-driven machine learning models. This new methodology can enhance the availability of large datasets, thereby facilitating the application of high-performance machine learning predictive models for optimal structural design.

Keywords: predictive modeling; optimization; structural design

check for updates

Citation: Bekdaş, G.; Cakiroglu, C.; Kim, S.; Geem, Z.W. Optimization and Predictive Modeling of Reinforced Concrete Circular Columns. *Materials* **2022**, *15*, 6624. https://doi.org/10.3390/ma15196624

Academic Editor: Ayman El-Zohairy

Received: 7 August 2022
Accepted: 19 September 2022
Published: 23 September 2022

1. Introduction

Structural optimization aims at designing structures with the best possible dimensions that minimize cost without any impact on structural performance. In recent years metaheuristic optimization techniques have been increasingly applied to the optimization of different structures such as cylindrical reinforced concrete walls [1,2], retaining walls [3–5], plate girders [6], laminated composite plates [7–9], concrete-filled steel tubes [10,11], truss systems [12], timber structures [13], and liquid mass dampers [14–16]. These algorithms can be divided into evolutionary, physics-based, swarm-based, and population-based algorithms [17–19]. A detailed classification of the state-of-the-art metaheuristic optimization techniques can be found in Figure 1 [20–35]. Among the metaheuristic optimization algorithms applied to structural optimization, the harmony search algorithm stands out as one of the most widely used techniques. Besides structural optimization, the harmony search technique has been applied to various areas of science and engineering such as transportation engineering [36–39], environmental engineering [40,41], healthcare systems [42], bioinformatics [43–45], and cloud computing [46].

Reinforced concrete (RC) circular columns have been used in broad applications in structural engineering. The total amount of longitudinal reinforcement determines their load-carrying capacity of them. Therefore, the accurate determination of the right amount of reinforcement in these structural members under axial forces and bending moments bears the utmost importance. Figure 2 shows a general description of an RC circular column where the lateral and longitudinal reinforcements can be seen. The outer diameter D and the total length L describe the geometry of the column in addition to the longitudinal reinforcement area A_s.

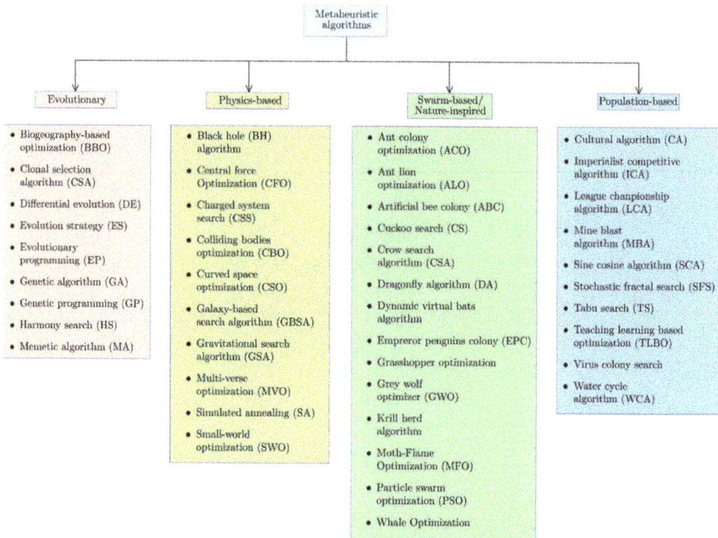

Figure 1. Classification of metaheuristic optimization algorithms.

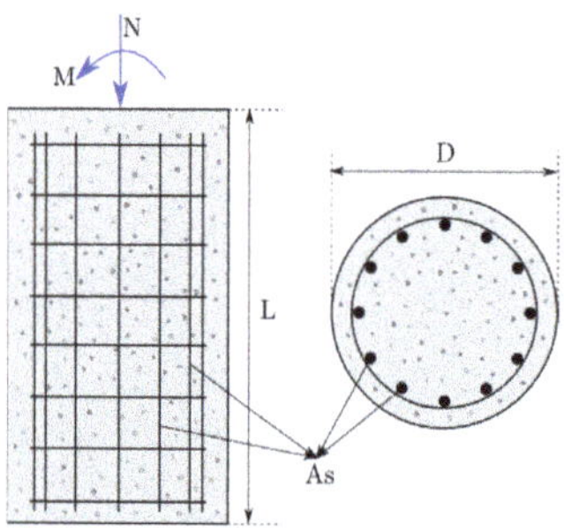

Figure 2. Cross-section and longitudinal section of a circular RC column.

This paper presents a novel data-driven technique for the prediction of the area of the longitudinal reinforcement (A_s) in RC circular columns. To this end, four different ensemble learning algorithms have been utilized to obtain predictive models. The performances of these algorithms in terms of predicting A_s accurately have been compared using the coefficient of determination (R^2), mean absolute error (MAE), and root mean square error (RMSE) as the metrics of accuracy. The datasets needed to train these predictive models have been generated using the harmony search algorithm such that each data sample corresponds to an optimal design configuration that satisfies certain load-carrying requirements defined by the design codes. A combination of axial force and the bending moment was applied in each data sample. A dataset of 3125 samples was generated where each sample consists of six input variables and the corresponding output variable. The input variables in this dataset consist of the outer diameter of the column (D), the unit cost of concrete used in the construction (C_c), the unit cost of steel (C_s), the total length of the column (L), the bending moment acting on the column (M), and the axial force acting on the column (N). The corresponding output variable is the optimal longitudinal reinforcement area (A_s). These input variables were selected in order to have a description of the geometry, material properties, and external loading for each data sample. In this regard, the unit costs of concrete and steel quantify the material properties whereas the column length and outer diameter describe the geometry corresponding to a data sample. To clarify the impact of each input variable on the output of the predictive models and to show the dependencies between different variables, the SHAP algorithm has been utilized. Furthermore, a four-level factorial analysis has been carried out to visualize the variation of the output variable for the different levels of each input variable [47]. Based on the dependencies of the input variables, a predictive equation has been proposed for the reinforcement area. The equation has been developed using the harmony search algorithm to minimize the difference between the predicted and true optimal reinforcement areas. An R^2 score of 0.998 could be achieved by the developed equation.

The current paper investigates the optimal design of circular RC columns under combined loading, which is an area of structural engineering that has not been investigated using data-driven machine learning techniques to the best of the authors' knowledge. The work related to failure mode classification and capacity prediction of RC columns using an ensemble machine learning algorithm AdaBoost by Feng et al. [48] can be counted among the recent machine learning-related research works in the field of conventional RC columns. The ensemble learning algorithm was developed based on a data set consisting of 254 data samples collected from cyclic loading tests. Also, Dogan et al. [49] investigated the damage levels of RC columns under cyclic lateral loading conditions using machine learning methods for classification such as support vector machines, K-nearest neighbors and discriminant analysis. The machine learning models were trained on a set of 390 damage images. However, the research activity in the area of RC columns using machine learning methods has been limited compared to other areas such as concrete columns with fiber-reinforced polymer wrappings [50,51], or crack propagation and corrosion in RC structures [52–54]. Nasrollahzadeh and Nouhi [50] proposed fuzzy inference system models to predict the strength and strain capacity of square concrete columns wrapped with fiber-reinforced polymer. Experimental data sets consisting of 261 and 112 test samples were used for the prediction of compressive strength and ultimate strain respectively. Naderpour et al. [51] utilized artificial neural network and gene expression programming techniques to predict the compressive strength of columns confined with fiber reinforced polymers based on a data set consisting of 95 data samples. However, despite the benefits of using composite materials as reinforcement or confinement for columns reported in the literature, the overwhelming majority of new constructions and existing infrastructure rely on conventional reinforced concrete.

One of the reasons for the lack of machine learning related research in the field of RC columns is the difficulty of obtaining large datasets. Machine learning-based predictive models need to be trained using large and comprehensive data sets in order to be relevant in general-purpose structural design. On the other hand, experimental research in the field of RC members is generally costly and experimental programs usually deliver a limited number of data points. Therefore, alternative techniques need to be devised for the training of machine learning models in order to use these powerful techniques in the field of RC design.

An important current issue in previous literature pertaining to machine learning applications in structural engineering is the size of the data sets used during the model training process. Evidently, most studies in this field depend on data sets with less than a thousand data samples. However, the reliability of a machine learning model heavily depends on the size and quality of the data used in its training. The aim of the current study is to present a methodology for the generation of large datasets related to the optimal design of RC columns. The novelty of this current work shows the applicability of newly developed techniques of artificial intelligence to the design process of RC columns. The availability of quality data is a major requirement in this process. However, large datasets are needed for the development of accurate predictive machine learning models, which is one of the limiting factors in the process of machine learning model development. The current paper proposes a novel technique for the generation of large datasets. This technique is highly valuable since it can remove one of the major bottleneck points in the application of machine learning techniques to structural design.

2. Dataset Generation and Analysis

A large dataset consisting of 3125 unique configurations has been generated with the help of the harmony search optimization algorithm. Each sample in this dataset corresponds to a design configuration that minimizes the total cost associated with the construction process while keeping the load-carrying capacity above a certain level. To this end, the axial load and moment capacities were kept above the applied load which can be described in Equation (1):

$$\phi P_n \geq P_u, \quad \phi M_n \geq M_u \tag{1}$$

In Equation (1), P_u and M_u are the applied loads, P_n and M_n are the nominal strengths of the column cross-section, and ϕ is the strength reduction factor. This process starts with the generation of a randomly populated harmony memory matrix (HM) as shown in Equation (2) where f denotes the cost function, HMS is the size of the solution candidate population, V_c and W_s are the total volume of concrete and the total weight of steel respectively and x^i is a solution candidate vector containing the variables D^i, C_c^i, C_s^i, L^i, M^i, N^i, A_s^i. The cost function f determines the performance of each population member and the solution candidates can be ranked accordingly.

$$HM = \begin{bmatrix} D^1 & C_c^1 & C_s^1 & L^1 & M^1 & N^1 & A_s^1 & f(x^1) \\ D^2 & C_c^2 & C_s^2 & L^2 & M^2 & N^2 & A_s^2 & f(x^2) \\ \vdots & \vdots & \vdots & \vdots & \vdots & \vdots & \vdots & \vdots \\ D^{HMS} & C_c^{HMS} & C_s^{HMS} & L^{HMS} & M^{HMS} & N^{HMS} & A_s^{HMS} & f(x^{HMS}) \end{bmatrix}$$
$$f(x^i) = (C_c)^i V_c + (C_s)^i W_s \tag{2}$$

In Equation (2), C_c^i, C_s^i, L^i, M^i, N^i, and A_s^i stand for the concrete cost per unit volume, steel cost per unit weight, length of the column, bending moment, axial force and the total area of the longitudinal reinforcement in the i-th solution candidate respectively. The harmony search technique obtains an optimum solution that minimizes the total cost by an evolutionary process in which the solution candidates improve incrementally and eventually converge to an optimum solution. The evolutionary process of incremental improvement of the harmony memory matrix can be described in Equations (3)–(6).

$$k = \text{int}(\text{rand} \cdot \text{HMS}), \ \text{rand} \in (0, 1) \tag{3}$$

$$x_{i,new} = x_{i,\,min} + \text{rand} \cdot (x_{i,\,max} - x_{i,\,min}), \ \text{if HMCR} + \text{rand} \tag{4}$$

$$x_{i,new} = x_{i,k} + \text{rand} \cdot \text{PAR} \cdot (x_{i,\,max} - x_{i,\,min}), \ \text{if HMCR} + \text{rand} \tag{5}$$

$$\text{HMCR} = 0.5\left(1 - \frac{i}{\max(i)}\right), \ \text{PAR} = 0.05\left(1 - \frac{i}{\max(i)}\right) \tag{6}$$

In Equations (3)–(6), HMCR, PAR, $x_{i,\,min}$ and $x_{i,\,max}$ stand for the harmony memory consideration rate, the pitch adjustment rate, the minimum and the maximum values of the i-th input variable in the population respectively. After each harmony search iteration, the updated solution candidates replace the old ones if they have superior performance and satisfy the structural design code requirements. For a more detailed review of the harmony search technique, the reader is referred to [55].

The variable ranges of the dataset generated using the harmony search algorithm can be seen in Figure 3 and Table 1. In Figure 3, the variable ranges have been divided into four partitions and the ranges of these partitions have been shown on the horizontal bars whereas the number of samples in each partition is shown inside the horizontal bars. The length of each partition in Figure 3 is in proportion to the number of samples belonging to that partition. Figure 3 shows that the largest partition for the outer diameter D consists of 1213 samples ranging between 0.574 m and 0.661 m. The second largest partition for this variable with 1000 samples ranges between 0.661 m and 0.747 m. The remaining two partitions ranging between the lower bound of the outer diameter of 0.4 m and 0.574 m constitute 29% of the entire dataset. The partitions for the variables C_c, C_s, L, M, and N are evenly distributed. The largest partition for the column length L consists of 1250 samples ranging between 3 m and 4 m, which is 40% of the entire dataset. The horizontal bars for C_c and C_s in Figure 3 show the unit prices of concrete and steel in USD/m^3 and USD/ton respectively.

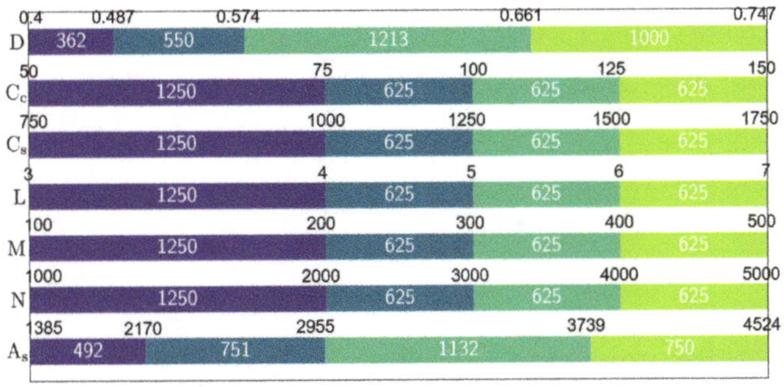

Figure 3. Design variable ranges in the dataset.

Table 1. Variable ranges and statistical properties.

Variable	Min		Max		Average		Standard Deviation		Variance	
	Actual	Normalized	Actual	Normalized	Actual	Normalized	Actual	Normalized	Actual	Normalized
D [m]	0.4	0.647	0.747	1.209	0.618	1	0.087	0.141	0.0076	0.02
Cc [USD/m^3]	50	0.5	150	1.5	100	1	35.4	0.354	1250	0.125
Cs [USD/ton]	750	0.6	1750	1.4	1250	1	354	0.283	125,000	0.08
L [m]	3	0.6	7	1.4	5	1	1.41	0.283	2	0.08
M [kNm]	100	0.333	500	1.667	300	1	141	0.471	20,000	0.222
N [kN]	1000	0.333	5000	1.667	3000	1	1414	0.471	2,000,000	0.222
As [mm^2]	1385	0.447	4524	1.459	3101	1	799	0.258	639,080	0.067

Table 1 includes the upper and lower bounds as well as statistical properties of the design variables represented in the dataset. These statistical properties are the average value, standard deviation, and variance of each variable inside the dataset. Furthermore, for each variable, the corresponding boundaries and statistical properties have been listed after normalizing the variables by their average values. These normalized values are used at a later stage for the development of a predictive equation. Also, the partitions presented in Figure 3 are used as the basis of a four-level factorial analysis to determine the variation of A_s for each design variable. In addition to the partition plot in Figure 3, also a correlation plot has been generated for the dataset (Figure 4). For each input variable and the output variable A_s, Figure 4 shows the Pearson correlation coefficient between any two of these variables in the upper right portion of the diagram. Pearson correlation values close to 1 indicate a high correlation between two variables. The highest correlation coefficient in Figure 4 can be observed between A_s and D which indicates that the outer diameter has a significant impact on the reinforcement area. The second highest correlation can be observed between A_s and N with a Pearson correlation value of 0.79. Another relatively high correlation is observed between D and N with a correlation value of 0.77. Finally, the column length L is correlated to A_s and D with correlation coefficients of 0.53 and 0.54 respectively. Greater correlation between variables is represented by larger font size and stars inside the tiles of the matrix. In Figure 4, each variable occupies one of the diagonal tiles and the scale of this variable is shown in one of the horizontal axes and one of the vertical axes. Furthermore, each diagonal tile contains a histogram showing the distribution of the variable in it. The lower left portion of the correlation matrix contains bivariate scatter plots with regression lines. The equation for the computation of the Pearson correlation coefficient as well as the other metrics of accuracy used in this paper can be found in Appendix A.

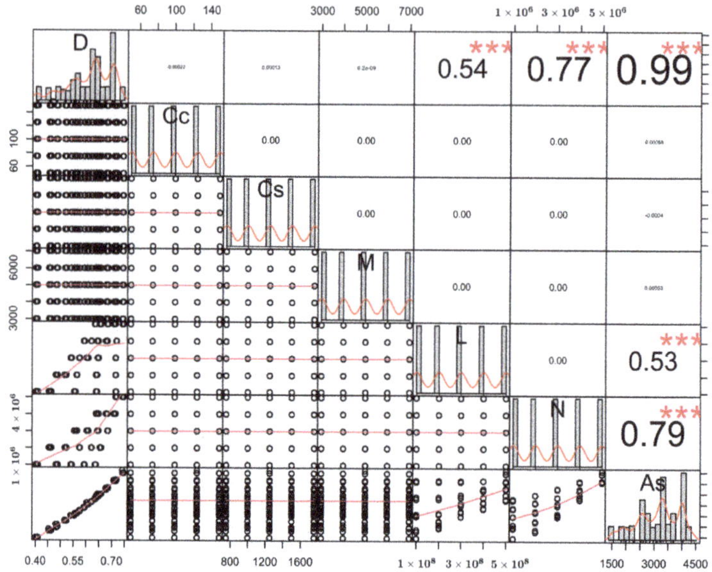

Figure 4. Correlation matrix of the dataset.

Using the dataset whose properties have been shown in Figure 3, Table 1, and Figure 4, four different data-driven predictive models have been trained using the ensemble learning algorithms XGBoost, LightGBM, Random Forest, and CatBoost. The results of them and their interpretations using the SHapley Additive exPlanations (SHAP) technique have been presented in the next section. The theoretical background of ensemble learning and SHAP algorithms can be found in [3,56–62]. In addition to the SHAP analysis also, a four-level factorial analysis has been carried out to further investigate the sensitivity of A_s to the variations in different design variables. Afterward, a closed-form equation has been proposed for the prediction of A_s as a function of the six design variables shown in Figures 3 and 4. The overall process of dataset generation, training of the machine learning models, and the development of a predictive closed-form equation have been summarized in a flow chart in Figure 5.

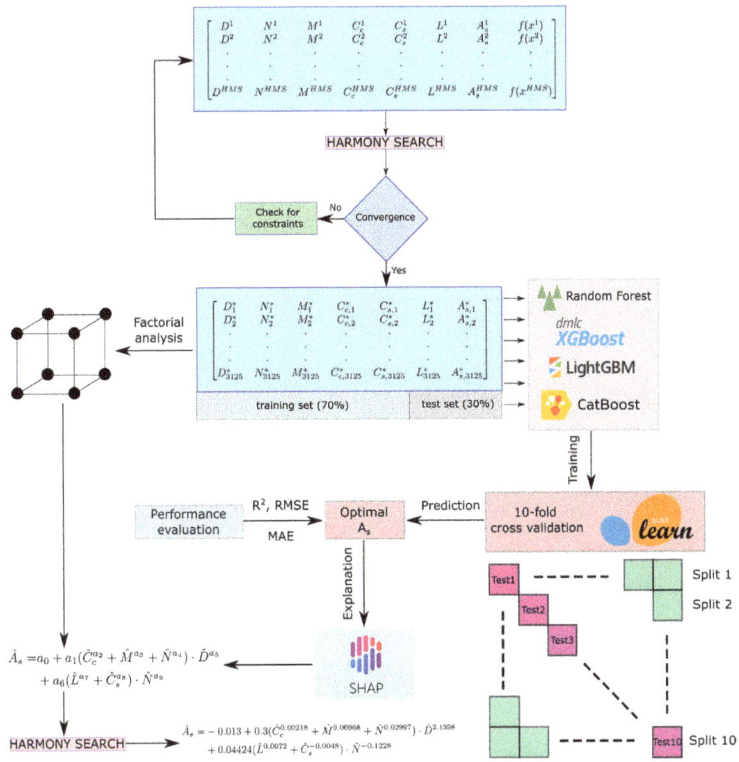

Figure 5. Machine learning and equation development process.

3. Results

This section presented the comparison of the optimal reinforcement areas predicted by the ensemble learning algorithms with the true optimal values obtained through the harmony search algorithm. The performances of each predictive model have been measured by the metrics of the coefficient of determination, mean absolute error and root mean squared error. The outcome of the ensemble learning models has been analyzed with the SHAP algorithm to determine the variables with the highest impact on the model predictions. Furthermore, a four-level factorial analysis has been performed to visualize the variation of the optimal reinforcement cross-section for each input variable. Based on the outcome of the SHAP and factorial analyses, a predictive equation format has been proposed. This predictive equation has been developed using the harmony search algorithm, and the accuracy of the obtained equation has been demonstrated by the same accuracy metrics applied to the ensemble learning models.

3.1. Ensemble Learning Model Predictions

The ensemble learning models have been trained by splitting the entire dataset into a training set and a test set in 70% to 30% proportions. This division was made based on past machine learning studies in the area of structural engineering. Particularly, the study of Nguyen et al. [63] demonstrated that among the 10/90, 20/80, 30/70, 40/60, 50/50, 60/40, 70/30, 80/20, and 90/10 training set to test set ratios, the 70/30 ratio delivered the best performance. The models have been trained on the training set using ten-fold cross-validation. After the completion of the model training, the test set was used to measure the model performances. The performances of the ensemble learning models have been visualized by plotting the true optimal A_s values of the test set against the

A_s values predicted by the ensemble learning models. Figure 6 shows the comparison of the predicted and actual optimal A_s values for each of the four predictive models. In Figure 6, the diagonal solid lines represent the case when the actual and predicted values are equal, whereas the dotted lines represent $\pm$ 10% deviation from a perfect prediction. The performances of these predictive models are compared to each other in Table 2. According to Table 2, the Random Forest algorithm demonstrated the best performance in terms of both prediction accuracy on the test set and the speed of execution (3.71 s).

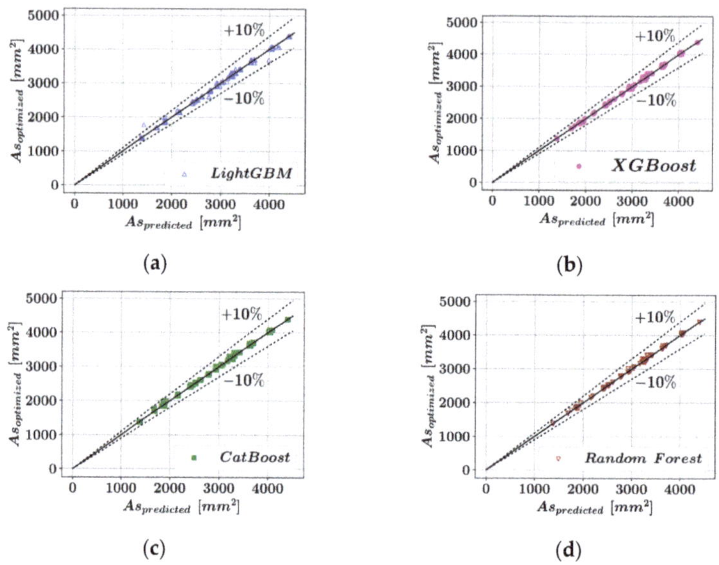

Figure 6. Comparison of the predicted and optimized dimensions. (**a**) LightGBM. (**b**) XGBoost. (**c**) CatBoost. (**d**) Random Forest.

Table 2. Prediction accuracy of the machine learning models.

Algorithm	Variable	R^2		MAE		RMSE		Duration [s]
		Train	Test	Train	Test	Train	Test	
XGBoost	As	0.9999	0.9995	1.998	7.523	3.839	17.072	5.14
Random Forest	As	0.9999	0.9996	2.593	7.111	6.095	15.929	3.71
LightGBM	As	0.9994	0.9988	9.962	12.767	19.673	27.157	6.07
CatBoost	As	0.9998	0.9994	7.579	10.788	12.440	18.940	28.23

3.2. SHAP Analysis

The SHAP analysis visually describes the contribution of each design variable to the prediction of a machine learning model. The SHAP summary plot and feature dependence plots in this section are based on the Random Forest algorithm selected due to its superior performance on this dataset. The SHAP summary plot in Figure 7 ranks the six input variables according to their impact on the predictive model output. In Figure 7, each data sample is represented by a dot and positive SHAP values indicate an increasing effect of a variable on the model output whereas negative SHAP values indicate a decreasing effect on the model output. The impact of a variable on model output for a particular data sample is a function of the position of a dot along the horizontal axis. On the other hand, the numerical values of the input parameters are represented with color so that high parameter values are shown with shades of red and the low parameter values are shown with shades of blue. According to Figure 7, the outer diameter D has the greatest impact on the model output. It can be deduced that an increase in the value of D also leads to an increase in the model prediction. Conversely, decreasing the D value leads to lower model predictions. On the other hand, the impacts of the remaining parameters on the model output are an order of magnitude smaller than the impact of D according to Figure 7.

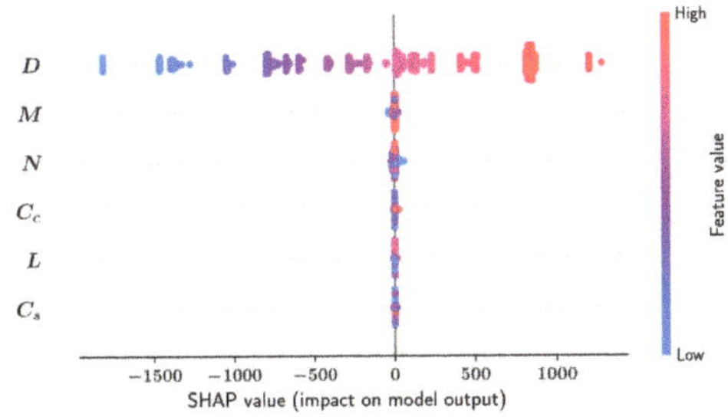

Figure 7. SHAP summary plot for A_s (Random Forest).

The feature dependence plots in Figure 8 can help better understand the interdependencies between different variables and their effects on the model output. Figure 8a clearly shows that as the D value increases, so does the SHAP value for this variable. This confirms the inference from Figure 7 that increased D values lead to increased reinforcement area. The colors of the dots in Figure 8a indicate the numerical values of C_c which is the variable most dependent on D. On the other hand, the feature dependence plots of C_c, C_s and L show that the SHAP values of these variables have a horizontal trend as the variable values increase. For these three variables, most of the SHAP values stay in the range of −5 to 5 for the entire dataset. For a significant portion of the dataset, the SHAP values are clustered around zero. Also, for all three of these variables, N is the variable most dependent on them. The feature dependence plots of M and N exhibit certain patterns for different levels of these variables. Figure 8e,f show that the feature dependence plots for M and N are fragmented and can be investigated separately for different levels of these variables. For each value of M and N, the SHAP values of these variables are concentrated around different levels depending on the value of D which is the parameter most dependent on M and N. It can be observed that M has the greatest impact on the model output when $M = 3 \times 10^8$Nmm and D have large values shown in red and the least impthe act when $M = 10^8$Nmm and D have small values shown in blue. Similarly, N has the greatest impact when $N = 10^6$N and D have small values shown in blue and the least impact when $N = 3 \times 10^6$N and D have large values shown in red.

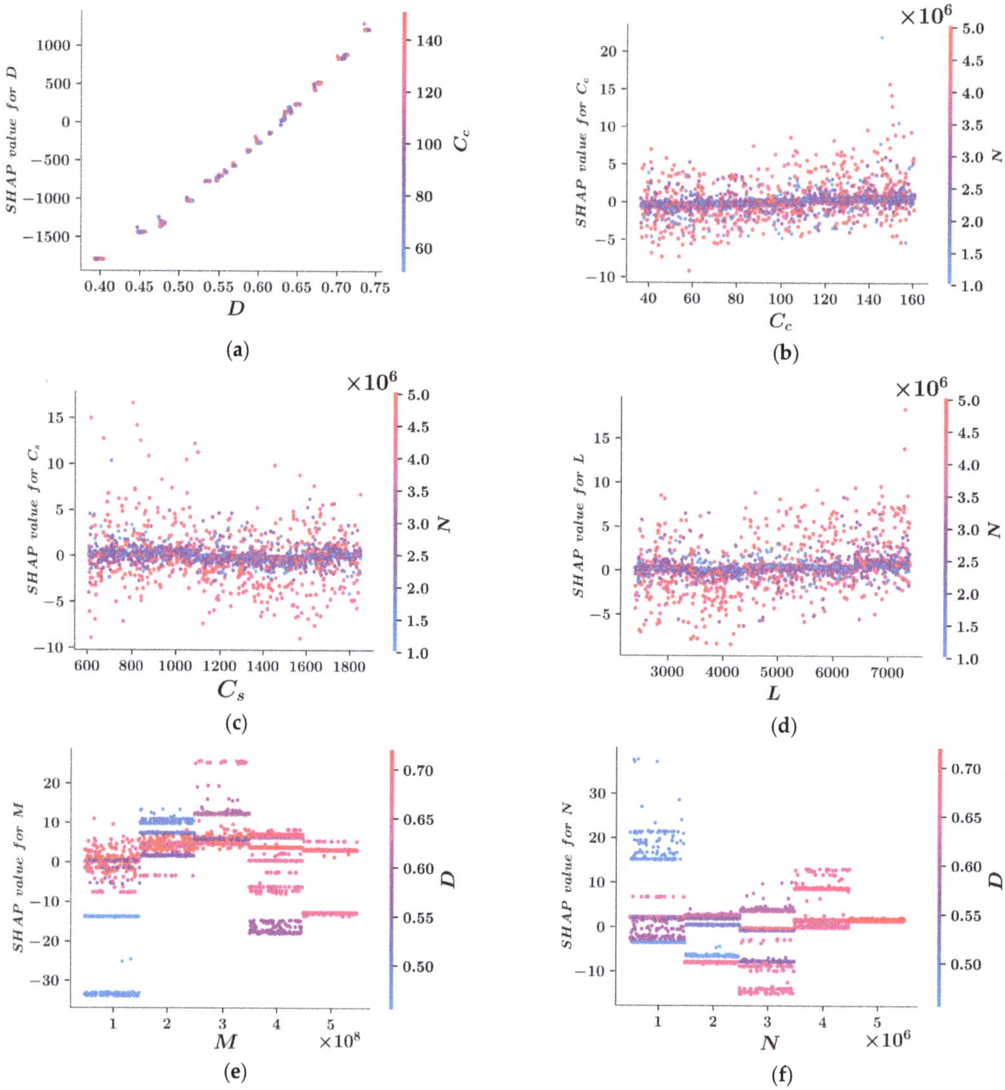

Figure 8. Feature dependence plots for the variables (**a**) D, (**b**) C_c, (**c**) C_s, (**d**) L, (**e**) M, (**f**) N.

3.3. Four-Level Factorial Analysis

The factorial analysis technique is widely used for gaining insights into the response and sensitivity of a system that depends on multiple variables to the variations in a single variable. The factorial analysis technique is particularly useful when the variables of a system can be broken down into different levels. In this paper, the cross-section of reinforcement is predicted as a function of six variables. Each of these variables has been broken down into four levels as shown in Figure 3. The four-level factorial analysis enables the visualization of the nonlinear variations in the target variables or curvatures in the system response [64]. Afterward, for each level of each variable, the average value of the area of reinforcement has been calculated. These average values are plotted for different levels of each variable in Figure 9. A significant variation of A_s in Figure 9 with respect to changes in a certain variable indicates the high sensitivity of A_s to the changes in this variable. According to Figure 9, the greatest change in the value of A_s is observed when D increases from its lowest level (level 0) to its highest level (level 3). A total increase from 1641 mm^2 to 3993 mm^2 can be observed which corresponds to a 143% increase. The second largest increase in the average value of A_s can be observed when N increases from its lowest level to its highest level. In this case, the increase of the area from 2459 mm^2 to 4119 mm^2 can be observed which corresponds to a 68% increase. The third largest percentage-wise increase in the average A_s value can be observed when M increases from its lowest level to its highest level. An increase from an average area of 2661 mm^2 to an average area of 3750 mm^2 can be observed which corresponds to a 41% increase. For the remaining three variables C_c, C_s and L, the changes in the average value of A_s was negligible in comparison to D, N, and M.

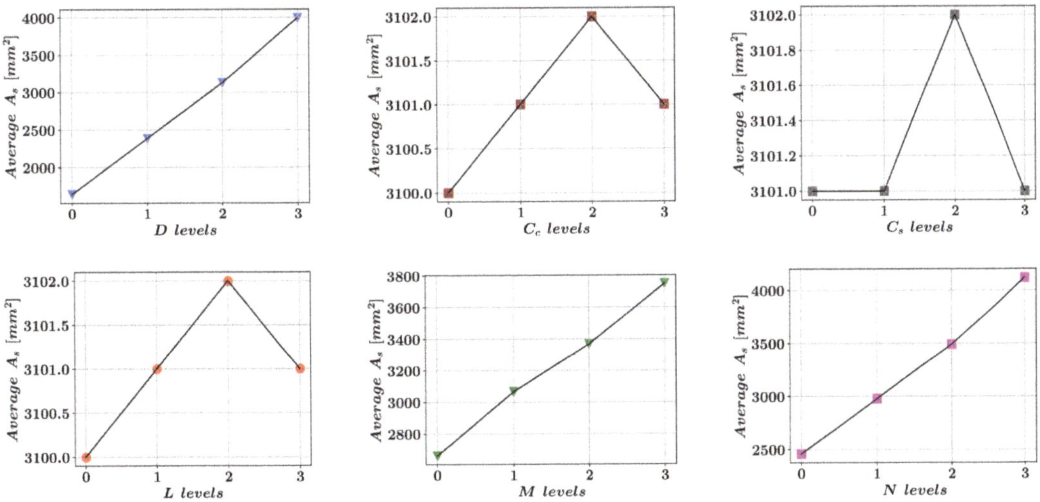

Figure 9. Four-level factorial analysis.

3.4. *Development of an Equation for the Prediction of the Optimum Reinforcement Area*

In light of the results presented in the previous sections, the formula in Equation (7) has been proposed for the prediction of the reinforcement area in an optimal design. In Equation (7) the variables A_s, $\hat{C}_c$, $\hat{M}$, $\hat{N}$, $\hat{D}$, $\hat{L}$, $\hat{C}_s$ are normalized by the average value of each variable in the dataset consisting of 3125 samples.

$$\hat{A}_s = a_0 + a_1\left(\hat{C}_c^{a_2} + \hat{M}^{a_3} + \hat{N}^{a_4}\right) \cdot \hat{D}^{a_5} + a_6\left(\hat{L}^{a_7} + \hat{C}_s^{a_8}\right) \cdot \hat{N}^{a_9} \tag{7}$$

The coefficients a_0 to a_9 in Equation (7) have been adjusted using harmony search iterations. This process necessitates the declaration of a new harmony memory matrix that contains the coefficients of Equation (7) as shown in Equation (8) where the population consists of 30 different solution candidates.

$$HM = \begin{bmatrix} a_0^1 & a_1^1 & a_2^1 & a_3^1 & a_4^1 & a_5^1 & a_6^1 & a_7^1 & a_8^1 & a_9^1 \\ a_0^2 & a_1^2 & a_2^2 & a_3^2 & a_4^2 & a_5^2 & a_6^2 & a_7^2 & a_8^2 & a_9^2 \\ \vdots & \vdots & \vdots & \vdots & \vdots & \vdots & \vdots & \vdots & \vdots & \vdots \\ a_0^{30} & a_1^{30} & a_2^{30} & a_3^{30} & a_4^{30} & a_5^{30} & a_6^{30} & a_7^{30} & a_8^{30} & a_9^{30} \end{bmatrix} \tag{8}$$

After every harmony search iteration, the performances of the solution candidates are measured by comparing their predictions of $\hat{A}_s$ with the actual $\hat{A}_s$ values for the entire dataset. The prediction error is represented by the Euclidean norm of the vector containing the differences between the actual and predicted optimal $\hat{A}_s$ values. The development of these vector norms for the best- and worst-performing members of the harmony memory population is presented in Figure 10.

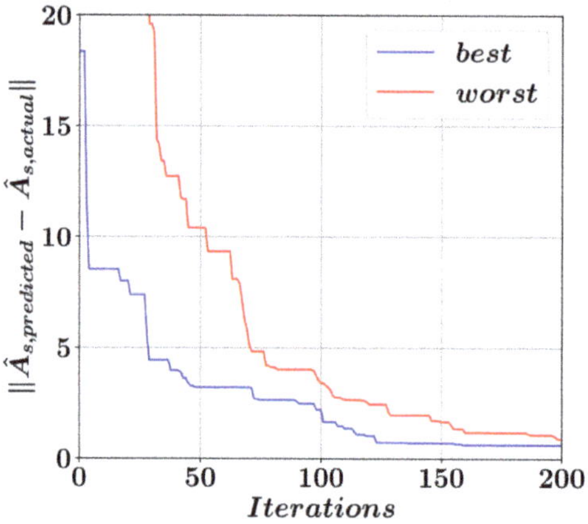

Figure 10. Equation performance throughout the harmony search iterations.

A total of 5000 harmony search iterations was carried out to obtain the best possible equation coefficients with the smallest possible error norm. Figure 10 shows the development of the best and worst solution candidates in the initial 200 iterations. It should be noted that the largest improvements in the solution candidates take place during the initial phases of the harmony search iterations. Figure 11 shows the process of obtaining the coefficients a_0 to a_9 that minimize the difference between the optimal $\hat{A}_s$ values predicted by Equation (7) and the actual optimal $\hat{A}_s$ values.

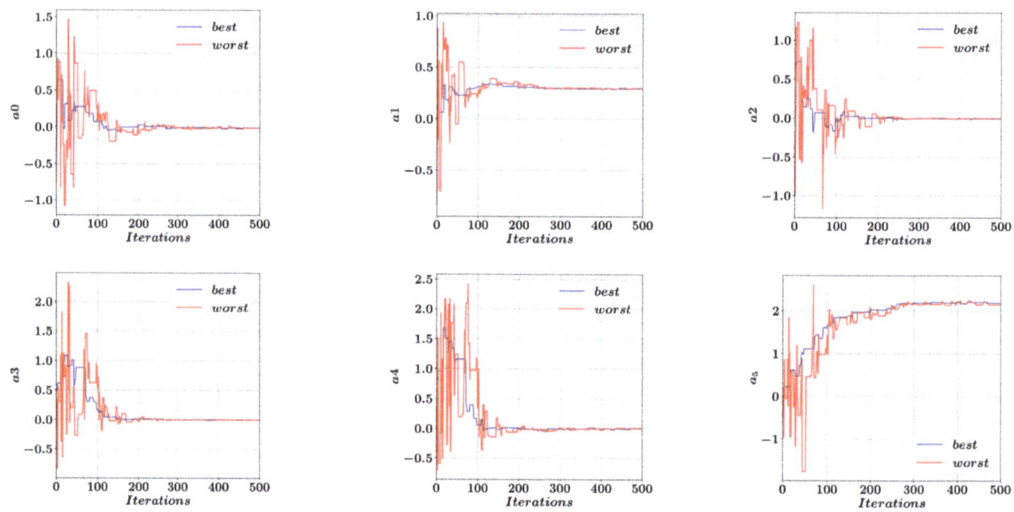

Figure 11. *Cont.*

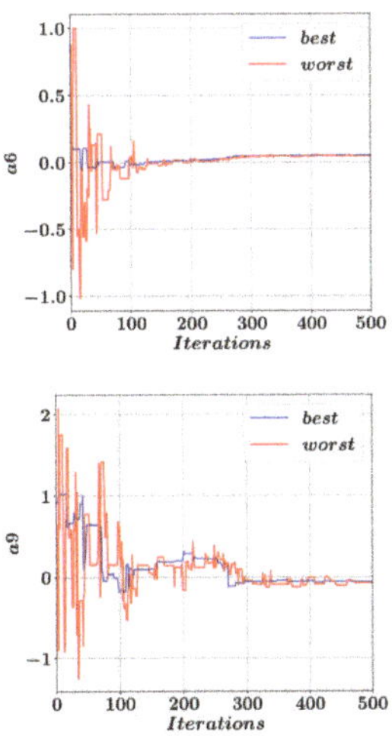

Figure 11. Development of the equation coefficients in Equation (7).

Figure 11 shows the values of the coefficients a_0 to a_9 in the first 500 harmony search iterations. The coefficient values corresponding to the best- and worst-performing members of the harmony memory population are shown in blue and red respectively. It can be observed that after the initial fluctuations, these coefficients tend to converge to their optimal limit values. Inserting these limit values of the coefficients a_0 to a_9 in Equation (7), we obtain Equation (9) for the prediction of the normalized reinforcement area $\hat{A}_s$ from which the actual reinforcement area A_s can be obtained after multiplication with the average value of A_s from Table 1.

$$\begin{aligned}\hat{A}_s = &-0.013 + 0.3\left(\hat{C}_c^{0.00218} + \hat{M}^{0.00968} + \hat{N}^{0.02997}\right) \cdot \hat{D}^{2.1398} \\ &+ 0.04424\left(\hat{L}^{0.0072} + \hat{C}_s^{-0.0048}\right) \cdot \hat{N}^{-0.1228}\end{aligned} \tag{9}$$

A coefficient of determination of 0.9985, mean absolute error of 0.0076, and root mean square error of 0.0099 could be achieved using Equation (9). The $\hat{A}_s$ values predicted by Equation (9) are plotted against the actual $\hat{A}_s$ values in Figure 12 where the dotted lines indicate the ±10% deviation from a perfect match between the predicted and actual values.

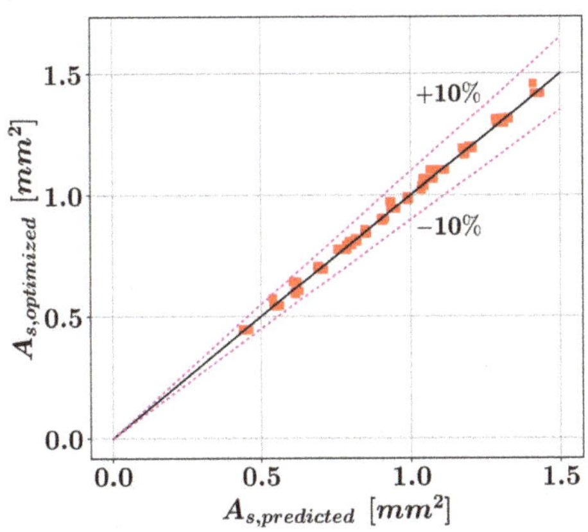

Figure 12. Comparison of the predicted and actual optimal reinforcement areas.

4. Discussion

Circular RC columns are ubiquitous in the structural design of buildings, bridges, and ports. Therefore, the optimal design of these structural members can have enormous economic and environmental benefits. This paper demonstrates the applicability of machine learning models to the structural optimization of RC circular columns. The total cross-sectional area of the longitudinal reinforcement was selected as the decisive parameter in the design of these structures. A novel technique has been applied in the training phase of these machine learning models. The harmony search algorithm was used to generate a large dataset consisting of 3125 samples where each sample represents an optimum configuration of column geometry and external loading. XGBoost, LightGBM, Random Forest, and CatBoost algorithms are used to generate four different ensemble learning predictive models. The performances of these models have been compared using the coefficient of determination (R^2), mean absolute error (MAE) and root mean squared error (RMSE) as the metrics of accuracy. It was found that the Random Forest algorithm performed better than the other three ensemble learning algorithms in terms of both the accuracy of predicted optimal cross-sectional areas and the computational speed. However, it should be noted that all ensemble learning algorithms demonstrated high accuracy with an R^2 score greater than 0.99. On the other hand, the execution speed of the CatBoost algorithm was significantly slower than the remaining three algorithms. The output of the Random Forest algorithm has been further analyzed using the SHAP algorithm to better understand the impact of each input variable on the model prediction and the interdependencies between the input variables. Furthermore, a four-level factorial analysis has been carried out to visualize the sensitivity of the reinforcement cross-sectional area (A_s) to various input parameters.

The results of the four-level factorial analysis showed that the column diameter and applied external loads largely determine the cross-sectional area of steel reinforcement necessary for a safe design. The outer diameter of the column was also singled out by the SHAP summary plot as the most impactful design variable that determines the output of the random forest model. The outcome of the SHAP analysis and the factorial analysis can be interpreted as the external loading and column size being an order of magnitude more significant than the unit costs of steel and concrete in terms of their effect on the necessary amount of reinforcement.

In the design of a predictive equation, both the factorial analysis and the SHAP feature dependence plots have been decisive. In addition to a bias term, the predictive equation consists of two other terms. The first term after the bias term is chosen to be a product of the normalized diameter with the sum of the three parameters most dependent on D according to the feature dependence plots. The second term is chosen to be a product of the axial load with the sum of the normalized values of the two design variables most dependent on the axial load. In order to capture the nonlinear variations in the optimum reinforcement area, each term in the equation was raised to a power and multiplied by a coefficient. Afterward, the optimal coefficient values were determined through harmony search optimization. A coefficient of determination greater than 0.99 could be obtained by the predictive equation thus developed. However, it should be noted that more comprehensive studies including a larger number of design variables and greater ranges for the variable values should be performed to enhance the reliability of the developed equation.

Since machine learning techniques are data-driven, the applicability of these techniques to structural design depends on the availability of quality datasets. Furthermore, the accuracy of the obtained predictive models depends largely on the size of the dataset. However, most of the recent machine learning related research in the field of structural engineering is based on data sets not large enough to be statistically significant. This issue is being addressed in this study by proposing a harmony search-based novel technique for generating large data sets. The application of the harmony search methodology to the problem of data generation can solve the problem of limited data availability in structural engineering. Using optimization techniques such as harmony search, large datasets can be generated where the structural performances of the generated optimal samples are controlled according to the requirements in the existing design codes.

5. Conclusions

The performance of any predictive model depends mostly on the size and quality of the dataset used in its training. The current paper demonstrated a novel technique for the generation of large datasets using the harmony search optimization methodology. The generated dataset was used in the prediction of the optimal amount of longitudinal reinforcement for circular RC columns. Four different ensemble learning models were demonstrated to perform well in the prediction process. Furthermore, a closed-form equation was proposed that predicts the optimal amount of reinforcement that minimizes the cost associated with the construction process without violating the design code requirements. The most important results of this research work can be listed as follows:

- Four different machine learning models were developed using the XGBoost, LightGBM, Random Forest, and CatBoost algorithms. All of these algorithms performed well on the dataset with an R^2 score greater than 0.99. Among these models, the Random Forest algorithm performed best in terms of both accuracy and computational speed whereas the CatBoost algorithm was nearly an order of magnitude slower than the rest of the algorithms.
- The results of the SHAP analysis showed that the outer diameter of the circular column has the greatest impact on the machine learning model predictions. The impacts of the applied axial loading (N) and bending moment (M) were found to be dependent on the value of D. At smaller values of D, N was shown to have a larger impact on the model output.
- After dividing the dataset into four segments for each variable the four-level factorial analysis showed that a 59% increase in the outer diameter can lead to a 143% increase in the optimal value of A_s. A_s was also found to be highly sensitive to variations in N and M. Doubling the magnitude of N was observed to cause a 68% increase in the optimal value of A_s whereas doubling the magnitude of M led to a 41% increase in the optimal value of A_s.
- A closed-form equation with an R^2 score of 0.9985 was proposed which predicts the optimal value for A_s as a function of column outer diameter, axial loading, bending moment, column length, and the unit prices of concrete and steel.

The availability of closed-form equations that deliver optimal dimensions for structural design can greatly facilitate and accelerate the design process for practicing engineers. With the help of these equations, the most favorable design combinations can be obtained without the need for complex optimization methodologies. However, it should be noted that the proposed equation and machine learning models in this paper are limited by the range of variables that constitute the dataset. Therefore, further research needs to be carried out for the development of more comprehensive predictive models. Furthermore, the scope of the variables included in the dataset could be enhanced to include variables such as the number and diameter of longitudinal reinforcing bars. In its current form, the output of the predictive equation could be evenly distributed to determine the area and a total number of the longitudinal reinforcing bars. Also, the spacing and dimensions of the lateral reinforcements can be included in the database. Future research towards the design of RC columns using machine learning methodologies can include composite materials such as carbon fiber or glass fiber reinforced polymers as the material of reinforcement.

Author Contributions: Methodology, G.B.; formal analysis (coding), C.C.; writing—original draft preparation, C.C. and G.B.; writing—review & editing, S.K. and Z.W.G.; visualization, C.C.; supervision, G.B., S.K. and Z.W.G.; funding acquisition, Z.W.G. All authors have read and agreed to the published version of the manuscript.

Funding: This research was supported by the Energy Cloud R&D Program through the National Research Foundation of Korea (NRF) funded by the Ministry of Science, ICT (2019M3F2A1073164).

Data Availability Statement: Data will be made available upon reasonable request.

Conflicts of Interest: The authors declare no conflict of interest.

Appendix A

In Table A1, x_i and y_i denote the values of two different data series, $\tilde{y}_i$ denotes the predicted values of the data series y_i, and n is the total number of data points in the series.

Table A1. Metrics of Model Accuracy.

Metric	Formula
Root mean square error (RMSE)	$\mathrm{RMSE} = \sqrt{\frac{\sum_{i=1}^{n}(y_i - \tilde{y}_i)^2}{n}}$
Coefficient of determination (R^2):	$R^2 = \left(\frac{n\sum_{i=1}^{n} y_i\tilde{y}_i - \sum_{i=1}^{n} y_i \sum_{i=1}^{n} \tilde{y}_i}{\sqrt{n\sum_{i=1}^{n} y_i^2 - (\sum_{i=1}^{n} y_i)^2}\sqrt{n\sum_{i=1}^{n} \tilde{y}_i^2 - (\tilde{y}_i)^2}}\right)^2$
Mean absolute error (MAE):	$\mathrm{MAE} = \frac{\sum_{i=1}^{n}\left\lvert y_i - \tilde{y}_i\right\rvert}{n}$
Pearson correlation coefficient:	$r_{xy} = \frac{n\sum_{i=1}^{n} x_i y_i - \sum_{i=1}^{n} x_i \sum_{i=1}^{n} y_i}{\sqrt{n\sum_{i=1}^{n} x_i^2 - (\sum_{i=1}^{n} x_i)^2}\sqrt{n\sum_{i=1}^{n} y_i^2 - (\sum_{i=1}^{n} y_i)^2}}$

References

1. Bekdaş, G.; Cakiroglu, C.; Islam, K.; Kim, S.; Geem, Z.W. Optimum Design of Cylindrical Walls Using Ensemble Learning Methods. *Appl. Sci.* **2022**, *12*, 2165. [CrossRef]
2. Bekdaş, G. Harmony Search Algorithm Approach for Optimum Design of Post-Tensioned Axially Symmetric Cylindrical Reinforced Concrete Walls. *J. Optim. Theory Appl.* **2015**, *164*, 342–358. [CrossRef]
3. Bekdaş, G.; Cakiroglu, C.; Kim, S.; Geem, Z.W. Optimal Dimensioning of Retaining Walls Using Explainable Ensemble Learning Algorithms. *Materials* **2022**, *15*, 4993. [CrossRef]
4. Arama, Z.A.; Kayabekir, A.E.; Bekdaş, G.; Kim, S.; Geem, Z.W. The Usage of the Harmony Search Algorithm for the Optimal Design Problem of Reinforced Concrete Retaining Walls. *Appl. Sci.* **2021**, *11*, 1343. [CrossRef]
5. Kayabekir, A.E.; Arama, Z.A.; Bekdaş, G.; Nigdeli, S.M.; Geem, Z.W. Eco-Friendly Design of Reinforced Concrete Retaining Walls: Multi-objective Optimization with Harmony Search Applications. *Sustainability* **2020**, *12*, 6087. [CrossRef]
6. Cakiroglu, C.; Bekdaş, G.; Kim, S.; Geem, Z.W. Optimisation of Shear and Lateral–Torsional Buckling of Steel Plate Girders Using Meta-Heuristic Algorithms. *Appl. Sci.* **2020**, *10*, 3639. [CrossRef]
7. Cakiroglu, C.; Bekdaş, G. Buckling Analysis and Stacking Sequence Optimization of Symmetric Laminated Composite Plates. In *Advances in Structural Engineering—Optimization. Studies in Systems, Decision and Control*; Nigdeli, S.M., Bekdaş, G., Kayabekir, A.E., Yucel, M., Eds.; Springer: Cham, Switzerland, 2021; Volume 326. [CrossRef]
8. Cakiroglu, C.; Bekdaş, G.; Geem, Z.W. Harmony Search Optimisation of Dispersed Laminated Composite Plates. *Materials* **2020**, *13*, 2862. [CrossRef]
9. Cakiroglu, C.; Islam, K.; Bekdaş, G.; Kim, S.; Geem, Z.W. Metaheuristic Optimization of Laminated Composite Plates with Cut-Outs. *Coatings* **2021**, *11*, 1235. [CrossRef]
10. Cakiroglu, C.; Islam, K.; Bekdaş, G.; Billah, M. CO_2 emission and cost optimization of concrete-filled steel tubular (CFST) columns using metaheuristic algorithms. *Sustainability* **2021**, *13*, 8092. [CrossRef]
11. Cakiroglu, C.; Islam, K.; Bekdaş, G.; Kim, S.; Geem, Z.W. CO_2 Emission Optimization of Concrete-Filled Steel Tubular Rectangular Stub Columns Using Metaheuristic Algorithms. *Sustainability* **2021**, *13*, 10981. [CrossRef]
12. Toklu, Y.C.; Temur, R.; Bekdaş, G. Teaching learning based optimization algorithm for analyses of trusses considering elastoplastic behavior. In *Fluids, Heat and Mass Transfer, Mechanical and Civil Engineering*; Recent Advances in Mechanical Engineering Series 17; WSEAS Press: Athens, Greece, 2015; ISBN 978-1-61804-358-0.
13. Ulusoy, S. Optimum design of timber structures under fire using metaheuristic algorithm. *Građevinar* **2022**, *74*, 115–124. [CrossRef]
14. Ocak, A.; Nigdeli, S.M.; Bekdaş, G.; Kim, S.; Geem, Z.W. Adaptive Harmony Search for Tuned Liquid Damper Optimization under Seismic Excitation. *Appl. Sci.* **2022**, *12*, 2645. [CrossRef]
15. Ocak, A.; Bekdaş, G.; Nigdeli, S.M.; Kim, S.; Geem, Z.W. Optimization of Tuned Liquid Damper Including Different Liquids for Lateral Displacement Control of Single and Multi-Story Structures. *Buildings* **2022**, *12*, 377. [CrossRef]
16. Ulusoy, S.; Nigdeli, S.M.; Bekdaş, G. Introduction and Review on Active Structural Control. In *Optimization of Tuned Mass Dampers. Studies in Systems, Decision and Control*; Bekdaş, G., Nigdeli, S.M., Eds.; Springer: Cham, Switzerland, 2022; Volume 432. [CrossRef]
17. Khalilpourazari, S.; Khalilpourazary, S. An efficient hybrid algorithm based on Water Cycle and Moth-Flame Optimization algorithms for solving numerical and constrained engineering optimization problems. *Soft Comput.* **2019**, *[illegible]*, 1699–1722. [CrossRef]

18. Mirjalili, S. SCA: A sine cosine algorithm for solving optimization problems. *Knowl. -Based Syst.* **2016**, *96*, 120–133. [CrossRef]
19. Harifi, S.; Mohammadzadeh, J.; Khalilian, M.; Ebrahimnejad, S. Giza Pyramids Construction: An ancient-inspired metaheuristic algorithm for optimization. *Evol. Intel.* **2021**, *14*, 1743–1761. [CrossRef]
20. Holland, H.; Reitman, J.S. Cognitive systems based on adaptive algorithms. *ACM SIGART Bull.* **1977**, *63*, 49. [CrossRef]
21. Storn, R.; Price, K. Differential Evolution—A Simple and Efficient Heuristic for global Optimization over Continuous Spaces. *J. Glob. Optim.* **1997**, *11*, 341–359. [CrossRef]
22. Fogel, D.B. Artificial Intelligence through Simulated Evolution. In *Evolutionary Computation: The Fossil Record;* IEEE: Piscataway, NJ, USA, 1998; pp. 227–296. [CrossRef]
23. Koza, J.R. *Genetic Programming: On the Programming of Computers by Means of Natural Selection;* MIT Press: Cambridge, MA, USA, 1992; ISBN 0-262-11170-5.
24. Simon, D. Biogeography-based optimization. *IEEE Trans. Evol. Comput.* **2008**, *12*, 702–713. [CrossRef]
25. Rashedi, E.; Nezamabadi-Pour, H.; Saryazdi, S. GSA: A gravitational search algorithm. *Inf. Sci.* **2009**, *179*, 2232–2248. [CrossRef]
26. Kaveh, A.; Mahdavi, V.R. Colliding bodies optimization method for optimum discrete design of truss structures. *Comput. Struct.* **2014**, *139*, 43–53. [CrossRef]
27. Hatamlou, A. Black hole: A new heuristic optimization approach for data clustering. *Inf. Sci.* **2013**, *222*, 175–184. [CrossRef]
28. Kirkpatrick, S.; Gelatt, C.D.; Vecchi, M.P. Optimization by simmulated annealing. *Science* **1983**, *220*, 671–680. [CrossRef]
29. Mirjalili, S.; Mirjalili, S.M.; Hatamlou, A. Multi-verse optimizer: A nature-inspired algorithm for global optimization. *Neural Comput. Appl.* **2016**, *27*, 495–513. [CrossRef]
30. Du, H.; Wu, X.; Zhuang, J. Small-World Optimization. In *Advances in Natural Computation. ICNC 2006: Lecture Notes in Computer Science;* Jiao, L., Wang, L., Gao, X., Liu, J., Wu, F., Eds.; Springer: Berlin/Heidelberg, Germany, 2006; Volume 4222. [CrossRef]
31. Mirjalili, S. The ant lion optimizer. *Adv. Eng. Soft.* **2015**, *83*, 80–98. [CrossRef]
32. Harifi, S.; Khalilian, M.; Mohammadzadeh, J. Emperor Penguins Colony: A new metaheuristic algorithm for optimization. *Evol. Intel.* **2019**, *12*, 211–226. [CrossRef]
33. Li, M.D.; Zhao, H.; Weng, X.W.; Han, T. A novel nature-inspired algorithm for optimization: Virus colony search. *Adv. Eng. Soft.* **2016**, *92*, 65–88. [CrossRef]
34. Salimi, H. Stochastic fractal search: A powerful metaheuristic algorithm. *Knowl. -Based Syst.* **2015**, *75*, 1–18. [CrossRef]
35. Kashan, A.H. League Championship Algorithm (LCA): An algorithm for global optimization inspired by sport championships. *Appl. Soft Comput.* **2014**, *16*, 171–200. [CrossRef]
36. Geem, Z.W.; Lee, K.S.; Park, Y. Application of harmony search to vehicle routing. *Am. J. Appl. Sci.* **2005**, *2*, 1552–1557. [CrossRef]
37. Zhang, Y.; Su, R.; Zhang, Y.; Gammana Guruge, N.S. A Multi-Bus Dispatching Strategy Based on Boarding Control. *IEEE Trans. Intell. Transp. Syst.* **2022**, *23*, 5029–5043. [CrossRef]
38. Villalobos, O.A.R.; Rojas, E.J.R.; Velandia, J.B.; Cuchango, H.E.E. Application of the Harmonic Search Algorithm for Identification of Model Parameters of Traffic Lights for a High Way of Bogotá. *Int. J. Eng. Res. Technol.* **2020**, *13*, 240–246. [CrossRef]
39. Ganeshkumar, N.; Kumar, S. QoS Aware Modified Harmony Search Optimization for Route Selection in VANETs. *Indian J. Comput. Sci. Eng. (IJCSE)* **2022**, *13*, 2. [CrossRef]
40. You, C.-H.; Suh, S.-H.; Jung, W.; Kim, H.-J.; Lee, D.-I. Dual-Polarization Radar-Based Quantitative Precipitation Estimation of Mountain Terrain Using Multi-Disdrometer Data. *Remote Sens.* **2022**, *14*, 2290. [CrossRef]
41. Loy-Benitez, J.; Li, Q.; Nam, K.; Nguyen, H.T.; Kim, M.; Park, D.; Yoo, C. Multi-objective optimization of a time-delay compensated ventilation control system in a subway facility—A harmony search strategy. *Build. Environ.* **2021**, *190*, 107543. [CrossRef]
42. Abdulkhaleq, M.T.; Rashid, T.A.; Alsadoon, A.; Hassan, B.A.; Mohammadi, M.; Abdullah, J.M.; Vimal, S. Harmony search: Current studies and uses on healthcare systems. *Artif. Intell. Med.* **2022**, *131*, 102348. [CrossRef]
43. Taghipour, S.; Zarrineh, P.; Ganjtabesh, M. Improving protein complex prediction by reconstructing a high-confidence protein-protein interaction network of Escherichia coli from different physical interaction data sources. *BMC Bioinform.* **2017**, *18*, 10. [CrossRef]
44. Li, X.; Wang, B.; Lv, H.; Yin, Q.; Zhang, Q.; Wei, X. Constraining DNA Sequences With a Triplet-Bases Unpaired. *IEEE Trans. NanoBiosci.* **2020**, *19*, 299–307. [CrossRef]
45. Mohsen, A.M.; Khader, A.T.; Ramachandram, D. HSRNAFold: A harmony search algorithm for RNA secondary structure prediction based on minimum free energy. In Proceedings of the 2008 International Conference on Innovations in Information Technology, Al Ain, United Arab Emirates, 16–18 December 2008; pp. 11–15. [CrossRef]
46. Bibartiu, O.; Dürr, F.; Rothermel, K. Optimal Refinement for Component-based Architectures. In Proceedings of the 2021 IEEE 25th International Enterprise Distributed Object Computing Conference (EDOC), Gold Coast, Australia, 25–29 October 2021; pp. 142–151. [CrossRef]
47. Schofield, J.S.; Evans, K.R.; Hebert, J.S.; Marasco, P.D.; Carey, J.P. The effect of biomechanical variables on force sensitive resistor error: Implications for calibration and improved accuracy. *J. Biomech.* **2016**, *49*, 786–792. [CrossRef]
48. Feng, D.C.; Liu, Z.T.; Wang, X.D.; Jiang, Z.M.; Liang, S.X. Failure mode classification and bearing capacity prediction for reinforced concrete columns based on ensemble machine learning algorithm. *Adv. Eng. Inform.* **2020**, *45*, 101126. [CrossRef]
49. Dogan, G.; Arslan, M.H.; Baykan, O.K. Determination of damage levels of RC columns with a smart system oriented method. *Bull. Earthquake Eng.* **2020**, *18*, 3223–3245. [CrossRef]

50. Nasrollahzadeh, K.; Nouhi, E. Fuzzy inference system to formulate compressive strength and ultimate strain of square concrete columns wrapped with fiber-reinforced polymer. *Neural Comput. Appl.* **2018**, *30*, 69–86. [CrossRef]
51. Naderpour, H.; Nagai, K.; Fakharian, P.; Haji, M. Innovative models for prediction of compressive strength of FRP-confined circular reinforced concrete columns using soft computing methods. *Compos. Struct.* **2019**, *215*, 69–84. [CrossRef]
52. Bossio, A.; Lignola, G.P.; Fabbrocino, F.; Monetta, T.; Prota, A.; Bellucci, F.; Manfredi, G. Nondestructive assessment of corrosion of reinforcing bars through surface concrete cracks. *Struct. Concr.* **2017**, *18*, 104–117. [CrossRef]
53. Bossio, A.; Fabbrocino, F.; Monetta, T.; Lignola, G.P.; Prota, A.; Manfredi, G.; Bellucci, F. Corrosion effects on seismic capacity of reinforced concrete structures. *Corros. Rev.* **2019**, *37*, 45–56. [CrossRef]
54. Fabbrocino, F.; Funari, M.F.; Greco, F.; Lonetti, P.; Luciano, R.; Penna, R. Dynamic crack growth based on moving mesh method. *Compos. Part B Eng.* **2019**, *174*, 107053. [CrossRef]
55. Kayabekir, A.E.; Bekdaş, G.; Yücel, M.; Nigdeli, S.M.; Geem, Z.W. Harmony Search Algorithm for Structural Engineering Problems. In *Nature-Inspired Metaheuristic Algorithms for Engineering Optimization Applications*; Springer Tracts in Nature-Inspired Computing, Carbas, S., Toktas, A., Ustun, D., Eds.; Springer: Singapore, 2021.
56. Chen, T.; Guestrin, C. XGBoost: A scalable tree boosting system. In Proceedings of the 22nd ACM SIGKDD International Conference on Knowledge Discovery and Data Mining, San Francisco, CA, USA, 13–17 August 2016; Association for Computing Machinery: New York, NY, USA, 2016; pp. 785–794.
57. Feng, D.C.; Wang, W.J.; Mangalathu, S.; Taciroglu, E. Interpretable XGBoost-SHAP machine-learning model for shear strength prediction of squat RC walls. *J. Struct. Eng.* **2021**, *147*, 04021173. [CrossRef]
58. Rahman, J.; Ahmed, K.S.; Khan, N.I.; Islam, K.; Mangalathu, S. Data-driven shear strength prediction of steel fiber reinforced concrete beams using machine learning approach. *Eng. Struct.* **2021**, *233*, 111743. [CrossRef]
59. Degtyarev, V.V.; Naser, M.Z. Boosting machines for predicting shear strength of CFS channels with staggered web perforations. *Structures* **2021**, *34*, 3391–3403. [CrossRef]
60. Dorogush, A.V.; Ershov, V.; Gulin, A. Catboost: Gradient boosting with categorical features support. *arXiv* **2018**, arXiv:1810.11363.
61. Lee, S.; Vo, T.P.; Thai, H.T.; Lee, J.; Patel, V. Strength prediction of concrete-filled steel tubular columns using Categorical Gradient Boosting algorithm. *Eng. Struct.* **2021**, *238*, 112109. [CrossRef]
62. Lundberg, S.M.; Lee, S.I. A unified approach to interpreting model predictions. In Proceedings of the 31st Conference on Neural Information Processing Systems (NIPS 2017), Long Beach, CA, USA, 4–9 December 2017.
63. Nguyen, Q.H.; Ly, H.B.; Ho, L.S.; Al-Ansari, N.; Le, H.V.; Tran, V.Q.; Prakash, I.; Pham, B.T. Influence of data splitting on performance of machine learning models in prediction of shear strength of soil. *Math. Probl. Eng.* **2021**, *2021*, 4832864. [CrossRef]
64. Montgomery, D.C. *Design and Analysis of Experiments*; John Wiley & Sons: Hoboken, NJ, USA, 2017.

materials

MDPI

Article

Optimal Dimensioning of Retaining Walls Using Explainable Ensemble Learning Algorithms

Gebrail Bekdaş [1,*], Celal Cakiroglu [2], Sanghun Kim [3] and Zong Woo Geem [4,*]

1 Department of Civil Engineering, Istanbul University-Cerrahpasa, Istanbul 34320, Turkey
2 Department of Civil Engineering, Turkish-German University, Istanbul 34820, Turkey; cakiroglu@tau.edu.tr
3 Department of Civil and Environmental Engineering, Temple University, Philadelphia, PA 19122, USA; sanghun.kim@temple.edu
4 Department of Smart City & Energy, Gachon University, Seongnam 13120, Korea
* Correspondence: bekdas@iuc.edu.tr (G.B.); geem@gachon.ac.kr (Z.W.G.)

Abstract: This paper develops predictive models for optimal dimensions that minimize the construction cost associated with reinforced concrete retaining walls. Random Forest, Extreme Gradient Boosting (XGBoost), Categorical Gradient Boosting (CatBoost), and Light Gradient Boosting Machine (LightGBM) algorithms were applied to obtain the predictive models. Predictive models were trained using a comprehensive dataset, which was generated using the Harmony Search (HS) algorithm. Each data sample in this database consists of a unique combination of the soil density, friction angle, ultimate bearing pressure, surcharge, the unit cost of concrete, and six different dimensions that describe an optimal retaining wall geometry. The influence of these design features on the optimal dimensioning and their interdependence are explained and visualized using the SHapley Additive exPlanations (SHAP) algorithm. The prediction accuracy of the used ensemble learning methods is evaluated with different metrics of accuracy such as the coefficient of determination, root mean square error, and mean absolute error. Comparing predicted and actual optimal dimensions on a test set showed that an R^2 score of 0.99 could be achieved. In terms of computational speed, the LightGBM algorithm was found to be the fastest, with an average execution speed of 6.17 s for the training and testing of the model. On the other hand, the highest accuracy could be achieved by the CatBoost algorithm. The availability of open-source machine learning algorithms and high-quality datasets makes it possible for designers to supplement traditional design procedures with newly developed machine learning techniques. The novel methodology proposed in this paper aims at producing larger datasets, thereby increasing the applicability and accuracy of machine learning algorithms in relation to optimal dimensioning of structures.

Keywords: machine learning; optimization; structural design

Citation: Bekdaş, G.; Cakiroglu, C.; Kim, S.; Geem, Z.W. Optimal Dimensioning of Retaining Walls Using Explainable Ensemble Learning Algorithms. *Materials* **2022**, *15*, 4993. https://doi.org/10.3390/ma15144993

Academic Editor: Ayman El-Zohairy

Received: 30 June 2022
Accepted: 15 July 2022
Published: 18 July 2022

1. Introduction

Retaining walls are a ubiquitous element in structural design. Due to their relatively large dimensions, optimizing their dimensions can lead to significant gains with construction costs. Furthermore, designing with minimum dimensions has certain advantages for CO_2 emissions because of using the minimum amount of cement. Therefore, the application of advanced methodologies of optimization in the design of retaining walls has economic and environmental benefits.

Many newly developed optimization techniques have been used for structural optimization in recent years. Gomes [1] applied the particle swarm optimization technique to the mass optimization of steel trusses under frequency constraints. Similarly, Dede [2] analyzed the weight minimization of steel trusses using the teaching–learning-based optimization algorithm. Bekdaş et al. [3] used several metaheuristic optimization algorithms in the minimum total potential energy analysis of steel trusses. Bekdaş [4] applied the applications of harmony search, flower pollination, and teaching–learning-based optimization algorithms to minimizing the total construction cost associated with axially symmetric cylindrical reinforced concrete walls. Ocak et al. [5,6] optimized a tuned liquid damper device, which was used for lateral displacement control of structures using the adaptive harmony search algorithm. Ulusoy [7] applied the teaching–learning-based optimization algorithm to the problem of the fire-resistant design of timber-based roof structures. Cakiroglu et al. [8,9] showed that the social spider algorithm could affect the cost minimization problem for concrete-filled steel tubular columns. The optimum design of retaining walls has been investigated using various metaheuristic algorithms including the non-dominated sorting genetic algorithm (NSGA-II) [10], flower pollination algorithm [11], gravitational search algorithm [12], and harmony search algorithm [13–15].

In addition to various optimization algorithms, the application of machine learning techniques in structural design has been increasingly reported in the literature. Feng et al. [16] developed an XGBoost-SHAP machine learning model which estimates the shear strength of reinforced concrete shear walls. In their study, the database, which consisted of 434 samples, was split into a training and a test set in a 70% to 30% ratio. This split ratio is adopted by the majority of the studies in this field and it is based on the optimum split ratio established by Mangalathu et al. [17]. Somala et al. [18] showed that in the fundamental period estimation of masonry infilled reinforced concrete frames, Ensemble Learning Techniques such as Random Forest and XGBoost could outperform the existing empirical predictive models available in the literature. Ahmed et al. [19] developed a novel long short-term memory network with overlapping data for the accurate prediction of earthquake-induced damage in ductile and non-ductile frame structures. Ni et al. [20] generated fragility curves for buried pipelines using Lasso Regression Analysis. Bekdaş et al. [21] demonstrated the high accuracy of different Ensemble Learning Algorithms in predicting the optimal wall thickness of reinforced concrete cylindrical walls. Cakiroglu et al. [22] developed predictive models using Ensemble Learning Algorithms to estimate the axial load-carrying capacity of FRP-reinforced concrete columns.

The current paper deals with optimizing six key dimensions which define the dimensioning of a retaining wall. These dimensions are the length of the heel (X1), length of the toe (X2), the thickness of the stem at the top of the wall (X3), the thickness of the stem at the bottom of the wall (X4), the thickness of the foundation of the wall (X5), and the stem height of the wall (H). For each of them, a separate predictive model has been developed using four different Ensemble Learning Algorithms. Ensemble Learning Techniques have been demonstrated to have superior performance in terms of prediction accuracy in recent years in comparison to traditional methods of structural performance prediction. The dataset needed to train the predictive models has been created using the Harmony Search Algorithm. More than seventy thousand data samples have been created, where each one of these data points corresponds to an optimum design configuration. Every data sample in this dataset contains, in addition to the six geometric variables which define the retaining wall geometry, the soil density (γ), surcharge loading (q), soil friction angle (ϕ), the unit cost of concrete (C_c), and the soil bearing capacity (q_z).

Optimal dimensioning of retaining walls can lead to significant gains in terms of cost and environmental protection. In recent years, various optimization techniques have been demonstrated for minimizing the construction cost associated with retaining walls. On the other hand, machine learning algorithms are increasingly being used in the prediction of structural performance. However, it is necessary to train these predictive algorithms using large datasets for their accuracy. The availability of large-enough datasets has been a major bottleneck in the development of accurate predictive machine learning models for structural design in the recent years. Most of the research in this area has been conducted using datasets in the order of magnitude of a thousand data samples or fewer. To overcome this limitation, the current paper demonstrates a novel technique to generate significantly larger datasets with the help of optimization algorithms. The current paper is unique in its combination of metaheuristic optimization with machine learning models to obtain predictive models that can determine optimal dimensions of a retaining wall under various loading and soil conditions. The novelty of the paper is the usage of a well-established optimization methodology for the generation of large datasets that can be used in the training of machine learning models.

2. Methods of Optimization and Predictive Model Development

The current paper demonstrates the application of the harmony search algorithm in generating large datasets consisting of optimum design configurations. These design configurations consist of six key geometric variables which define the geometry of a retaining wall in addition to soil properties, concrete unit cost, and applied surcharge load. The variables of retaining wall geometry are shown in Figure 1. After generating a large dataset with more than seventy thousand combinations of design variables, four different machine learning models are trained based on this dataset. The following sections describe optimization and machine learning techniques.

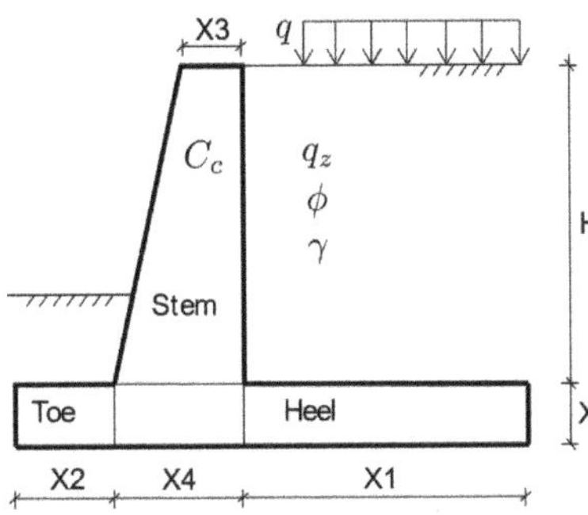

Figure 1. Retaining wall dimensions.

2.1. Harmony Search Algorithm

The application of metaheuristic optimization algorithms to structural optimization is an active area of research. Among a large number of metaheuristic algorithms, the harmony search (HS) algorithm is one of the most widely used and established techniques, and applied to numerous areas such as structural design [23], water network design [24], flood model calibration [25], economic load dispatch [26], concrete mix proportion design [27], chaotic systems [28], timetabling [29], weapon target assignment [30], stock price prediction [31], mobile network security [32], COVID-19 detection from CT scans [33], and subway ventilation [34].

The technique is based on the incremental improvement of an initial population of randomly generated solution candidates, also called the harmony memory matrix. In the case of cost optimization of the retaining wall, the solution candidates are vectors consisting of variables such as the wall geometry, soil properties, unit cost of material used in the retaining wall construction, and the external loads, as shown in Equation (1) where harmony memory size (HMS) denotes the size of the population of candidate solution vectors.

$$HM = \begin{bmatrix} X1^1 & X2^1 & X3^1 & X4^1 & X5^1 & H^1 & q_z^1 & q^1 & \gamma^1 & \phi^1 & C_c^1 & f(x^1) \\ X1^2 & X2^2 & X3^2 & X4^2 & X5^2 & H^2 & q_z^2 & q^2 & \gamma^2 & \phi^2 & C_c^2 & f(x^2) \\ \vdots & \vdots & \vdots & \vdots & \vdots & \vdots & \vdots & \vdots & \vdots & \vdots & \vdots & \vdots \\ X1^{HMS} & X2^{HMS} & X3^{HMS} & X4^{HMS} & X5^{HMS} & H^{HMS} & q_z^{HMS} & q^{HMS} & \gamma^{HMS} & \phi^{HMS} & C_c^{HMS} & f(x^{HMS}) \end{bmatrix} \quad (1)$$

In Equation (1), each row of the harmony memory matrix (HM) contains the components of a candidate solution vector. The last column of the HM contains the output of a function f that takes a candidate solution vector as its argument and returns the performance of the solution vector. In the case of cost optimization, the output of f is the total cost of material used in constructing the retaining wall. Based on their performances, the solution vectors are ranked, and the best- and worst-performing members of the population are determined. In each HS iteration, the solution vectors are updated according to the steps shown in Equations (2)–(5).

$$k = \text{int}(\text{rand}\cdot\text{HMS}),\ \text{rand} \in (0,1) \tag{2}$$

$$x_{i,new} = x_{i,min} + \text{rand}\cdot(x_{i,max} - x_{i,min}),\ \text{if HMCR} > \text{rand} \tag{3}$$

$$x_{i,new} = x_{i,k} + \text{rand}\cdot\text{PAR}\cdot(x_{i,max} - x_{i,min}),\ \text{if HMCR} \le \text{rand} \tag{4}$$

$$\text{HMCR} = 0.5\left(1 - \frac{i}{\max(i)}\right), \text{PAR} = 0.05\left(1 - \frac{i}{\max(i)}\right) \tag{5}$$

HMCR and PAR in Equations (2)–(5) are the harmony memory consideration rate and the pitch adjustment rate, respectively. After each modification step, the newly generated solution vectors are ranked against the existing vectors. Among these vectors, the ones that perform better than the vectors of the previous iteration replace those worse-performing vectors. In the process of generating new solution candidate vectors, the constraints of optimization are regarded based on design codes for retaining walls so that the new design has enough capacity to resist the applied loads. The details of the HS algorithm and its different variants can be found in [35].

2.2. Machine Learning Methodologies

The database of optimum design combinations generated through the HS algorithm has been used in training predictive models. The design variables included in this dataset and their ranges are shown in Figure 2, where the values that each design variable takes are split into four different subgroups. For each one of these groups, the total number of samples belonging to that group is written inside the horizontal bars and the subgroup ranges are written above the subgroup boundaries. In Figure 2, the length of a subgroup indicates the percentage of the samples belonging to that group inside of the entire dataset.

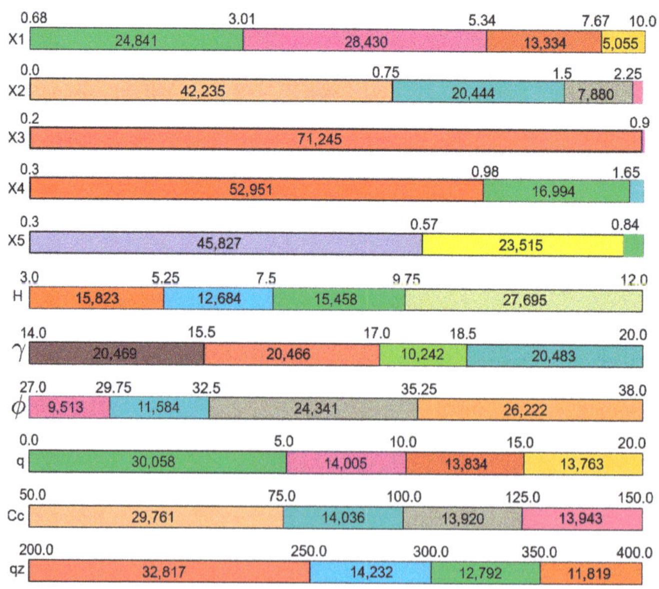

Figure 2. Design variable ranges in the dataset.

Figure 2 shows the concrete unit price (C_c) ranging between 50 and 150 USD/m^3. It can be observed that the majority of cases were within the 75–150 USD/m^3 range. The entire range of unit prices for concrete corresponds to a compressive strength of 16 to 50 MPa, which includes the compressive strengths of most commonly used concrete classes, excluding high strength concrete [36,37]. For the details of the correlation between the soil friction angles included in this study with the other soil properties and the soil classification, the reader is referred to [38].

The predictive models in this paper were generated using the XGBoost, Random Forest, LightGBM, and CatBoost algorithms. These models are further analyzed using the SHapley Additive exPlanations (SHAP) methodology. The following sections show a summary of the theoretical background of these methods.

2.2.1. Extreme Gradient Boosting (XGBoost)

The XGBoost algorithm is a decision tree-based method that has the capability of scaling to large datasets with billions of samples. The decision tree technique starts with testing a root criterion and recursively branches into leaf nodes, testing further criteria, ultimately reaching a terminal node that contains the prediction. The algorithm controls overfitting by using a special regularization technique. The objective of the algorithm is to obtain mapping between the input vectors x^i and the output values y^i as shown in Equations (6) and (7), where L is the loss function, f_k is a weak learner, α_k is the learning rate, T is the number of leaves, w_k are the leaf weights, and γ and λ are the penalty coefficients [16,39].

$$\hat{y}^i = \phi\left(x^i\right) = \sum_{k=1}^{K} \alpha_k f_k\left(x^i\right) \tag{6}$$

$$L(\phi) = \sum_i \left(y^i - \hat{y}^i\right)^2 + \sum_k \gamma T + \frac{1}{2}\lambda||w_k|| \tag{7}$$

2.2.2. Random Forest

The Random Forest technique combines the predictions of an ensemble of single decision trees. The algorithm implements bagging and random feature selection techniques such that every decision tree in the ensemble is built using a bootstrap sample of the training set and the mean value of the individual tree predictions determines the overall predictive model prediction. In every node split, a random subset of features is selected for tree building. The random forest model can be summarized as in Equation (8), where $\hat{m}_j$ stands for a single decision tree [40–42].

$$\hat{m}(x) = \frac{1}{M} \sum_{i=1}^{N} \hat{m}_j(x) \tag{8}$$

2.2.3. Light Gradient Boosting Machine (LightGBM)

LightGBM is another decision tree-based algorithm, where the leaf-wise generation of the predictive model enables the creation of more complex trees. This is a version of the gradient boosting algorithm with improved computational speed and better accuracy. Using the Gradient-based One-Side Sampling (GOSS) method, LightGBM can handle large datasets. The Exclusive Feature Bundling (EFB) method makes it possible to handle datasets with a large number of design features in a more efficient way compared to the basic gradient boosting decision tree [43–45].

2.2.4. Categorical Gradient Boosting (CatBoost)

CatBoost differentiates itself from the basic gradient boosting decision tree in that it is capable of dealing with categorical input features more efficiently. The built-in one-hot encoding capability of CatBoost can obtain target statistics from categorical features. Furthermore, the ordered boosting technique allows the CatBoost algorithm to overcome the gradient bias. Let $X_i = (x_{i,1}, \ldots, x_{i,m})$ be an input vector consisting of m design features and $Y = (Y_1, \ldots, Y_n)$, the vector of labels. Let $\sigma = (\sigma_1, \ldots, \sigma_n)$ be a permutation. To reduce overfitting and use the entire dataset, the CatBoost algorithm uses a random permutation by substituting $x_{\sigma_p,k}$ with the expression in Equation (9), where P is a prior value and $a > 0$ is its weight [46,47].

$$\frac{\sum_{j=1}^{p-1}\left[x_{\sigma_j,k} = x_{\sigma_p,k}\right] Y_{\sigma_j} + a \cdot P}{\sum_{j=1}^{p-1}\left[x_{\sigma_j,k} = x_{\sigma_p,k}\right] + a} \tag{9}$$

2.2.5. SHapley Additive exPlanations (SHAP)

The SHAP analysis is a great contribution to the explainability of the machine learning models, in that it enables a visual representation of the impact of each input variable on the predictive model outcome. Furthermore, Shapley values can measure the interdependencies between different input variables. The algorithm is based on game theory and uses the additive feature attribution, method where an output model is defined as a linear combination of simplified input vectors, as shown in Equation (10) [48].

$$f(\mathbf{x}) = g(x') = \phi_0 + \sum_{i=1}^{M} \phi_i x_i' \tag{10}$$

In Equation (10), the functions f and g are the original predictive model and the explanation model, respectively. M is the total number of input variables, $\mathbf{x}$ is a vector of input variables, x_i are the simplified input vectors, and ϕ_i are the Shapley values. The Shapley values are calculated using Equation (11). Further details of the SHAP algorithm can be found in [49].

$$\phi_i(f, \mathbf{x}) = \sum_{\mathbf{z}' \subseteq \mathbf{x}'} \frac{|\mathbf{z}'|!(M - |\mathbf{z}'| - 1)!}{M!}\left[f_x(\mathbf{z}') - f_x(\mathbf{z}' \backslash i)\right] \tag{11}$$

3. Results

In this section, the predictions of four different Ensemble Learning Algorithms are compared to the actual optimum dimensions obtained through the harmony search method.

3.1. Comparison of the Model Predictions

The comparisons have been visualized for all of the six key dimensions that describe the retaining wall geometry. For each algorithm and each dimension that is being predicted, the accuracy of the predictive models has been measured using three different metrics and listed in Table 1. In Figures 3–6, the predicted optimum dimensions are plotted against the actual optimized dimensions. It can be observed that in the plots showing the predictions for X1, X5, and H, the points representing the different configurations are within a relatively narrow band, which indicates higher accuracy of prediction. In each one of these plots, the dotted ±10% lines can be seen, which indicates a 10% deviation from a perfect match between the predicted and actual optimal dimensions.

Table 1 shows that five out of the six parameters defining the wall geometry could be accurately predicted using the ensemble learning models. Among these models in Table 1, low R^2 values were obtained for X3, since the database used in the training of the predictive models is mostly populated with samples where the X3 value is 0.2. This distribution of the design variable values was obtained after eliminating the design configurations for which the harmony search method did not converge to an optimum result within design constraints. Taking the average value of the metrics of accuracy corresponding to different dimensions, it can be seen from Table 2 that the CatBoost model has the best performance in terms of all three accuracy metrics. CatBoost was followed by Random Forest and LightGBM, whose performances were close to each other. Lastly, the XGBoost models had the lowest accuracy among the four predictive models.

Table 1. Prediction accuracy of the machine learning models.

Algorithm	Variable	R^2	MAE	RMSE	Duration [s]
XGBoost	X1	0.977	0.0697	0.2988	16.49
	X2	0.958	0.0573	0.1319	19.12
	X3	0.562	0.0092	0.0759	16.89
	X4	0.967	0.0197	0.0708	17.88
	X5	0.988	0.0075	0.0192	17.62
	H	0.998	0.0907	0.1351	14.98
Random Forest	X1	0.997	0.0279	0.1091	65.02
	X2	0.958	0.0378	0.1220	62.47
	X3	0.559	0.0083	0.0762	94.74
	X4	0.960	0.0162	0.0776	66.19
	X5	0.989	0.0052	0.0188	61.61
	H	0.997	0.0702	0.1525	51.24
LightGBM	X1	0.998	0.0463	0.0989	5.86
	X2	0.947	0.0719	0.1383	5.68
	X3	0.566	0.0100	0.0756	6.27
	X4	0.966	0.0208	0.0725	5.59
	X5	0.989	0.0075	0.0186	7.36
	H	0.997	0.1051	0.1517	6.28
CatBoost	X1	0.998	0.0281	0.0860	85.31
	X2	0.960	0.0505	0.1189	73.05
	X3	0.642	0.0093	0.0687	74.85
	X4	0.971	0.0167	0.0660	85.76
	X5	0.991	0.0056	0.0170	90.40
	H	0.999	0.0524	0.0890	75.43

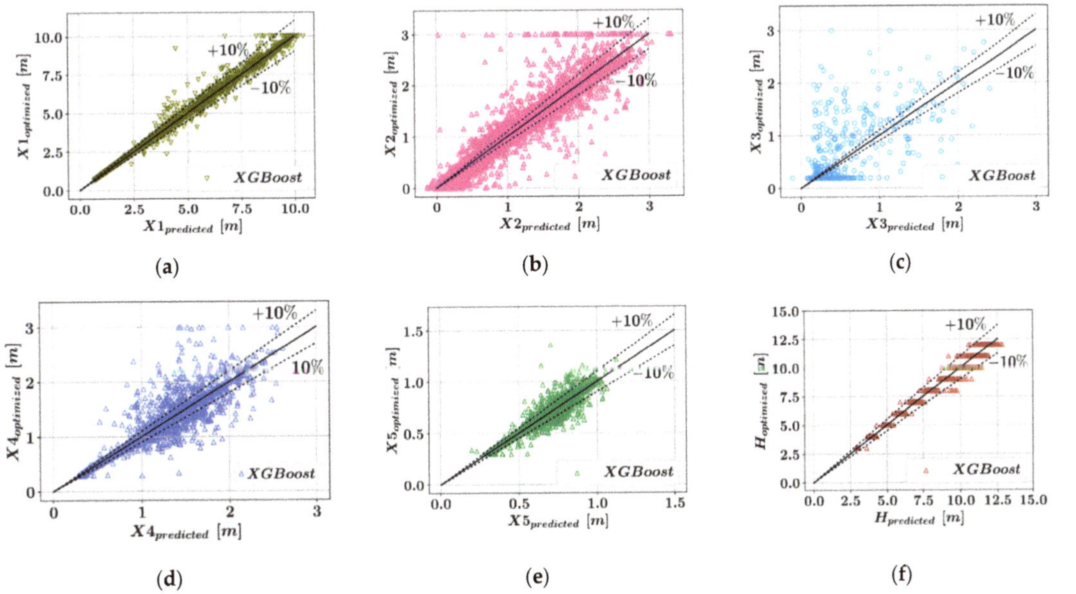

Figure 3. Comparison of the predicted and optimized dimensions using XGBoost. (**a**) X1, (**b**) X2, (**c**) X3, (**d**) X4, (**e**) X5, (**f**) H.

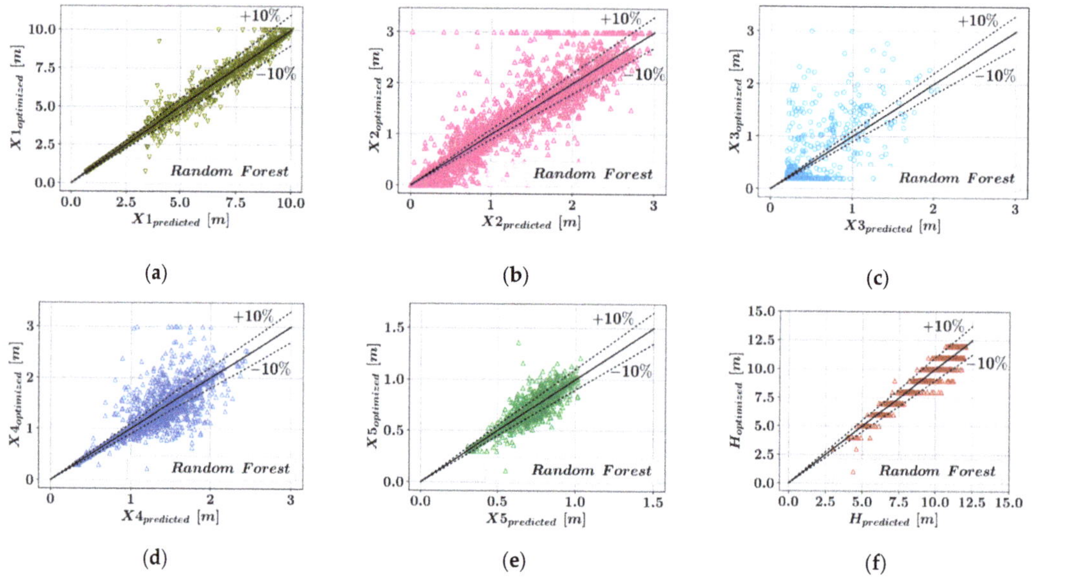

Figure 4. Comparison of the predicted and optimized dimensions using Random Forest. (**a**) X1, (**b**) X2, (**c**) X3, (**d**) X4, (**e**) X5, (**f**) H.

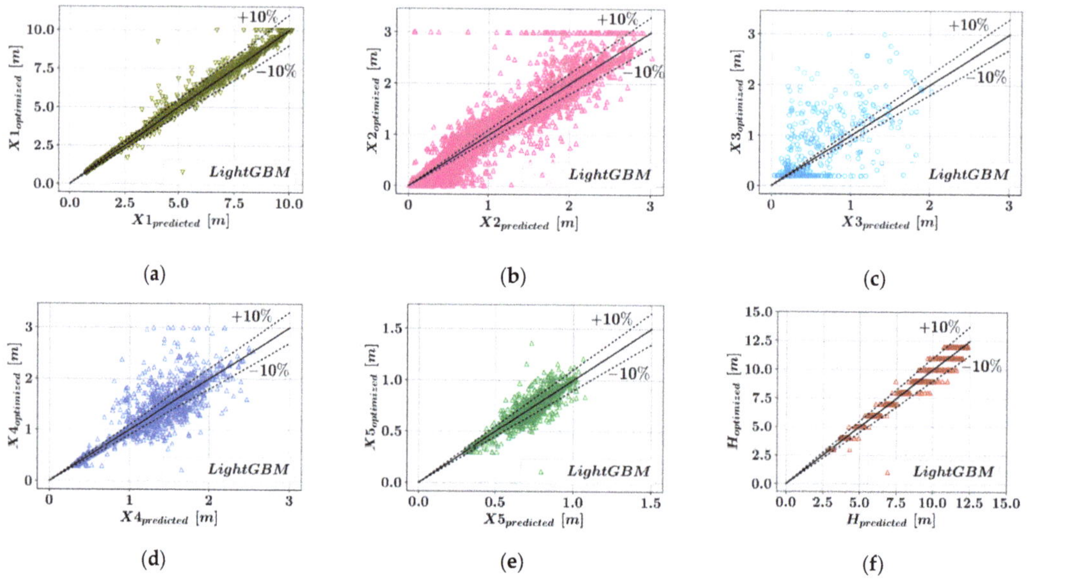

Figure 5. Comparison of the predicted and optimized dimensions using LightGBM. (**a**) X1, (**b**) X2, (**c**) X3, (**d**) X4, (**e**) X5, (**f**) H.

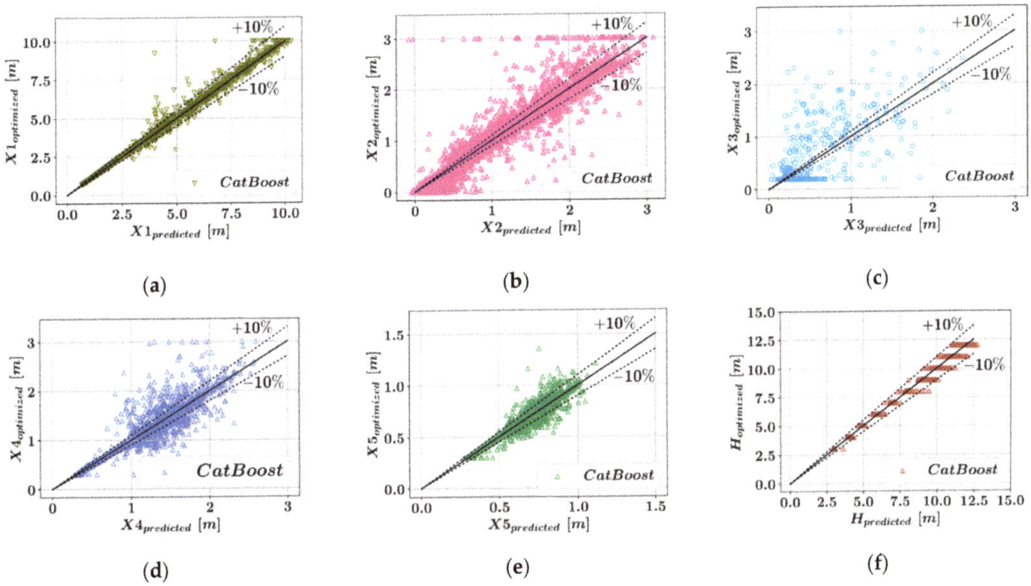

Figure 6. Comparison of the predicted and optimized dimensions using CatBoost. (**a**) X1, (**b**) X2, (**c**) X3, (**d**) X4, (**e**) X5, (**f**) H.

Table 2. Average predictive model accuracy and performance.

Algorithm	R^2	MAE	RMSE	Duration [s]
XGBoost	0.9083	0.04235	0.12195	17.16
Random Forest	0.91	0.0276	0.0927	66.88
LightGBM	0.9105	0.0436	0.0926	6.17
CatBoost	0.92683	0.0271	0.07427	80.80

The Taylor diagrams in Figure 7 show the model quality by using the Pearson correlation coefficient as the metric of accuracy. The equation for the calculation of the Pearson correlation is given in Appendix A. The prediction of each model is shown with a circle and the corresponding correlation coefficient is shown on the radial grid, which ranges from 0 to 1. Furthermore, for each predictive model as well as the original dataset, the corresponding standard deviation values are calculated and shown on both the horizontal and vertical axes. From Figure 7, it can be seen that for the design variables X1, X5, and H, the correlation coefficients were greater than 0.99 for all predictive models, which indicates excellent accuracy of prediction. For X2 and X4, the correlation values were 0.98 for all predictive models. The lowest correlation values were observed for X3 in the interval from 0.75 to 0.80, where the highest correlation values could be obtained through the CatBoost model. A summary of the predictive model performance can be observed in Table 2, where the average values of the error metrics are listed for all the models. According to Table 2, the CatBoost models have the highest accuracy on average.

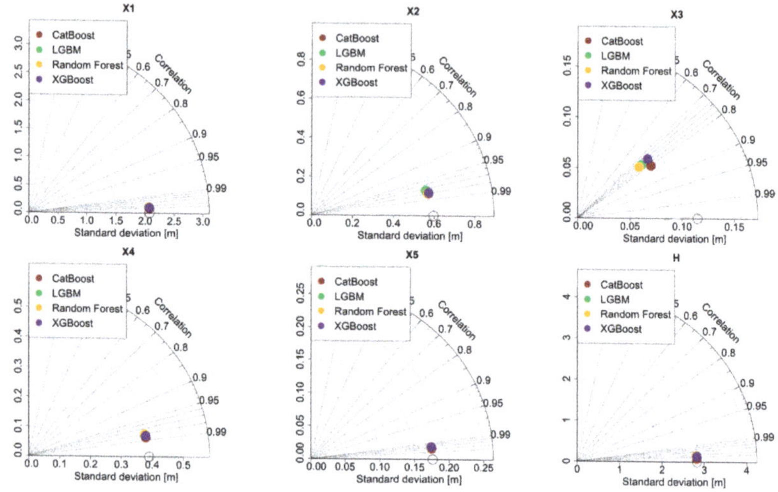

Figure 7. Taylor diagrams for the design variables.

3.2. SHAP Analysis

The SHAP summary plots and feature dependence plots presented in this section provide an effective way of visualizing the impact of various design variables on the overall predictions of the machine learning models. The summary plot shown in Figure 8 is an information-rich representation of how ten different input variables affected the CatBoost model outcome for the prediction of the wall stem thickness at the bottom (X4). In Figure 8, each dot corresponds to a different sample in the database. The dot positions along the horizontal axis are related to the SHAP value of the variable such that greater positive values indicate an increasing effect on the model prediction and negative SHAP values indicate a decreasing effect on the model output. Furthermore, the magnitude of a variable in a sample is represented with color such that greater magnitudes are shown with the shades of blue and lower values are shown with the shades of red. According to Figure 8, the thickness of the wall foundation (X5) has the greatest impact on X4 such that increasing the X5 value also increases X4.

The feature dependence plots in Figure 9 present further information about the interdependencies of the different input variables. Figure 9a shows that as the value of X1 increases, the SHAP value decreases. Therefore, its impact on the model output tends to decrease. Particularly when the length of the toe (X2) has a high value, this relationship between X1 and its impact is more pronounced. From Figure 9b, it can be observed that up to a certain value as X2 increases, its SHAP value increases regardless of the value of X5, which is the variable most dependent on X2. For X2 > 1.5, the impact of these variables decreases with its size when X5 has higher values shown with the shades of red. The relationship between the values of X5 and the impact of this variable on the model output can be observed in Figure 9d. For X5 < 1, the value of X5 and its impact are linearly proportional regardless of the value of C_c, which is the most dependent parameter on X5.

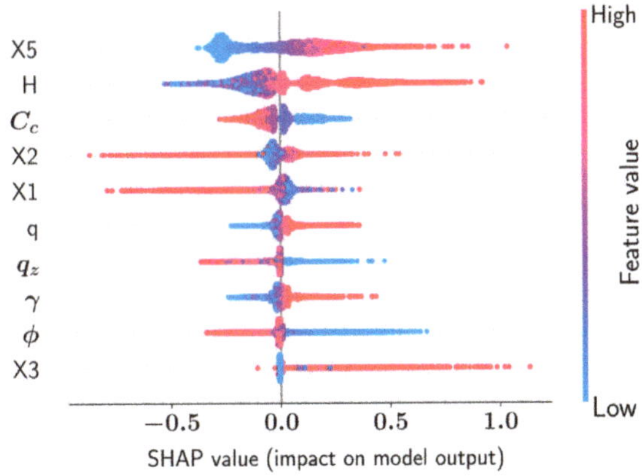

Figure 8. SHAP summary plot for X4.

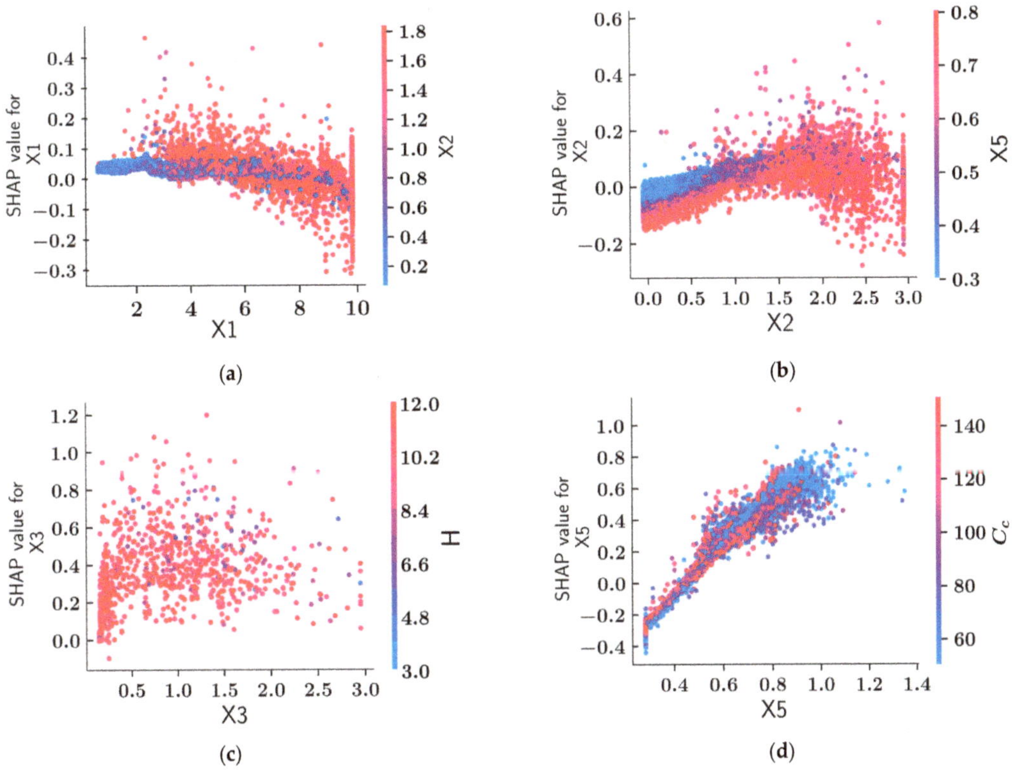

Figure 9. Feature dependence plots for the geometric variables. (**a**) CatBoost dependence plot for X1, (**b**) CatBoost dependence plot for X2, (**c**) CatBoost dependence plot for X3, (**d**) CatBoost dependence plot for X5.

4. Discussion

The current paper presents a novel technique for the optimum dimensioning of retaining walls. A data-driven approach is presented using four different ensemble learning techniques. Predictive machine learning models have been generated using a large dataset obtained through optimization. The thickness of the retaining wall stem at the bottom has been used as the decisive parameter that determines the overall size and cost of the structure. The database necessary to develop the predictive models has been generated using the HS optimization technique. Using this technique, a large database with over seventy thousand data samples was generated where each data sample consists of an optimum combination of eleven design variables and the total construction cost associated with them. The prediction accuracy of the different models has been presented using root mean square error (RMSE), mean absolute error (MAE), coefficient of determination (R^2), and Pearson correlation as the metrics of model performance. The highest prediction accuracy could be achieved by the CatBoost models followed by LightGBM, Random Forest, and XGBoost. The focus of this analysis was the optimization of the geometric dimensions of a retaining wall. The overall wall size and shape were described using six key geometric dimensions.

Previous studies in the area of optimal structural dimensioning mostly attempted to minimize structural cost or weight for a single load case [8,9]. More recent studies in the area attempted to develop general-purpose predictive models based on a dataset of structural configurations with known structural behavior [21,22]. However, the availability of experimental or numerical data describing the structural behavior is a major limiting factor in the training of robust predictive models since the size of the database used in the training of these predictive models is a decisive factor that effects to what extent these models could be used reliably. Furthermore, the range of design variables included in the dataset determines the accuracy of the predictions on new data samples. The current study differentiates itself from the previous ones by generating comprehensive predictive models that incorporate a large number of samples and design variables. Both the size of the dataset and the ranges of the design variables were selected so that these ranges would include most load cases with practical relevance. As a result, a dataset of 71,660 data samples was generated using the harmony search optimization algorithm, which is significantly larger than the datasets previously used in this field. The current paper demonstrates the possibility of generating significantly larger datasets using optimization techniques. This novel approach has the potential to overcome the data availability limitations associated with training machine learning models for structural dimensioning. Using this approach, the applicability of machine learning algorithms to the field of engineering design can be greatly enhanced.

From the SHAP summary plot, it could be observed that all geometric variables, except for the length of the heel (X1), have an increasing effect on the wall stem thickness at the bottom (X4). On the other hand, variables such as concrete unit cost, soil friction angle, and soil bearing capacity have a decreasing effect on X4 as their values increase. Furthermore, increasing the magnitude of the soil density (γ) was observed to have an increasing effect on X4. The thickness (X5) of the wall foundation was found to have the greatest impact on X4, whereas X3 was the variable with the least impact. The low impact of X3 on the model output can be attributed to the concentration of the X3 values around a single value in the entire database.

5. Conclusions

The availability of large datasets is crucial for the development of accurate predictive models in machine learning and particularly in structural dimensioning. The current paper shows the generation of a database consisting of 71,660 unique optimal combinations of six different geometric variables and five parameters that describe the material properties and external loads. The harmony search algorithm has been utilized to obtain these optimal configurations. The major outcomes of this paper can be summarized as follows:

- Among the four ensemble learning models developed in this paper, the highest overall prediction accuracy could be achieved by the CatBoost model, with a maximum coefficient of determination score of 0.999 for the prediction of the optimum stem height and an average R^2 score of 0.927, while the XGBoost models demonstrated, on average, the lowest prediction accuracy.

- In terms of computational speed, the LightGBM models demonstrated the best performance, with an average duration of 6.17 s for the training and testing, whereas the CatBoost models were an order of magnitude slower than the LightGBM models.
- The results of the SHAP analysis showed that the thickness of the retaining wall foundation (X5), the unit cost of concrete (C_c), and the stem height of the wall have the greatest impact on the optimal design.
- The foundation thickness and concrete unit cost were found to be highly dependent on each other and a linear proportionality could be observed between the foundation thickness and the impact of this parameter on the optimal design configuration.

Further research in this area can be carried out by setting different material properties such as the compressive strength of concrete or the yield strength of steel as the optimization objective. Furthermore, the arrangement of the steel reinforcement can be included in future studies as a design variable or optimization objective. One of the limitations of the current study is that a certain range of unit prices is assumed during the generation of the dataset which represents the quality of concrete. However, fluctuations in concrete unit prices have not been considered. Furthermore, it should be noted that the developed ensemble learning models are only applicable within variable ranges included in the training dataset. For variable values outside these ranges, detailed structural analysis and optimization techniques should be applied on a case-by-case basis.

Author Contributions: Methodology, G.B.; formal analysis (coding), C.C.; writing—original draft preparation, C.C. and G.B.; writing—review and editing, S.K. and Z.W.G.; visualization, C.C.; supervision, G.B., S.K. and Z.W.G.; funding acquisition, Z.W.G. All authors have read and agreed to the published version of the manuscript.

Funding: This work was supported by the National Research Foundation of Korea (NRF) grant funded by the Korean government (MSIT) (2020R1A2C1A01011131).

Institutional Review Board Statement: Not applicable.

Informed Consent Statement: Not applicable.

Data Availability Statement: The data presented in this study are available on request from the corresponding author.

Conflicts of Interest: The authors declare no conflict of interest.

Appendix A

Table A1. Metrics of Model Accuracy.

Root mean square error (RMSE):

$$RMSE = \sqrt{\frac{\sum_{i=1}^{n}\left(y_i - \tilde{y}_i\right)^2}{n}}$$

Coefficient of determination (R^2):

$$R^2 = \left(\frac{n\sum_{i=1}^{n} y_i\tilde{y}_i - \sum_{i=1}^{n} y_i \sum_{i=1}^{n} \tilde{y}_i}{\sqrt{n\sum_{i=1}^{n} y_i^2 - \left(\sum_{i=1}^{n} y_i\right)^2}\sqrt{n\sum_{i=1}^{n} \tilde{y}_i^2 - \left(\tilde{y}_i\right)^2}}\right)^2$$

Mean absolute error (MAE):

$$MAE = \frac{\sum_{i=1}^{n}\left|y_i - \tilde{y}_i\right|}{n}$$

Pearson correlation coefficient:

$$r_{xy} = \frac{n\sum_{i=1}^{n} x_i y_i - \sum_{i=1}^{n} x_i \sum_{i=1}^{n} y_i}{\sqrt{n\sum_{i=1}^{n} x_i^2 - \left(\sum_{i=1}^{n} x_i\right)^2}\sqrt{n\sum_{i=1}^{n} y_i^2 - \left(\sum_{i=1}^{n} y_i\right)^2}}$$

References

1. Gomes, H.M. Truss optimization with dynamic constraints using a particle swarm algorithm. *Expert Syst. Appl.* **2011**, *38*, 957–968. [CrossRef]
2. Dede, T. Application of teaching-learning-based-optimization algorithm for the discrete optimization of truss structures. *Ksce J. Civ. Eng.* **2014**, *18*, 1759–1767. [CrossRef]
3. Bekdaş, G.; Kayabekir, A.E.; Nigdeli, S.M.; Toklu, Y.C. Advanced energy-based analyses of trusses employing hybrid metaheuristics. *Struct. Des. Tall Spec. Build.* **2019**, *28*, e1609. [CrossRef]
4. Bekdaş, G. New improved metaheuristic approaches for optimum design of posttensioned axially symmetric cylindrical reinforced concrete walls. *Struct. Des. Tall Spec. Build.* **2018**, *27*, e1461. [CrossRef]
5. Ocak, A.; Nigdeli, S.M.; Bekdaş, G.; Kim, S.; Geem, Z.W. Adaptive Harmony Search for Tuned Liquid Damper Optimization under Seismic Excitation. *Appl. Sci.* **2022**, *12*, 2645. [CrossRef]
6. Ocak, A.; Bekdaş, G.; Nigdeli, S.M.; Kim, S.; Geem, Z.W. Optimization of Tuned Liquid Damper Including Different Liquids for Lateral Displacement Control of Single and Multi-Story Structures. *Buildings* **2022**, *12*, 377. [CrossRef]
7. Ulusoy, S. Optimum design of timber structures under fire using metaheuristic algorithm. *Građevinar* **2022**, *74*, 115–124. [CrossRef]
8. Cakiroglu, C.; Islam, K.; Bekdaş, G.; Billah, M. CO_2 emission and cost optimization of concrete-filled steel tubular (CFST) columns using metaheuristic algorithms. *Sustainability* **2021**, *13*, 8092. [CrossRef]
9. Cakiroglu, C.; Islam, K.; Bekdaş, G.; Kim, S.; Geem, Z.W. CO_2 Emission Optimization of Concrete-Filled Steel Tubular Rectangular Stub Columns Using Metaheuristic Algorithms. *Sustainability* **2021**, *13*, 10981. [CrossRef]
10. Kaveh, A.; Kalateh-Ahani, M.; Fahimi-Farzam, M. Constructability optimal design of reinforced concrete retaining walls using a multi-objective genetic algorithm. *Struct. Eng. Mech.* **2013**, *47*, 227–245. [CrossRef]
11. Mergos, P.E.; Mantoglou, F. Optimum design of reinforced concrete retaining walls with the flower pollination algorithm. *Struct. Multidisc. Optim.* **2020**, *61*, 575–585. [CrossRef]
12. Khajehzadeh, M.; Taha, M.R.; Eslami, M. Efficient gravitational search algorithm for optimum design of retaining walls. *Struct. Eng. Mech.* **2013**, *45*, 111–127. [CrossRef]
13. Kayabekir, A.E.; Yücel, M.; Bekdaş, G.; Nigdeli, S.M. Comparative study of optimum cost design of reinforced concrete retaining wall via metaheuristics. *Chall. J. Concr. Res. Lett.* **2020**, *11*, 75–81. [CrossRef]
14. Kayabekir, A.E.; Arama, Z.A.; Bekdaş, G.; Nigdeli, S.M.; Geem, Z.W. Eco-friendly design of reinforced concrete retaining walls: Multi-objective optimization with harmony search applications. *Sustainability* **2020**, *12*, 6087. [CrossRef]
15. Arama, Z.A.; Kayabekir, A.E.; Bekdaş, G.; Kim, S.; Geem, Z.W. The usage of the harmony search algorithm for the optimal design problem of reinforced concrete retaining walls. *Appl. Sci.* **2021**, *11*, 1343. [CrossRef]
16. Feng, D.C.; Wang, W.J.; Mangalathu, S.; Taciroglu, E. Interpretable XGBoost-SHAP machine-learning model for shear strength prediction of squat RC walls. *J. Struct. Eng.* **2021**, *147*, 04021173. [CrossRef]
17. Mangalathu, S.; Jeon, J.S.; DesRoches, R. Critical uncertainty parameters influencing seismic performance of bridges using Lasso regression. *Earthq. Eng. Struct. Dyn.* **2018**, *47*, 784–801. [CrossRef]
18. Somala, S.N.; Karthikeyan, K.; Mangalathu, S. Time period estimation of masonry infilled RC frames using machine learning techniques. *Structures* **2021**, *34*, 1560–1566. [CrossRef]
19. Ahmed, B.; Mangalathu, S.; Jeon, J.S. Seismic damage state predictions of reinforced concrete structures using stacked long short-term memory neural networks. *J. Build. Eng.* **2022**, *46*, 103737. [CrossRef]
20. Ni, P.; Mangalathu, S.; Liu, K. Enhanced fragility analysis of buried pipelines through Lasso regression. *Acta Geotech.* **2020**, *15*, 471–487. [CrossRef]
21. Bekdaş, G.; Cakiroglu, C.; Islam, K.; Kim, S.; Geem, Z.W. Optimum Design of Cylindrical Walls Using Ensemble Learning Methods. *Appl. Sci.* **2022**, *12*, 2165. [CrossRef]
22. Cakiroglu, C.; Islam, K.; Bekdaş, G.; Kim, S.; Geem, Z.W. Interpretable Machine Learning Algorithms to Predict the Axial Capacity of FRP-Reinforced Concrete Columns. *Materials* **2022**, *15*, 2742. [CrossRef] [PubMed]
23. Hasançebi, O.; Erdal, F.; Saka, M.P. Adaptive harmony search method for structural optimization. *J. Struct. Eng.* **2010**, *136*, 419–431. [CrossRef]
24. Geem, Z.W.; Cho, Y.H. Optimal design of water distribution networks using parameter-setting-free harmony search for two major parameters. *J. Water Resour. Plan. Manag.* **2011**, *137*, 377–380. [CrossRef]
25. Geem, Z.W. Parameter estimation of the nonlinear Muskingum model using parameter-setting-free harmony search. *J. Hydrol. Eng.* **2011**, *16*, 684–688. [CrossRef]
26. Geem, Z.W. Economic dispatch using parameter-setting-free harmony search. *J. Appl. Math.* **2013**, *2013*, 427936. [CrossRef]
27. Lee, J.H.; Yoon, Y.S. Modified harmony search algorithm and neural networks for concrete mix proportion design. *J. Comput. Civ. Eng.* **2009**, *23*, 57–61. [CrossRef]
28. dos Santos Coelho, L.; de Andrade Bernert, D.L. An improved harmony search algorithm for synchronization of discrete-time chaotic systems. *Chaos Solitons Fractals* **2009**, *41*, 2526–2532. [CrossRef]
29. Al-Betar, M.A.; Khader, A.T. A harmony search algorithm for university course timetabling. *Ann. Oper. Res.* **2012**, *194*, 3–31. [CrossRef]
30. Chang, Y.Z.; Li, Z.W.; Kou, Y.X.; Sun, Q.P.; Yang, H.Y.; Zhao, Z.Y. A new approach to weapon-target assignment in cooperative air combat. *Math. Probl. Eng.* **2017**, *2017*, 2936279. [CrossRef]

31. Aghakhani, K.; Karimi, A. A new approach to predict stock big data by combination of neural networks and harmony search algorithm. *Int. J. Comput. Sci. Inf. Secur.* **2016**, *14*, 36.
32. Fahad, A.M.; Muniyandi, R.C. Harmony search algorithm to prevent malicious nodes in mobile ad hoc networks (MANETs). *Inf. Technol. J.* **2016**, *15*, 84–90. [CrossRef]
33. Basu, A.; Sheikh, K.H.; Cuevas, E.; Sarkar, R. COVID-19 detection from CT scans using a two-stage framework. *Expert Syst. Appl.* **2022**, *193*, 116377. [CrossRef] [PubMed]
34. Loy-Benitez, J.; Li, Q.; Nam, K.; Nguyen, H.T.; Kim, M.; Park, D.; Yoo, C. Multi-objective optimization of a time-delay compensated ventilation control system in a subway facility—A harmony search strategy. *Build. Environ.* **2021**, *190*, 107543. [CrossRef]
35. Kayabekir, A.E.; Bekdaş, G.; Yücel, M.; Nigdeli, S.M.; Geem, Z.W. Harmony Search Algorithm for Structural Engineering Problems. In *Nature-Inspired Metaheuristic Algorithms for Engineering Optimization Applications*; Springer Tracts in Nature-Inspired Computing; Carbas, S., Toktas, A., Ustun, D., Eds.; Springer: Singapore, 2021. [CrossRef]
36. National Bureau of Statistics of China, Market Prices of Important Means of Production in Circulation, 1–10 June 2022. Available online: http://www.stats.gov.cn/english/PressRelease/202206/t20220614_1858099.html (accessed on 7 July 2022).
37. Zhang, H. (Ed.) *Building Materials in Civil Engineering*; Woodhead Publishing: Sawston, UK, 2011; ISBN 978-1-84569-955-0.
38. Carter, H.; Bentley, S.P. *Correlations of Soil Properties*; Pentech: London, UK, 1990.
39. Chen, T.; Guestrin, C. XGBoost: A scalable tree boosting system. In Proceedings of the 22nd ACM SIGKDD International Conference on Knowledge Discovery and Data Mining, San Francisco, CA, USA, 13–17 August 2016; Association for Computing Machinery: New York, NY, USA, 2016; pp. 785–794.
40. Mangalathu, S.; Jeon, J.-S. Machine Learning–Based Failure Mode Recognition of Circular Reinforced Concrete Bridge Columns: Comparative Study. *J. Struct. Eng.* **2019**, *145*, 04019104. [CrossRef]
41. Scikit-Learn Documentation. Available online: https://scikit-learn.org/stable/modules/ensemble.html#forest (accessed on 26 June 2022).
42. Feng, D.C.; Wang, W.J.; Mangalathu, S.; Hu, G.; Wu, T. Implementing ensemble learning methods to predict the shear strength of RC deep beams with/without web reinforcements. *Eng. Struct.* **2021**, *235*, 111979. [CrossRef]
43. Degtyarev, V.V.; Naser, M.Z. Boosting machines for predicting shear strength of CFS channels with staggered web perforations. *Structures* **2021**, *34*, 3391–3403. [CrossRef]
44. Mangalathu, S.; Jang, H.; Hwang, S.H.; Jeon, J.S. Data-driven machine-learning-based seismic failure mode identification of reinforced concrete shear walls. *Eng. Struct.* **2020**, *208*, 110331. [CrossRef]
45. Ke, G.; Meng, Q.; Finley, T.; Wang, T.; Chen, W.; Ma, W.; Ye, Q.; Liu, T.Y. Lightgbm: A highly efficient gradient boosting decision tree. In Proceedings of the 31st Conference on Neural Information Processing Systems (NIPS 2017), Long Beach, CA, USA, 4–9 December 2017.
46. Dorogush, A.V.; Ershov, V.; Gulin, A. Catboost: Gradient boosting with categorical features support. *arXiv* **2018**, arXiv:1810.11363.
47. Lee, S.; Vo, T.P.; Thai, H.T.; Lee, J.; Patel, V. Strength prediction of concrete-filled steel tubular columns using Categorical Gradient Boosting algorithm. *Eng. Struct.* **2021**, *238*, 112109. [CrossRef]
48. Mangalathu, S.; Hwang, S.H.; Jeon, J.S. Failure mode and effects analysis of RC members based on machine-learning-based SHapley Additive exPlanations (SHAP) approach. *Eng. Struct.* **2020**, *219*, 110927. [CrossRef]
49. Lundberg, S.M.; Lee, S.I. A unified approach to interpreting model predictions. In Proceedings of the 31st Conference on Neural Information Processing Systems (NIPS 2017), Long Beach, CA, USA, 4–9 December 2017.

materials

Article

Numerical Model for Flexural Analysis of Precast Segmental Concrete Beam with Internal Unbonded CFRP Tendons

Wutong Yan [1,2,*], Liangjiang Chen [1], Bing Han [2], Huibing Xie [2] and Yue Sun [3]

[1] China Railway Economic and Planning Research Institute Co., Ltd., Beijing 100038, China; chenliangjiang66@126.com
[2] School of Civil Engineering, Beijing Jiaotong University, Beijing 100044, China; bhan@bjtu.edu.cn (B.H.); hbxie@bjtu.edu.cn (H.X.)
[3] China Building Technique Group Co., Ltd., Beijing 100013, China; 17121105@bjtu.edu.cn
* Correspondence: yanwutong@bjtu.edu.cn

Abstract: A new construction scheme was recently developed for precast segmental concrete beams by replacing steel tendons with internal unbonded carbon-fiber-reinforced polymer tendons. The discontinuous behaviors of the opening joints and unbonded phenomenon of tendons made their flexural behaviors more complicated than those of monolithic beams and members with bonded tendons. Currently, the knowledge on the structural performance of precast segmental concrete beams with internal unbonded carbon-fiber-reinforced polymer tendons is still limited. An efficient numerical model is urgently needed for the structural analysis and performance evaluation of this new construction scheme. In this paper, a new beam–cable hybrid model was proposed accounting for the mechanical behaviors of open joints and unbonded tendons. The numerical model was implemented in the OpenSees software with the proposed modeling method for joint elements and a newly developed element class for internal unbonded tendons. The effectiveness of the proposed model was verified by comparisons against two simply supported experimental tests. Then, the numerical model was employed to evaluate the flexural performance of a full-scale bridge with a span of 37.5 m. Compared with the precast segmental concrete beam with external steel tendons, the scheme with internal unbonded carbon-fiber-reinforced polymer tendons significantly improved the flexural capacity and ductility by almost 54.6% and 8.9%, respectively. The span-to-depth ratio and prestressing reinforcement ratio were the main factors affecting the flexural behaviors. With the span-to-depth ratio increasing by 23%, the flexural capacity decreased by approximately 38.6% and the tendon stress increment decreased by approximately 15.7%. With the prestressing reinforcement ratio increasing by 65.4%, the flexural capacity increased by 88.7% and the tendon stress increment decreased by approximately 25.2%.

Keywords: precast concrete segmental beam; internal unbonded tendons; CFRP; numerical model; flexural capacity; ductility

Citation: Yan, W.; Chen, L.; Han, B.; Xie, H.; Sun, Y. Numerical Model for Flexural Analysis of Precast Segmental Concrete Beam with Internal Unbonded CFRP Tendons. *Materials* **2022**, *15*, 4105. https://doi.org/10.3390/ma15124105

Academic Editor: Ayman El-Zohairy

Received: 29 April 2022
Accepted: 8 June 2022
Published: 9 June 2022

1. Introduction

Precast segmental concrete beams (PSCBs) have become a popular construction option in recent decades due to their significant advantages in construction speed and quality control [1–3]. As the most critical component to ensure the integrity of the assembled concrete segments, the reliable prestressing scheme is of great concern for PSCBs.

Nowadays, steel tendons constitute the most widely adopted prestressing solution in PSCBs, including internal fully bonded steel tendons, external unbonded steel tendons, and hybrid tendons [4]. Numerous studies have been performed on the flexural and shear behaviors of steel tendon prestressed PSCBs [5–10]. The internal bonded steel tendons are beneficial to improving structural load-carrying capacity and ductility, but they are vulnerable to corrosion at the joint locations [11,12]. Once corrosion occurs, deterioration is inevitable, and the embedded steel tendons are almost impossible to be replaced. The application of external steel tendons is a solution option due to their property of being easier to handle, monitor, and replace during their service life [13]. However, the flexural capacity of PSCBs with external steel tendons is reduced due to the decreased tendon effective depth at the ultimate states [14–17]. A better prestressing scheme thoroughly considering both the tendon durability and structural load-carrying capacity is greatly needed for PSCBs.

Fiber-reinforced polymer (FRP) composites have the remarkable advantages of high strength and free corrosion [18,19], which are favorable for use in prestressing applications as alternatives to steel tendons. Previously, carbon-fiber-reinforced polymer (CFRP) tendons were usually applied to monolithic beams [20]. Le et al. first proposed the scheme of the precast segmental concrete beam with internal carbon-fiber-reinforced polymer tendons (PSCB-IUCFRPs) [21,22]. This new system has two main expected advantages compared with previous ones. First, the pipe grouting and epoxy joints are no longer needed, due to the noncorrosion property of the CFRP material, contributing to rapid construction and maintenance [23]. Second, the flexural capacity can be improved due to the better-maintained tendon eccentricity and high tensile strength of CFRP tendons. However, the structural performance of the PSCB-IUCFRP was only tested in two scaled experiments by Le et al. [21,22]. The knowledge on the flexural behaviors of the full-scale PSCB-IUCFRP is still limited. There is a need to develop the applicative numerical model and verify the structural performance of the full-scale PSCB-IUCFRP.

Compared with monolithic beams [20,24–27], the PSCB-IUCFRP system showed more complicated flexural behaviors, including the discontinuous behaviors of open joints and the unbonded phenomenon of the internal unbonded CFRP tendons. Le et al. [28] and Tran et al. [29] conducted the only numerical model for the PSCB-IUCFRP. In their models, concrete segments and internal tendons were modeled using the 3D solid finite elements (FE) model. The joint connection and unbonded behavior of tendons were modeled using contact elements. The proposed 3D FE method was verified against the experiments but came at a high computational cost. It was inconvenient for application to the full-scale beam. Lou et al. [30,31] and Pang et al. [32] proposed a higher-efficiency numerical model, in which the concrete beam was modeled by layered beam elements, and FRP tendons were modeled by the equivalent loads. However, their models could only be used for monolithic concrete beams, and two inadequacies were revealed. First, the discontinuous behaviors of the open joints were not considered. Then, the unbonded CFRP tendon was only modeled by equivalent loads by neglecting the stiffness, which decreased the numerical convergence for highly nonlinear solutions. To date, there is still no available high-efficiency numerical model for the structural analysis of the full-scale PSCB-IUCFRP.

In this paper, a new high-efficiency nonlinear analysis model is proposed for the analysis of the PSCB-IUCFRP, accounting for the discontinuous behaviors of open joints and the unbonded phenomenon of internal CFRP tendons. The concrete segment is modeled by the fiber beam element. The segmental joints are modeled by the "plain concrete" fiber beam element using the smeared strain method. The unbonded CFRP tendons are modeled by the proposed multi-node slipping cable element with complete formulations of the stiffness matrix and resistance vector, which improves the numerical convergency. The numerical model is implemented in the OpenSees software (version 2.5.0) [33] with newly developed element types and verified against experimental tests. Then, the numerical model is applied to evaluate the flexural capacity, ductility, and failure mode of a 37.5 m full-scale PSCB-IUCFRP. The effects of the main factors, including loading types, span-to-depth ratio, and prestressing reinforcement ratio, on the flexural behaviors of the PSCB-IUCFRP are discussed. To the best of the author's knowledge, this is the first time to evaluate the flexural performance of the PSCB-IUCFRP in full-scale bridges.

2. Numerical Model Description

2.1. Modeling Schemes

The PSCB-IUCFRP system mainly comprises three components, namely, concrete segments, segmental joints, and internal unbonded CFRP tendons, as shown in Figure 1. The discontinuity of the opening joints and the unbonded phenomenon of prestressing tendons lead to more complicated flexural behaviors than monolithic or bonded prestressed concrete beams.

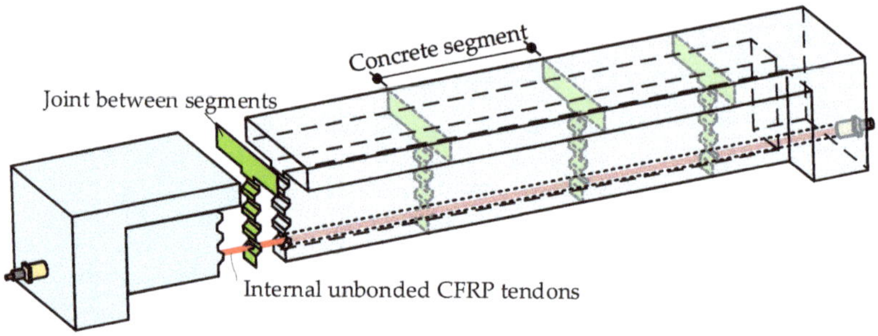

Figure 1. Schematic diagram of a PSCB prestressed with internal unbonded CFRP tendons.

For the segmental joints, once decompressing at the dry joints or reaching the cracking strain at epoxy joints, the joints will open and the joint width will increase quickly due to the lack of bonded reinforcements. The segments adjacent to the open joints become discontinuous, which is the first issue that should be considered in the numerical model.

For the internal unbonded tendons, the friction forces between the unbonded tendons and the adjacent concrete are smaller than tendon traction [16,34]. Therefore, the frictionless assumption is usually adopted for unbonded tendons to simplify analysis [28–30]. With this assumption, internal unbonded tendons behave as a slipping cable with equal traction anywhere between two anchorages. The elongation of the unbonded tendons is member-dependent rather than exhibiting only section-dependent behavior. The interaction between internal unbonded tendons and concrete beams exists within the entire beam length. They have the same vertical deformation but different longitudinal deformation, which is the second issue to consider.

Tran et al. conducted the first numerical model study for the PSCB-IUCFRP using the 3D solid element FE model [29]. The CFRP tendon and concrete beam were modeled by solid elements and specified no-friction contact elements between them to model the unbonded phenomenon. The joints were simulated using the contact elements with a friction coefficient of 0.7 to model the discontinuous behaviors of the opening joints. The solid FE model showed good accuracy but came at a high computational cost due to the large amounts of elements and complex contact interactions that existed, decreasing the computational efficiency and convergence for the analysis of full-scale beams.

In this paper, we propose a one-dimensional beam–cable hybrid element model to improve the analysis efficiency and convergence. Based on the experimental tests and observations [21,22], the following reasonable assumptions are considered:

- The structural deformation and capacity are dominated by flexural behaviors. The effects of the shear deformation on the beam are neglected.
- The direct shear failure of joints will not occur before the structural flexural failure.
- Previous tests [16,34] showed that the friction between the unbonded tendons and surrounding concrete is usually small. For simplification, we neglect the friction effects in this work and assume that the traction of the internal unbonded tendon is constant between the anchorages [28–30].

With the above assumptions, the proposed modeling scheme is established, as shown in Figure 2. The main characteristics of the proposed model are concluded as follows:

(1) The segmental concrete beam is modeled by the nonlinear fiber beam element to improve the analysis efficiency. In particular, the segmental joints are modeled by the "plain concrete" fiber beam element with a specified element length l_{se}. The discontinuous joint behaviors are converted into the deformation of equivalent continuum elements based on the smeared strain method, which is described in Section 2.2.

(2) The newly developed multi-node slipping cable element is employed to model the unbonded behavior of internal unbonded tendons. The deformation compatibility is satisfied and the strain is equal anywhere within the element, as presented in Section 2.3.

(3) The interaction between the concrete beam and unbonded tendon is simulated by the rigid beam, which transfers the translational and rotational deformations of beam nodes to the translation of tendon nodes, as presented in Section 2.4.

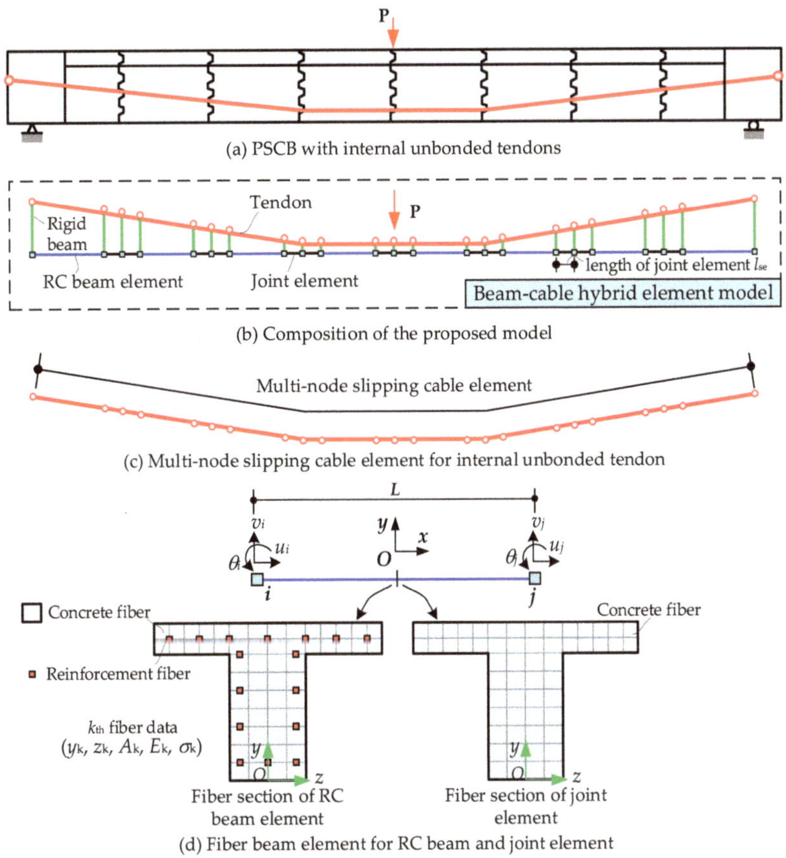

Figure 2. Modeling schemes of the proposed model.

The proposed modeling method and derived FE formulations are implemented in the OpenSees software [33] for simulation. The mechanical characteristics of the PSCB-IUCFRP, including nonlinear material behaviors, discontinuous joint behaviors, and the unbonded phenomenon of CFRP tendons, are all thoroughly considered using one-dimensional element types, improving the analysis efficiency.

2.2. Fiber Beam Element for Segmental Concrete Beam

The modeling method for the segmental concrete beam in this paper is shown in Figure 3, in which each concrete beam segment can be generally modeled by three elements. Two joint elements are arranged at two ends of the segment and one RC beam element is placed in the middle. Both the joint element and the RC beam element are modeled by the fiber beam element in the OpenSees software [35]. The cross-section of this element is divided into many small regions, which are called "fibers" in the OpenSees software [35]. The region for concrete is called "concrete fiber" and the region for reinforcement is called "reinforcement fiber", as shown in Figure 2d. Each fiber contains a uniaxial material constitutive model, an area, and centroid coordinates (y, z). For the Euler–Bernoulli beam, the axial deformations of the fibers within the cross-section satisfy the plane section assumption.

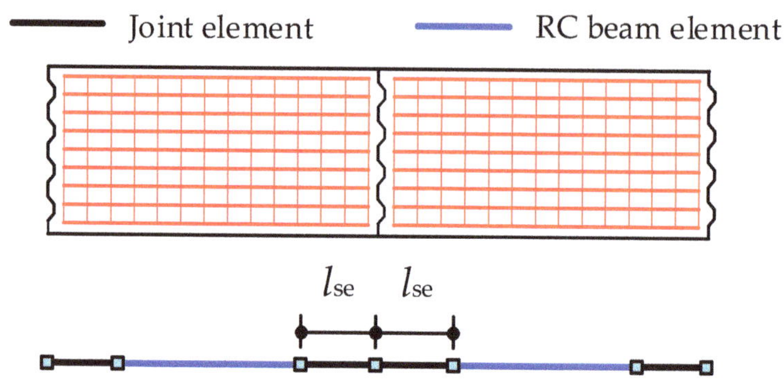

Figure 3. Modeling method of concrete beam segments.

The nonlinear state updating process of the element can be concluded as follows. With the specified element deformation, the axial deformation of each fiber in the cross-section is first calculated. Then, the stress and tangent modulus of each fiber are updated according to the uniaxial nonlinear constitutive model. Then, the sectional resistance and stiffness can be calculated by the sum of all of the fibers' axial forces and tangent stiffness. The element resistance and stiffness are subsequently calculated with the sectional resistance and stiffness using the Gauss–Lobatto numerical integration method. The detailed FE formulations and algorithmic implementation of this element can be seen in [36]. The corotational coordinate transformation is adopted to consider the geometrically nonlinear behavior of the element.

The main issue that should be emphasized here is the difference in fiber section definition between the joint and RC beam elements, as shown in Figure 2d. The fiber section of the RC beam element includes concrete fibers and reinforcement fibers, while in joint elements, only concrete fibers are included as reinforcements are cut-off at the joint locations. With the increasing load up to ultimate states, the bottom fiber of the joint will be open, and the compression region will be crushed. In the numerical model, we define the length of the joint element as l_{se}. The actual deformations of the joint and the adjacent concrete (as shown in Figure 4a) are averagely smeared into the joint element. The concrete constitutive model in the joint element is equivalently modified according to the fracture and crushing band theory [37,38], as shown in Figure 4b. Based on this equivalent smeared strain method, the effects of the joint element length can be normalized. Generally, specifying l_{se} as no greater than one-fourth of the segment length is applicable.

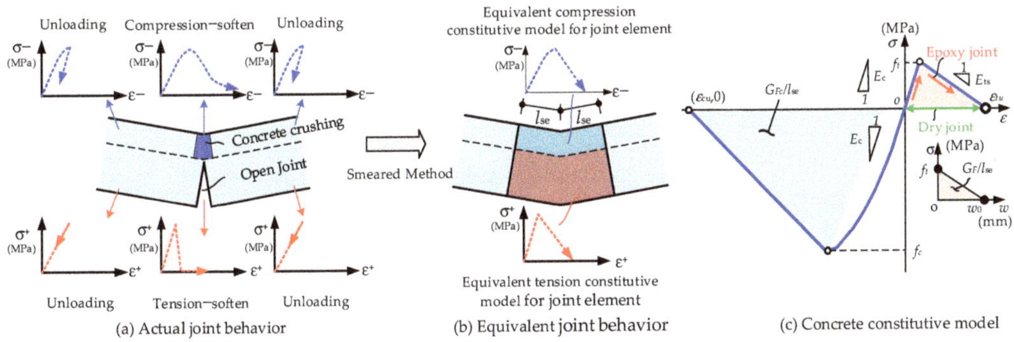

Figure 4. Smeared strain method and concrete constitutive model for joint element.

According to the concept of smeared strain, the unified concrete constitutive model is adopted, as shown in Figure 4c. For epoxy joints, the tensile strength of the fiber in the joint element is set to be equal to that of the concrete segments. For dry joints, the tensile strength of the fiber is set to be zero. The shadows under the tensile and compressive curves in Figure 4c denote fracture energy G_F (N/mm) and crushing energy G_{Fc} (N/mm) of the concrete per unit length, respectively. The slopes of the compression-softening and tension-stiffening curves are adjusted according to G_{Fc} (N/mm) and G_F (N/mm), as shown in Equations (1) and (2). The values of G_{Fc} (N/mm), G_F (N/mm), and G_{F0} (N/mm) are calculated according to the specifications of CEB-FIP [39]. Then, the crack and crushing bands are smeared into the whole element length l_{se} (mm) to reach normalization. Furthermore, the joint opening width can be approximatively predicted by the product of tensile strain and joint element length l_{se} (mm). In this way, the discontinuous joint behaviors are converted to the equivalent "continuous behavior".

$$\varepsilon_{cu} = 2G_{Fc}/(f_c l_{se}) + \varepsilon_{c0},\ G_{Fc} = 8.8\sqrt{f_c} \tag{1}$$

$$\varepsilon_{tu} = 2G_F/(l_{se} f_t),\ G_F = G_{F0}(f_c/10)^{0.7} \tag{2}$$

2.3. *Multi-Node Slipping Cable Element for Internal Unbonded Tendons*

In the traditional FE model, the prestressing tendons are usually modeled by the truss element, which is a 2-node line element. For internal unbonded prestressed members, the tendon needs to be divided into many truss elements to apply the interaction with the beam along the whole span. However, the strain compatibility cannot be satisfied by the traditional truss elements, which are not suitable for the simulation of the free-slip behaviors of internal unbonded tendons.

To model the free-slip effects of internal unbonded tendons, a new type of element called the multi-node slipping cable element is proposed by the authors. The schematic diagram of the proposed element is shown in Figure 5. The element consists of several nodes and segments, and the nodal as well as segmental numbers are arbitrary without limitation. In this paper, we assume the slipping cable element as a configuration with n_{p} nodes and $n_{\mathrm{p}-1}$ segments. Each node has two translational DOFs, and each segment can only transmit only axial force. The axial forces in the segments are equal anywhere.

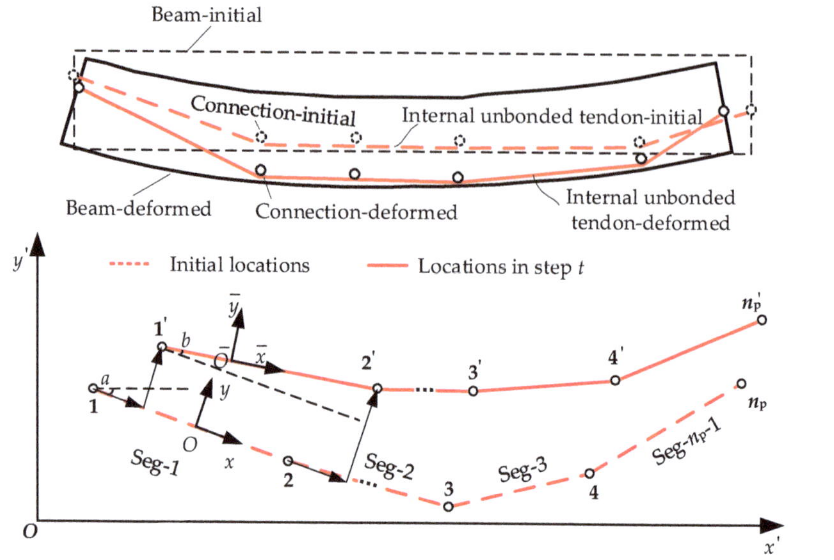

Figure 5. Schematic diagram of multi-node slipping cable element for internal unbonded tendon.

At step t, we express the element displacement vector $\mathbf{U}_t$ as Equation (3).

$$\mathbf{U}_t = \begin{bmatrix} \mathbf{u}_{1,t} & \mathbf{u}_{i,t} & \cdots & \mathbf{u}_{n_p,t} \end{bmatrix}^T_{2n_p \times 1}, \tag{3}$$

where $\mathbf{u}_{i,t} = [u_{i,t} \quad v_{i,t}] (i = 1 \sim n_p)$ is the displacement vector of node i at step t; $u_{i,t}$ and $v_{i,t}$ denote the displacement along the x and y direction of the global coordinate system, respectively.

For each segment, we define the element local coordinate system $\overline{x}\overline{O}\overline{y}$ along the deformed element axis based on the corotational theory. Then, the element displacement vector $\mathbf{U}_t^l$ in the local coordinate system can be denoted as Equation (4).

$$\begin{gathered}
\mathbf{U}_t^l = \mathbf{R}\mathbf{P}_t\mathbf{U}_t \\
\mathbf{P}_t = \begin{bmatrix} -1 & 0 & 1 & & & & \\ & -1 & 0 & 1 & & & \\ & & -1 & 0 & 1 & & \\ & & & \ddots & \ddots & \ddots & \\ & & & & -1 & 0 & 1 \end{bmatrix}_{(2n_p-2)\times(2n_p)}, \ \mathbf{R} = diag\left(\begin{bmatrix} \mathbf{r}_1 & \mathbf{r}_j & \cdots & \mathbf{r}_{n_p-1} \end{bmatrix}\right)_{(2n_p-2)\times(2n_p-2)} \\
\mathbf{r}_j = \begin{bmatrix} \cos X & \cos Y \\ -\cos Y & \cos X \end{bmatrix}, \ \cos X = \frac{x_{j+1,0} - x_{j,0}}{l_{j,0}}, \ \cos Y = \frac{y_{j+1,0} - y_{j,0}}{l_{j,0}}
\end{gathered} \tag{4}$$

Then, the tendon strain at step t can be calculated using the definition of the engineering strain, as shown in Equation (5).

$$\begin{gathered}
\varepsilon^t = L_t/L_0 = \mathbf{U}_t^T\mathbf{P}_t^T\Upsilon_t\left[\mathbf{P}_t\mathbf{U}_t + \mathbf{R}^T\mathbf{L}_0\right]/L_0 \\
\Upsilon_t = diag\left(\begin{bmatrix} l_{1,t}^{-1}\mathbf{I}_{2\times 2} & l_{j,t}^{-1}\mathbf{I}_{2\times 2} & \cdots & l_{n_p-1,t}^{-1}\mathbf{I}_{2\times 2} \end{bmatrix}\right) \\
\mathbf{L}_0 = \begin{bmatrix} \mathbf{L}_{1,0}^T & \mathbf{L}_{j,0}^T & \cdots & \mathbf{L}_{n_p-1,0}^T \end{bmatrix}^T, \ \mathbf{L}_{j,0} = \begin{bmatrix} l_{j,0} & 0 \end{bmatrix}^T
\end{gathered} \tag{5}$$

where L_0 denotes the initial sum length of all segments and L_t denotes the sum length of all segments at step t.

Through the virtual work equations and variational derivation, the stiffness matrix $\mathbf{K}_t$ and resistance vector $\mathbf{f}_t$ of the slipping cable element can be obtained and expressed as

Equations (6) and (7), respectively. $\mathbf{k}_t^g$ and $\mathbf{k}_t^m$ denote the geometric and material stiffness matrix of the element, respectively. E_p is the tangent modulus of the material and A_p is the cross-sectional area of the tendon. σ^t denotes the stress in current analysis step t.

$$\begin{aligned}\mathbf{K}_t &= \mathbf{k}_t^g + \mathbf{k}_t^m \\ \mathbf{k}_t^g &= \sigma^t A_p \mathbf{P}_t^T (\Upsilon_t - \Gamma_t)\mathbf{P}_t \\ \mathbf{k}_t^m &= E_p A_p / L_0 \mathbf{P}_t^T \Gamma_t \mathbf{R}^T \mathbf{L}_0 \mathbf{L}_0^T \mathbf{R} \Upsilon_t \mathbf{P}_t\end{aligned} \tag{6}$$

$$\mathbf{f}_t = \sigma^t A_p \mathbf{L}_0^T \mathbf{R} \Gamma_t \mathbf{P}_t \quad \mathbf{f}_t = \sigma^t A_p \mathbf{L}_0^T \mathbf{R} \Gamma_t \mathbf{P}_t \tag{7}$$

In the current OpenSees version, there are no suitable element types for the simulation of internal unbonded tendons. Based on the above-derived equation, the slipping cable element is added into the OpenSees software by the authors as the newly developed element type. Meanwhile, the *Tcl* command interpreters are also programmed into OpenSees for the definition of the element. It should be pointed out that the element and developed procedure have no limitation on the tendon shape profiles.

2.4. Rigid Beam Connection

The interactions between the concrete beam and tendons are modeled by a rigid beam connection, as shown in Figure 6. The node of the concrete beam has three degrees of freedom (DOFs), and the nodal displacement vector is denoted as $\mathbf{u}_c = [u_c, v_c, \theta_c]^T$. The node of the tendon element has two DOFs, whose displacement vector is denoted as $\mathbf{u}_t = [u_t, v_t]^T$. Taking the concrete beam node as the master node and the tendon node as the slave node, the rigid beam connection builds a transformation of displacements of the beam node to the displacements of the tendon node. The transformation relation of the rigid beam is presented in Equation (8). Δx and Δy denote the projection distances from the master node to the slave node in the x and y-directions, respectively.

$$\mathbf{u}_t = \begin{bmatrix} 1 & 0 & \Delta y \\ 0 & 1 & -\Delta x \end{bmatrix} \mathbf{u}_c \tag{8}$$

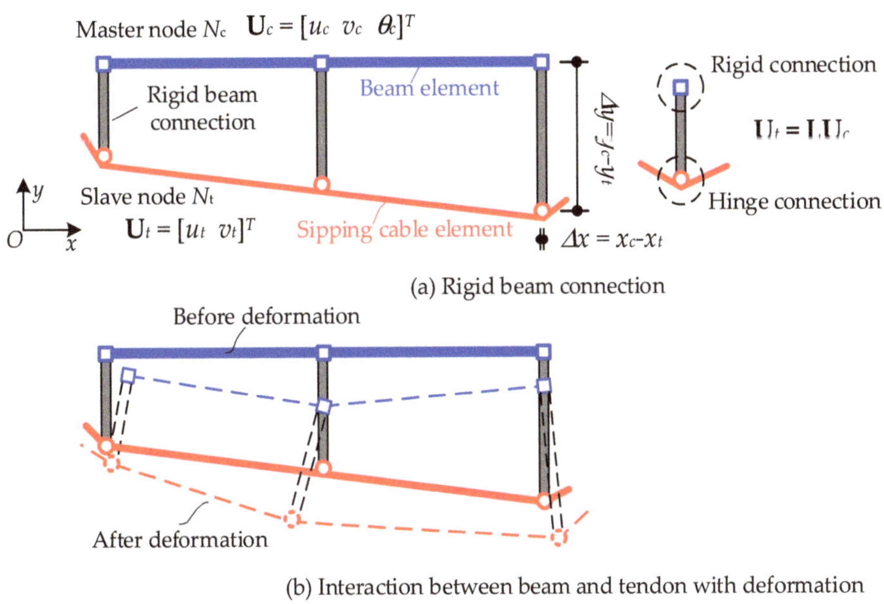

Figure 6. Rigid beam connection between beam and tendon.

2.5. Constitutive Model

The nonlinear constitutive models of materials used in this paper are presented as follows:

(1) Concrete Model

The concrete model for the RC beam element and joint element uses a parabolic-ascending linear-descending form for the compressive stress–strain relation. The tensile stress–strain relation is modeled by a linear-elastic before cracking and linear-descending form after cracking. The diagram can be seen in Figure 4c. f_t (MPa) and f_c (MPa) are the tensile and compression strength of concrete, respectively. ε_{tu} and ε_{cu} denote the ultimate tensile and compression strain of concrete, respectively. E_c (MPa) represents the initial modulus of concrete. For dry joint elements, the tensile strength of concrete is set to be zero and the built-in material Concrete01 in OpenSees without tension is adopted. For the epoxy joint, the built-in material Concrete02 in OpenSees is used and the tension softening stiffness is adjusted by the fracture energy, as presented in Equation (2). The compression-soften slope is adjusted by the crushing energy, as presented in Equation (1).

(2) Reinforcements Model

The Steel01 Model in the OpenSees software is adopted to model the material behaviors of reinforcements, as shown in Figure 7a. The model is linear-elastic before reaching the yield strain and then linear strain-hardening up to ultimate stress. E_s (MPa) denotes the initial elastic modulus and f_y (MPa) is the yield strength of reinforcement. b represents the strain-hardening ratio and is set to be 0.005 in this paper. The reinforcements model is only used in the RC beam element.

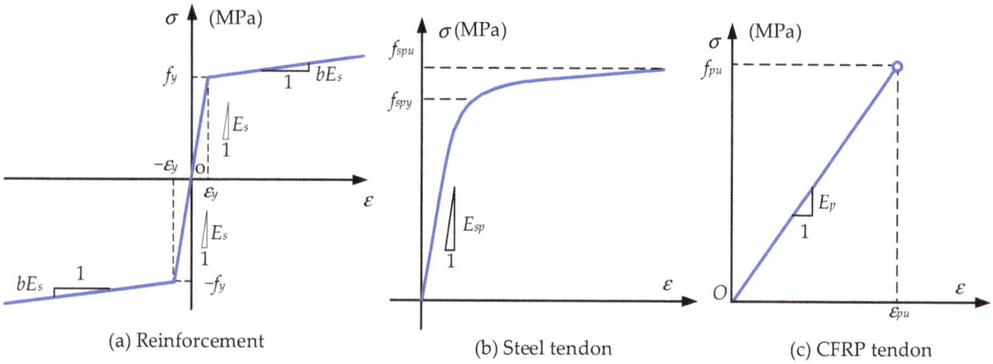

Figure 7. Constitutive model of reinforcement and tendons.

(3) Steel tendon

The material behavior of the prestressing steel tendon is modeled by the Steel02 model in OpenSees, as shown in Figure 7b. f_{spy} (MPa) is the yield strength of the steel tendon. E_{sp} (MPa) denotes the initial elastic modulus and b_{sp} represents the strain-hardening ratio. In this paper, b_{sp} = 0 and R_0 = 18 are employed in the Steel02 model definition and the default values are used for the other parameters (see http://opensees.berkeley.edu/ (accessed on 1 March 2020)). At the same time, the initial prestress σ_{pe} (MPa) is defined in the Steel02 model.

(4) CFRP tendon

The CFRP tendon has no yield states and behaves linear-elastically until rupture. The adopted material model for the CFRP is shown in Figure 7c. In OpenSees, there is no available material model for the CFRP. Therefore, we develop the new material constitutive model for the CFRP, which is linear elastic for a strain less than ε_{pu} and stress, and zero otherwise.

2.6. Novelty of the Proposed Model

Compared with previous studies, the novelty of the proposed model can be divided into three aspects as follows.

- This work proposes the first one-dimensional numerical model for the newly developed precast segmental concrete beam with internal unbonded CFRP tendons. Compared with the solid element model, the analysis efficiency and numerical convergence can be improved.
- Accounting for the discontinuous behaviors of the opening joint, a novel modeling idea, the "plain concrete" fiber beam element with the smeared strain method, is proposed.
- Accounting for the unbonded phenomenon of the internal unbonded tendon, a multi-node slipping cable element is proposed. In contrast to the equivalent load method, the complete formulations containing resistance and stiffness matrices are strictly derived, improving the numerical convergence for highly nonlinear solutions. The derived formulations are further developed in the OpenSees software as a newly developed element class.

3. Verification of Numerical Model

3.1. Verification for Tests of Monolithic Beam Prestressed with Internal Unbonded CFRP Tendon

Heo et al. tested the flexural behaviors of the monolithic beam prestressed with the internal unbonded CFRP tendon [24,40], in which the specimens RU50 and RO50 were designed with the same dimensions but different prestressing reinforcement ratios. These two specimens are employed here to verify the effectiveness of the proposed model for the simulation of internal unbonded tendons. Figure 8 shows the structural details and material parameters of RU50 and RO50. In the numerical model, the concrete beam is modeled by 18 fiber beam elements of 150 mm in length, and the internal tendon is modeled by only one slipping cable element. Prestressing and self-weight are applied to the beam first, and then the external load is applied up to failure.

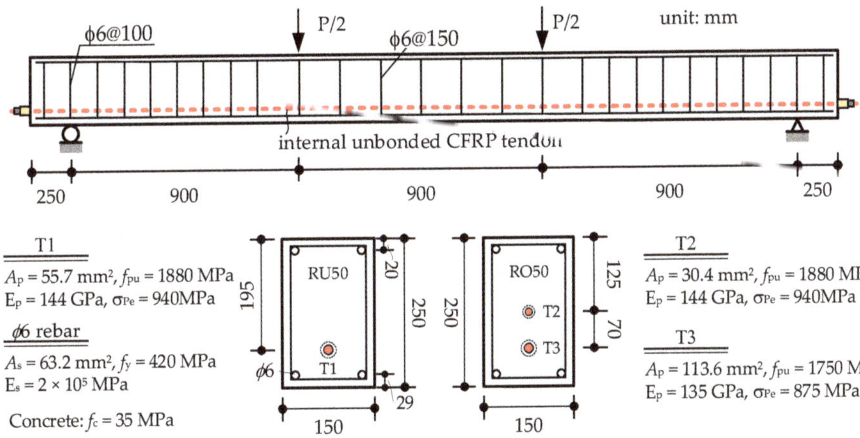

Figure 8. Design details of specimens RU50 and RO50 (unit: mm).

The analysis results and the comparisons with the tests are shown in Figure 9. Figure 9a shows the load versus mid-span deflection curves. The specimen BO50, with larger prestressing reinforcement ratios, shows greater cracking load and flexural capacity than BU50. The predicted deviations in the flexural capacity for BU50 and BO50 are −7.3% and −0.15%, respectively. The numerical model shows fairly good agreements with the tests overall. Meanwhile, it should be noted that the load–deflection curves of the monolithic beam with internal unbonded CFRP tendons show three feature points, namely, cracking, the yield of rebar, and concrete crushing. The sectional curvature at the ultimate state is extracted from the numerical results and the distributions along the beam length are shown in Figure 9b. It can be seen that a plastic region with a certain length is formed at the mid-span of both RU50 and RO50. The numerical model shows similar failure modes to the tests.

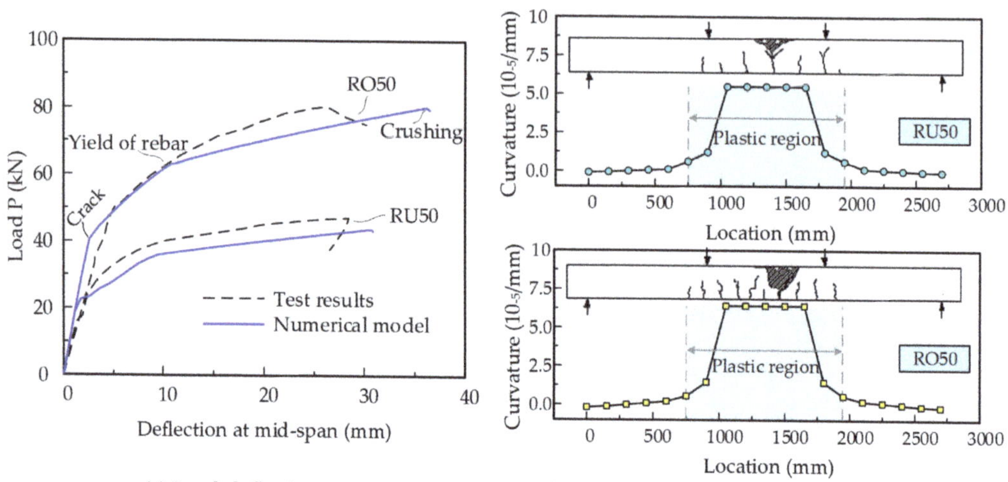

(a) Load-deflection curves (b) Curvature distribution at the ultimate state

Figure 9. Analysis results for specimens RU50 and RO50.

3.2. *Verification for Tests of PSCBIU-CFRP*

Le et al. [21,22] conducted the first experiments on the flexural behavior of a precast segmental concrete beam with internal unbonded CFRP tendons. The specimens named C2 and C4 are analyzed to verify the numerical model. The structural dimensions are shown in Figure 10 and the material parameters of the rebar and tendon are listed in Table 1. Two ϕ12 rebars are employed for bottom longitudinal reinforcement. Four ϕ10 rebars are placed at the top slab. ϕ10 reinforcements are used as the steel stirrups, with a 100 mm spacing for the middle segments and 75 mm for the end segments. The cylinder compressive strength of the concrete is 44.0 MPa on the testing day. The specimens C2 and C4 adopt the multiple shear-keyed epoxied joints, and the pre-tension control stress is set as $0.4f_{pu}$ for the CFRP tendon. After the completion of post-tensioning, specimen C2 is grouted as the bonded members and specimen C4 is not grouted as the unbonded PSCBs.

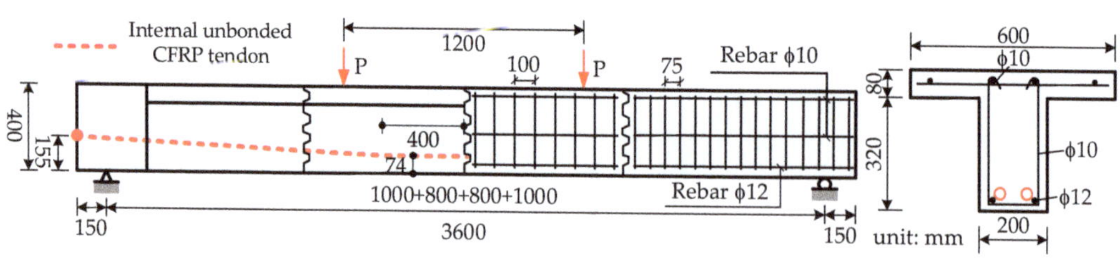

Figure 10. Structural dimensions of specimens C2 and C4 (unit: mm).

Based on the structural dimensions, two different models are built using OpenSees software. In the numerical model, the concrete segment element is modeled by the joint and RC beam element presented in Section 2.2. The length of the joint element is set as 100 mm and the concrete constitutive model is modified by the mentioned smeared strain method. The main differences between the two models are the modeling method for the CFRP tendons. For specimen C2, the bonded CFRP tendon is modeled by several traditional truss elements, and the tendon nodes are connected to the beam element by the rigid beam. Under loading states, the forces of the tendon elements in C2 are different and the maximum tendon forces usually occur in the tendon element adjacent to the failure region. For specimen C4, the unbonded CFRP tendon is modeled by the proposed slipping cable element. The tendon force is equal anywhere in the whole tendon. The analysis results and comparisons with the tests are shown in Figure 11. Figure 11a shows the curves of the applied load versus the mid-span deflection. Figure 11b shows the curves of the applied load versus the tendon stress increment. The tendon stress increment is equal within the internal unbonded tendon in specimen C4. For specimen C2, the tendon stress increment is obtained from the middle segment where the tendon stress is the maximum. Figure 11c shows the curves of the applied load versus the joint opening width at mid-span. The schematic diagrams of the numerical models for C2 and C4 are shown in Figure 11d.

Table 1. Properties of material used in specimens C2 and C4.

Type	Diameter (mm)	Area (mm^2)	f_y (MPa)	f_u (MPa)	E (GPa)
Steel rebar	12.0	113.0	534	587	200
Steel rebar	10.0	78.5	489	538	200
CFRP tendon	12.9	126.7	—	2450	145

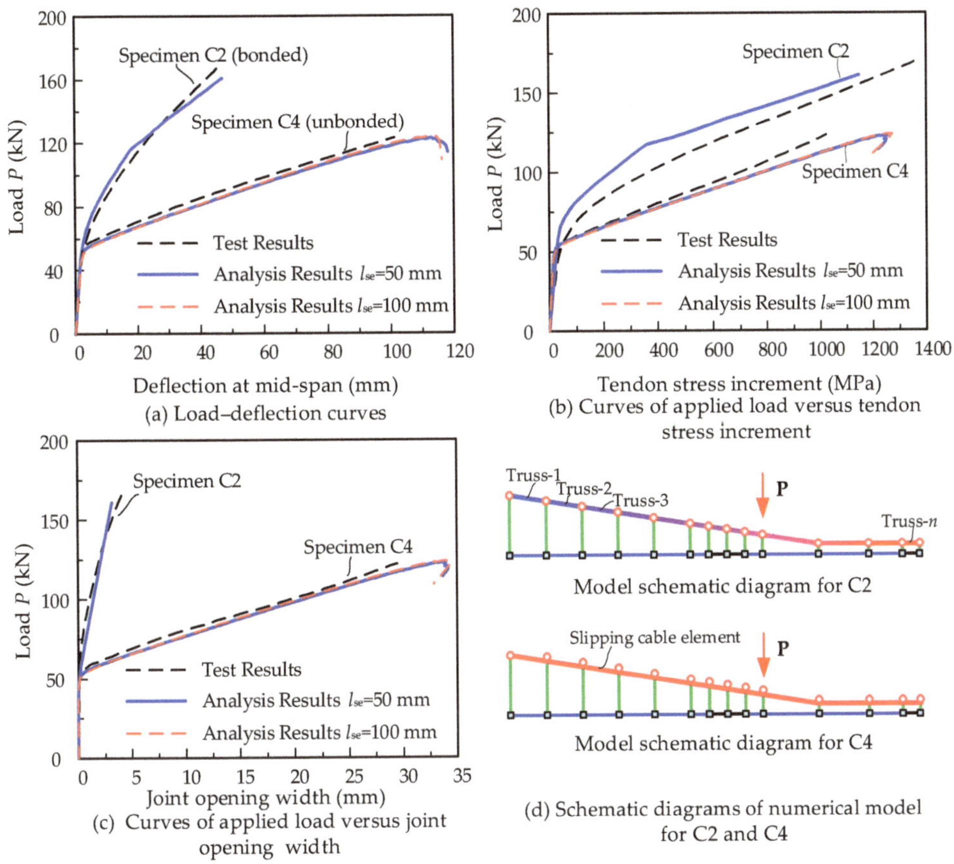

Figure 11. Analysis results for specimens C2 and C4.

Overall, the analysis results show good agreements with the tests of structural capacity, deformation, tendon stress increments, and joint opening width. The bonded prestressed member C2 has a larger capacity but the unbonded prestressed member C4 shows better ductility. The conventional truss element could only be used for the simulation of the bonded CFRP tendon and the proposed slipping cable element could be employed to analyze the effects of the unbonded phenomenon. Moreover, the joint element length is taken as a discussed parameter. With the normalized constitutive model based on the fracture and crushing energy, the analysis results of models with joint element lengths of l_{se} = 50 mm and l_{se} = 100 mm are almost the same. It can be concluded that the proposed method can be effectively used to model the joint behavior and the effects of the joint element length can be eliminated.

To compare the failure mode differences, the curvature distributions along the beam axial line are extracted from the last analysis step and plotted in Figure 12. The comparisons show that the bonded member C2 has a certain length of the plastic region at mid-span, as with the monolithic beam in Figure 8. However, this is quite different for the unbonded beam C4. Figure 12b shows that a concentrated plastic hinge is formed at the mid-span of C4 and the other regions are almost elastic. The PSCBIU-CFRP shows different behaviors to the bonded beam.

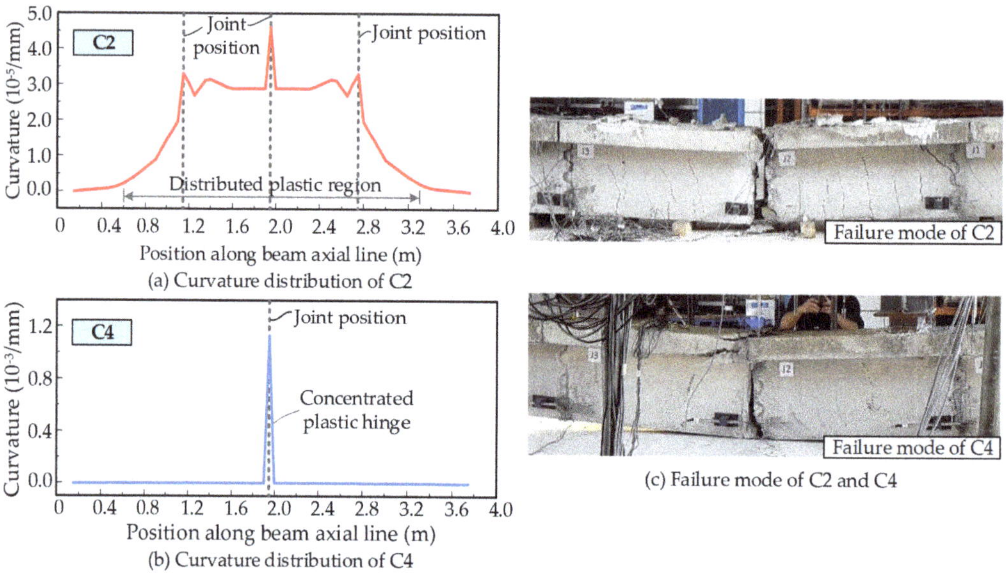

Figure 12. Failure mode comparisons for specimens C2 and C4.

However, the specimens by Le et al. [21,22] have only one joint at mid-span, which is different from the practice bridge. The flexural performances should be further evaluated in the full-scale bridge and the proposed model could be employed as a highly efficient analysis tool.

4. Flexural Performance Evaluation of PSCBIU-CFRP in Full-Scale Bridge

4.1. Flexural Performance Evaluation

Compared with the scaled beam, the full-scale PSCB consists of more segments with larger dimensions. The scale of the shear keys is very small compared with the concrete segment length, requiring significant numbers of elements in the 3D solid model during the mesh generation. The application of the 3D solid element model to the full-scale bridge is sometimes limited. The verified and high-efficiency numerical model in this paper provides an optional analysis tool for evaluating the flexural performance of the full-scale PSCBIU-CFRP system.

In this section, three different prestressing schemes for PSCBs are discussed, namely, the widely used external unbonded tendons, internal fully bonded tendons, and the newly developed internal unbonded CFRP tendons. The models are loaded by four-point loading (as shown in Figure 13f) up to failure; then, the flexural capacity and ductility of the different schemes are compared.

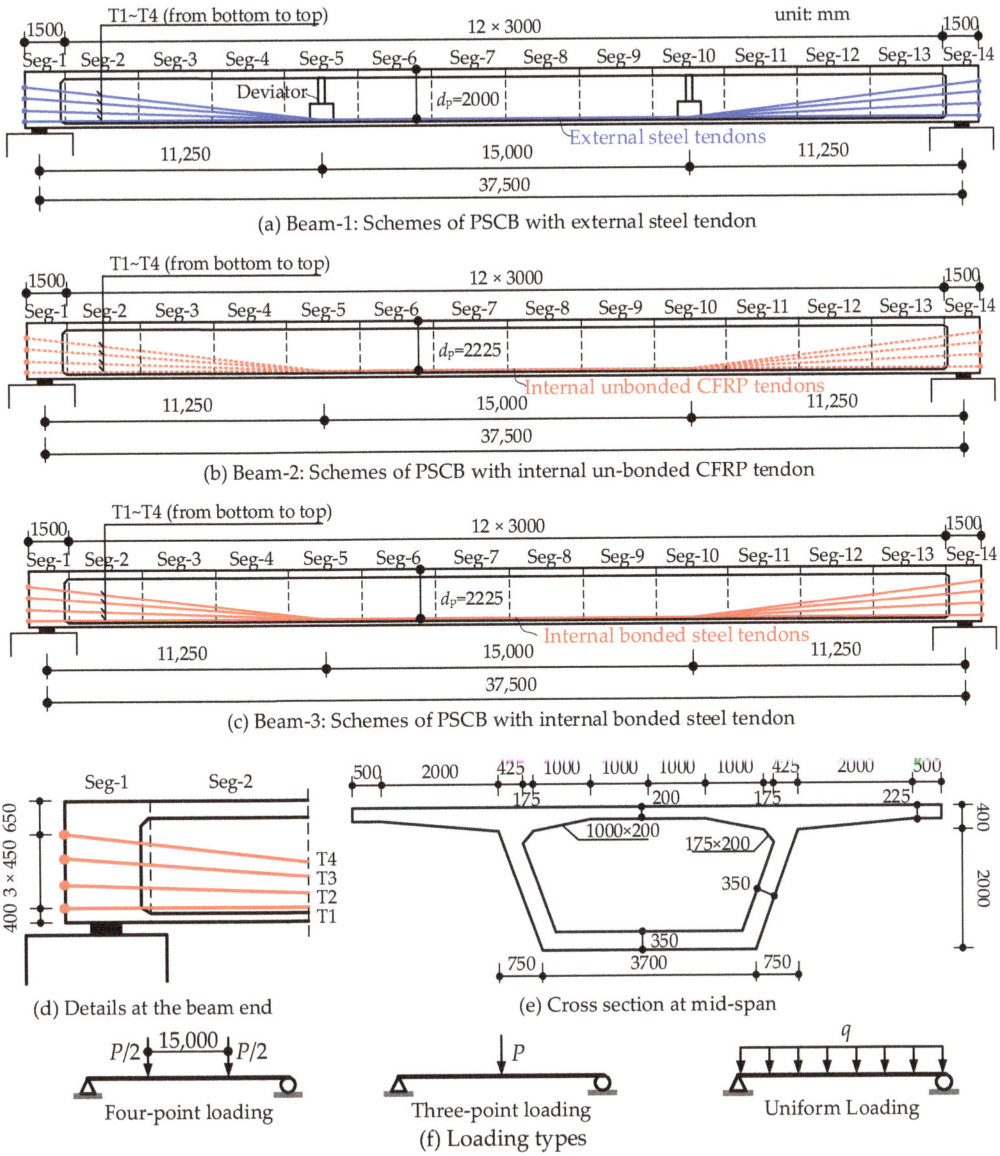

Figure 13. Structural dimensions of the numerical beam (unit: mm).

Figure 13a,b present the structural design details of PSCBs with external steel tendons (called Beam-1) and internal unbonded CFRP tendons (called Beam-2), respectively. Meanwhile, Beam-3 is designed to have the same structural dimensions as Beam-2, but replaces the internal unbonded CFRP tendons with internal bonded steel tendons (as shown in Figure 13c). The supported span is 37.5 m, consisting of 14 concrete box beam segments and prestressed by 8 tendons with symmetrical layouts. The dimensions of the cross-section are shown in Figure 13e. There are 120 and 60 conventional rebars ϕ12 in the top and bottom flanges, respectively. Furthermore, ϕ10 reinforcements are placed in the segment at 150 mm spacing to increase the shear strength of the segments. The concrete strength is 40.0 MPa, and the other material properties are listed in Table 2.

Table 2. Material properties of Beam-1 and Beam-2.

Type	Area (mm^2)	f_y (MPa)	f_u (MPa)	E (GPa)	Post-Tensioning Stress (MPa)
Steel rebar ϕ12	113.0	350	380	200	-
Steel tendon	2100	1748	1860	195	1000
CFRP tendon	2100	-	2450	145	1000

The calculated flexural capacity, deformation, and tendon stress of the three models are compared in Figure 14, and the results are listed in Table 3. Some observations are concluded below:

(1) As shown in Figure 14a, the load–deflection curves of these beams show different shape styles. For Beam-3, prestressed with internal fully bonded steel tendons, the load–deflection curve is a three-broken-line shape with feature points of joint opening, steel tendon yield, and concrete crushing. However, for Beam-2 and Beam-1, the load–deflection curve shape is a double broken line with feature points of joint opening and concrete crushing.

(2) The beam with external steel tendons (Beam-1) shows a minimum flexural capacity of 22.9×10^3 kN·m. The beam with internal fully bonded steel tendons (Beam-3) shows a maximum flexural capacity of 47.2×10^3 kN·m. The application of the CFRP in the PSCB as internal unbonded prestressing significantly improves the flexural capacity. The ultimate capacity of 35.4×10^3 kN·m increases by 54.6% compared with Beam-1 and is close to that of Beam-3, with only a 25% decrease. The comparisons show that the PSCB-IUCFRP system can be taken as the substitute for an external prestressed PSCB to improve structural capacity.

(3) Following the adopted ductility indices defined in the references [21,22], $\mu = \Delta_u / \Delta_y$, the structural ductility is evaluated and compared. Δ_y is the deflection at mid-span when the joints are open. Δ_u is the deflection at mid-span when reaching the maximum load. The calculated ductility indices are listed in Table 3. All of the beams show good structural ductility, and the PSCB-IUCFRP system increases ductility indices by 8.9% compared with the external prestressed beam (Beam-1).

(4) The tendon stress along with the increasing applied load is shown in Figure 14b, in which the results of Beam-3 are from the tendon at mid-span. The tendon stress increments are all small before joint opening and then increase quickly once reaching the cracking load. At the ultimate state, the steel tendon in Beam-3 has been yielded and tendons in the unbonded beam Beam-1 and Beam-2 are still inelastic. The tendon stress increments of Beam-2 are larger than those of Beam-1. Meanwhile, the effective depth of the tendon in Beam-1 reduces by 201.7 mm due to the second-order effects, leading to a decrease in flexural capacity compared with Beam-2.

(5) Figure 14c shows the curvature distribution curves along the beam length in the ultimate state. In Beam-1, the curvature is mainly concentrated at mid-span joints and the other regions are almost still elastic. However, for Beam-2 and Beam-3, the deformations spread to the other joints, increasing the tendon stress and further flexural capacity.

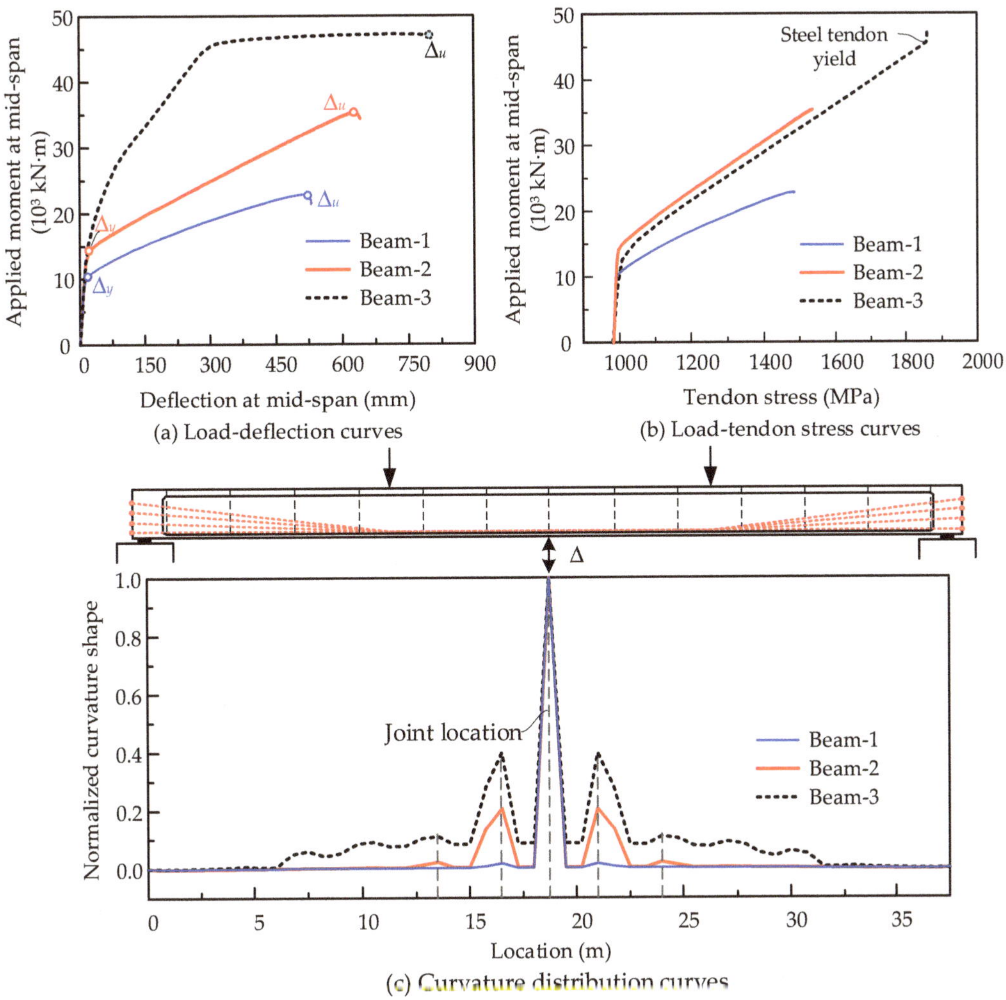

Figure 14. Performance comparisons of PSCB with different prestressing schemes.

Table 3. Comparisons of indices for PSCB.

Beam ID	$\Delta\sigma_p$ (MPa)	σ_{pu} (MPa)	M_y (10^3 kN·m)	M_u (10^3 kN·m)	Δ_y (mm)	Δ_u (mm)	$\mu = \Delta_u/\Delta_y$
Beam-1	505	1486	10.2	22.9	18	524	29.1
Beam-2	552	1536	14.1	35.4	20	633	31.7
Beam-3	879	1860	45.5	47.2	20	724	36.2

4.2. Parametric Study

Taking Beam-2 in Figure 13b as the reference beam, the effects of loading types, span-to-depth ratio L/d_p, and prestressing reinforcement ratio $\rho_p = A_p/A_c$ on the flexural performance are discussed.

(1) Loading types

Three loading types as shown in Figure 13e are considered in the simulation and the analysis results are shown in Figure 15. We find that the loading types have certain effects on the flexural capacity. The model under three-point loading obtains the minimum flexural capacity. The maximum capacity can be carried under four-point loading states. This finding differs from the conclusions of Le et al. [21,22]. In their analysis model, there was only one joint placed at mid-span, and no additional joint opened under any loading types. However, more joints exist in the practice bridge, and the joint opening may not only occur at one joint. In this simulation, under four-point loading, more joints open in the pure bending region, as shown in Figure 15c. The joint opening width significantly influences the tendon stress increment (Figure 15b) and further increases the capacity.

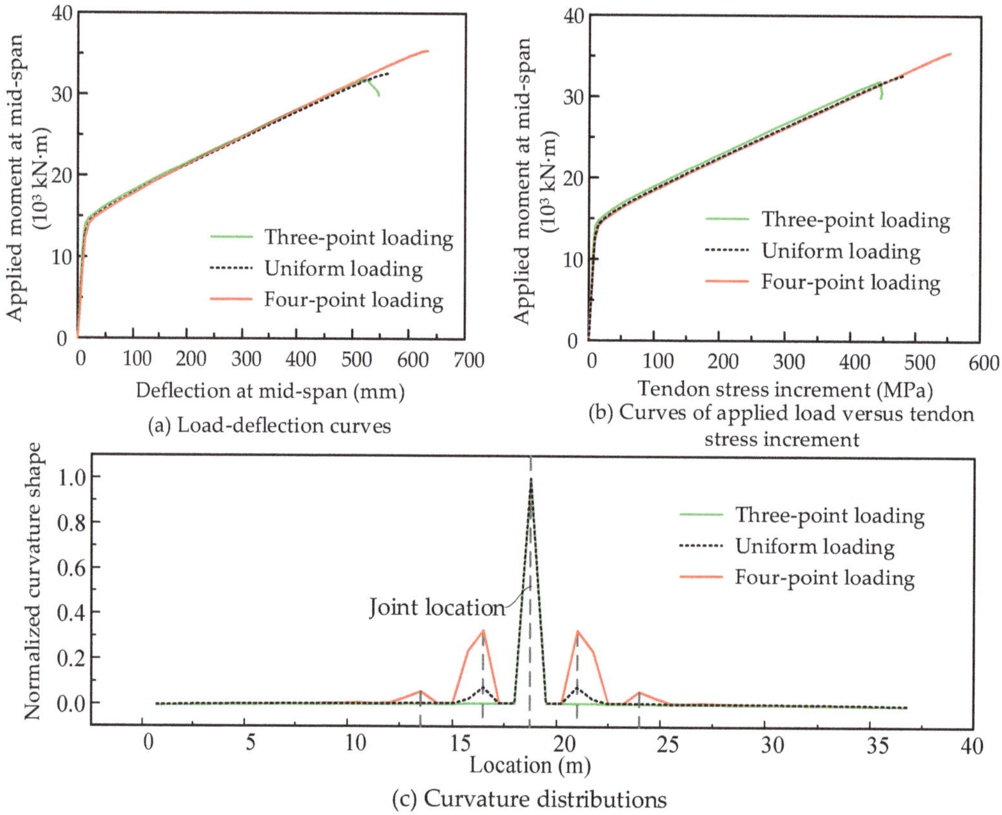

Figure 15. Effects of loading types on the flexural behaviors of PSCBIU-CFRP.

(2) Span-to-depth ratio L/d_p

Based on the design details of Beam-2 in Figure 13b, the tendon effective depth d_p varies from 2.025 and 2.225 to 1.8 m; then, beams with a span-to-depth ratio L/d_p of 16.85, 18.5, and 20.8 are obtained to discuss the effects on the flexural performances. All of the models are applied with three-point loading up to failure and the results are shown in Figure 16. With L/d_p increasing by 23%, the flexural capacity decreases by about 38.6% and the tendon stress increment decreases by about 15.7%. This reveals that a smaller span-to-depth ratio L/d_p is conducive to flexural performance.

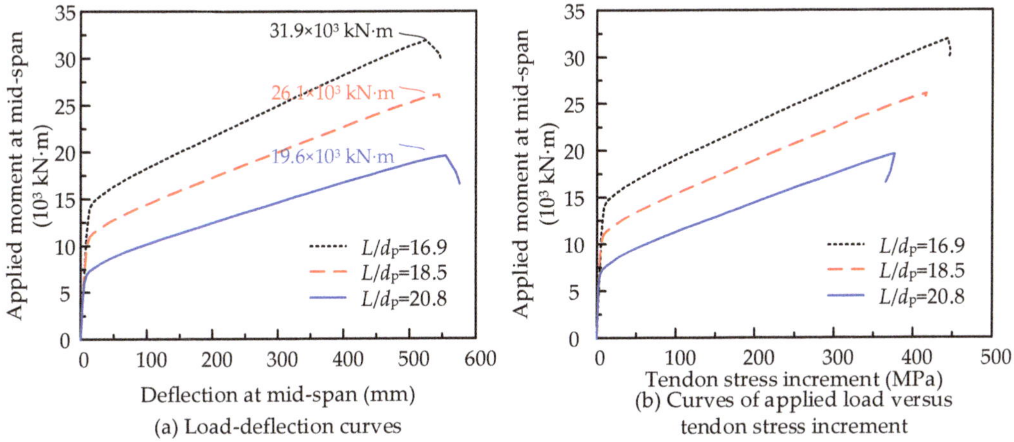

Figure 16. Effects of span-to-depth ratio L/d_p on the flexural behaviors of PSCBIU-CFRP.

(3) Prestressing reinforcement ratio ρ_p

Based on the referenced beam in Figure 13b, the area of the tendon is taken as the variable to discuss the effects of ρ_p on the flexural performance. The cross-sectional area per tendon is set to 1575, 2100, and 2625 mm^2, respectively. Three models are built with prestressing reinforcement ratios ρ_p of 0.26%, 0.35%, and 0.43%. Under three-point loading states, the flexural behaviors are analyzed and the results are presented in Figure 17. With ρ_p increasing by 65.4%, the flexural capacity increases by 88.7%. However, we also observe that the ultimate stress increment and the structural ductility decrease by 25.2% and 29.6%, respectively.

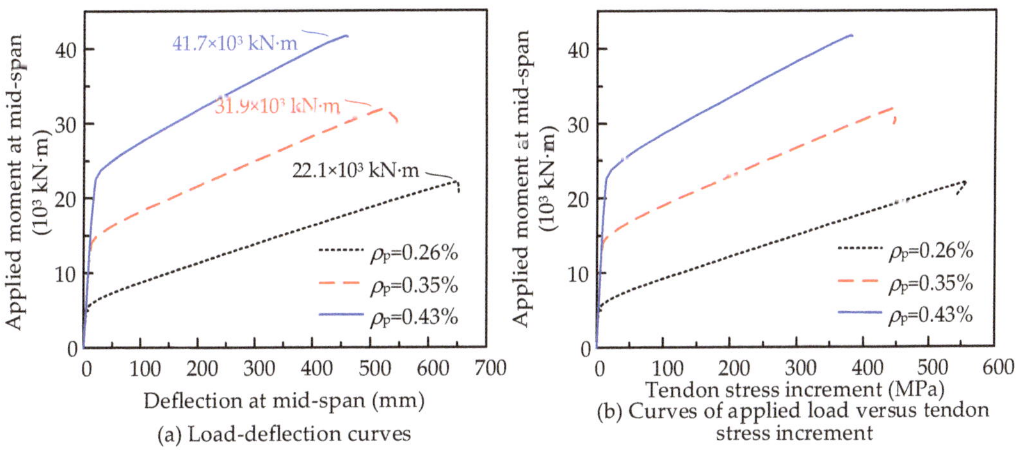

Figure 17. Effects of ρ_p on the flexural behaviors of PSCBIU-CFRP.

5. Conclusions

In this study, a new numerical model was proposed for the flexural analysis of a precast segmental concrete beam prestressed with internal unbonded CFRP tendons (PSCB-IUCFRPs). The modeling scheme, finite element formulations, and element implementation were presented. Two scaled experimental tests, namely, two monolithic beams and two segmental beams with internal unbonded CFRP tendons, were performed to verify the effectiveness of the proposed model. The verified numerical model was then applied to evaluate the flexural performance of a full-scale PSCB-IUCFRP with a span of 37.5 m. The influences of loading types, span-to-depth ratios, and prestressing reinforcement ratios on the flexural behaviors were further discussed. The following conclusions could be drawn:

1. The comparisons of two scaled experimental tests showed that the opening joint and the bonded condition of the tendons had significant influences on the flexural behaviors of the beam. The opening joint led to the concentrated plastic deformation at the joint location. The unbonded phenomenon of tendons caused a lower flexural capacity but higher ductility. The proposed modeling method for segmental joints and the developed muti-node slipping cable element for unbonded tendons can efficiently capture their mechanical behaviors. The effectiveness of the proposed model for the predictions of tendon stress increments, flexural capacity, and failure modes was fully verified against the experimental tests.
2. The flexural performances of the full-scale 37.5 m span PSCBs prestressed with three different prestressing schemes were compared. The PSCB with internal bonded steel tendons had the highest flexural capacity and ductility. The results for the PSCB with externally prestressed steel tendons were lowest. Compared with the externally prestressing scheme, with the same prestressing reinforcement ratio, the PSCB with internal unbonded CFRP tendons improved the flexural capacity by 54.6% and ductility indices by 8.9%. The internal unbonded CFRP tendon was a competitive alternative solution to the external steel tendon for the PSCB.
3. The parametric study showed that the loading types, span-to-depth ratios, and prestressing reinforcement ratios all had certain influences on the flexural behaviors of the PSCB with internal unbonded CFRP tendons. Under four-point loading states, more joints opened at the ultimate states, which was different from the analysis results of scaled specimens. The tendon stress increments and the flexural capacity were greater than those of the other loading conditions. The larger span-to-depth ratio L/d_p was not conducive to the flexural capacity. With L/d_p increasing by 23%, the flexural capacity decreased by about 38.6% and the tendon stress increment decreased by about 15.7%. With ρ_p increasing by 65.4%, the flexural capacity increased by 88.7%. However, the ultimate stress increment and the structural ductility decreased by 25.2% and 29.6%, respectively.

Author Contributions: Conceptualization, methodology, software, validation, writing—original draft preparation, W.Y.; project administration, L.C.; supervision, B.H.; data curation, writing—review and editing, H.X.; visualization, Y.S. All authors have read and agreed to the published version of the manuscript.

Funding: This research was funded by the Funds of China Railway Economic Planning and Research Institute, funding number 2021BSH01; Fundamental Research Funds for the Central Universities, funding number 2022JBMC047; Project of Science and Technology Research Development Plan of China Railway, funding numbers N2021G046 and K2021G013; National Natural Science Foundation of China, funding number 51708020; Project from Key Laboratory of Transport Industry of Bridge Detection Reinforcement Technology (Beijing), funding number C21M00030.

Institutional Review Board Statement: Not applicable.

Informed Consent Statement: Not applicable.

Data Availability Statement: Not applicable.

Conflicts of Interest: The authors declare no conflict of interest.

References

1. Li, H.; Li, J.; Xin, Y.; Hao, H.; Le, T.D.; Pham, T.M. Prestress Force Monitoring and Quantification of Precast Segmental Beams through Neutral Axis Location Identification. *Appl. Sci.* **2022**, *12*, 2756. [CrossRef]
2. Zhu, Y.; Zhang, Y.; Shi, J. Finite Element Analysis of SSFlexural Behavior of Precast Segmental UHPC Beams with Prestressed Bolted Hybrid Joints. *Eng. Struct.* **2021**, *238*, 111492. [CrossRef]

3. Rettinger, M.; Hueckler, A.; Schlaich, M. Technologies and Developments in Precast Segmental Bridge Construction. *Beton Und Stahlbetonbau* **2021**, *116*, 12–23. [CrossRef]
4. Ahmed, G.H.; Aziz, O.Q. Shear Strength of Joints in Precast Posttensioned Segmental Bridges during 1959–2019, Review and Analysis. *Structures* **2019**, *20*, 527–542. [CrossRef]
5. Ahmed, G.H.; Aziz, O.Q. Stresses, deformations and damages of various joints in precast concrete segmental box girder bridges subjected to direct shear loading. *Eng. Struct.* **2020**, *206*, 110151. [CrossRef]
6. Chai, S.; Guo, T.; Chen, Z.; Yang, J. Flexural Behavior of Precast Concrete Segmental Box-Girders with Dry Joints. *Adv. Civ. Eng.* **2020**, *2020*, 8895180. [CrossRef]
7. Jiang, H.; Chen, M.; Sha, Z.; Xiao, J.; Feng, J. Numeric Analysis on Shear Behavior of High-Strength Concrete Single-Keyed Dry Joints with Fixing Imperfections in Precast Concrete Segmental Bridges. *Materials* **2020**, *13*, 2914. [CrossRef]
8. Yan, W.-T.; Han, B.; Xie, H.-B.; Li, P.-F.; Zhu, L. Research on numerical model for flexural behaviors analysis of precast concrete segmental box girders. *Eng. Struct.* **2020**, *219*, 110733. [CrossRef]
9. Yuan, A.; Zhao, X.; Lu, R. Experimental investigation on shear performance of fiber-reinforced high-strength concrete dry joints. *Eng. Struct.* **2020**, *223*, 111159. [CrossRef]
10. Jiang, H.; Huang, C.; Feng, J.; Gao, X.; Wang, T.; Zhong, Z. Direct shear behavior of castellated dry RPC joints in precast concrete segmental bridges. *Structures* **2021**, *33*, 4579–4595. [CrossRef]
11. Anania, L.; Badalà, A.; D'Agata, G. Damage and collapse mode of existing post tensioned precast concrete bridge: The case of Petrulla viaduct. *Eng. Struct.* **2018**, *162*, 226–244. [CrossRef]
12. Wouters, J.P.; Kesner, K.; Poston, R.W. Tendon Corrosion in Precast Segmental Bridges. *J. Transp. Res. Board* **1999**, *1654*, 128–132. [CrossRef]
13. Xue, X.; Wang, X.; Hua, X.; Wu, M.; Wu, L.; Ma, Z.; Zhou, J. Experimental Investigation of the Shear Behavior of a Concrete Beam without Web Reinforcements Using External Vertical Prestressing Rebars. *Adv. Civ. Eng.* **2019**, *2019*, 3452056. [CrossRef]
14. Peng, F.; Xue, W.; Tan, Y. Design Approach for Flexural Capacity of Prestressed Concrete Beams with External Tendons. *J. Struct. Eng.* **2018**, *144*, 04018215. [CrossRef]
15. Yuan, A.; He, Y.; Dai, H.; Cheng, L. Experimental Study of Precast Segmental Bridge Box Girders with External Unbonded and Internal Bonded Posttensioning under Monotonic Vertical Loading. *J. Bridg. Eng.* **2015**, *20*, 04014075. [CrossRef]
16. Jiang, H.; Cao, Q.; Liu, A.; Wang, T.; Qiu, Y. Flexural behavior of precast concrete segmental beams with hybrid tendons and dry joints. *Constr. Build. Mater.* **2016**, *110*, 1–7. [CrossRef]
17. Yang, X.; Zohrevand, P.; Mirmiran, A.; Arockiasamy, M.; Potter, W. Effect of Elastic Modulus of Carbon Fiber-Reinforced Polymer Strands on the Behavior of Posttensioned Segmental Bridges. *J. Compos. Constr.* **2016**, *20*, 04016030. [CrossRef]
18. Karimipour, A.; Farhangi, V. Effect of EBR- and EBROG-GFRP laminate on the structural performance of corroded reinforced concrete columns subjected to a hysteresis load. *Structures* **2021**, *34*, 1525–1544. [CrossRef]
19. Karimipour, A.; de Brito, J.; Ghalehnovi, M.; Gencel, O. Torsional behaviour of rectangular high-performance fibre-reinforced concrete beams. *Structures* **2021**, *35*, 511–519. [CrossRef]
20. Wu, T.; Sun, Y.; Liu, X.; Cao, Y. Comparative study of the flexural behavior of steel fiber-reinforced lightweight aggregate concrete beams reinforced and prestressed with CFRP tendons. *Eng. Struct.* **2021**, *233*, 111901. [CrossRef]
21. Le, T.D.; Pham, T.M.; Hao, H.; Hao, Y. Flexural behaviour of precast segmental concrete beams internally prestressed with unbonded CFRP tendons under four-point loading. *Eng. Struct.* **2018**, *168*, 371–383. [CrossRef]
22. Le, T.D.; Pham, T.M.; Hao, H.; Yuan, C. Performance of precast segmental concrete beams posttensioned with carbon fiber-reinforced polymer (CFRP) tendons. *Compos. Struct.* **2018**, *208*, 56–69. [CrossRef]
23. Tang, L. Maintenance and Inspection of Fiber-Reinforced Polymer (FRP) Bridges: A Review of Methods. *Materials* **2021**, *14*, 7826. [CrossRef] [PubMed]
24. Heo, S.; Shin, S.; Lee, C. Flexural Behavior of Concrete Beams Internally Prestressed with Unbonded Carbon-Fiber-Reinforced Polymer Tendons. *J. Compos. Constr.* **2013**, *17*, 167–175. [CrossRef]
25. Bakar, M.B.C.; Rashid, R.S.M.; Amran, M.; Jaafar, M.S.; Vatin, N.I.; Fediuk, R. Flexural Strength of Concrete Beam Reinforced with CFRP Bars: A Review. *Materials* **2022**, *15*, 1144. [CrossRef]
26. Rogowski, J.; Kotynia, R. Comparison of Prestressing Methods with CFRP and SMA Materials in Flexurally Strengthened RC Members. *Materials* **2022**, *15*, 1231. [CrossRef]
27. Alqam, M.; Alkhairi, F.; Naaman, A. An Improved Methodology for the Prediction of the Stress at Ultimate in Unbonded Internal and External Steel Tendons. *Arab. J. Sci. Eng.* **2020**, *45*, 7915–7954. [CrossRef]
28. Le, T.D.; Pham, T.M.; Hao, H. Numerical study on the flexural performance of precast segmental concrete beams with unbonded internal steel tendons. *Constr. Build. Mater.* **2020**, *248*, 118362. [CrossRef]
29. Tran, D.T.; Pham, T.M.; Hao, H.; Chen, W. Numerical investigation of flexural behaviours of precast segmental concrete beams internally post-tensioned with unbonded FRP tendons under monotonic loading. *Eng. Struct.* **2021**, *249*, 113341. [CrossRef]
30. Lou, T.; Lopes, S.M.; Lopes, A.V. Response of continuous concrete beams internally prestressed with unbonded FRP and steel tendons. *Compos. Struct.* **2016**, *154*, 92–105. [CrossRef]
31. Lou, T.; Liu, M.; Lopes, S.M.; Lopes, A.V. Effect of bond on flexure of concrete beams prestressed with FRP tendons. *Compos. Struct.* **2017**, *173*, 168–176. [CrossRef]

32. Pang, M.; Li, Z.; Lou, T. Numerical Study of Using FRP and Steel Rebars in Simply Supported Prestressed Concrete Beams with External FRP Tendons. *Polymers* **2020**, *12*, 2773. [CrossRef] [PubMed]
33. Mckenna, F.; Fenves, G. *Opensees 2.5.0, Computer Software*; UC Berkeley: Berkeley, CA, USA, 2015; Available online: http://opensees.berkeley.edu (accessed on 1 March 2020).
34. Tao, X.; Du, G. Ultimate Stress of Unbonded Tendons in Partially Prestressed Concrete Beams. *PCI J.* **1985**, *30*, 72–91. [CrossRef]
35. Scott, M.H.; Filippou, F.C. Response Gradients for Nonlinear Beam-Column Elements under Large Displacements. *J. Struct. Eng.* **2007**, *133*, 155–165. [CrossRef]
36. Kenawy, M.; Kunnath, S.; Kolwankar, S.; Kanvinde, A. Fiber-Based Nonlocal Formulation for Simulating Softening in Reinforced Concrete Beam-Columns. *J. Struct. Eng.* **2018**, *144*, 04018217. [CrossRef]
37. Wang, J.-J.; Liu, C.; Fan, J.-S.; Hajjar, J.F.; Nie, X. Triaxial Concrete Constitutive Model for Simulation of Composite Plate Shear Wall–Concrete Encased: THUC3. *J. Struct. Eng.* **2019**, *145*, 04019088. [CrossRef]
38. Lignola, G.; Prota, A.; Manfredi, G.; Cosenza, E. Non-linear modeling of RC rectangular hollow piers confined with CFRP. *Compos. Struct.* **2009**, *88*, 56–64. [CrossRef]
39. CEB-FIP. *Model Code 2010*; Thomas Telford: Lausanne, Switzerland, 2010.
40. Lee, C.; Shin, S.; Lee, H. Balanced Ratio of Concrete Beams Internally Prestressed with Unbonded CFRP Tendons. *Int. J. Concr. Struct. Mater.* **2016**, *11*, 1–16. [CrossRef]

materials MDPI

Article

Optimization of the Structural Performance of Buried Reinforced Concrete Pipelines in Cohesionless Soils

Odey Alshboul [1,*], Ghassan Almasabha [1], Ali Shehadeh [2], Omar Al Hattamleh [1] and Ali Saeed Almuflih [3]

1 Department of Civil Engineering, Faculty of Engineering, The Hashemite University, P.O. Box 330127, Zarqa 13133, Jordan; ghassans@hu.edu.jo (G.A.); hattam@hu.edu.jo (O.A.H.)
2 Department of Civil Engineering, Hijjawi Faculty for Engineering Technology, Yarmouk University, P.O. Box 566, Irbid 21163, Jordan; ali.shehadeh@yu.edu.jo
3 Department of Industrial Engineering, King Khalid University, Fahad St, Guraiger, Abha 62529, Saudi Arabia; asalmuflih@kku.edu.sa
* Correspondence: odey.shboul@hu.edu.jo

Abstract: Pipelines are widely used to transport water, wastewater, and energy products. However, the recently published American Society of Civil Engineers report revealed that the USA drinking water infrastructure is deficient, where 12,000 miles of pipelines have deteriorated. This would require substantial financial investment to rebuild. Furthermore, the current pipeline design practice lacks the guideline to obtain the optimum steel reinforcement and pipeline geometry. Therefore, the current study aimed to fill this gap and help the pipeline designers and practitioners select the most economical reinforced concrete pipelines with optimum steel reinforcement while satisfying the shear stresses demand and serviceability limitations. Experimental testing is considered uneconomical and impractical for measuring the performance of pipelines under a high soil fill depth. Therefore, a parametric study was carried out for reinforced concrete pipes with various diameters buried under soil fill depths using a reliable finite element analysis to execute this investigation. The deflection range of the investigated reinforced concrete pipelines was between 0.5 to 13 mm. This indicates that the finite element analysis carefully selected the pipeline thickness, required flexural steel reinforcement, and concrete crack width while the pipeline does not undergo excessive deformation. This study revealed that the recommended optimum reinforced concrete pipeline diameter-to-thickness ratio, which is highly sensitive to the soil fill depth, is 6.0, 4.6, 4.2, and 3.8 for soil fill depths of 9.1, 12.2, 15.2, and 18.3 m, respectively. Moreover, the parametric study results offered an equation to estimate the optimum pipeline diameter-to-thickness ratio via a design example. The current research outcomes are imperative for decision-makers to accurately evaluate the structural performance of buried reinforced concrete pipelines.

Keywords: reinforced concrete pipelines; optimum diameter-to-thickness ratio; buried pipelines; deep embankment soil

check for updates

Citation: Alshboul, O.; Almasabha, G.; Shehadeh, A.; Al Hattamleh, O.; Almuflih, A.S. Optimization of the Structural Performance of Buried Reinforced Concrete Pipelines in Cohesionless Soils. *Materials* **2022**, *15*, 4051. https://doi.org/10.3390/ma15124051

Academic Editor: Ayman El-Zohairy

Received: 27 April 2022
Accepted: 5 June 2022
Published: 7 June 2022

1. Introduction

Pipelines are widely used in infrastructures to transport water, wastewater, and energy products. The USA uses 2.2 million miles of underground pipes to deliver drinking water. However, the recently released report by ASCE [1] classifies the USA drinking water infrastructure as a poor category "-C", as the pipelines are aged. Financing shortages result in an estimated water leakage of 6 billion gallons per day of drinking water. The estimated deteriorated pipelines length was 12,000 miles in 2019 [1]. Pipelines and culverts are considered the most economical options compared with highway bridges, where the utilities of water, wastewater, or energy products intersect with roadways and barriers of high soil depths. However, limited studies were conducted to investigate the performance of pipelines and culverts under deep soil burial since it is difficult to measure the response of such structures experimentally, and it costs a significant amount of money to conduct the experimental setup. On the other hand, the successful use of validated finite element analysis in recent years has significantly helped researchers and designers to measure the structural response of such infrastructural utilities.

The interaction between the pipeline structure and the compacted soil around them is critical in their structural performance. The interface layer between the pipeline structure and the surrounding soil requires a unique estimation. Analyzing such a soil–structure interaction is not easy to estimate, as no closed-form solution can predict this interaction accurately; however, this interaction can be effectively analyzed using the finite element analysis (FEA) [2–6]. In addition, the soil–structure interaction usually is not considered in pipeline and box culverts designs, where the applied pressure above the pipeline and the box culvert is assumed to be equivalent to the geostatic load, which is the earth prism weight measured above the pipeline or the box culvert (Pimentel et al., 2009 [7]). As the soil–structure interaction is challenging to experimentally estimate, several studies have efficiently exploited numerical tools such as finite element modeling and artificial intelligence to predict the effect of the soil–structure interaction for various structural types [3–19]. Therefore, the soil–structural interaction should be included in the pipeline and culvert design in order to estimate the structural response of such infrastructures accurately. Excluding this interaction may result in the erroneous structural design of these elements.

Various studies have investigated pipelines' stress redistribution and performance under a deep soil fill depth. For example, Marston and Anderson (1913) [20] were the pioneer researchers who studied loading distribution on buried pipes, whereas Marston (1930) [21] found that installation conditions and pipe properties influence the loading distribution. Likewise, Spangler (1950 [22], 1968 [23]) revealed that vertical pressure on the steel buried pipes developed by the surrounding compacted soil layers is coupled to the relative settlement between the soil barrier and the original soil. Similarly, Abuhajar et al. (2015) [24] indicated that soil arching is significantly affected by the soil fill depth, culvert thickness, elastic modulus of soil, and poison ratio.

Bashir (2000) [25] investigated several steel pipelines with diameter-to-thickness ratios ranging from 50 to 150, subject to soil fill depths ranging from 0.6 m to 3.7 m. The study detected that regardless of the pipe diameter-to-thickness ratio, the live load of vehicles was insignificant for soil depths more than 2.5 m. Similarly, Orton et al. (2015) [26] investigated 10 reinforced concrete box culverts, and the soil fill depth ranged from 0.76 to 4.1 m. The study found that the live load of vehicles was not significant when the soil fill depth increased. In addition, the experimental results indicate that the deflection of culverts subjected to soil fill of less than 2.4 m is overestimated by the American Association of State Highway and Transportation Officials (AASHTO) load resistance design factors (LRFD) bridge design specifications (AASHTO 2012 [27]).

The experimental investigation of the structural performance of pipelines and culverts under a high soil fill depth is uneconomical and difficult to conduct. Finite element modeling is an effective alternative tool. For instance, Shatnawi et al. (2017) [28] effectively used the finite element analysis (CANDE software) to explore the behavior of various box culvert geometries under a high soil fill depth. This paper used CANDE-2019 [29] as the FEA tool to perform the parametric study.

In this study, FEA was used to investigate the structural performance of reinforced concrete pipelines under various soil fill depths (9.1, 12.2, 15.2, and 18.3 m) with different pipe diameters (610, 915, 1220, and 1524 mm). A soil depth of 18 m may be reached in deep embankment soil; an example of deep-buried culverts is the Peter Smith Brook Culvert with a 23.4 m deep embankment soil, which is a part of a highway upgrade near Longs Creek, New Brunswick, Canada [30]. This study provides a guideline for engineers to estimate the most efficient pipeline geometry for various soil filling depths. In addition, this research also explored the optimum diameter-to-thickness ratio for various pipeline conditions. The shear reinforcement in this study was excluded, as Yee (2003) [31] found that the provided shear reinforcement in reinforced concrete culverts did not improve the load capacity of the culverts, but the culvert section became over-reinforced. Shatnawi et al., 2017 [28], previously published work related to the investigated topic, and thus this study aimed to compare the obtained results with the previously gained ones.

2. Methodology

Using finite element analysis (FEA), this study investigated reinforced concrete (RC) buried pipelines under various soil depths using finite element analysis (FEA). Culvert Analysis and Design (CANDE-2019 software [29]) was used to conduct the nonlinear FEA. The description of CANDE-2019 [29] is discussed in Section 3.

The following points represent the most critical features of the executed FEA modeling. Various pipe diameters, as shown in Table 1, were investigated, where Pipe 24, Pipe 36, Pipe 48, and Pipe 60 were 610, 915, 1220, and 1524 mm, respectively. The geometric properties of the RC pipelines are illustrated in Figure 1.

Table 1. Dimensions of the pipes used in the analysis.

Pipe Designation	Pipe 24	Pipe 36	Pipe 48	Pipe 60
Pipe diameter, mm (in.)	610 (24)	915 (36)	1220 (48)	1524 (60)

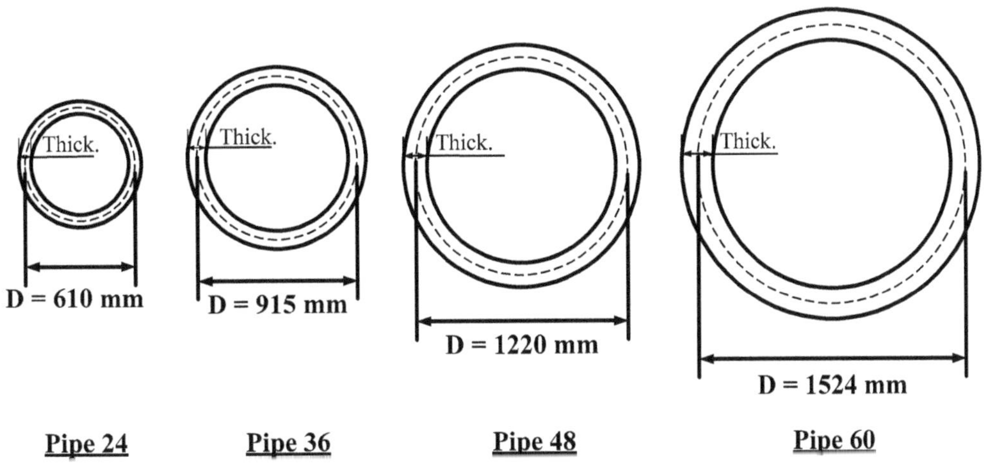

Figure 1. Geometric dimensions of the investigated pipeline culverts in this study.

- Varying soil fill depths (h) were investigated: 9.1, 12.2, 15.2, and 18.3 m.
- Flexural steel reinforcement (FSR) was provided in the RC pipelines. However, no shear reinforcement was provided.
- Duncan et al. (1980 [32]) and Selig (1988 [33])'s hyperbolic stress–strain parameters were used for the nonlinear FEA soil properties.
- The parametric FEA study investigated various variables (e.g., soil fill depth, pipeline diameter, and pipeline thickness).

Fifty-eight FEA runs were executed to calculate the FSR for each pipeline model, as indicated in Table 2. In each FEA model run, the following process was carried out:

1. The soil fill depth was defined considering the values in Table 2.
2. The diameter of the RC pipeline was selected from the dimensions specified in Table 1.
3. The thickness of the RC pipeline was reduced in each FEA model until the ratio of applied-to-capacity shear stress (τ_a/τ_c) was equal to or less than the unity to exclude the shear reinforcement and reduce the cost of constructing the pipelines. In addition, the thickness reduction process was stopped once the maximum crack width (w_{max})

reached 0.25 mm, which allowed the pipeline to have small crack width at the ultimate loading state.

4. The FEA results were summarized and analyzed to investigate the effect of soil depth on the diameter and thickness of the reinforced concrete pipeline.
5. The total required FSR and the maximum deflection (Δ_{max}) of each pipe were recorded.

Table 2. Thickness and soil fill depth of the analyzed reinforced concrete pipes.

Soil Fill Height h, m (ft)	Reinforced Concrete Pipe Thickness mm (in)			
	Pipe 24	Pipe 36	Pipe 48	Pipe 60
9.1 (29.9)	102 (4) 152 (6) 203 (8) –	102 (4) 152 (6) 203 (8) –	102 (4) 152 (6) 203 (8) 254 (10)	109 (4.3) 152 (6) 203 (8) 254 (10)
12.2 (40)	102 (4) 152 (6) 203 (8) –	114 (4.5) 152 (6) 203 (8) –	102 (4) 152 (6) 178 (7) 254 (10)	152 (6) 254 (10) 305 (12) –
15.2 (49.9)	102 (4) 152 (6) 203 (8) – –	203 (8) 254 (10) 312 (12.3) – –	190 (7.5) 305 (12) 363 – –	152 (6) 215 (8.5) 254 (10) 317 (12.5) 363 (14.3)
18.3 (60)	89 (3.5) 127 (5) 152 (6) 203 (8) –	152 (6) 254 (10) 312 (12.3) – –	152 (6) 203 (8) 254 (10) 284 (11.2) 363 (14.3)	152 (6) 210 (8.3) 262 (10.3) 318 (12.5) 363 (14.3)

As discussed in Step 3 in the procedure mentioned above, the shear reinforcement was omitted. Furthermore, it was excluded because the ratio of applied-to-capacity shear stress was equal to or less than unity. The flowchart summary of the methodology in this study is shown in Figure 2.

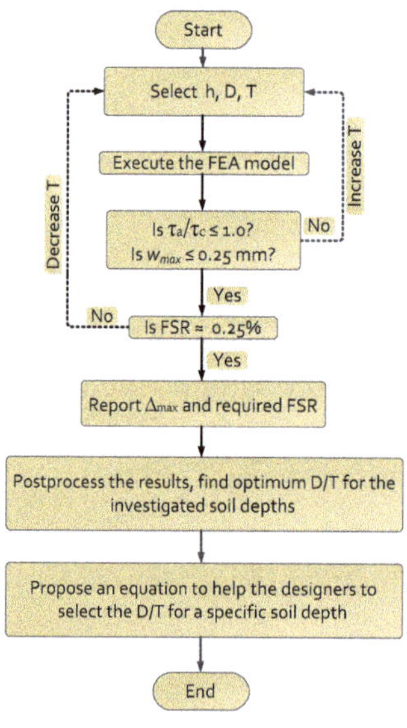

Figure 2. Flow chart to summarize the methodology of this study.

3. FEA Model and Construction Stages

Figure 3 represents the FEA model's geometry, boundary conditions, and mesh. The soil layers are the in situ soil, bedding, and fill soil. As the pipeline is symmetric about the pipe centerline, one-half of the pipe and soil layers were modeled. The pipeline was simulated using 10 equal-length beam elements connected between 11 nodes. The 2D quadrilateral elements of the modeled soil layers were defined below and above the pipeline up to three times the pipeline radius. The load factor of 1.42 was defined according to AASHTO 2020 [34]. Therefore, the load pressure equals the factor 1.42 multiplied by the soil fill unit weight and soil fill depth. The live load was excluded from the FEA as all of the investigated cases in this study were subjected to a soil depth more than the 2.44 m-threshold proposed by AASHTO 2020 [34], as the effect of live load on culverts diminishes for soil depths of more than 2.44 m. The bottom layer of the in situ soil was restrained in the horizontal and vertical directions.

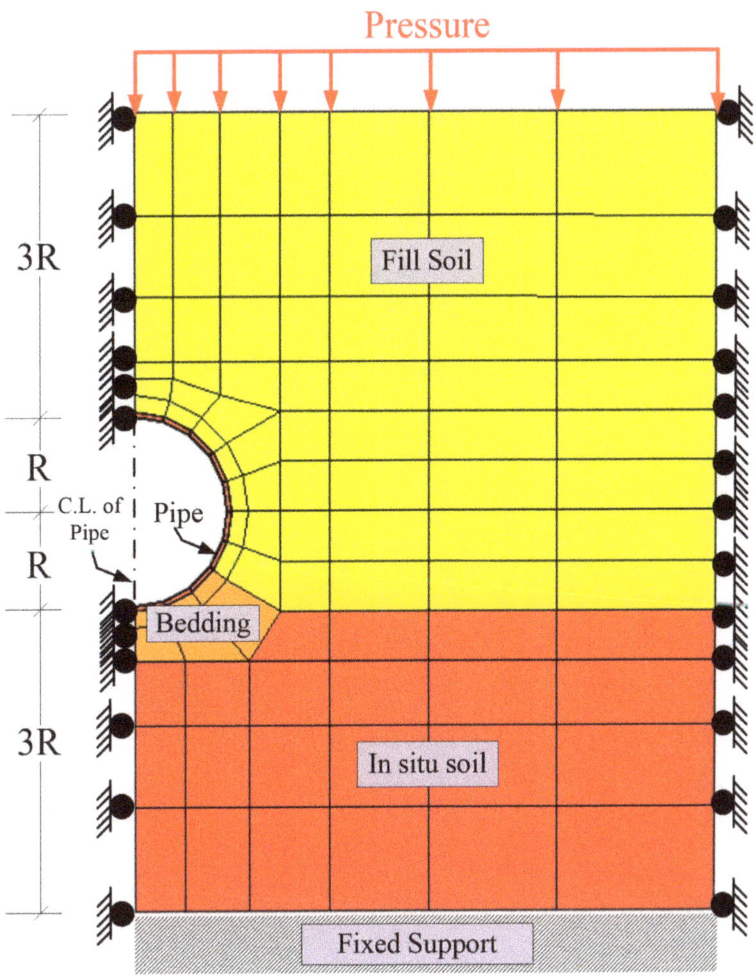

Figure 3. FEA model soil layers, boundary conditions, and loading pressure.

In contrast, the left and right sides of the model were restrained in the horizontal direction only because the model was symmetric at the pipeline's centerline. The FEA model considered the soil–structure interaction, as shown in Figure 3. It is worth mentioning that an extensive investigation of the literature was conducted. However, the literature lacked experimental results related to this study. On the other hand, the used model was reliable and was validated using the project developed under the National Cooperative Highway Research Project NCHRP 15–28 [35]. In this study, the Level 2 option (automatic FEA) automatically created the nodes and the optimum mesh size of the FEA model using simple input parameters; this option made the modeling process easy to implement and reduced the required time to generate debug for the mesh of the FEA models. The automatic mesh selected coarse mesh size in the soil layers and fine mesh size in the pipeline geometry, as we were more concerned about the response of the pipeline.

It is imperative to simulate the construction stages of the reinforced concrete pipelines in order to model the structural responses of the pipelines accurately. Therefore, the pipeline, bedding, and in situ soil layer were modeled in the first stage, while the soil fill was added gradually in the second, third, fourth, and fourth stages, respectively. On the other hand, to increase the analysis accuracy and avoid solution divergence, the loading pressure was divided into several equal increments applied to the model in stages 6 to 20, as shown in Figure 4.

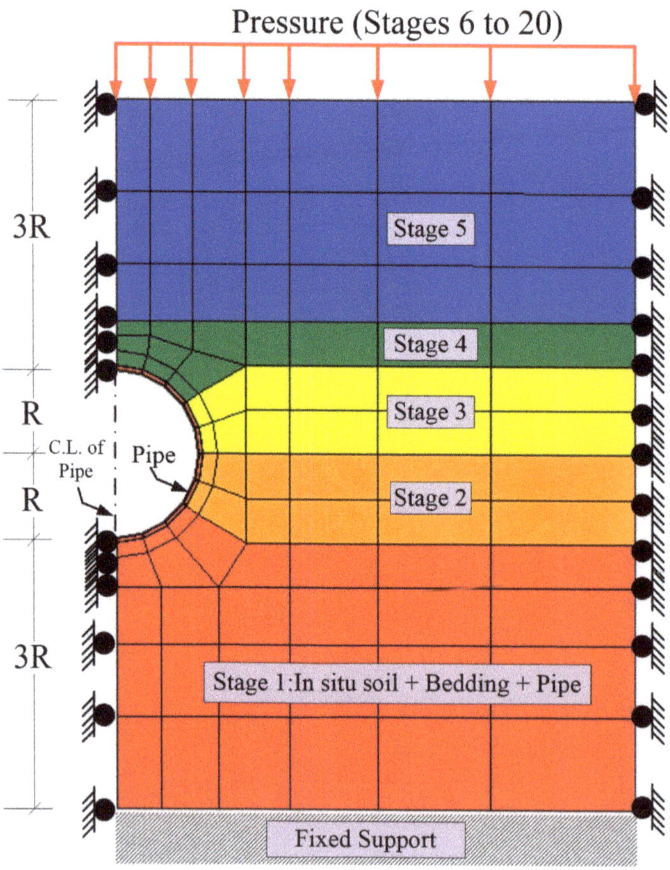

Figure 4. Construction and pressure stages of the FEA model.

3.1. FEA Soil Properties

We utilized three types of soil layers in the FEA model. The Duncan/Selig parameters were proposed by Duncan et al. (1980) [32] and were modified by Selig (1988) [33], where SW95 and SW90 gravely sand with 95% compaction was used to define the FEA modeling. The SW95 and SW90 soil were used in the fill soil and bedding. The library of CANDE-2019 contains the hyperbolic stress–strain parameters of SW95 and SW90, as summarized in Table 3 [36]. The in situ soil was considered isotropic, where the soil's young's modulus was 34.5 MPa, Poisson's ratio was 0.4, and soil density was 18.9 kg/m^3. CANDE-2019 offers various culvert installations, including barrier and trench installations. In general, the installation type refers to the location of the culvert relative to the ground level. In the embankment installation, the bottom face of the culvert coincides with the original ground level, where soil excavation is not needed. While trench installation refers to the case when the bottom face of the culvert is below the original soil level, in this case, soil excavation is needed. In this study, embankment installation was considered in the FEA modeling.

Table 3. Duncan/Selig soil parameters used in the FEA analysis.

Parameter	Parameter Symbol	Soil Properties	
		SW95 (Fill Soil)	SW90 (Bedding)
Young's tangent	K (kPa)	950	640
Modulus parameters	n	0.6	0.43
	C (kPa)	0	0
	ϕ_0 (Degree)	48	42
	$\Delta\phi$ (Degree)	8	4
	R_f	0.7	0.75
Bulk parameters	B_i/P_a	74.8	40.8
	ε_u	0.02	0.05
Soil unit weight	γ (kN/m^3)	18.9	18.9

Note: K = modulus number; n = modulus exponent; R_f = failure ratio; C = cohesion; ϕ_0 = initial internal friction angle; $\Delta\phi$ = reduction in friction angle; B_i = initial tangent bulk modulus; P_a = atmosphere pressure; and ε_u = ultimate volumetric strain.

3.2. FEA Steel and Concrete Properties

Normal strength concrete with a compressive stress of 34.5 MPa and unit weight of 23.6 kg/m^3 was utilized, the 448 MPa tensile strength steel reinforcement with a concrete cover of 32 mm was used, and more details of the concrete and steel properties are summarized in Table 4.

Table 4. Concrete and steel reinforcement properties.

Concrete Properties	
Unit weight (kN/m^3)	23.6
Compressive strength (MPa)	34.5
Young's modulus (MPa)	29,557
Poison's ratio	0.17
Steel Reinforcement Properties	
Yield stress (MPa)	448
Young's modulus (MPa)	200,000
Poison's ratio	0.3
Concrete cover to the centerline of steel rebar cage, mm	32

4. FEA Results and Discussions

The parametric study was conducted using CANDE 2019 software. The deformation response and required flexural steel reinforcement (FSR) were recorded and analyzed. The following two sections illustrate the FEA modeling outputs and the detailed analysis.

4.1. Deformed FEA Model

Figure 5 shows the FEA model's deformed shape representing a reinforced concrete pipeline with a diameter of 1524 mm, pipeline thickness of 152 mm, and soil fill depth of 18.3 m. Nodes A, B, C, D, E, F, G, H, L, M, and N represent various in situ soil, bedding, and soil fill, while points 1 to 11 connect the pipeline elements. The vertical displacement (Y-displacement) of points A, B, C, and D was 0, −6.1, −10.3, and −14.4 mm, respectively. It is worth mentioning that the vertical displacement was cumulative, e.g., the vertical displacement at point C was equal to the vertical displacement at point B plus the local deflection at point B, which resulted from the stress demand at point C. Similarly, the net deflection that occurred in the pipe can be calculated using the relative displacement between points 1 and 11, which represent the extreme top and bottom points of the pipeline. In the case of Figure 5, the pipeline experienced a deflection of −39.2 mm − (−31.6 mm) = −7.6 mm. The summary of the vertical displacement for points A to N and 1 to 11 are tabulated in Figure 5.

(b) Deformation of pipeline

Node	X-displacement (mm)	Y-displacement (mm)
1	0.0	−39.2
2	0.1	−38.7
3	0.7	−37.5
4	1.9	−36.2
5	3.1	−35.5
6	3.6	−35.4
7	3.1	−35.3
8	1.8	−34.6
9	0.7	−33.4
10	0.1	−32.1
11	0.0	−31.6

(c) Deformation of soil layers

Node	Y-displacement (mm)
A	0
B	−6.1
C	−10.3
D	−14.4
E	−42.6
F	−52.9
G	−65
H	−77.1
L	−41.7
M	−31.9
N	−20.3

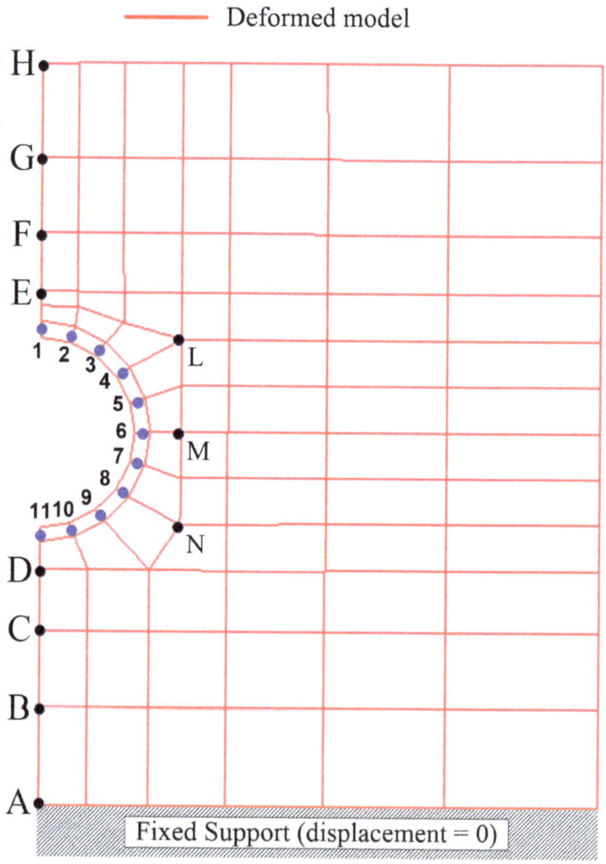

(a) Model deformation

Figure 5. Sample of FEA model results for a soil fill depth of 18.3 m, diameter of 1524 mm, and pipe thickness of 152 mm: (**a**) deformed model, which includes the pipeline elements with nodes 1 to 11 and selected soil nodes A to N, (**b**) nodal displacement of the pipeline structural elements, and (**c**) Y-displacement of selected soil nodes.

The pipeline deflection for the investigated parametric study can be plotted using the procedure mentioned above, as shown in Figure 6, for soil fill depths of 9.1, 12.2, 15.2, and 18.3 m. Figure 6a shows that if Pipe 60 with a thickness of 254 mm was decreased to 203 mm, 152 mm, or 109 mm, the pipeline deflection increased from 5.5 mm to 6.3 mm, 9.2 mm, or 13 mm, respectively. The same pipeline response trend was noticeable for Pipe 24, Pipe 36, and Pipe 60. The FEA model considers three different variables during the analysis process: pipeline deflection, maximum crack width, and FSR. The maximum crack width must not exceed the 0.25 mm limit, and if this limitation cannot be achieved, the model increases the required FSR, which reduces the crack width and pipeline deflection.

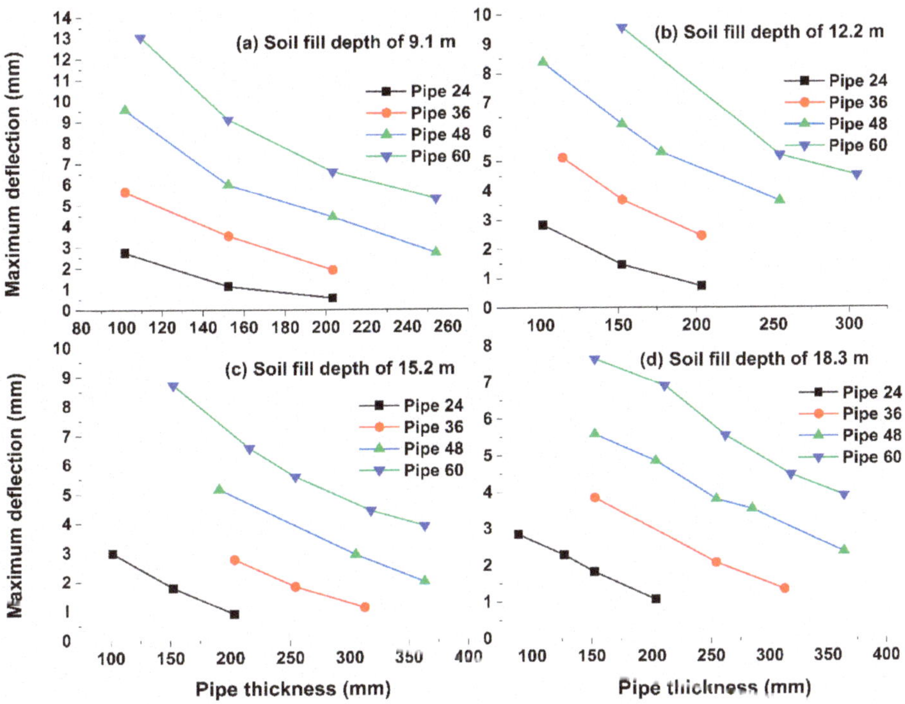

Figure 6. Maximum deflection versus reinforced concrete pipe thickness for various soil fill depths.

4.2. Flexural Steel Reinforcement (FSR)

The parametric study was performed using CANDE-2019 according to the procedure described in the Methodology. A total of 58 FEA model runs were conducted, as specified in Table 2. In each model run, the required FSR and maximum deflection were reported. Figure 7 illustrates the required FSR corresponding to various reinforced concrete pipeline thicknesses for the soil depths of 9.1, 12.2, 15.2, and 18.3 m. Figure 7a shows the effect of reducing the pipe thickness on the FSR for soil fill depths of 9.1 m. In contrast, the required FSR for Pipe 60 with a cross-section thickness of 254 mm required an FSR of 0.26%; however, if the cross-section of Pipe 60 was reduced to 203, 152, and 109 mm, the required FSR increased to 0.35%, 0.5%, and 0.65%, respectively. Therefore, considering that the most expensive component of the pipe is the steel reinforcement, the optimum thickness of Pipe 60 under soil fill of 9.1 m was 254 mm, as it requires the least amount of steel reinforcement.

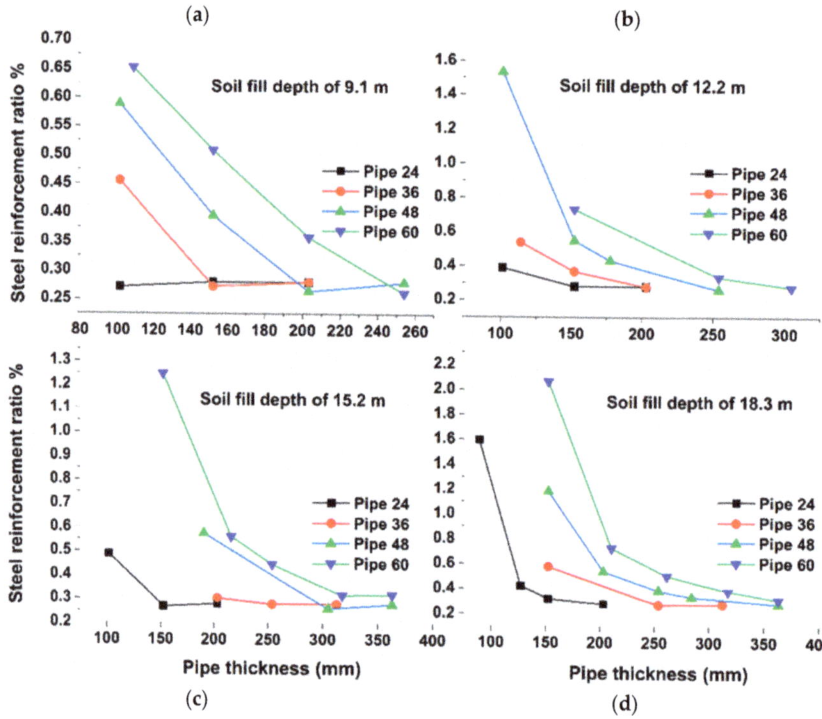

Figure 7. Required steel ratio versus reinforced concrete pipe thickness for various soil fill depths.

Similarly, for Pipe 48, changing the pipe thickness from 254 to 203, 152, or 102 mm, increased the required FSR to 0.26%, 0.39%, and 0.59%, respectively. As a thickness of 203 and 254 mm required the same amount of reinforcement (FSR = 0.26%), the thickness of 203 mm was considered to be optimum, as it reduced the amount of RC. The same criteria can be applied to Pipe 36 and Pipe 24, where the optimum thicknesses that required the least amount of steel reinforcement were 152 and 102 mm, respectively. The same procedure was applied for soil fill depths of 12.2, 15.2, and 18.3 m, as illustrated in Figure 7b–d. The summary of the optimum thickness for all cases is listed in Table 5.

Table 5. The optimum thickness of the analyzed reinforced concrete pipes.

Soil Depth (m)	The Optimum Thickness of the Reinforced Concrete Pipe (mm)			
	Pipe 24	Pipe 36	Pipe 48	Pipe 60
9.1	102	152	203	254
12.2	152	203	254	304
15.2	152	203	305	363
18.3	203	254	284	363

4.3. Optimum RC Pipeline Diameter-to-Thickness Ratio

In Section 4.2, identifying the optimum pipeline thickness is illustrated. The summary of the optimum pipeline thickness for various investigated pipeline diameters and soil fills is listed in Table 5. Moreover, the normalized optimum diameter-to-thickness (D/T) ratios from Table 5 were calculated and are listed in Table 6. For example, the optimum thickness of Pipe 24 under a soil fill depth of 9.1 m was 102 mm. Thus, the optimum D/T is 6.0. This procedure was performed for the remaining pipeline geometries and soil fill depths, as shown in Table 6. The last column in Table 6 lists the average optimum D/T ratios for various pipeline diameters for soil fill depths of 9.1, 12.2, 15.2, and 18.3 m. For instance, the average optimum D/T at a soil depth of 15.2 m is the average of 4, 4.5, 4, and 4.2, which equals 4.2.

Table 6. Optimum diameter-to-thickness ratio (*D/T*) of the analyzed reinforced concrete pipes.

Soil Depth (m)	The Optimum *D/T* Ratio of the Reinforced Concrete Pipe				Average *D/T*
	Pipe 24	Pipe 36	Pipe 48	Pipe 60	
9.1	6.0	6.0	6.0	6.0	6.0
12.2	4.0	4.5	4.8	5.0	4.6
15.2	4.0	4.5	4.0	4.2	4.2
18.3	3.0	3.6	4.3	4.2	3.8

Figure 8 illustrates the average optimum D/T ratio for various soil fill depth values; the data were based on the last column in Table 6. The figure reveals that the optimum D/T ratio linearly decreased with the soil fill depth, and the optimum D/T could be related to the soil depth using Equation (1). This equation is very important for pipeline designers to select the optimum pipeline geometry for various soil fill depths.

$$\text{D/T ratio} = -0.232 \times \text{soil fill depth (m)} + 7.8 \quad (1)$$

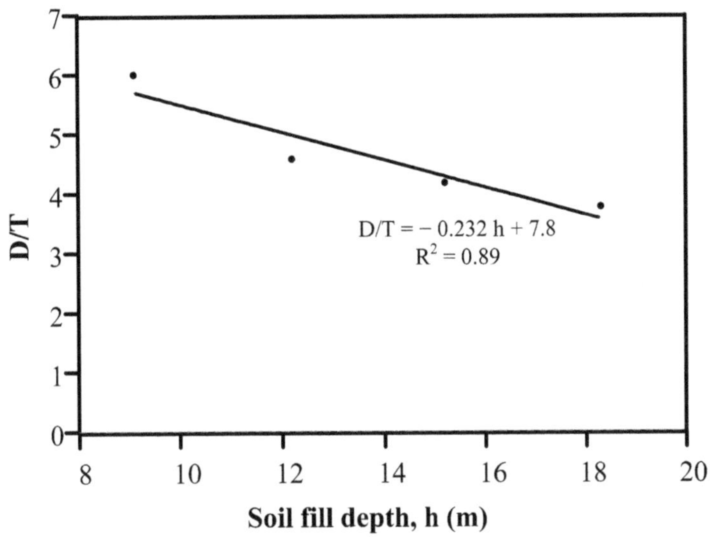

Figure 8. Soil fill depth versus optimum D/T ratio.

4.4. Designers Aid in Selecting Optimum RC Pipeline Thickness

This section includes a practical example to assist designers and practitioners in benefitting from this study and ensuring that the findings are generalizable. The example predicted the best thickness for pipes with diameters of 500, 750, 1000, 1250, and 1500 mm, with soil fill depths ranging from 2.4 m to 20 m taken into account, as shown in Figure 9. The optimal pipeline thickness was determined using Equation (1). According to Figure 9, the ideal thickness was determined by the pipe diameter and soil fill depth. For the examined pipe diameters up to 10 m, the optimal reinforced concrete pipe thickness was directly proportional to the soil depth.

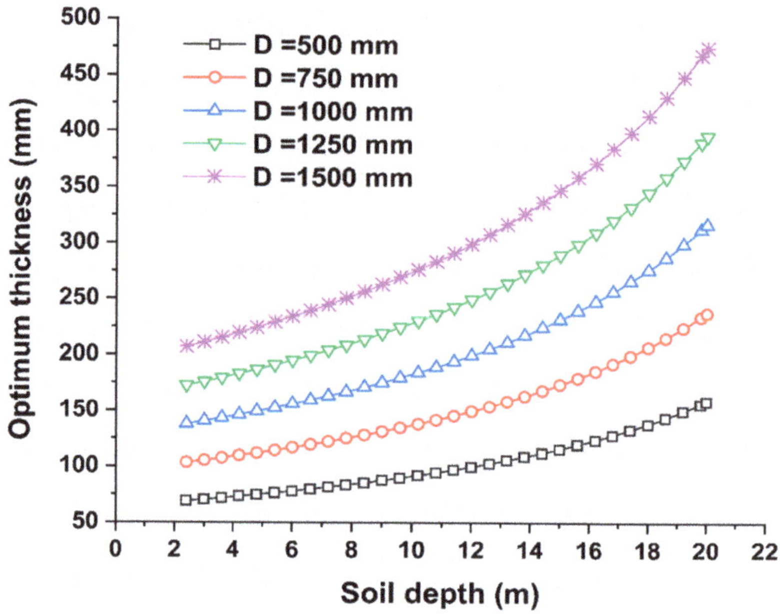

Figure 9. Optimum pipeline thickness under various soil fill depths and pipeline diameters.

However, beyond 10 m of soil depth, the ideal pipeline thickness was nonlinearly related to the soil depth, with the increasing rate in this stage (i.e., for soil depth greater than 10 m) being more significant than the slope of the first stage (i.e., for soil depth less than 10 m). Furthermore, the ideal thickness for a pipeline diameter of 500 mm at a soil depth of 2.4 m was 70 mm, while the optimum thickness for the soil fill when raised to 20 m was 160 mm. However, for pipe diameters of 1500 mm, the optimal thickness was 210 mm and 475 mm at soil depths of 2.4 and 20 m, respectively, indicating that the optimum thickness was much higher for big pipe diameters under a high burial soil depth. For example, the D/T ratio for a 1500-mm pipe diameter at a soil depth of 20 m was 1500 mm/475 mm = 3.15, but the D/T ratio at a soil depth of 2.4 m was 1500 mm/210 mm = 7.14. As previously indicated in Figure 8, the D/T ratio had an inverse relationship with the soil fill depth. The same holds for pipe diameters of 500, 750, 1000, 1250, and 1500 mm, where the optimal D/T decreased from 7.14 to 3.15 for soil depths of 2.4 and 20 m, respectively. As a result, regardless of the pipe diameter, the optimal pipe thickness rose (nearly doubles) as the soil depth increases from 2.4 m to 20 m. This example shows how designers and practitioners utilize Equation (1) to determine the best reinforced concrete pipeline for a given pipe geometry and soil depth. Thus, in construction projects, providing decision-support tools using machine and deep learning approaches is vital [37–48].

5. Summary and Conclusions

The current practice does not include guidelines for using the most efficient reinforced concrete pipeline geometry under deep soil embankments. To fill this gap and propose a guideline for designing reinforced concrete pipelines, this study investigated the performance of reinforced concrete pipelines of various diameters buried under soil fill depths of 9.1 to 18.3 m. High costs accompany testing or inspecting pipelines. Thus, a tool such as finite element analysis is critical for reducing such costs; CANDE 2019 software was used here as the FEA tool to conduct the parametric study. As a result, 58 FEA runs were executed, whereas the flexural steel reinforcement (FSR) and the maximum pipeline deflection in each FEA run were reported. In addition, the optimum pipeline diameter-to-thickness (D/T) ratio that had the least required FSR and crack width did not exceed 0.25 mm, which was specified for various pipeline diameters and soil fill depths. The Duncan/Selig soil parameters SW90 and SW95 were used to define the soil properties for the in situ bedding and fill layers. The analysis results revealed that the optimum reinforced concrete pipeline diameter-to-thickness (D/T) ratio was 6.0, 4.6, 4.2, and 3.8 for soil fill depths of 9.1, 12.2, 15.2, and 18.3 m, respectively. In addition, the deflection range of the investigated reinforced concrete pipelines was between 0.5 to 13 mm, which indicates that the finite element analysis carefully selected the pipeline thickness, required FSR, and concrete crack width. In contrast, the pipeline did not undergo excessive deformation.

A detailed example was presented to help designers and practitioners select the optimum thickness for various soil depths (2.4 to 20 m) and different pipeline diameters (500, 750, 1000, 1250, and 1500 mm), which ensures the generalization of the proposed equation for a variety of pipeline conditions. The study proposed an equation to select the optimum diameter-to-thickness ratio for various soil fill depths, which will help practitioners select cost-effective pipeline geometry and steel reinforcement at a specific soil fill depth. The optimum pipeline thickness was significantly sensitive to the soil fill depth. In contrast, the optimum reinforced concrete pipeline thickness was substantially increased (almost doubled) when the soil fill depth increased from 2.4 to 20 m. Field data collection of pipelines is required for further research to reach a solid stage in order for more analytical investigations. It is thus recommended to use deep and machine learning algorithms to improve the analysis accuracy and pave the road for researchers to deeply validate the proposed model performance.

Author Contributions: Conceptualization, O.A., A.S. and O.A.H.; Formal analysis, O.A.H. and A.S.A.; Investigation, G.A.; Methodology, O.A., G.A., A.S. and A.S.A.; Resources, A.S.A.; Supervision, O.A., G.A. and A.S.; Writing—original draft, O.A. and O.A.H. All authors have read and agreed to the published version of the manuscript.

Funding: The authors extend their appreciation to the Deanship of Scientific Research at King Khalid University for funding this work through the Large Groups Project under grant number (RGP. 2/178/43).

Conflicts of Interest: The authors declare no conflict of interest.

References

1. ASCE's 2021 Infrastructure Report Card | GPA: C-. 2021. Available online: http://www.infrastructurereportcard.org. (accessed on 12 September 2021).
2. Kang, J. Finite element analysis for deeply buried concrete pipes in proposed imperfect trench installations with expanded polystyrene (EPS) foams. *Eng. Struct.* **2019**, *189*, 286–295. [CrossRef]
3. Kang, J.; Parker, F.; Yoo, C.H. Soil–structure interaction for deeply buried corrugated steel pipes Part I: Embankment installation *Eng. Struct.* **2008**, *30*, 384–392. [CrossRef]
4. Kang, J.; Parker, F.; Yoo, C.H. Soil–structure interaction for deeply buried corrugated steel pipes Part II: Imperfect trench installation. *Eng. Struct.* **2008**, *30*, 588–594. [CrossRef]
5. Phuor, T.; Harahap, I.S.H.; Ng, C.Y.; Al-Bared, M.A.M. Development of the skew boundary condition for soil-structure interaction in three-dimensional finite element analysis. *Comput. Geotech.* **2021**, *137*, 104264. [CrossRef]
6. Zanni, A.A.; Spyridis, M.S.; Karabalis, D.L. Discrete model for circular and square rigid tanks with concentric openings—Seismic analysis of a historic water tower. *Eng. Struct.* **2020**, *211*, 110433. [CrossRef]
7. Pimentel, M.; Costa, P.; Félix, C.; Figueiras, J. Behavior of reinforced concrete box culverts under high embankments. *J. Struct. Eng.* **2009**, *135*, 366–375. [CrossRef]
8. Boksmati, J.I.; Madabhushi, G.S.; Thusyanthan, I.N. Dynamic soil-structure interaction of a shallow founded shear frame and a frame equipped with viscous dampers under seismic loading. *Eng. Struct.* **2021**, *227*, 111388. [CrossRef]
9. Bolisetti, C.; Whittaker, A.S. Numerical investigations of structure-soil-structure interaction in buildings. *Eng. Struct.* **2020**, *215*, 110709. [CrossRef]
10. Chee, K.H.; Krauthammer, T.; Astarlioglu, S. Shallow buried RC structures behavior under airblast in the time and P-I domains. *Eng. Struct.* **2020**, *203*, 109885. [CrossRef]
11. Dancygier, A.N.; Yankelevsky, D.Z. A Soft Layer to Control Soil Arching above a Buried Structure. *Eng. Struct.* **1996**, *18*, 378–386. [CrossRef]
12. Ding, H.; Kaup, A.; Wang, J.-T.; Lu, L.-Q.; Altay, O. Real-time hybrid simulation framework for the investigation of soil-structure interaction effects on the vibration control performance of shape memory alloys. *Eng. Struct.* **2021**, *243*, 112621. [CrossRef]

13. Fernanda, D.B.; Domizio, M.; Ambrosini, D.; Curadelli, O. Influence of various parameters in the seismic soil-structure interaction response of a nuclear power plant. *Eng. Struct.* **2020**, *217*, 110820.
14. Gong, J.; Zou, D.; Kong, X.; Liu, J.; Qu, Y. An approach for simulating the interaction between soil and discontinuous structure with mixed interpolation interface. *Eng. Struct.* **2021**, *237*, 112035. [CrossRef]
15. Huang, Y.; Gu, M.; El Naggar, M.H. Effect of soil-structure interaction on wind-induced responses of supertall buildings with large pile groups. *Eng. Struct.* **2021**, *243*, 112557. [CrossRef]
16. Pham, T.A.; Tran, Q.-A.; Villard, P.; Dias, D. Geosynthetic-reinforced pile-supported embankments—3D discrete numerical analyses of the interaction and mobilization mechanisms. *Eng. Struct.* **2021**, *242*, 112337. [CrossRef]
17. Rachedi, M.; Matallah, M.; Kotronis, P. Seismic behavior & risk assessment of an existing bridge considering soil-structure interaction using artificial neural networks. *Eng. Struct.* **2021**, *232*, 111800. [CrossRef]
18. Terzi, V.G.; Athanatopoulou, A. Influence of soil structure interaction effects on the real elastic axis of asymmetric buildings. *Soil Dyn. Earthq. Eng.* **2021**, *146*, 106775. [CrossRef]
19. Venanzi, I.; Salciarini, D.; Tamagnini, C. The effect of soil–foundation–structure interaction on the wind-induced response of tall buildings. *Eng. Struct.* **2014**, *79*, 117–130. [CrossRef]
20. Marston, A.; Anderson, A.O. *The Theory of Loads on Pipes in Ditches and Tests of Cement and Clay Drain Tile and Sewer Pipes*; Iowa Engineering Experiment Station: Ames, IA, USA, 1913.
21. Marston, A. *The Theory of External Loads on Closed Conduits in the Light of the Latest Experiments*; Iowa Engineering Experiment Station: Ames, IA, USA, 1930.
22. Sprangler, M.G. *Field Measurements of the Settlement Ratios of Various Highway Culverts*; Iowa Engineering Experiment Station: Ames, IA, USA, 1950.
23. Sprangler, M.G. Discussion to the modification of the pressures on rigid culverts with fill procedures. *Highw. Res. Rec.* **1968**, *249*, 41–43.
24. Abuhajar, O.; El Naggar, H.; Newson, T. Static soil culvert interaction the effect of box culvert geometric configurations and soil properties. *Comput. Geotech.* **2015**, *69*, 219–235. [CrossRef]
25. Bashir, R. Analysis and Design of Buried Pipelines. Master's Thesis, King Fahd University of Petroleum and Minerals, Dhahran, Saudi Arabia, 2000.
26. Orton, S.L.; Loehr, J.E.; Boeckmann, A.; Havens, G. Live load effect in reinforced concrete box culverts under soil fill. *J. Bridge Eng.* **2015**, *20*, 04015003. [CrossRef]
27. AASHTO. *LRFD Bridge Design Specifications*, 7th ed.; AASHTO: Washington, DC, USA, 2012.
28. Shatnawi, A.; Almasabha, G.; Tarawneh, B. Structural behavior of concrete box culverts under deep burial. *J. Pipeline Syst. Eng. Pract.* **2017**, *8*, 04017025. [CrossRef]
29. *CANDE-2019*, Computer Software; Transportation Research Board: Washington, DC, USA, 2019.
30. Parker, B.A.; McAffee, R.P.; Valsangkar, J. Field performance and analysis of 3-m-diameter induced trench culvert under a 19.4-m soil cover. In *Transportation Research Record 2045*; TRB, National Research Council: Washington, DC, USA, 2008; pp. 68–76.
31. Yee, R.A. Shear Behaviour of Concrete Box Culverts. Master's Thesis, University of Toronto, Toronto, ON, Canada, 2013.
32. Duncan, J.M.; Byrne, P.; Wong, K.S.; Mabry, P. *Strength, Stress-Strain and Bulk Modulus Parameters for Finite Element Analysis of Stresses and Movements in Soil Masses*; Geotechnical Engineering Rep. No. UCB/GT/80-01; Univesity of California: Berkeley, CA, USA, 1980.
33. Selig, E.T. Soil parameters for design of buried pipelines. In Proceedings of the Pipeline Infrastructure Conference, Boston, MA, USA, 6–7 June 1988; ASCE: Reston, VA, USA, 1988; pp. 99–116.
34. AASHTO. *LRFD Bridge Design Specifications*, 9th ed.; AASHTO: Washington, DC, USA, 2020.
35. National Academies of Sciences, Engineering, and Medicine. *Modernize and Upgrade CANDE for Analysis and LRFD Design of Buried Structures*; The National Academies Press: Washington, DC, USA, 2008.
36. Katona, M.G.; Mlynarski, M.; McGrath, T.J. *CANDE-2019 Culvert Analysis and Design Solution Methods and Formulations*; Transportation Research Board. Washington, DC, USA, 2019.
37. Alshboul, O.; Shehadeh, A.; Tatari, O.; Almasabha, G.; Saleh, E. Multiobjective and multivariable optimization for earthmoving equipment. *J. Facil. Manag.* **2022**, *ahead-of-print*. [CrossRef]
38. Shehadeh, A.; Alshboul, O.; Tatari, O.; Alzubaidi, M.A.; Hamed El-Sayed Salama, A. Selection of heavy machinery for earthwork activities: A multi-objective optimization approach using a genetic algorithm. *Alex. Eng. J.* **2022**, *61*, 7555–7569. [CrossRef]
39. Shehadeh, A.; Alshboul, O.; al Mamlook, R.E.; Hamedat, O. Machine learning models for predicting the residual value of heavy construction equipment: An evaluation of modified decision tree, LightGBM, and XGBoost regression. *Autom. Constr.* **2021**, *129*, 103827. [CrossRef]
40. Alshboul, O.; Shehadeh, A.; Al-Kasasbeh, M.; al Mamlook, R.E.; Halalsheh, N.; Alkasasbeh, M. Deep and machine learning approaches for forecasting the residual value of heavy construction equipment: A management decision support model. *Eng. Constr. Archit. Manag.* **2021**, *ahead-of-print*. [CrossRef]
41. Alshboul, O.; Shehadeh, A.; Hamedat, O. Development of integrated asset management model for highway facilities based on risk evaluation. *Int. J. Constr. Manag.* **2021**, 1–10. [CrossRef]

42. Shehadeh, A.; Alshboul, O.; Hamedat, O. A Gaussian mixture model evaluation of construction companies' business acceptance capabilities in performing construction and maintenance activities during COVID-19 pandemic. *Int. J. Manag. Sci. Eng. Manag.* **2022**, *17*, 112–122. [CrossRef]
43. Alshboul, O.; Shehadeh, A.; Hamedat, O. Governmental Investment Impacts on the Construction Sector Considering the Liquidity Trap. *J. Manag. Eng.* **2022**, *38*, 04021099. [CrossRef]
44. Almasabha, G. Gene expression model to estimate the overstrength ratio of short links. *Structures* **2022**, *37*, 528–535. [CrossRef]
45. Almasabha, G.; Alshboul, O.; Shehadeh, A.; Almuflih, A.S. Machine Learning Algorithm for Shear Strength Prediction of Short Links for Steel Buildings. *Buildings* **2022**, *12*, 775. [CrossRef]
46. Shehadeh, A.; Alshboul, O.; Hamedat, O. Risk Assessment Model for Optimal Gain-Pain Share Ratio in Target Cost Contract for Construction Projects. *J. Constr. Eng. Manag.* **2022**, *148*, 04021197. [CrossRef]
47. Alshboul, O.; Alzubaidi, M.A.; Mamlook, R.E.A.; Almasabha, G.; Almuflih, A.S.; Shehadeh, A. Forecasting Liquidated Damages via Machine Learning-Based Modified Regression Models for Highway Construction Projects. *Sustainability* **2022**, *14*, 5835. [CrossRef]
48. Alshboul, O.; Shehadeh, A.; Almasabha, G.; Almuflih, A.S. Extreme Gradient Boosting-Based Machine Learning Approach for Green Building Cost Prediction. *Sustainability* **2022**, *14*, 6651. [CrossRef]

materials

MDPI

Article

Mechanical, Physico-Chemical and Morphological Characterization of Energy Optimised Furnace (EOF) Steel Slag as Coarse Aggregate in Concrete

Arivoli Masilamani [1,*], Malathy Ramalingam [1], Parthiban Kathirvel [2,*], Gunasekaran Murali [3] and Nikolai Ivanovich Vatin [3]

1 Department of Civil Engineering, Sona College of Technology, Salem 636005, India; malathycivil@sonatech.ac.in
2 School of Civil Engineering, SASTRA Deemed University, Thirumalaisamudram, Thanjavur 613401, India
3 Peter the Great St. Petersburg Polytechnic University, 195251 St. Petersburg, Russia; murali_22984@yahoo.com (G.M.); vatin@mail.ru (N.I.V.)
* Correspondence: arivolisenthil@gmail.com (A.M.); parthiban@civil.sastra.edu (P.K.)

Abstract: This research tests energy optimised furnace (EOF) steel slag as substitution for natural coarse aggregate in concrete. Steel slag's usefulness as a substitute for natural coarse aggregate in concrete is the primary goal of this research. According to IS:2386-1963, the characterization of EOF steel slag, as coarse, is done by examining the shape and size of a particle, mechanical properties, physical properties, soundness, and alkali-aggregate reactivity. Tests for detection of staining material in steel slag and hardness of inter-facial transition zone in hardened cement paste were also carried out. The chemical analysis of the steel slag reveals the stability of oxides present in the steel slag. Microstructural characterization by SEM (scanning electron microscope) analysis of steel slag aggregate was also employed to support the characterization and XRD analysis, and it was found that the EOF steel slag is crystalline. The digital image processing technique (DIP) is adopted to study the shape indices, circularity, sphericity, shape factor, and roundness of natural and EOF steel slag aggregate. According to the characterization and strength investigation, steel slag aggregate outperforms natural coarse aggregate.

Keywords: steel slag; chemical composition; physical property; mechanical property; scanning electron microscope; digital image processing

check for updates

Citation: Masilamani, A.; Ramalingam, M.; Kathirvel, P.; Murali, G.; Vatin, N.I. Mechanical, Physico-Chemical and Morphological Characterization of Energy Optimised Furnace (EOF) Steel Slag as Coarse Aggregate in Concrete. *Materials* **2022**, *15*, 3079. https://doi.org/10.3390/ma15093079

Academic Editor: Ayman El-Zohairy

Received: 31 March 2022
Accepted: 19 April 2022
Published: 24 April 2022

1. Introduction

Concrete is a versatile composite material used for the worldwide constructions that is made up of coarse granular material known as aggregates that are strongly linked together with the assistance of cement to form a strong structure. The quality of aggregate used therein is the major contributing factor to the quality of concrete; proper proportioning and aggregate mixing influence the concrete strength and mechanical properties [1]. As the mechanical forces are received and conducted by the aggregate, they can be referred to as the skeleton of the concrete. Raw steel slag is produced as a by-product in the process of steel-making during the change of iron into steel [2]. In a blast furnace, molten slag and steel are formed by oxidation of molten iron, metals, scrap, and flux to remove impurities [3]. The slow cooling of molten slag converts it into a dense rock-like material. As per the Indian scenario, the steel industry generates a huge quantity of waste, which will increase enormously with increasing steel production. According to the Ministry of Environment and Forests and Climate Change (MOEFCC) and Central Pollution Control Board, India, steel slag is produced at a rate of 4 million tons annually, which are continuously discarded in open spaces. A typical blast furnace slag is produced at the rate of 300–400 kg per tonne of liquid iron produced in India. The quantity of steel slag produced has increased from 35.35 million tonne (out of 101.4 million tonne of steel production) in 2017–2018 to 53 million tonne (out of 150 million tonne of steel production) in 2019–2020, which tends to strengthen the utilization of 100% slag effectively and economically viable [4].

Nippon slag association [5] has been developing technology to resolve difficulties related to the consumption of steel slag. The mining, transportation, and processing of natural resources will be reduced if steel slag is utilized in place of natural aggregate. Additionally, there will be less disruption to the land, less energy consumption, less pollution of air, land, water, and less emission of greenhouse gases. Reducing the amount of natural aggregate in the concrete mix by replacing it with steel slag can enhance the sustainability of natural resources, since the extraction and transportation phases in the production of natural aggregate are eliminated. Steel slag, a waste product, forms the process of steel-making, requires a little energy to produce. The steel industry will be benefited, since their by-products will be used beneficially, instead of degrading the environment by unsafe dumping. The American National Slag Association (NSA) [6] established that iron and steel-making slag poses no health or environmental hazards. The great mechanical resistance of steel slag makes it superior to natural aggregate when used as an aggregate. Steel slag as a filler ingredient in concrete has been the subject of several investigations.

It has been identified that the free calcium oxide (CaO) present in the steel slag is the dominant cause of expansion as a result of the formation of calcium hydroxide and calcium carbonate when reacted with water [7]. By minimizing the free calcium oxide content in steel slag, the stress resulting from the expansion of free calcium oxide will be lowered and higher will be the stability of cement concrete. Steel slag undergoes mineral change at high temperatures, resulting in irregular volumetric expansion that reflected adversely on slag-based concrete mixtures' residual properties [8]. The harmful effect of free CaO could be minimized to acceptable limits by aging. During this aging, the steel slag is stockpiled for 4–6 months and hydration of free lime (CaO) takes place with atmospheric moisture. Aging for a long time mainly reduces the presence of harmful expansive materials in steel slag [9]. By studying the morphological character of and electric arc furnace steel slag and basic oxygen furnace (BOF) slag, it can be seen that slag particles have high sphericity, solid structure, and favorable frictional properties [10]. The scanning electron microscope test reveals the steel slags rough texture, which enhances the mechanical strength of concrete [11].

In building road foundations and anti-slippery blacktop, steel slag may be substituted for the limestone aggregates entirely to generate lightweight concrete, due to its hydraulic property [12]. The compressive, split tensile, and flexure strength were enhanced by between 4 and 8 percent in concrete made using steel slag as coarse aggregate, compared to the control mixes. The density of concrete is increased by 5 to 7% when using coarse aggregate that is 100 percent slag, rather than natural aggregate, as a reference [13]. Recent researchers reported that the split tensile, flexural, and compressive strength of steel slag aggregate concrete were higher than the control mix [7,14,15]. This shows that adding slag could satisfy the role of the accelerator at the initial stage, whereas, at 28 days, the effect diminished. Replacement of slag notches is the primary outcome [16]; additionally, the utilization of iron slag in concrete can augment the concrete strength properties [17]. Steel slag aggregate concrete has also inclined to decrease the water absorption characteristics and porosity after 28 days, compared to the reference mix with coarse aggregate in the form of stone chips [14].

The aggregate shape can profoundly characterize the aggregate quality, affecting concrete mixtures property. Daily, measuring the shape parameter using calliper devices is tiresome and time-consuming for quality control. Digital image processing (DIP) has been applied to characterize aggregate shape in recent years. The relationship between the DIP and experimental results was strong. It was inferred that the DIP method was a much faster and improved substitute for traditional particle shape measurement [18–20]. The surface roughness, shape, and surface area of fine aggregate were analyzed by data acquisition procedures [20]. The shape features of the aggregates were examined using

Image J, including elongation, flatness, roundness, aspect ratio, form factor, shape factor, and sphericity. The coarse aggregate's shape and texture greatly influenced the concrete slump. As the sphericity and form factor of aggregate increased, the compressive strength also increased [21,22]. Interfacial roughness, specific surface area, and aspect ratio of the lightweight aggregate significantly affected mechanical strength [23].

Many investigations have been conducted to see whether concrete aggregate may be replaced with industrial by-products. It is observed from the previous studies that the characteristics of steel slag largely depend on the production process of steel and the chemical composition of flux added [6,8,9]. EOF steel slag is a by-product of steel production industries. EOF steel slag is not used in cement production because it has no cementitious properties. The mechanical qualities of EOF steel slag are critical to its use, and they are intimately linked to concrete's physical, chemical, and morphological features. The chemical composition of the EOF steel slag influences the volumetric stability of the resulting material. Similarly, the EOF steel slag based concrete strength is a function of the mechanical properties of steel slag. This paper presents a thorough characterization of EOF steel slag to facilitate its use in concrete. In this context, an experimental investigation has been carried to evaluate the effectiveness of EOF steel slag on the fresh properties (such as the slump cone test), hardened properties (such as compressive strength), microstructural characteristics (such as scanning electron microscope), X-ray diffraction, and shape parameters (using digital image processing of concrete mixes) were carried out, and the results are summarized.

2. Material Properties

2.1. Cement and Sand (Fine Aggregate)

According to IS 8112-2013 [24], 43 grade ordinary Portland cement (OPC) was employed as a binder during the whole investigation. A sieve test was used to assess the cement fineness and observed that the residue on 90 μ sieve was 8%, which is less than 10%, as required for OPC. The cement has a specific gravity, standard consistency, and initial setting time were 3.16, 31%, and 35 min, respectively. The fine aggregate used in this investigation has been acquired from a riverbed, which has the advantage of a cubical shape with a smooth surface. The fine aggregate physical properties used in this investigation are detailed in Table 1, which satisfies the requirement of IS 383-2016 [25], since they have a more significant impact on the fresh concrete properties.

Table 1. Physical properties of Fine aggregates.

Description of Physical Properties	Obtained Value
Bulk density	1450 kg/m^3
Surface moisture	Nil
Fineness modulus	3.1 (zone II)
Water absorption	1.57%
Specific gravity	2.6

2.2. Coarse Aggregate and EOF Steel Slag

Construction-grade aggregate, made from locally sourced granite, was utilized to produce concrete. Aggregates of less than 20 mm in size were crushed and well-graded by machines. Steel-making operations produce this slag as a by-product when pig iron and steel scrap are transformed into steel. For its outstanding toughness and workability, this steel is highly prized. It is known that approximately 200 kg of slag is created for every ton of steel production. In JSW Steel Ltd., Salem Works (JSWS), 12,000–13,000 tons of energy optimized furnace (EOF) slag is produced per month and 200,000–300,000 m^3 of EOF slag is produced per annum. The utilization of steel slag in concrete delivers excellent mechanical properties.

2.3. Test Techniques of Coarse Aggregates

Scarcity in aggregate availability from natural sources is being faced in many situations. This problem can be overcome by using natural aggregate as minimum as possible by replacing it with industrial by-products, such as EOF steel slag. As aggregates are essential components for making concrete, and their properties are greatly affected by various characteristics of the aggregate. For this reason, it is vital to investigate the materials characteristics that are utilized in concrete production. This knowledge will help in choosing appropriate ingredients for good concrete quality. EOF steel slag characterization requires the following tests, and these tests were employed according to ASTM C641-98 [26], ASTM C123-03 [27] and IS: 2386:1963 [28–33]. A comparison is made between the methods and outcomes and the characteristics of natural coarse aggregate. Figure 1 discusses the systematic research framework used to characterize steel slag for use as coarse aggregate in concrete step-by-step.

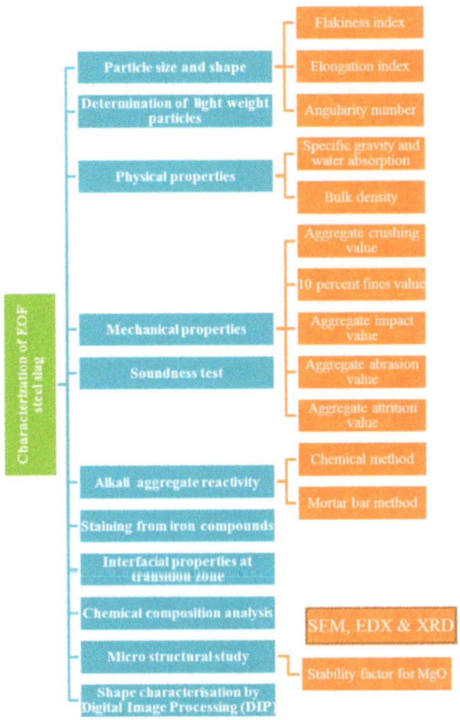

Figure 1. Framework of characterization study.

2.3.1. Particle Size and Shape

A suitable gradation of aggregate is a significant aspect of producing workable concrete, which contains all standard aggregate fractions in the required proportion. There are minimal voids that need little paste to fill the spaces between the aggregate in graded aggregates. It was necessary to reduce the nominal size of the raw and aged slag from steel mills to 20 mm, in order to make it easier to include in concrete mixes. Using sieve analysis, the gradation of EOF steel slag was determined according to IS 383-2016 [25], which indicates the coarse aggregate for concrete. The aggregate shape is an imperative characteristic because it dramatically influences the workability and consolidation of fresh concrete. EOF steel slag shape properties, such as the flakiness index, elongation index, and angularity number, are evaluated. With a greater length to thickness ratio than the specified values, the particles are elongated and flat. The flat and elongated particles tend to break easily, causing adverse effects in concrete. An aggregate's angularity, or the sharpness of its particles, is an essential feature, since it affects how easily an aggregate mixture may be handled. This supports strong interlocking, binding hardened concrete and stimulates workability of concrete at fresh state. The stability of a concrete mixture relies on the degree of particles interlocking.

2.3.2. Determination of Light Weight Particles

Lightweight particles affect durability and tend to cause pop-outs. Therefore, it is essential to determine the presence of lightweight particles by following the ASTM C 123 [27] guidelines. Zinc chloride salt was diluted in a sufficient quantity of distilled water, in order to get the required specific gravity of 2.4 for the solution. EOF steel slag sample was allowed to submerge in the prepared zinc chloride solution.

2.3.3. Physical Properties

All aggregates are porous, and the amount of liquid absorbed by an aggregate, when submerged in water, depends on its porosity. Aggregates water absorption value impacts water–cement ratio and fresh concrete workability [34]. EOF steel slag has a water absorption rate of around 3%, much greater than that of natural aggregate. This means that the EOF steel slag must be saturated before being mixed with concrete, in order to lower its water absorption capability.

2.3.4. Mechanical Properties

It was considered that concrete is an amalgamation of separate particles held together by a cementing agent. The concrete's strength is determined by the mechanical characteristics of aggregate, regardless of the cement paste quality and bonding with aggregate. According to IS: 2386 (Part IV)—1963 [32], experiments were conducted on a EOF steel slag and natural coarse aggregate to determine their mechanical properties, including crushing, 10% fines, impact, and the Los Angeles abrasion value. Several factors contribute to concrete's strength, including mechanical properties. Aggregates that are strong and durable are necessary for producing strong concrete.

2.3.5. Soundness Test

For concrete, the aggregates' resistance to the effects of weathering or their long-term durability is a significant consideration. In order to determine aggregate durability, sodium sulphate soundness testing is one of the most often used methods. When an aggregate is subjected to a chemical assault, it is utilized to calculate the percentage loss of the aggregate, according to IS: 2386 (Part V)—1963 [32]. The aggregate was sorted into size fractions: fractions crossing 20 mm retained 10 mm, fractions passing 10 mm retained 4.75 mm, and fractions passing 10 mm kept 4.75 mm. At a temperature of 23 °C, the aggregate sample was subjected to a sodium sulphate saturated solution. The aggregate fractions were subjected to 10 immersion cycles in the salt solution, with each cycle requiring 16 h of immersion time in the solution. Following each immersion cycle, 105 °C oven was used to dry the samples.

2.3.6. Alkali Aggregate Reactivity (AAR)

Chemical reactions of alkalis present in concrete and some alkaline minerals present in aggregate, are termed alkali aggregate reactivity (AAR). Alkali-silicate gel formed from AAR attracts moisture by osmosis or absorption, which increases volume, due to the surrounding confinement, internal pressure results, causes of the disruption of cement paste, and cracking of aggregate. There are two methods to determine alkali aggregate reactivity. One is the chemical method in which this test was performed, in order to identify the presence of reactive minerals in the aggregates that reacted with the alkalis present in the cement. Another method is the mortar bar method. Through the measurement of the expansion created in mortar bars, this test seeks to determine the potential expansive alkali reactivity of a cement-aggregate mixture. IS: 2386 [33] standard procedure was followed to cast, cure and store mortar bars of dimension 25 × 25 × 250 mm appropriately. For this test, the coarse aggregates were crushed into fine aggregates to meet the requirements.

2.3.7. Staining from Iron Compounds

Unsightly stains appeared in exposed concrete surfaces, due to the occasional presence of iron oxide and iron sulfide particles in aggregates. As per ASTM C641 [26], the test to know the presence of any insoluble iron components in steel slag was carried out. The presence of iron compounds in aggregate attributes the potential degree of staining, which is evaluated primarily using a visual classification method. If the aggregate has insoluble products from the decomposition of iron components, the insoluble products will deposit as black, red, and green strains.

2.3.8. Interfacial Transition Zone

Normal cement concrete has a weak point at the interfacial transition zone between hardened cement paste (HCP) and aggregate, roughly 100 μm wide. To further understand the performance of the HCP aggregate interface and difference between the steel slag aggregate HCP interfaces and natural coarse aggregate, micro-hardness variation in this area was evaluated. The micro-hardness was determined by standard Vickers hardness testing equipment. The hardness is defined as the ratio of applied load and area of the indentation. The test was performed on the surface of concrete specimens polished with emir sheets and tested with a pyramid diamond cone indentor of apical angle of 136°. It is necessary to measure the diagonal length of the indentation for a particular force and then use conventional tables to calculate the hardness. Alternatively, the following formula may be used to determine hardness.

$$\text{Hardness } (HV) = (1854.4 \times P)/D^2 \tag{1}$$

where, D is the mean diagonal length (mm) and P is the applied load (g).

2.3.9. Microstructural Study

Energy dispersive X-ray (EDX), SEM, and XRD analysis studied the microstructure properties of EOF steel slag. The microscopic examination of steel slag particles revealed their shape and surface texture. EOF steel slag particles were examined with a SEM (Jeol JSM 6390 scanning electron microscope). Elemental analysis was performed using EDX analysis.

2.3.10. Stability Factor for MgO

Free MgO hydrates take months or even years to form, resulting in a considerable volume change. The slag generated from molten steel making technology has low MgO content [6]. The hydration of MgO causes the autogenous expansion, forming $Mg(OH)_2$ crystals, which are larger than the sum of volumes of MgO and H_2O (IS: 2386 (Part V)—1963) [32]. The presence of unstable MgO content can be confirmed by determining the stability factor 'K'.

$$\text{Stability factor K} = \text{MgO}/(\text{FeO} + \text{MnO}) \tag{2}$$

if $K < 1$ MgO is stable or $K > 1$ MgO is unstable

2.3.11. Shape Parameters from Digital Image Processing

The shape aggregate can profoundly characterise the quality of aggregate, as it affects concrete mixtures property. Daily, measuring the shape parameter using calliper devices is tiresome and time-consuming for quality control. Digital image processing (DIP) has been applied to characterise aggregate shape in recent years. Four indices are chosen to characterize natural aggregate and EOF steel slag: sphericity, circularity, shape factor, and roundness (angularity index).

Sphericity is expressed as the ratio of the surface area of a sphere whose volume is as that of aggregate to the original surface area of aggregate [19]. To find sphericity, the following Equation (3) [35] is used:

$$\text{Sphericity} = \sqrt[3]{\frac{\text{bd}}{\text{l}^2}} \tag{3}$$

where, b is breadth area of particle, d is depth of the particle, and l is the length of the particle.

Circularity is expressed as the ratio between the perimeter of the equivalent circle of aggregate and perimeter of the aggregate. If the circularity is one, the shape of the aggregate is smooth and round [19]. Therefore, if the circularity is zero means, the shape of aggregate is less smooth and round. The circularity of the particles is calculated by Equation (4):

$$\text{Circularity} = \sqrt{\frac{4\pi A}{P^2}} \tag{4}$$

where, P is the perimeter of the particle, and A is the projected area of the particle.

Roundness is a measure of sphericity; the roundness value of a particle closer to one means the particle is more circular in shape [19], i.e., for a circle, the roundness value is 1. The higher roundness of an aggregate indicates lesser fracture faces on the aggregate surface [20]. Roundness is measured by the given formula [36]:

$$\text{Roundness} = \frac{P^2}{4\pi A} \tag{5}$$

Shape factor of aggregate is a dimensionless quantity that defines the aggregate shape. The shape factor is independent of the size of the aggregate. The ranges of the shape factor are 0–1. The following equation proposed by Kuo et al. (1996) [36] is used to calculate the shape factor.

$$\text{Shape factor} = \frac{\text{d}}{\sqrt{\text{bl}}} \tag{6}$$

A computerised technique to capture and digitalize the scene is a two-dimensional image expressed as digital image processing (DIP). A scanner or a video camera can be used to capture the scene. The produced video signals are stored as pixels; later, they are analysed to generate the required information. Various algorithms and techniques can be engaged to generate the required information. In this study, the DIP technique is applied in concrete technology to investigate aggregate shape parameters.

The algorithm adopted identifies shape variations in greyscale or in colour of an aggregate to define the edges of the aggregate and discriminate the aggregate form the background. Many public domain programs are available in DIP technique. During this investigation, Image J is employed as an image analysis application. The National Institute of Health in United States has developed Image J, a Java-based program for image processing. Image J can be downloaded in public domains. The working methodology of image processing is represented in Figure 2.

2.3.12. Slump Test

Fresh concrete workability is essential as the concrete needs to be adequately compacted, transported, readily placed, and adequately finished. A concrete mix having good workability will produce concrete with maximum density and less void space. An increase in the density of concrete improves the concrete strength. To measure the workability of concrete, a slump test is commonly carried out. The mix proportion for different water–cement ratios was arrived at by following the recommendations of IS: 10262-2009 [37] standard. Mild exposure circumstances and high-quality standards have been included in the design. Different mix proportions were used, i.e., 1:1.94–3.21, 1:1.45–2.61, and 1:1.42–2.62 ratio of cement, fine aggregate, and coarse aggregate for water–cement ratios of 0.55, 0.45, and 0.44, respectively.

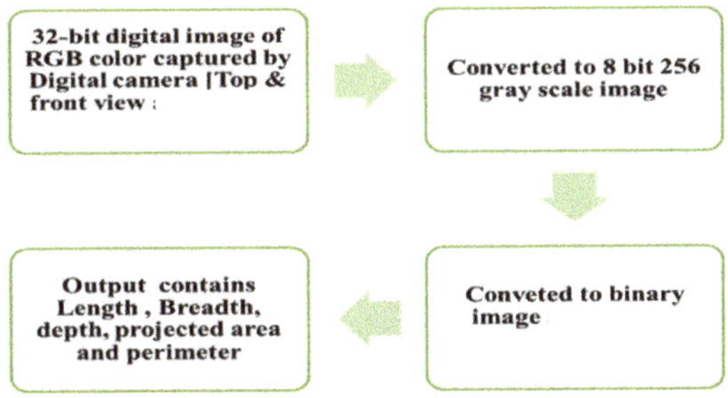

Figure 2. Image processing methodology of DIP.

2.3.13. Compressive Strength

The compressive strength of hardened concrete is a primary mechanical property to confirm whether the concrete has the requisite specified strength. Several factors, i.e., water–cement ratio, aggregates' grading, size, surface texture, shape, and superplasticizer dosage, influence the concrete strength. With EOF steel slag and natural coarse aggregate, 150 mm cubic concrete cubes were made. Twenty-four hours afterward, the specimens were de-molded and exposed to a 28-day water curing procedure. The compressive strength of all specimens was tested and evaluated, as per IS: 516-1959 [38].

3. Discussion of Results

3.1. Particle Size and Shape

Figure 3 depicts the particle size distribution for a graded aggregate with a nominal size of 20 mm, which is then compared to the limitations of percentage passing established by IS 383-2016 [25] for this aggregate. Table 2 demonstrates the properties of EOF steel slag and natural coarse aggregate. Compared to natural coarse aggregate, the EOF steel slag has a lower flakiness index, suggesting that the slag particles are thicker. Because EOF steel slag has a lower elongation index than natural coarse aggregate, it may be inferred that slag particles are shorter and rounder than the natural coarse aggregate. The natural coarse aggregate's flakiness and elongation index value is 33.2%. EOF steel slag is 24.4%, which should not exceed 40% for concrete mixes and 30% for bituminous non-bituminous mixes, as recommended in IS 383-2016 [25]. The lesser percentage of flat particles makes sure high stability in concrete. According to Table 2, EOF steel slag (20.83) and natural coarse aggregate (20.41) have the same angular number. EOF steel slag's angular and rougher surface texture will ensure a better degree of interlocking with cement matrix [39]. It can be observed from Table 2 that the EOF steel slag particles are thicker than natural coarse aggregate, and the angular and rougher surface texture of EOF steel slag will ensure a better degree of interlocking with the cement matrix.

Table 2. Shape properties of natural coarse aggregate and EOF steel slag.

Test	EOF Steel Slag	Natural Coarse Aggregate
Angularity number	20.41	20.83
Elongation index	7.7%	15%
Flakiness index	16.7%	18.2%

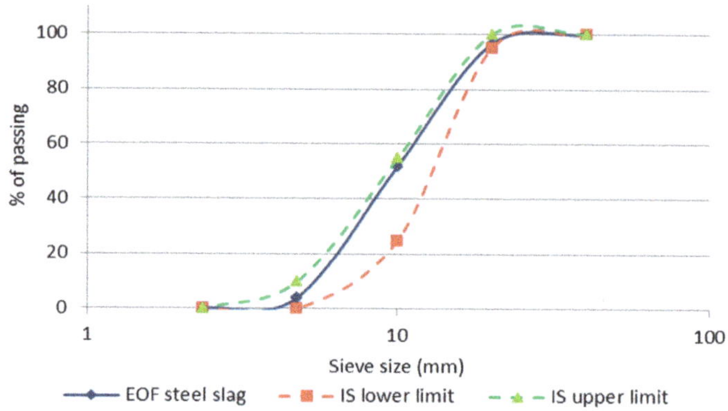

Figure 3. Granulometric curve.

3.2. Evaluation of Light Weight Particles

As shown in Figure 4, the lightweight particles in the EOF steel slag sample began to float on the solution's surface. The floating fraction is collected separately. The collected fraction of the sample is dried in the oven for 24 h and is weighed. The percent of lightweight particles is found to be 3%. This meets the prescribed limit of 3%. The effects of lightweight particles in concrete could be assessed by studying the durability properties of concrete. There are no traces of other deleterious materials, such as clay, coal, lignite, and particles, finer than 75 µm in the EOF steel slag. Therefore, the sample taken meets the limit of total deleterious content of 5%, as per IS: 383—2016 [25].

Figure 4. Lightweight particles floating in the zinc chloride solution.

3.3. Physical Properties

According to the criteria established in IS 2386 (part III) (1963), the physical characteristics of EOF steel slag and natural aggregate are investigated and shown in Table 3 [30]. EAF slag has a bulk density 4.6 percent higher than natural coarse aggregate, which is likewise the case with EOF steel slag [40]. Steel slag has a denser composition than natural coarse aggregate, with a density difference of 4% [34], due to high iron content [39]. Natural aggregates have a lower bulk density and specific gravity (see Table 3) than BOF and EAF steel slag (see Table 3).

Table 3. Physical properties of coarse aggregate.

Test	Natural Coarse Aggregate	EOF Steel Slag	EAF Steel Slag [34]	BOF Steel Slag [39]	EAF Steel Slag [39]	EAF Steel Slag [40]
Specific gravity	2.75	2.86	3.58	3.26	3.64	3.68
Bulk density (kg/m^3)	2886	3020	2581	2049	2092	3810
Water absorption	0.8%	3%	0.7%	2.31%	1.75%	2.4%

3.4. Mechanical Properties

Table 4 shows the mechanical parameters of EOF steel slag. For the strength tests, such as the crushing, 10% fines, and impact values, steel slag has greater values than natural aggregate. EOF steel slag has a greater crushing value because it is more resistant to crushing under progressively applied compressive stress. IS: 2386 [32] states that EOF steel slag may be utilized in concrete pavement wear surfaces and other concrete constructions because of its strong crushing strength of above 30%. The EOF steel slag's increased impact value results from its resistance to failure under sudden loading, as well as the greater the concrete's elastic modulus, due to the stronger aggregate [41]. Resistance against fragmentation of EOF steel slag aggregate is 7.11% higher in the Los Angeles abrasion test [34] and 14.88% higher in Deval's attrition test (resistance against wear) than that of the natural aggregate. The impartial resistance to wear and fragmentation of EOF slag particle qualifies its use as a replacement for natural aggregate in high strength concretes (50–60 MPa) [40].

Table 4. Mechanical properties of coarse aggregate.

Type of Test	EOF Steel Slag	Natural Aggregate	Recommended Value, as per IS: 2386
Deval attrition	3.68%	18.56%	-
Los Angeles abrasion	1.16%	8.27%	Not exceeding 50%
Impact value	29.57%	22.87%	Less than 45%
10 percent fines value	8.3%	7.8%	7.5% to 12.5%
Crushing value	8%	6.55%	Not exceeding 45%

3.5. Soundness Test

Figure 5 shows the results of the sodium sulphate soundness test for natural aggregate and EOF steel slag. Soundness is the resistance against volume change brought out by changes in physical condition. The EOF steel slag has non-hydrated lime and magnesium oxides derived from the steel-making process. The free calcium and magnesium are susceptible to expansion when exposed to moisture. The soundness test result is presented in Figure 5 and points out that the resistance of EOF steel slag against volume change is comparable to natural aggregate. This can be attributed to the aging process happening in the EOF steel slag during its storage. The EOF steel slag is left as a stockpile over 4–6 months. During this aging process, free lime and magnesium react with atmospheric moisture and are converted into magnesium hydroxide ($Mg(OH)_2$) and calcium hydroxide ($Ca(OH)_2$). Throughout the aging, the steel slag is free to expand in this aging process. The expansion of EOF steel slag is considerably minimized in this aging process; thereby, EOF steel slag becomes free from expandable materials when used in concrete [40].

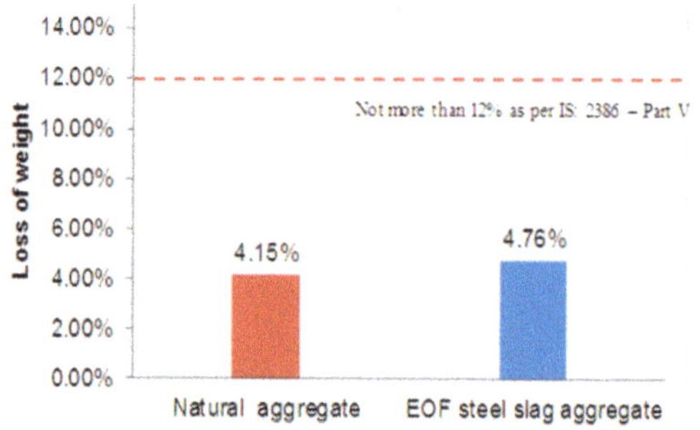

Figure 5. Percentage losses in soundness test.

3.6. Determination of Alkali Aggregate Reactivity

The chemical method's test results of alkali reactivity are incorporated in the graph of IS: 2386-1963 [33], as shown in Figure 6a. Figure 6a also represents the non-reactivity mechanism of EOF steel slag with alkali cement. The points lie to the left of the boundary line. It indicates that EOF steel slag causes mortar expansion of less than 0.1 percent in a year, under the same conditions. The EOF steel slag can very well be considered an innocuous aggregate. By the mortar bar method, the difference in the length of the mortar bars was recorded for 6 months. The percentage of change in length of bars is displayed in Figure 6b. It is observed from the figure that there is no expansion at the early ages (i.e., from day 1 to 7); from 7th day onwards, the expansion is reflected in an increase in length of all specimens. Beyond 30 days, the increment in the expansion percentage of natural aggregate is low, when compared to progressive increment in EOF steel slag aggregate. The permissible limits for the expansion of the mortar bar test at 38 °C are 0.05% at 90 days and 0.10% at 180 days, which are recommended by ASTM C641-98 [26]. Though the EOF steel slag aggregate meets the permissible limits, it is expansive in nature, compared to natural aggregate.

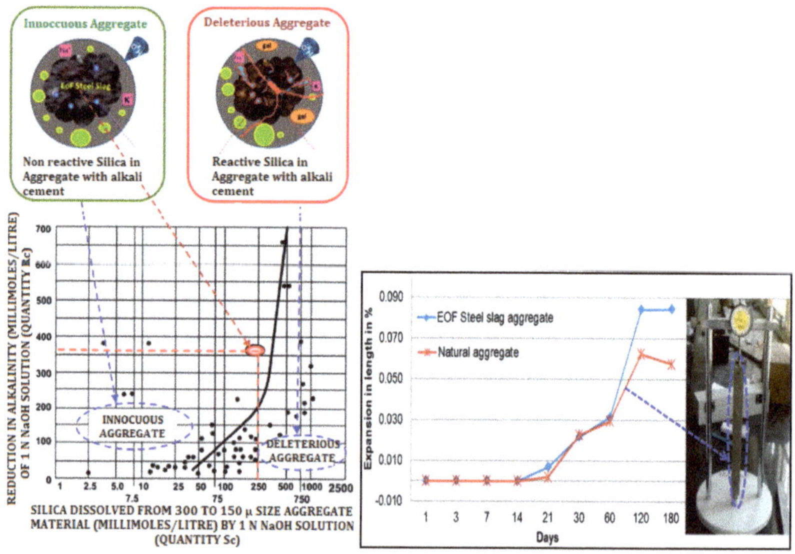

(**a**) (**b**)

Figure 6. Alkali aggregate reactivity of EOF steel slag (**a**) chemical method with chart (IS: 2386 (part VII)—1963); (**b**) mortar bar method.

3.7. Staining from Iron Compounds

Figure 7 shows the filter paper before and after being subjected to a steam bath. It is observed from Figure 7 that there are considerably no visible strains in the filter papers. From this, it is surmised that there is no decomposed iron component in the EOF steel slag. The absence of stain in the filter paper also proves that there are no soluble iron components present, which may tend to avoid the unpleasant appearance on the surface of concrete. In addition, the absence of soluble iron components tends to reduce the transformation of ferrous components to ferric components, thereby reducing the effect of the corrosion of the concrete reinforcement.

Figure 7. Filter papers without staining.

3.8. Properties of the Interfacial Transition Zone

Normal cement concrete has a weak point at the interfacial transition zone between the hardened cement paste (HCP) and aggregate, which is roughly 100 µm wide. Figure 8 shows that the microhardness was tested in the interfacial transition zone [41] as shown in Figure 8a. The results of micro-hardness are shown in Figure 8b, and it is evident that the hardness value of EOF steel slag aggregate HCP is greater than that of natural coarse aggregate. It is inferred that the hardened cement paste gets consolidated, due to hydration of EOF steel slag, i.e., the addition of steel slag to concrete increases the chemical reaction. Furthermore, the steel slag's rough surface texture improves the concrete's strength because of its strong adhesion [42].

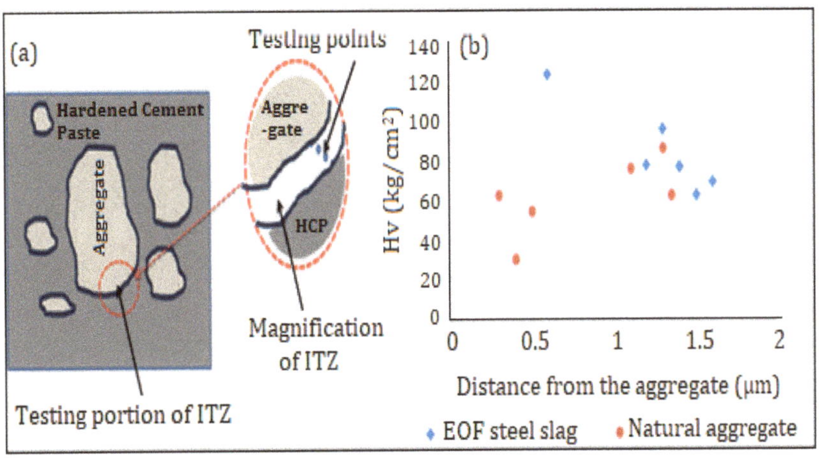

Figure 8. (**a**) Interfacial transition zone between hardened cement paste (HCP) and aggregate, (**b**) Micro hardness value of interfacial transition zone.

3.9. Stability Factor for MgO

By substituting the values of oxides from Table 5, in Equation (2), the value of stability factor K is determined as 0.322, which is less than one. This indicates that MgO exists in solid solution with FeO and MnO. Therefore, it is deduced that MgO is stable. Figure 9 shows the non-reactivity mechanism of EOF steel slag with stable MgO. The reactive aggregate will expand by forming $Mg(OH)_2$ and forming cracks because of $K > 1$, but EOF steel forms cracks because $K < 1$.

Table 5. Chemical composition of steel slag.

Chemical Compounds	EOF Steel Slag	EAF Steel Slag [11]	Steel Slag [43]	EAF Steel Slag [34]	Steel Slag [44]
CaO (%)	35.28	39.62	46.73	32.52	24.62
MgO (%)	9.27	9.23	6.27	4.56	-
SiO_2 (%)	16.69	17.23	14.77	17.17	17.79
Al_2O_3 (%)	6.20	4.08	5.52	7.96	7.82
MnO (%)	1.88	0.70	2.76	3.8	-
Fe_2O_3 (%)	26.91	20.40	18.42	30.8	35.3
P_2O_5 (%)	1.43	-	1.67	0.58	-
Na_2O (%)	0.16	-		0.16	0.74
K_2O (%)	0.03	-		0.03	-
SO_3 (%)	0.56	-		0.25	0.13

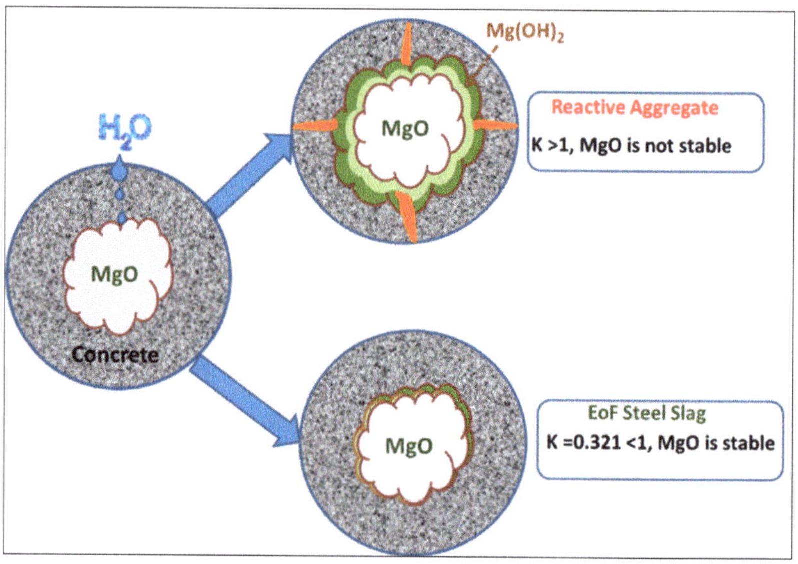

Figure 9. Stable Mechanism of Magnesium Oxide in EOF steel slag.

3.10. Microstructural Study

The EDX spectrum of the EOF steel slag sample that confirms the presence of oxides of Ca, Mg, Al, Si, and Fe appears in Figure 10. In general, the chemical composition of steel slag is mainly variable, depending upon materials, process, and location [45].

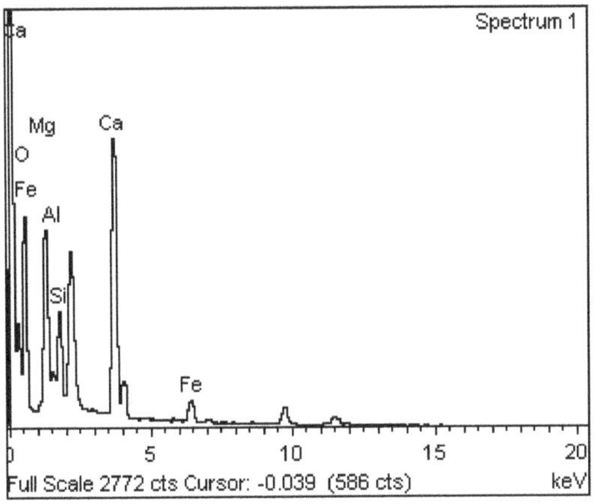

Figure 10. EDX spectrum of EOF steel slag.

Figure 11a,b shows the morphology of EOF steel slag. The microstructure of EOF steel slag exhibits subrounded and angular shapes [10,43]. Most of the particles are examined using SEM, which showed extremely crystalline structures and rough textures with platy. The platy particles examined by SEM are highly irregular in shape. This morphology provides a greater surface area to form a strong transition zone between the aggregate and hardened cement paste. The EOF steel slags chemical analysis is presented in Table 5, and the percentage of chemical composition is compared with other types of steel slag, such as EAF steel slag. The major chemical oxides identified are CaO, MgO, SiO_2, and Al_2O_3 [11,34]. The amount of MnO is insignificant. In transforming molten pig iron into steel, a certain percentage of iron (Fe) remains as hot metal, and this fraction cannot be streamlined into the produced steel. Chemical investigation of steel slag samples reveals the presence of this oxidized iron [44]. Because so much lime (flux) is needed in the iron-to-steel conversion process, EOF steel slag often has a high CaO component [11,43]. Figure 12 shows the XRD pattern of the EoF steel slag. The diffraction pattern of EOF steel slag shows the presence of hematite-Fe_2O_3 (H), calcite-CaCO (C), lanite-Ca_2SiO_2 (L), gehlenite- $Ca_2Al(AlSiO)_7$ (G), and wusite-FeO (W). The high crystalline nature of EOF steel slag shows its weak potential for any hydraulic activity that confirms dimensional stability. EOF steel slag does not have any pozzolanic activity, due to its high crystallinity [45]. This is highly attributed to the EOF steel slag's volume stability, which is major quality of aggregate [46,47].

Figure 11. SEM micro graph of EOF steel slag, (**a**) magnification ×55,000, and (**b**) magnification ×100,000.

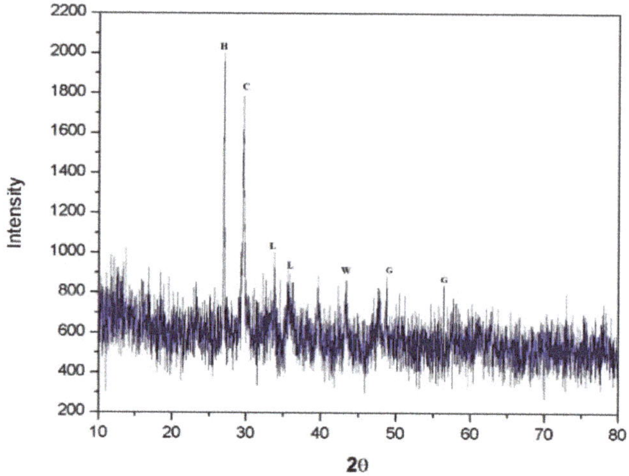

Figure 12. X-ray diffraction analysis of EOF steel slag.

3.11. Shape Parameters from Digital Image Processing

Image of natural aggregate and EOF steel slag are captured using a high-resolution digital camera. Eighty particles from each natural aggregate and EOF steel slag are chosen. The particles are placed on a white background to capture the top and front views. Then, the images are processed through Image J. The virtual quality of the original image is enhanced, and the particles are highlighted. Later, the chosen particles are segmented from the background by choosing the appropriate threshold value. Figures 13 and 14 show the natural aggregate and EOF steel slag's graphical and threshold image. In segmenting particles having lower intensity, threshold values are set to blue when the background is white. Figure 15 represents the binary image of the aggregate sample with numbering.

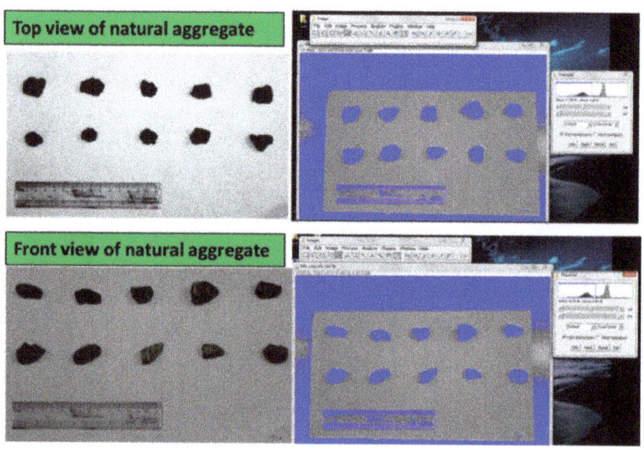

Figure 13. Top and front view of natural aggregate and thresholded images.

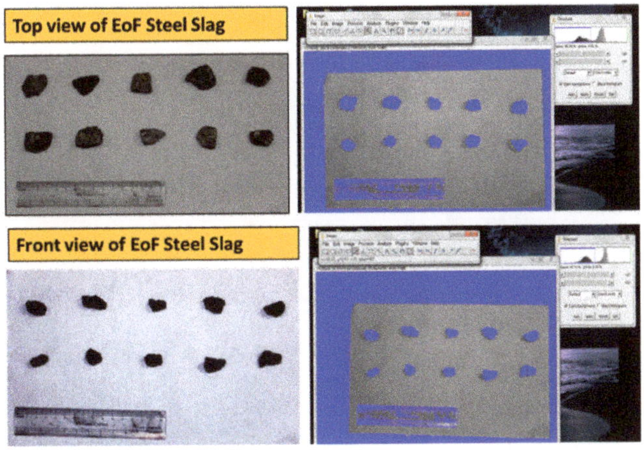

Figure 14. Top and front view of EOF steel slag and thresholded images.

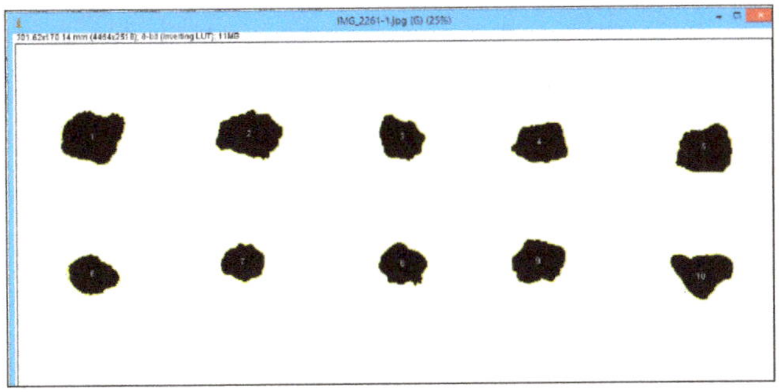

Figure 15. Binary image of aggregate with numbering.

The shape parameters of 80 particles each from matural coarse aggregate (NCA) and steel slag aggregate (SSA) are noted. Statistically, the results of the shape parameter obtained from DIP are significant with a 95% confidence interval (high precision). Table 6 gives the arithmetic mean of the shape parameters of natural aggregate and EOF steel slag. From the Table, it is inferred that the circularity of EOF steel slag is lesser than the natural coarse aggregate, which indicates EOF steel slag is less smooth and highly irregular in shape than the natural coarse aggregate. Sphericity and shape factor explains the form (complete shape) of the aggregate [19]. The sphericity and shape factor of EOF steel slag is marginally higher than natural coarse aggregate and comparable to natural coarse aggregate. The mean value of sphericity and shape factor shows that the natural coarse aggregate and EOF steel slag are non-spherical and highly angular. The mean roundness value indicates that the natural aggregate and steel slag aggregate are of other shapes. It also denotes that EOF steel slag aggregate has an irregular surface and rough edges, since the roundness of EOF steel slag is lesser than that of natural aggregate. From the lesser roundness value of the EOF steel slag, it can be inferred that the EOF steel slag has more fractured faces. More fractured faces mean more surfaces contact with cement paste, which ensures better interlocking of aggregate [20].

Table 6. Shape indices of natural coarse aggregate and EOF steel slag.

Description	Natural Coarse Aggregate		EOF Steel Slag Aggregate	
	Mean	Standard Deviation	Mean	Standard Deviation
Circularity	0.692	0.0844	0.627	0.1078
Sphericity	0.992	0.122	1.034	0.0985
Shape factor	0.759	0.1136	0.770	0.1264
Roundness	0.679	0.239	0.602	0.4026

3.12. *Workability of Concrete with EOF Steel Slag*

For a moderate environment, IS 456: 2000 [48] recommends that the degree of workability of concrete should be in the range of 50 mm to 100 mm of a slump. From Table 7, it is perceived that irrespective w/c ratio, all concrete mixes with EOF steel slag meet the least slump value of 50 mm. The EOF steel slag aggregates with irregular shapes and rough surfaces necessitate more cement volume to cover the surface than natural aggregates require [39,49]. The decrease in the workability is highly significant in the lower water–cement ratio because of higher water absorption [44,50].

Table 7. Workability of concrete comprising EOF steel slag.

	w/c Ratio	Slump (mm)
EOF steel slag	0.55	72
EOF steel slag	0.45	63
EOF steel slag	0.4	89
Abu-Eishah et al.	0.52 & 0.38	190 & 220
B. Pang et al.	0.42	185
Manso et al.	0.6	70
S. Brand et al. (BOF & EAF)	0.56	76 & 32
G. I. Sezer et al.	0.40, 0.55 & 0.7	110, 100 & 95

Another factor is the higher water absorption capacity of EOF steel slag. As the slump value lies within 50 mm, EOF steel slag can be utilized as a partial substitution/replacement for natural coarse aggregate in normal and heavily reinforced concrete, with compaction and uniform dispersal through manual vibration [9,48].

3.13. *Compressive Strength of EOF Steel Slag Based Concrete*

Figure 16 shows a comparison of the compressive strength of concrete containing EOF steel slag and natural aggregate. The specimens comprising of EOF steel slag as a coarse aggregate exhibited a significant improvement in compressive strength, compared to the specimens comprising of a natural coarse aggregate at the same water–cement ratio. This is attributed to rough surface roughness, which promotes good adhesion between EOF steel slag the cement matrix. Interfacial transition zone integrity is an important factor in determining concrete's strength [44]. The hydration of CaO in the EOF steel slag has resulted in a thick, continuous, and hardened transition zone in the aggregate concrete. Aside from the physical contribution, the formation of cement paste surrounding EOF steel slag is facilitated by the hydration of EOF steel slag [6,9,51]. In addition, the EOF steel slag's Los Angeles abrasion value is lower than natural aggregate. The lesser abrasion value means a higher aggregate strength, which enhances a compressive strength of concrete [11]. Figure 11 shows that the enhancement of the compressive strength of concretes with water–cement ratios of 0.55, 0.45, and 0.4 were 15%, 22%, and 12%, respectively.

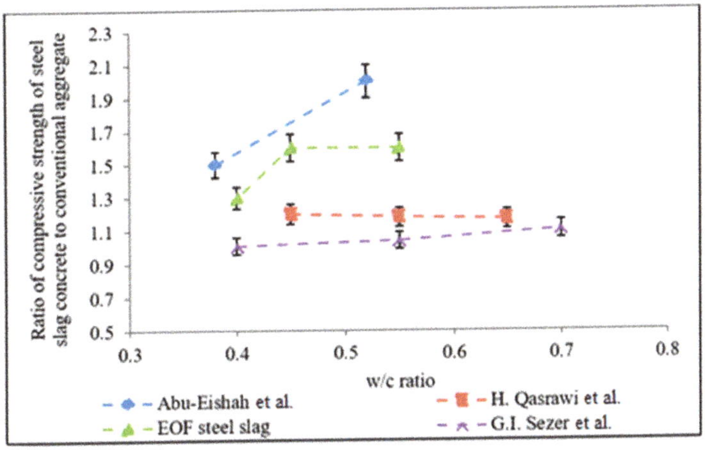

Figure 16. Concrete compressive strength versus water–cement ratios.

4. Conclusions

The current research was employed to compare the size of particle, shape, physical, mechanical, and chemical properties of EOF steel slag with conventional natural aggregates. Different ranges of graded EOF steel slag aggregate sizes facilitate good packing in concrete. The shape properties of EOF steel slag aggregate are more or less angular, making the workable concrete. Lightweight particles are present in EOF steel slag in insignificant quantities. The EOF steel slag had a greater water absorption capability than natural aggregate, which made the concrete more difficult to work. The mechanical properties show a remarkably better performance of EOF steel slag, in terms of crushing, impact, abrasion, and attrition. The soundness test and AAR test show the absence of expandable materials; additionally, there is no trace of decomposition of iron compounds in EOF steel slag. Steel slag aggregate concrete has a higher microhardness than natural aggregate because of its rough-textured surface and chemical reactivity with the cement matrix. From the chemical composition, it is noted that magnesium oxide is stable and inexpensive. The DIP analysis shows that EOF steel slag is considerable, similar to the natural aggregate, in the aspects of better form (sphericity and shape factor), roundness, and circularity. EOF steel slag aggregate morphology gives rise to a stronger interfacial transition zone, due to its rough-textured surface. The hardened interfacial zone and rough surface texture of EOF steel slag contributes to enhancing concrete strength.

Author Contributions: Conceptualization, A.M., M.R., P.K., G.M. and N.I.V.; methodology, A.M., M.R. and P.K.; investigation, A.M., M.R. and P.K.; resources, A.M., M.R. and P.K.; data curation, A.M., M.R. and P.K.; writing—original draft preparation, A.M., M.R., P.K., G.M. and N.I.V.; writing—review and editing, A.M., M.R., P.K., G.M. and N.I.V.; visualization, A.M., M.R., P.K., G.M. and N.I.V.; supervision, M.R.; project administration, M.R.; funding acquisition, G.M. and N.I.V. All authors have read and agreed to the published version of the manuscript.

Funding: The research is partially funded by the Ministry of Science and Higher Education of the Russian Federation, under the strategic academic leadership program 'Priority 2030' (agreement 075-15-2021-1333, dated 30 September 2021).

Institutional Review Board Statement: Not applicable.

Informed Consent Statement: Not applicable.

Data Availability Statement: Not applicable.

Conflicts of Interest: The authors declare no conflict of interest.

References

1. Demissew, A. Comparative Analysis of Selected Concrete Mix Design Methods Based on Cost-Effectiveness. *Adv. Civ. Eng.* **2022**, *2022*, 4240774. [CrossRef]
2. De Brito, J.; Francisco, A. *New Trends in Eco-Efficient and Recycled Concrete*; Woodhead Publishing: Cambridge, UK, 2019; pp. 169–190.
3. Beaumont, P.W.R.; Zweben, C.H. *Comprehensive Composite Materials II*; Elsevier: Amsterdam, The Netherlands, 2018; pp. 464–481.
4. Kumar, S.; Dhara, S.; Kumar, V.; Gupta, A.; Prasad, A.; Keshari, K.; Mishra, B. Recent trends in slag management & utilization in the steel industry. *Miner. Met. Rev.* **2019**, *2019*, 94–102.
5. Nippon Slag Association. *Iron and Steel Slag Products*; Nippon Slag Association: Tokyo, Japan, 2003.
6. National Slag Association. *Iron and Steel Making Slag Environmentally Responsible Construction Aggregates*; Environmental Technical Bulletin; Nippon Slag Association: Tokyo, Japan, 2003; p. 2.
7. Lin, W.-T.; Tsai, C.-J.; Chen, J.; Liu, W. Feasibility and Characterization Mortar Blended with High-Amount Basic Oxygen Furnace Slag. *Materials* **2019**, *12*, 6. [CrossRef] [PubMed]
8. Netinger, I. Concrete Containing Steel Slag Aggregate Performance after High Temperature Exposure. In *Concrete Repair, Rehabilitation and Retrofitting III*; Taylor & Francis Group: London, UK, 2012; pp. 1347–1352. ISBN 978-0-415-89952-9.
9. Jigar, P.P. Broader Use of Steel Slag Aggregates in Concrete. Master's Thesis, Cleveland State University, Cleveland, OH, USA, 2008; p. 401.
10. Yildirim, I.Z.; Prezzi, M. Chemical, Mineralogical, and Morphological Properties of Steel Slag. *Adv. Civ. Eng.* **2011**, *2011*, 463638. [CrossRef]
11. Abu-Eishah, S.I.; El-Dieb, A.S.; Bedir, M.S. Performance of concrete mixtures made with electric arc furnace (EAF) steel slag aggregate produced in the Arabian Gulf region. *Constr. Build. Mater.* **2012**, *34*, 249–256. [CrossRef]
12. Kirillidi, Y.; Frogoudakis, E. Electric arc furnace slag utilisation. In Proceedings of the 9th International Conference on Environmental Science and Technology, Rhodes Island, Greece, 1–3 September 2005.
13. Mohammed, N.; Arun, D.P. Experimental investigation of using slag as an alternative to normal aggregates (coarse and fine) in concrete. *Int. J. Civ. Struct. Eng.* **2012**, *3*, 117–127.
14. Hiraskar, K.G.; Chetan, P. Use of blast furnace slag aggregate in concrete. *Int. J. Sci. Eng. Res.* **2013**, *4*, 95–98.
15. Chinnaraju, K.; Ramkumar, V.R.; Lineesh, K.; Nithya, S.; Sathish, V. Study on concrete using steel slag as coarse aggregate replacement and ecosand as fine aggregate replacement. *Int. J. Res. Eng. Adv. Technol.* **2013**, *1*, 6.
16. Tarawneh, S.A.; Gharaibeh, E.; Saraireh, F.M. Effect of using steel slag aggregate on mechanical properties of concrete. *Am. J. Appl. Sci.* **2014**, *11*, 700–706. [CrossRef]
17. Khalid, R.; Apoorv, S.; Patel, R.D. Strength analysis of concrete by using iron slag as a partial replacement of normal aggregate (coarse) in concrete. *Int. J. Sci. Res.* **2014**, *3*, 190–193.
18. Maerz, N.H. Aggregate Sizing and Shape Determination Using Digital Image Processing. In Proceedings of the Center for Aggregates Research (ICAR) Sixth Annual Symposium, St. Louis, MI, USA, 19–20 April 1998; pp. 195–203.
19. Kwan, A.K.H.; Mora, C.F.; Chan, H.C. Particle shape analysis of coarse aggregate using digital image processing. *Cem. Concr. Res.* **1999**, *29*, 1403–1410. [CrossRef]
20. Bessa, I.S.; Branco, V.T.C.; Soares, J. Evaluation of different digital image processing software for aggregates and hot mix asphalt characterizations. *Constr. Build. Mater.* **2012**, *37*, 370–378. [CrossRef]
21. Polat, R.; Yadollahi, M.M.; Sagsoz, A.E.; Arasan, S. The correlation between aggregate shape and compressive strength of concrete: Digital image processing approach. *Int. J. Struct. Civ. Eng. Res.* **2013**, *2*, 63–80.
22. Seracettin, A.; Samet, H.A.; Suat, A. Shape Properties of Natural and Crushed Aggregate using Image Analysis. *Int. J. Civ. Struct. Eng.* **2010**, *1*, 221–233.
23. He, Y.; Zhang, X.; Zhang, Y.; Zhou, Y. Effects of particle characteristics of lightweight aggregate on mechanical properties of lightweight aggregate concrete. *Constr. Build. Mater.* **2014**, *72*, 270–282. [CrossRef]
24. *IS: 8112-2013*; Specification for 43 Grade Ordinary Portland Cement. Bureau of Indian Standards: New Delhi, India, 2013.
25. *IS: 383-1970*; Specification for Coarse and Fine Aggregates from Natural Sources for Concrete. Bureau of Indian Standards: New Delhi, India, 1970.
26. *ASTM C 641-98*; Standard Test Method for Iron Staining Materials in Lightweight Concrete Aggregate. American Society for Testing and Materials: West Conshohocken, PA, USA, 1998.
27. *ASTM C 123*; Standard Test Method for Lightweight Particles in Aggregate. American Society for Testing and Materials: West Conshohocken, PA, USA, 2011.
28. *IS 2386 (PT 1): 1963*; Methods of Test for Aggregates for Concrete: Part 1 Particle Size and Shape. Bureau of Indian Standards: New Delhi, India, 1963.
29. *IS 2386 (PT 2): 1963*; Methods of Test for Aggregates for Concrete: Part 2 Estimation of Deleterious Materials and Organic Impurities. Bureau of Indian Standards: New Delhi, India, 1963.
30. *IS 2386 (PT 3): 1963*; Methods of Test for Aggregates for Concrete: Part 3 Specific Gravity, Density, Voids, Absorption and Bulking. Bureau of Indian Standards: New Delhi, India, 1963.
31. *IS 2386 (PT 4): 1963*; Methods of Test for Aggregate for Concrete: Part 4 Mechanical Properties. Bureau of Indian Standards: New Delhi, India, 1963.

32. *IS 2386 (PT 5): 1963*; Methods of Test for Aggregate for Concrete: Part 5 Soundness. Bureau of Indian Standards: New Delhi, India, 1963.
33. *IS 2386 (PT 7): 1963*; Methods of Test for Aggregate for Concrete: Part 7 Alkali Aggregate Reactivity. Bureau of Indian Standards: New Delhi, India, 1963.
34. San-José, J.T.; Vegas, I.; Arribas, I.; Marcos, I. The performance of steel-making slag concretes in the hardened state. *Mater. Des.* **2014**, *60*, 612–619. [CrossRef]
35. Krumbein, C. Measurement of geological significance of shape and roundness of sedimentary particles. *Sediment. Petrol.* **1991**, *11*, 64–72. [CrossRef]
36. Kuo, C.Y.; Frost, J.D.; Lai, J.S.; Wang, L.B. Threedimensional image analysis of aggregate particles from orthogonal projections. *Transp. Res. Rec.* **1996**, *1526*, 98–103. [CrossRef]
37. *IS 10262-2009*; Concrete Mix Proportioning—Guidelines. Bureau of Indian Standards: New Delhi, India, 2009.
38. *IS 516: 1959*; Method of Test for Strength of Concrete. Bureau of Indian Standards: New Delhi, India, 1959.
39. Brand, A.S.; Roesler, J.R. Steel furnace slag aggregate expansion and hardened concrete properties. *Cem. Concr. Compos.* **2015**, *60*, 1–9. [CrossRef]
40. Rondi, L.; Bregoli, G.; Sorlini, S.; Cominoli, L.; Collivignarelli, C.; Plizzari, G. Concrete with EAF steel slag as aggregate: A comprehensive technical and environmental characterisation. *Compos. Part B* **2016**, *90*, 195–202. [CrossRef]
41. Winslow, D. A Rockwell Hardness Test for Concrete. *Cem. Concr. Aggreg.* **1984**, *6*, 137–141.
42. Ozturan, T.; Cecen, C. Effect of coarse aggregate type on mechanical properties of concretes with different strengths. *Cem. Concr. Res.* **1997**, *27*, 165–170. [CrossRef]
43. Pang, B.; Zhou, Z.; Xu, H. Utilization of carbonated and granulated steel slag aggregate in concrete. *Constr. Build. Mater.* **2015**, *84*, 454–467. [CrossRef]
44. Sezer, G.I.; Gulderen, M. Usage of steel slag in concrete as fine and/or coarse aggregate. *Indian J. Eng. Mater. Sci.* **2015**, *22*, 339–344.
45. Rojas, M.F.; Sanchez De Rojas, M.I. Chemical assessment of the electric arc furnace slag as construction material: Expansive compounds. *Cem. Concr. Res.* **2004**, *34*, 1881–1888. [CrossRef]
46. Pellegrino, C.; Cavagnis, P.; Faleschini, F.; Brunelli, K. Properties of concretes with black/oxidising electric arc furnace slag aggregate. *Cem. Concr. Compos.* **2013**, *37*, 232–240. [CrossRef]
47. Monosi, S.; Ruello, M.L.; Sani, D. Electric arc furnace slag as natural aggregate replacement in concrete production. *Cem. Concr. Compos.* **2016**, *66*, 66–72. [CrossRef]
48. *IS 456: 2000*; Code of practice for plain and reinforced concrete. Bureau of Indian Standards: New Delhi, India, 2000.
49. Manso, J.; Polanco, J.A.; Losañez, M.; Gonzalez, J.J. Durability of concrete made with EAF slag as aggregate. *Cem. Concr. Compos.* **2006**, *28*, 528–534. [CrossRef]
50. Qasrawi, H. The use of steel slag aggregate to enhance the mechanical properties of recycled aggregate concrete and retain the environment. *Constr. Build. Mater.* **2014**, *54*, 298–304. [CrossRef]
51. Arivoli, M.; Malathy, R. Optimization of Packing Density of M30 Concrete with Steel Slag as Coarse Aggregate Using Fuzzy Logic. *Arch. Met. Mater.* **2017**, *62*, 1903–1913. [CrossRef]

materials MDPI

Article

Enhanced Mechanical and Durability Properties of Cement Mortar by Using Alumina Nanocoating on Carbon Nanofibers

Huda Al Qader [1], Ahmed M. Jasim [2], Hani Salim [1,*], Yangchuan Xing [2] and David Stalla [3]

1 Department of Environmental and Civil Engineering, University of Missouri, Columbia, MO 65211, USA; noorih@mail.missouri.edu
2 Department of Biomedical, Biological, and Chemical Engineering, University of Missouri, Columbia, MO 65211, USA; amjtg6@mail.missouri.edu (A.M.J.); xingy@missouri.edu (Y.X.)
3 Electron Microscopy Core Center, University of Missouri, Columbia, MO 65211, USA; ds5x8@missouri.edu
* Correspondence: salimh@missouri.edu

Abstract: This study evaluated the effect of carbon nanofibers (CNFs) coated by aluminum oxide Al_2O_3 as a reinforcement on compressive strength, frost resistance, and drying shrinkage of cement mortars. Three weight ratios of 0.125%, 0.25%, and 0.5% of Al_2O_3/CNFs and bare CNF cement mortars were compared with reference cement mortar samples. The reactive porous and high surface area layer of alumina induced the hydration reaction and promoted the production of well-distributed hydration gel. Derivative thermal analysis–differential thermogravimetric (TGA-DTG) and X-ray powder diffraction (XRD) characterization showed that Al_2O_3/CNFs reinforcement led to greater hydration gel production than bare CNFs. Transmission electron microscopy (TEM) and scanning electron microscopy (SEM) were performed to study the coating and microstructure of the cement mortars evaluated in this paper. The results show that the optimum enhancement of the cement mortar properties was obtained at ratios of 0.125% for Al_2O_3/CNFs and 0.25% for CNFs. This enhancement was greater with Al_2O_3/CNFs-reinforced specimens in terms of high compressive strength, less compressive strength degradation after 150 cycles, and less drying shrinkage. The low use of the CNFs in Al_2O_3/CNFs samples indicates the coating is an economical and promising approach for improving the performance of cement mortars.

Keywords: carbon nanofibers; nanocoating; alumina; durability

Citation: Al Qader, H.; Jasim, A.M.; Salim, H.; Xing, Y.; Stalla, D. Enhanced Mechanical and Durability Properties of Cement Mortar by Using Alumina Nanocoating on Carbon Nanofibers. *Materials* **2022**, *15*, 2768. https://doi.org/10.3390/ma15082768

Academic Editor: F. Pacheco Torgal

Received: 7 March 2022
Accepted: 4 April 2022
Published: 9 April 2022

1. Introduction

Cementitious materials are the most prevalent construction materials nowadays, but they suffer from low tensile strength and are highly susceptible to cracking. Therefore, several research efforts have been devoted to enhancing the cement structure with macro-level or micro-level reinforcements [1–4] However, internal voids and cracks generally start from a nano size as embryos, then grow to a micro size, leading to a destructive level. If the voids' population is largely present in the cementitious structure, the water easily penetrates and precipitates them, which can lead to freeze–thaw damage [5]. Exposing the cementitious materials to repetitive freeze–thaw cycles severely deteriorates their structure. Therefore, nano-scale reinforcements are used to prevent voids' initiation as an effective way to protect the inherent mechanical and durability properties [6–9]. Researchers have found that nanomaterials are highly reactive to cementitious materials as they can act as seeding sites for the hydration products [10,11]. The hydration reaction is catalyzed by either oxygen functional groups (-OH, -COOH, -C=O) on carbon nanomaterials or the pozzolanic reactivity associated with some metal oxide nanoparticles (SiO_2, Al_2O_3) [12]. This mainly serves for porosity refinement, water impermeability, and structure bridging.

Recently, there has been a growing interest in integrating carbon nanomaterials, such as carbon nanotubes (CNTs), carbon nanofibers (CNFs), and graphene [13–15], in cement composites to increase their frost resistance. The efforts have focused on either eliminating the void formation or refining them to small sizes to reduce the permeability of the cementitious material, hence improving the durability. Others have reported allowing extra space for ice to expand by incorporating air bubbles in the composites [12]. Limited studies have explicitly reported on using CNFs in enhancing the cement frost resistance. Cwirzen et al. [16] studied the effect of the CNFs on the strength and residual cumulative strain of cement mortars after freeze–thaw cycles with a high water-to-cement ratio of 0.5. They found that the higher the content of the CNF composites, the lower the ultimate strain and the internal damage. Their study showed the bridging ability of CNFs with higher water ratios. Wang et al. [9] comparatively investigated the different CNFs' loading effect on frost resistance. They found a 0.3% CNF dosage to be an optimum choice that refined the microstructures, which promoted the durability. Metaxa et al. [17] demonstrated the ability of the CNFs to arrest crack growth in the cement matrix. According to their results, the CNFs performed well in reducing the crack propagation by a bridging phenomenon that provided a load transfer mechanism across cracks and pores. However, the above-mentioned properties of the addition of the CNFs to the cementitious materials are limited by the dispersion of CNFs [18] and by the bonding with the cement matrix. Therefore, surface treatment is essential to improve the solubility as well as the bonding with cement hydration products.

Metal oxide nanoparticles (MNPs) have been vastly utilized in reducing the porosity of cement composites. These materials possess a high surface area that makes them reactive, strengthening the microstructure of the cement with less porosity [19,20], especially those that possess a pozzolanic property, such as silica and alumina nanoparticles [20–23]. Those materials react with calcium hydroxide Ca $(OH)_2$, which has portlandite, forming a secondary reaction producing gels of calcium–silicate–hydrate (C-S-H) and calcium–aluminum–hydrate (C-A-H). The rate of such a reaction is controlled by the available surface area and the surface nucleation sites. Behfarnia et al. [21] discussed the feasibility of using both SiO_2 and Al_2O_3 nanoparticles in enhancing the frost resistance of the concrete mixes due to the pozzolanic and refining effects. Muzenski et al. [24] reported that using 0.25 wt.% of aluminum oxide nanofibers increased the compressive strength by 30%.

However, an inhomogeneous dispersion of these nanoparticles could devolve their functions, such as forming aggregates and inducing formation of large-size voids. Unlike CNFs or CNTs, MNPs lack a bridging phenomenon, which is highly needed to compact the cement microstructure. Another issue with MNPs is their cost, which prohibits it from being scaled up. Therefore, an innovative design for metal oxide nanoparticles with a high surface area, hydrophilicity, and low cost is needed.

Drying shrinkage is another property the community has sought to practically solve. Different nanomaterials have been applied, aiming to reduce shrinkage. The aforementioned nanomaterials have also been used to improve drying shrinkage response resulting from many effects, such as surface hydrophilicity of CNF [25], fast hydration reaction of SiO_2 [26], and formation of the secondary moieties of C-S-H and C-A-H [27].

Limited research exists on combining carbon nanomaterials and metal oxides as a promising approach in enhancing the mechanical properties and the durability of cement through improving the interfacial interaction of the carbon materials with cement, enhancing the dispersion and retaining the bridging [28–34]. Such combinations are achieved by incorporating nanoparticles or metal oxide nanocoating. Sanchez et al. [28] used silicon powder with CNFs to improve the dispersion of CNFs and promote the interfacial interaction between the cement paste and the CNFs. Kim et al. [33] reported an improvement in the CNTs' dispersion and in enhancing the cement mechanical properties by incorporating silica fumes with CNTs. Garg and Das [32] showed a significant improvement in carbon nanotubes' dispersion and bonding due to a pozzolanic reaction after silica nanoparticles' incorporation. Sikora et al. [31] has reported a comparison between CNTs and CNT/NS on the compressive strength of cement pastes under elevated temperatures. They found that adding silica in the form of a nanocoating with a dense layer has a favorable effect on increasing the bond and protecting the CNTs from being oxidized at high temperatures. Dong et al. [34] have reported a study on coating the CNTs with nickel nanoparticles and their role in the concrete. They studied different aspect ratios and volume loadings in the cement microstructure. They found coupling Ni/CNTs increased the dispersion and the compressive and flexural strength.

To our knowledge, no study has reported using aluminum oxide as a coating layer to improve mechanical properties and durability of cement mortars. Therefore, this paper reports on a core–shell structure of carbon nanofibers (CNFs) as the core and a porous, fluffy alumina Al_2O_3 as the shell, denoted Al_2O_3/CNFs, at a by-weight loading of 42% and 58% of Al_2O_3 and CNFs, respectively. The advantages of using such materials are a low cost, high specific surface area, high dispersion with pozzolanic property, and a bridging property. This study investigated the compressive strength, the freeze–thaw resistance, and drying shrinkage of the cement mortar. Three ratios of 0.125%, 0.25%, and 0.5% of Al_2O_3/CNFs and bare CNFs were compared with a standard sample. Our results show the 0.125% Al_2O_3/CNFs and 0.25 wt.% CNFs performed the best out of all tests. We deduce the reason for such a performance is that the Al_2O_3 nanolayer accelerated the production of the hydration gel. The gel production was confirmed by the thermal analysis (TGA) and XRD results.

2. Experimental Program

2.1. Materials

Ordinary Portland cement classified within the ASTM C150 standard type I was utilized to prepare the mortar mix. Natural river sand with a fineness modulus of 2.6 was used in all mixes. To enhance the workability, a water-reducing admixture was used. CNFs with a diameter ranging between 50 and 150 nm were used in this study. For coating, trimethylaluminum (TMA-1.0M) was used as a precursor.

2.2. Synthesis of Al_2O_3 Coating on CNFs

The synthesis of Al_2O_3 coating on CNFs was conducted by using a condensed layer deposition (CLD) [35]. The CNFs were surface-treated by a mixture of sulfuric acid and nitric acid at a volume ratio of 3:1 at a concentration of 6.0 M under a sonication probe for 2 h at 60 °C followed by deionized water washing to remove any acid residue. The coating process was prepared in batches where each batch of 400 mg of CNF was dispersed in 500 mL of heptane in two containers, 200 mg of CNFs in 250 mL each. After a 20 min period of sonication, 150 microliters of water were added to the mixture under both sonication and stirring. After another 20 min, a water film formed on the CNFs surface. Then, a 0.00789 mole of the TMA based on the reaction 2TMA + $3H_2O \rightarrow Al_2O_3 + 6CH_4$ was injected under a nitrogen blanket to prevent any atmosphere effect. The samples were dried under vacuum at 100 °C for 8 h. Subsequent annealing to remove any residues in the form of CHx was conducted at 350 °C for 2 h under air in a tubular furnace.

2.3. Fabrication of Mortars

A constant cement-to-sand ratio of 1:2 and a water-to-cement ratio of 0.35 were used. Three different ratios of CNFs and Al_2O_3/CNFs at weights of 0.125%, 0.25%, and 0.5% were prepared. In addition, a control specimen without nanomaterial, denoted as C0, was prepared (mixing properties are listed in Table 1). In a typical procedure, CNFs and Al_2O_3/CNFs suspensions can be prepared by sonicating them (at 800 W–20 kHz) in water while the mechanical stirring is ongoing for 30 min. This water can then be used as a mixing agent to prepare the composite mortars. To achieve fluidity, a certain ratio of a superplasticizer was added (see Table 1) to the mixture of water and nanomaterials or to water only. The superplasticizer, as a surfactant, was used only with the suspensions that had only CNFs. In the case of Al_2O_3/CNFs, 80% of water was used for Al_2O_3/CNF dispersion and 20% was mixed with superplasticizer. Figure 1 illustrates the mixing procedures for Al_2O_3/CNFs composites. Afterward, the fresh mixture was cast into different molds for the different testing procedures described in this paper.

Table 1. Ingredient proportions for cement mortar preparation.

Sample Designation	Water-to-Cement Ratio (w/c)	Sand-to-Cement Ratio (s/c)	Mix Proportion (% Weight)			Fluidity (mm)	CNF Content (%wt) *
			CNFs	Al_2O_3/CNFs	Superplasticizer		
C0	0.35	2	-	-	0.1	133	-
CNF-0.125	0.35	2	0.125	-	0.1	156	100
CNF-0.25	0.35	2	0.25	-	0.2	158	100
CNF-0.5	0.35	2	0.5	-	0.4	164	100
Al_2O_3/CNF-0.125	0.35	2	-	0.125	0.1	151	58
Al_2O_3/CNF-0.25	0.35	2	-	0.25	0.2	150	58
Al_2O_3/CNF-0.5	0.35	2	-	0.5	0.4	161	58

* Taken from TGA data.

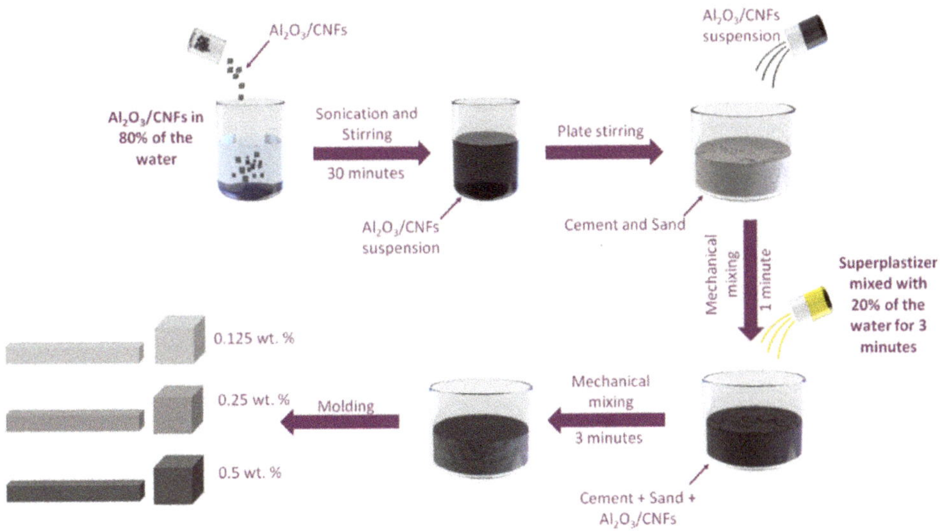

Figure 1. Schematic of preparation process.

2.4. Characterization Techniques

2.4.1. Surface Characterizations

Two types of surface characterizations were adopted. Transmission electron microscopy (TEM) was used to evaluate the Al_2O_3 coating on CNFs using Jeol-1400 at an acceleration voltage of 120 kV. Scanning electron microscopy (SEM) FEI Quanta 600 FEG SEM scope was used to investigate the cement internal structure. The SEM scope voltage was set at 5 kV to provide superior sensitivity to image surface structure. The surface area, whether that of Al_2O_3/CNFs or that of the cement composites, was measured using the Brunauer–Emmett–Teller method (BET).

2.4.2. Thermogravimetric Analysis

To exactly measure the Al_2O_3 coating layer loading on the CNFs and the sample overall, thermal analysis (TGA) model Q-500 (TA Instruments, Newcastle, DE, USA) was conducted in a range of 25–1000 °C with a scan rate of 10 °C/min. TGA-DTA was also performed to measure the Ca $(OH)_2$ content of the control and CNFs-0.25 and Al_2O_3/CNFs-0.125 samples at the ages of 7, 14, and 28 days. No special treatment, such as drying or pouring in any solvent, was used. Cement mortar slices of 5–8 mm were taken from the core of the sample cubes after being tested under compression to failure. The slices were crushed to powders and filtered by passing them through a 75 μm sieve. In each test, about 30 mg of sample was heated under nitrogen gas at a range of 25–1000 °C at a heating rate of 10 °C/min.

2.4.3. X-ray Diffraction Analysis

X-ray diffraction (XRD) (Philips X-Pert-Fisher Scientific, Hampton, NH, USA) equipped with CuKα was performed to analyze the crystalline phase of the samples. The data were collected for an angle range of 2θ = 5°–90° at a scan rate of 0.026° s^{-1}. The samples (50 mg in weight) were grained and sieved with a 75 μm sieve.

2.4.4. Fluidity and Compressive Strength Tests

To evaluate the effect of the addition of Al_2O_3/CNFs and CNFs on the cement mortars, fluidity and workability of different composite mixtures were examined based on ASTM C1437-13 standards [36]. After the mixing procedure was carried out, a specific amount of each mixture was poured into the flow table following ASTM standards. The compressive strength after 7, 14, and 28 days was determined following the procedure of the ASTM C109 standards [37]. Sixty-three specimens with dimensions of 50 mm × 50 mm × 50 mm were tested. The specimens were removed from the lime water solution and were wiped dry 24 h before testing.

2.4.5. Freeze–Thaw Tests

To study the durability of the bare and reinforced mortars, freeze–thaw cycling process was conducted at a temperature range of −18 °C to 4 °C based on ASTMC666, rapid freeze–thaw cycling in water [38]. The specimens were prisms of 25 mm × 25 mm × 280 mm dimensions. Before the freeze–thaw test, the specimens were demolded and moist-cured for 28 days. The samples were subjected to at least 300 freeze–thaw cycles. The loss of mass of all composite mortars and the reduction in compressive strength of composites with optimum dosage of nanomaterials was measured after a specified number of freeze–thaw cycles.

2.4.6. Drying Shrinkage Tests

Cement mortar samples with w/c ratio of 0.4 with dimensions of 25 mm × 25 mm × 280 mm were used for drying shrinkage measurements in accordance with ASTM C596 [39]. Samples were demolded after 48 h after casting and placed in lime water. After 72 h had passed since casting, the specimens were removed from the lime water and the initial dial gauge measurement of the length were recorded. The length measurements were conducted using length comparator according to the ASTM C490/C490M-17 [40]. The specimens were stored in a dry room at 22 ± 4 °C and 90% relative humidity. The length change measurements were taken at the ages of 5, 7, 14, 28, and 56 days.

3. Results and Discussion

Dispersion of bare CNFs and Al_2O_3/CNFs dispersed in water before and after surface functionalization is shown in Figure 2a, and Al_2O_3/CNFs before and after the annealing at 350 °C under air is shown in Figure 2b. The observations were made after the 30 min sonicating process without the addition of any surfactant, and the different solutions were left for 30 days. It could be seen that the untreated CNF solution showed no dispersion, while the treated CNFs solution showed relatively good dispersion. However, after 30 days, some CNF clusters were clearly visible at the bottom of the vials. The fresh Al_2O_3/CNFs settled down while in water, which was due to the organic ligands left over from the coating precursor molecule. The scenario was totally different after the fresh Al_2O_3/CNFs were exposed to air at a temperature of 350 °C for one hour. Therefore, annealed Al_2O_3/CNFs suspensions had much better long-term stability as compared with other suspensions. The removal of organic leftovers was confirmed by the TGA analysis, as shown in Figure 2b.

The better dispersion is attributed to the formation of pores due to oxidizing the organic content, converting them into highly hydrophilic surfaces.

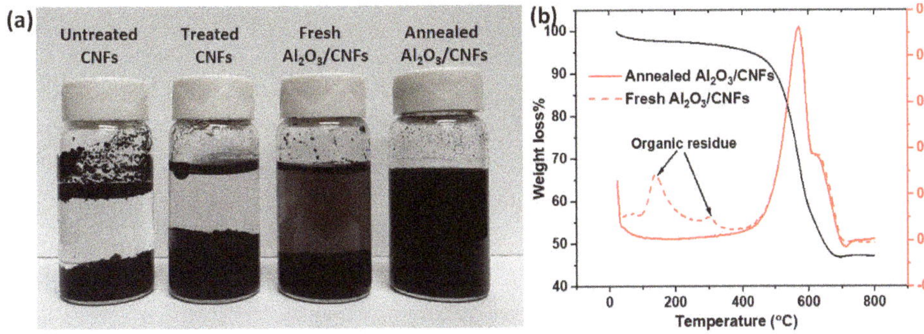

Figure 2. (**a**) CNF and Al_2O_3/CNF dispersion in water after one month, and (**b**) TGA-DTA profile for Al_2O_3/CNFs before and after annealing at 350 °C under air.

The findings of this paper provide fundamental evidence that could lead to practical implications in the future. The methodology and results obtained at a small scale can be advanced further with additional research to potentially impact current engineering practice. Specific discussion is provided next for the various results obtained in this paper.

3.1. Transmission Electron Microscopy (TEM)

The nanocoating of Al_2O_3 on CNFs, as characterized by TEM at both low and high resolution, is shown in Figure 3. Bare CNFs were examined by TEM as well, as shown in Figure 3a. Figure 3b shows a uniform coating with a porous morphology, which plays a role in the cement hydration. TGA analysis showed 42% and 58% weight percent for Al_2O_3 and CNFs, respectively. Figure 3c displays the adsorption scan of the nitrogen gas for surface area measurement. The surface area was found to be 274.36 m^2/g as well as the nanocoating layer pore size distribution. Figure 3d shows the TGA analysis for the mass loading of Al_2O_3 on CNFs, which was found to be 42 wt.%.

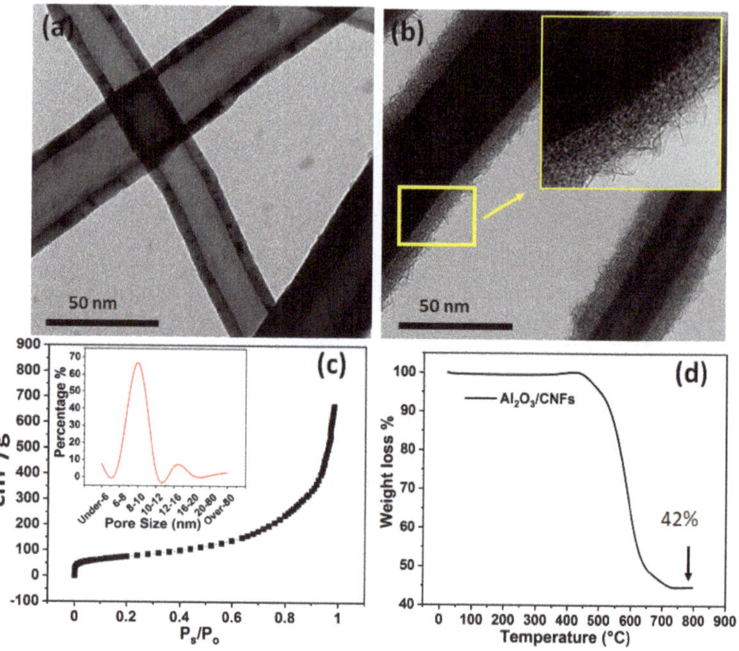

Figure 3. Bright-field transmission electron microscopy (TEM): (**a**) uncoated CNFs, (**b**) Al_2O_3 coating on CNFs showing a close-up of the layer morphology, (**c**) BET analysis for Al_2O_3/CNFs with its pore size distribution, and (**d**) TGA analysis showing mass loading of the Al_2O_3 on CNFs.

3.2. Microstructure Analysis

The microstructure of the cement mortar specimens was characterized by scanning electron microscopy (SEM). Figure 4 shows the SEM images for CNFs and Al_2O_3/CNFs at the age of 28 days, where the hydration products such as crystalline CH, needle-shaped ettringite, and amorphous calcium–silicate–hydrate (C-S-H) are clearly distinguished. The SEM session was performed at a voltage of 5 kV to avoid any surface charging under the beam exposure. Figure 4 shows the CNFs and Al_2O_3/CNF-embedded cement mortar (marked by arrows). Al_2O_3/CNFs was found to be well-anchored in the hydration products and was encapsulated by uniform and compacted C-S-H hydration products. The increased C-S-H gel formation is attributed to the contribution of the Al_2O_3 nanocoating as an effective layer with a high surface area that provides preferential nucleation sites for hydration products' growth. In addition, based on the observation of the surface morphology, a lower CH content was found when compared with CNFs-mortar composites. This could be due to the consumption of CH formed during the hydration of the cement [21]. On the other hand, the fracture surface for the CNFs composites showed a good interaction with the mortar matrix with a high CH content.

It is predicted that the porous morphology of Al_2O_3 enhances the hydration reaction as the porous sites adsorb the water, which becomes a nucleation site for the cement. The ratio of Ca/Si was used as a metric to determine the consumption of calcium converted to a hydrated silicate moiety [41]. The EDS analysis found the Ca/Si ratio was the highest in the case of C0 specimens, whereas it had the lowest value for Al_2O_3/CNFs-0.125% composite specimens, which further supports the findings from TGA and XRD data [42].

Figure 4. SEM images at different magnifications (**a**,**b**), CNFs embedded in cement mortar, and (**c**,**d**) Al_2O_3/CNFs embedded in cement mortar, all at the age of 28 days.

3.3. Cement Hydration by Thermal Analysis

Thermogravimetric (TGA) and differential thermogravimetric (DTG) curves were obtained to detect the influence of CNFs and Al_2O_3/CNFs in the hydration reaction of cement mortars at a water/cement ratio (w/c) of 0.35. In general, the decomposition of cement hydrates can be divided into three major regions. The first peak between 25 and 400 °C represents the evaporable water and dehydration of hydrated calcium silicate (C-S-H). The second one between 400 and 600 °C represents the dihydroxylation of portlandite (CH). As the CNFs started breaking down in this temperature window [43], this could be attributed to some overlapping between the CH and CNFs' thermal oxidation signal. The last peak due to decarbonation of $CaCO_3$ is between 600 and 800 °C [44].

The TGA and DTA results for the cement mortar composites for the standard C0, CNFs-0.25%, and Al_2O_3/CNFs-0.125% were collected at the ages of 7, 14 and 28 days, as shown in Figures 5–7, respectively. Compared with C0, the samples with nanomaterials showed an increase in the total mass loss. This is an indicator that the addition of the nanomaterial promoted the degree of hydration, but the Al_2O_3/CNF nanomaterials showed more significant effects at 14 and 28 days of curing age. One of the important metrics in assessing the extent of the hydration progress is determining the CH content. At 7 days, the largest amount of CH was observed in CNFs-0.25% and Al_2O_3/CNFs-0.125% at 17.1 and 16.82%, respectively. When increasing the curing age to 14 days, the CH content in C0 and CNFs-0.25% reached 17.75% and 18.6%, indicating that the degree of hydration increases with age. However, the CH content in the Al_2O_3/CNFs-0.125% composite was the lowest among the samples. This could also be explained due by the reactivity of the porous layer of alumina, which functions as an effective seeding site for hydration products' increase in CH and C-S-H. More uniform C-S-H can densify and compact the mortar matrix and restrict the growth of the crystal CH. The content of CH at the age of 28 days was also reduced in the case of CNF composites. Based on these results, one can conclude that the addition of CNFs and Al_2O_3/CNFs increases the degree of hydration by providing preferential nucleation sites at its oxygen functional groups. The effect of the covalent bonding provided by surface functional groups on CNFs or by the porous alumina govern its performance [45].

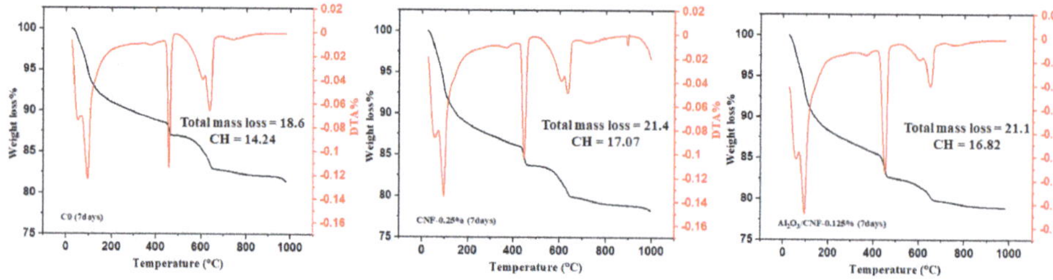

Figure 5. TGA analysis for C0 (**left**), CNFs-0.25 (**middle**), and Al_2O_3/CNFs-0.125 (**right**) at a curing age of 7 days.

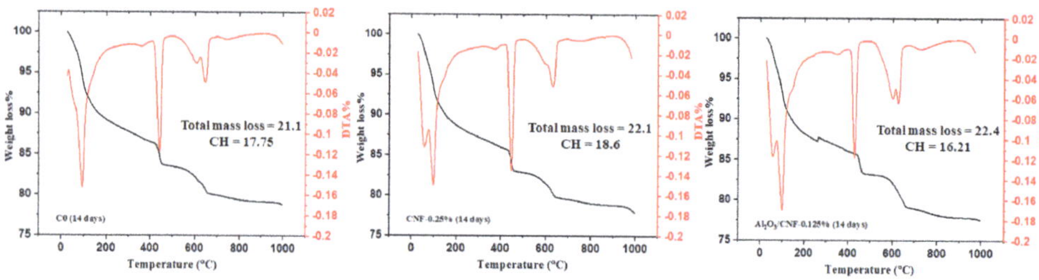

Figure 6. TGA analysis for C0 (**left**), CNFs-0.25 (**middle**), and Al_2O_3/CNFs-0.125 (**right**) at a curing age of 14 days.

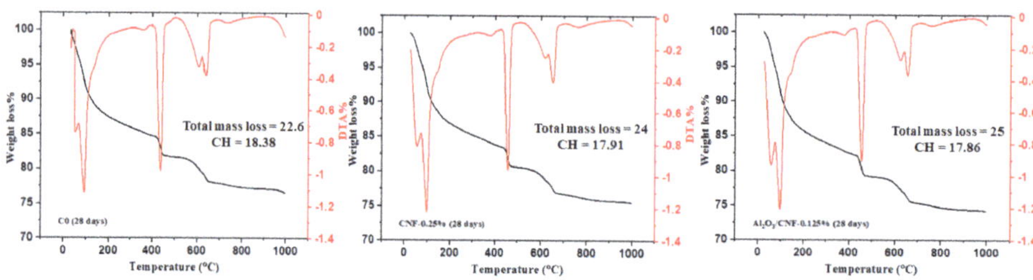

Figure 7. TGA analysis for C0 (**left**), CNFs-0.25 (**middle**), and Al_2O_3/CNFs-0.125 (**right**) at a curing age of 28 days.

3.4. X-ray Diffraction (XRD)

To qualitatively investigate the mineralogical compositions for the mortar composites, XRD patterns were performed. The results for samples C0, CNFs-0.25%, and Al_2O_3/CNFs-0.125% at the age of 28 days are displayed in Figure 8. It can be seen that the XRD patterns for all samples show identical diffraction angles. The phases found in all mixtures were ettringite (Aft), portlandite (CH), tricalcium silicate (C_3S), dicalcium silicate (C_2S), and calcium carbonate $CaCO_3$. The diffraction angle of 2θ = 9.1° belongs to ettringite, whereas the peak at 2θ = 22.5° is for $CaCO_3$. The peaks at 2θ = 18°, 28°, 47°, 51°, and 54° are attributed to calcium hydroxide (CH). The diffraction peaks at 2θ = 29.5°, 32°, 34°, and 39.3° were assigned to the anhydrate C_3S and C_2S [46]. The cement clinker was composed of about 75% chemical components of C_3S and C_2S. These two components react with water to produce crystalline calcium hydroxide or portlandite (CH) and the amorphous form of calcium–silicate–hydrate (C-S-H). In addition, these two hydration products comprise over

60% of hydration products. The chemical formulas expressing the hydration reaction of C_3S and C_2S are represented by the equations below:

$$C_3S + H_2O \rightarrow C\text{-}S\text{-}H + CH$$

$$C_2S + H_2O \rightarrow C\text{-}S\text{-}H + CH$$

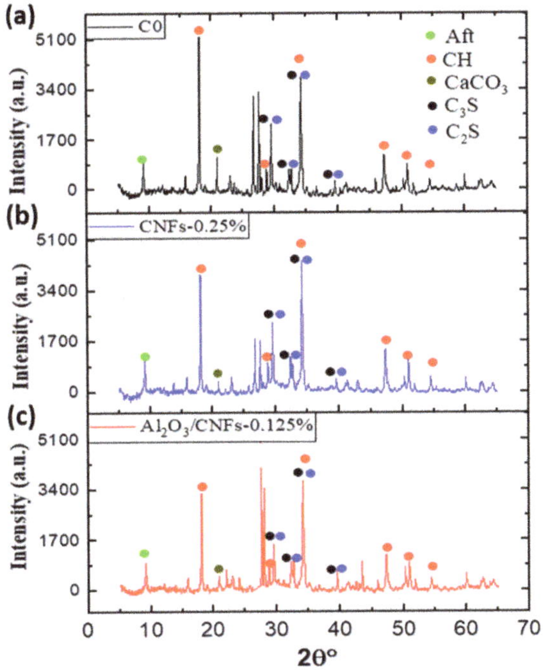

Figure 8. XRD patterns' total spectrum: (**a**) C0, (**b**) CNFs-0.25, and (**c**) Al_2O_3/CNFs-0.125%.

Therefore, measuring the peak intensities of C_2S and C_3S gave us a sense of how much C_2S and C_3S had been consumed during the hydration reaction. The intensity of the peaks of C_3S and C_2S at $2\theta = 29.5°$ was higher in the control sample. The peak intensity decreased when CNFs-0.25% was embedded, and a significant reduction in the peak was observed when Al_2O_3/CNFs-0.125% was in the cement matrix. The results indicate that hydration proceeds more efficiently in Al_2O_3/CNFs followed by CNFs-0.25% and finally C0. Another noted observation from the XRD data is the CH peak at $2\theta = 18°$, as shown in Figure 8. It was observed that the peak decreased in the case of Al_2O_3/CNFs-0.125% relative to the other two samples. This indicates the influence on the mortar hydration by accelerating the dissolution of C_3S and C_2S, leading to an increase in and better distribution of gel hydration C-S-H. This growth may restrict the growth of CH, as can be observed in the SEM morphology images (Figure 4d). The lesser CH intensity may also be an indicator of formation of an additional C-A-H gel [21–23,47]. The hydration gels are amorphous in nature and are not detectable in XRD. Therefore, an indirect measurement can be adapted by focusing on the CH intensity and anhydrate C_3S and C_2S phases. In conclusion, the addition of the CNFs and Al_2O_3/CNFs determined the degree of the hydration, and this promotion was significant in the case of Al_2O_3's presence. These results are in line with the TGA findings described earlier.

3.5. Compressive Strength

The compressive strength results of cement mortars reinforced by CNFs and Al_2O_3/CNFs are presented in Figure 9. These results are displayed based on the nanomaterials' weight percentage at the ages of 7, 14, and 28 days. These samples were compared with C0 at all ages. It was observed that with increasing curing age, all samples experienced an increase in their strength. The obtained compressive strength varies at all ages with different contents of CNFs and Al_2O_3/CNFs. The compressive strength for samples that contained CNFs at 0.125%, 0.25%, and 0.5% by weight of cement increased by about 15%, 36%, and 23%, respectively, at the age of 7 days. At the age of 14 days, the compressive strength of these composites increased by 8%, 24%, and 14%, respectively. When these samples reached a curing age of 28 days, the improvement in compressive strength was 6%, 15%, and 3%, respectively.

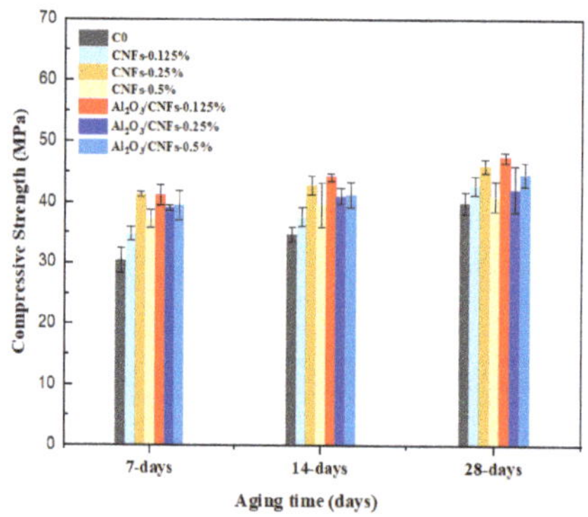

Figure 9. Compressive strength for a period of 7, 14, and 28 days.

The improvement in the compressive strength with different dosages of CNFs was due to the combined effects of the crack bridging and microstructure densifying due to refining the porosity [48]. In addition, the surface-functionalized CNFs possess active surface sites such as -COOH, -OH, -C=O. These groups have an ability to improve the bond with Ca^{2+} and to accelerate cement hydration, improving the mechanical strength as a result [49]. Out of the CNFs samples, CNFs-0.25% showed the highest strength due to the structure compacting and refinement of porosity. The filling effect of the CNFs can be clearly observed from the SEM images (Figure 4b). There was a reduction in the compressive strength when the ratio of the CNFs increased to 0.5%. This could be due to the agglomeration of CNFs inside the cement mortar matrix. The agglomeration creates large cavities which lower the strength [18].

On the other hand, the composites that contained Al_2O_3/CNFs showed progressive improvement. Relative to C0, the addition of 0.125% of Al_2O_3/CNFs increased the strength by 36%, 27%, and 19% at the ages of 7, 14, and 28 days, respectively. In the case of Al_2O_3/CNFs-0.25, the strength exceeded the compressive strength of the control sample by about 29%, 19%, and 6% at the ages of 7, 14, and 28 days. When increasing the dosage of Al_2O_3/CNFs powder to 0.5%, the strength of the samples was increased by 30%, 19%, and 12%, respectively. The reason behind the good performance of the Al_2O_3/CNFs nanomaterials is the stronger interfacial bonding between the porous layer of the alumina on CNFs with cement matrix. Interestingly, the quantity of the CNFs in Al_2O_3/CNFs samples was reduced by 42% compared to CNF samples. The effect of the coating layer promoted the mechanical properties of the composites significantly. This layer possesses a highly reactive surface area that promotes the nucleation effect. The reactivity of the surface is defined as highly populated active sites that can efficiently adsorb and convert the calcium to calcium–silicate hydrate products. This layer may also contribute to producing extra hydration products, which further enhance the pore refinement and densify the cement matrix. Compressive strength enhancements obtained in this paper are compared to those available in the literature (Table 2).

Table 2. Compressive strength comparison obtained in this paper with existing research.

Property	Enhancement in Properties	Material Concentration	Reference
Compressive strength	36%	0.125% Al_2O_3/CNFs	This study
	19%	0.5 wt.% MWCNTs	Li et al. [50]
	22%	0.1 wt.% MWCNTs	Bharj et al. [51]
	42.7%	0.16 wt.% CNFs	Gao et al. [52]
	25.4%	0.5 wt.% SiO_2/MWCNTs	Shaojie et al. [53]
	20%	0.125 wt.% MWCNTs/NSs	Pawel et al. [31]
	40%	0.5 wt.% CNT/TNT	Liu et al. [54]

3.6. Fluidity

The flow table diameter results are presented in Table 1. With different percentages of the nanomaterials and superplasticizer (SP), the mortar showed different fluidity performances. The control samples' fluidity was only 133 mm. The fluidity diameters increased remarkably with the addition of CNFs-SP and Al_2O_3\CNFs-SP at various ratios. However, the fresh mortars reinforced with Al_2O_3/CNFs showed a noticeable reduction in flow spread compared with the mortars with different CNF ratios. This reduction in the workability was due to the large surface area that characterized the Al_2O_3-coated CNFs. The high surface area of the nanomaterial needs more water to wet its surface; this need can reduce the free water in the mixture [10]. Increasing the dosages of the SP is important to offset the reduction in the fluidity of mortars reinforced with Al_2O_3/CNFs at different ratios.

3.7. Freeze–Thaw Effects

The effect of the addition of the nanomaterials on the freeze–thaw performance of cement composites is evaluated in this section. The freeze–thaw performance in terms of physical and mechanical properties is described next.

3.7.1. Compressive Strength

Figure 10 shows compressive strength versus the number of freeze–thaw cycles at 50, 100, and 150 cycles for C0, CNFs-0.25%, and Al_2O_3/CNFs-0.125% samples. Overall, samples with nanomaterials showed less compressive reduction when compared to C0 control samples. C0 displayed a decrease of 45% in compressive strength after 150 cycles, whereas specimens containing CNFs-0.25% showed a decrease of 22%. Interestingly, Al_2O_3/CNFs-0.125% had a significant enhancement in mortars' frost resistance, showing the lowest compressive strength reduction of 14%. This performance was confirmed by the Barrett–Joyner–Halenda (BJH) analysis, as demonstrated in the section pore structure below. The BJH indicates the ability of Al_2O_3/CNFs to further refine the small and medium capillary pores, resulting in more compacted and durable composites.

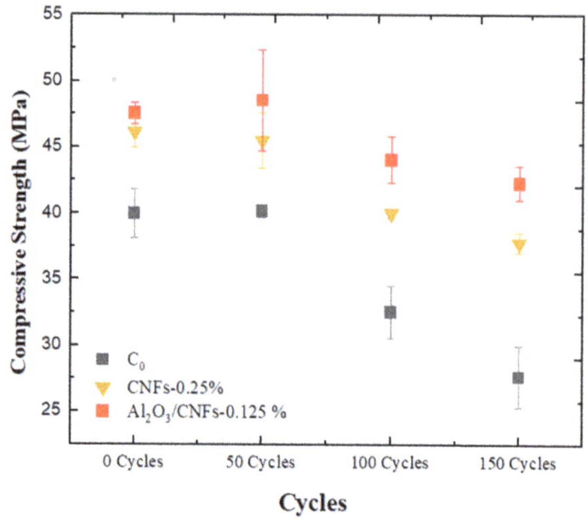

Figure 10. Compressive strength for C0, CNFs-0.125%, and Al_2O_3 /CNFs-0.125% composites after exposure to freezing–thawing cycles.

3.7.2. Mass Loss

The weight changes were monitored during the freeze–thaw cycles, which were conducted after 50, 100, 150, 200, 250, and 300 cycles. As shown in Figure 11, the mass loss for all mixtures increased with increasing the number of freeze–thaw cycles. With the addition of nanomaterials, the mass loss in the samples was significantly reduced. In the case of the control sample C0, the average mass loss was about 9% after 300 cycles, while the addition of CNFs at different ratios always lowered mass loss, with the lowest being observed for the case of CNFs-0.25%. This may indicate that the addition of the functionalized CNFs worked to reduce porosity and increase the density of the CNF/cement composites. However, the mass loss increased when CNFs-0.5% was embedded in cement mortars. This was expected as these composites showed lower strength due to the large pores that were created in the matrix because of the agglomeration of the CNFs bundles.

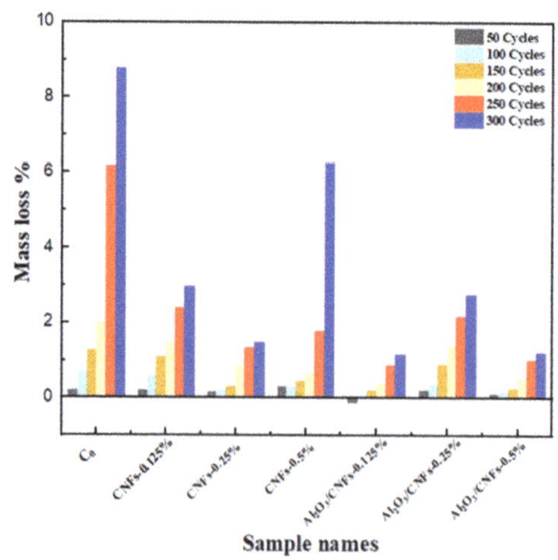

Figure 11. Mass loss ratios of mortar samples exposed to freezing–thawing cycles.

Following the same trend, the addition of Al_2O_3/CNFs showed a remarkable performance in resisting frost. Overall, the addition of different ratios of Al_2O_3/CNFs enhanced the frost resistance. The addition of Al_2O_3/CNFs-0.125% and Al_2O_3/CNFs-0.5% lead to a significantly lower mass loss after 300 cycles by only 1%. This improvement is explained by the decreasing porosity resulting from modifying the internal voids. This might be due to the refinement of the pores' structure by producing additional hydrated gels, which prevent intrusion of water molecules into the mixture's microstructure. Furthermore, this refinement could be due to the reactivity of the porous Al_2O_3 layer, which accelerates the hydration process, forming more well-distributed C-S-H hydrated gels in the matrix. Fewer CH crystals and more amorphous C-S-H in Al_2O_3/CNFs-0.125%, as shown in SEM, XRD, and TGA/DTA, is a testament of this speculation. The small-size voids could minimize the water settling time and subsequently retard the ice volume enlargement. This is expected since a small-size void would hold a high internal pressure and surface tension based on physics principles. The physical appearance of all samples after 300 cycles is presented in Figure 12. Clearly, the addition of the nanomaterials with good dispersion in the mortar matrix enhanced the frost resistance. The high frost resistance was observed in Al_2O_3/CNFs composites.

Figure 12. Physical appearance of the samples after 300 cycles.

3.7.3. Pore Structure

The cement mortar powder was grained and filtered by a 300 μm sieve, and then it was dried and vacuumed at 200 °C for 2 h during the degassing process. The pore volume of the samples was calculated based on the BJH method [55]. The overall BJH pore volume was also found to be 0.07805, 0.06783, and 0.06278 mL/g for C0, CNFs-0.25%, and Al_2O_3/CNFs-0.125%, respectively. Figure 13a shows a type IV isotherm hysteresis for adsorption–desorption [56]. Figure 13b shows the pore measurement statistics for the three samples. The pore diameter statistic is divided into four regions. It can be seen that the Al_2O_3/CNFs-0.125% had pores with a greater volume in the range of 1–10 nm, which are classified as gel pores [57]. The gel pores are related to the formation of C-S-H, which is often produced by the hydration reaction of C_3S and C_2S. In other ranges, where the capillary pores are present, the Al_2O_3/CNFs-0.125% pore volume was significantly reduced, indicating the refinement was enhanced. The pore refinement was also induced when CNFs were incorporated; however, Al_2O_3/CNFs-0.125% further reduced the pore volume, which may be due to the higher C-S-H content, as described in TGA and XRD results. It may also be due to the pozzolanic reaction offered by the Al_2O_3 nanocoating, which helps produce extra C-A-H. Recently, a study has demonstrated the effect of different aluminum oxide phases ($Al(OH)_3$, γ-Al_2O_3, α-Al_2O_3) on the calcium aluminate cement hydration [58]. In another study, it was found that $Al(OH)_3$ significantly accelerates hydration due to the high surface area and the ability to serve as a nucleation center for C-A-H moieties [59]. This is in line with the results obtained in this study, as aluminum oxide was in the boehmite phase, as confirmed by the XRD shown in this reference [35].

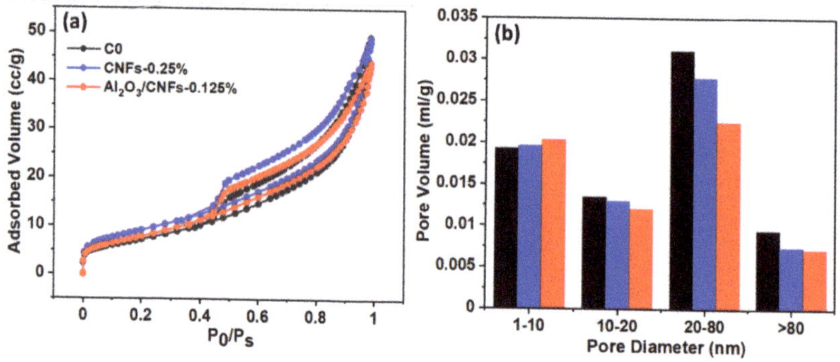

Figure 13. (**a**) BET isotherm analysis, (**b**) pore volume distribution.

3.8. Drying Shrinkage

The effect of the addition of the nanomaterials on the drying shrinkage was studied at a water/cement ratio of 0.40. The results, as shown in Figure 14, indicate that the cement mortars with CNFs and Al_2O_3-coated CNFs had lower shrinkage compared to the C0 control mortar at early and late ages. In comparison with C0, the average early drying shrinkage of composites that were reinforced by CNFs at ratios of 0.125%, 0.25%, and 0.5% showed noticeable decreases of about 19%, 33%, and 22%, respectively, at the age of 7 days, while after 56 days, the reduction was about 16%, 20%, and 14%. In addition, the results show a remarkable reduction in the drying shrinkage for the composites that contained Al_2O_3/CNFs at various ratios. It is clear that the addition of the Al_2O_3/CNFs-0.125% significantly reduced the shrinkage relative to other samples, whether in the early or late ages. At age of 7 days and 56 days, Al_2O_3/CNFs-0.125% reduced the shrinkage by 62% and 47%, respectively, compared to the control C0 mortar. The Al_2O_3/CNFs-0.25% mortar showed 36% and 27% shrinkage reduction during the same time periods. On the other hand, the composites that contained 0.5% of Al_2O_3/CNFs exhibited low shrinkage by 30% and 17% at the age of 7 and 56 days, respectively.

Drying shrinkage of cementitious materials is due to capillary surface tension because of the meniscus formation in the capillary pores [60]. The incorporation of both CNFs and Al_2O_3/CNFs in the mortar matrix reduced the total overall porosity. In addition, increasing the stiffness due to the microcrack bridging effect of these nanomaterials can be an effective way to reduce the shrinkage deformation. Furthermore, the Al_2O_3 porous layer worked as a reactive site to increase the nucleation, which worked to densify and to compact the microstructure of the matrix. The refinement in capillary pores served as an obstacle impeding the water loss from the cement skeleton [60,61]. The mitigation of the drying shrinkage of the Al_2O_3/CNFs composites could be due to the high hydrophilicity of these nanomaterials. The Al_2O_3 nanocoating is in a boehmite phase [35], and thus has a strong ability to absorb and retain water as long as it is in equilibrium with surrounding cement compounds. When the shrinkage takes place, the water is released for the gel to achieve the internal hydration equilibrium [62].

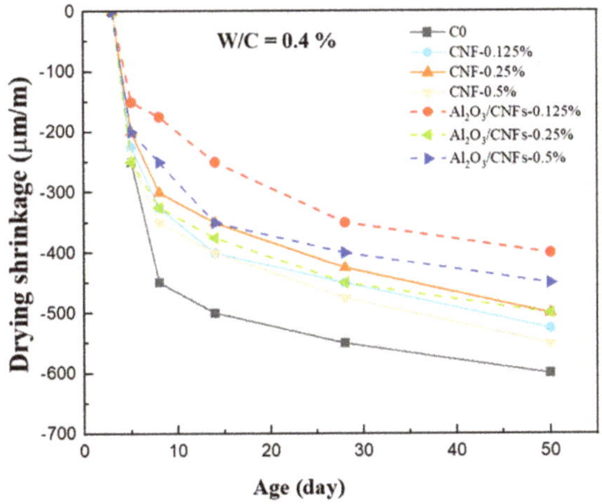

Figure 14. Drying shrinkage data for all the samples.

4. Limitations of the Study

This study investigated the effects of using Al_2O_3/CNFs on compressive strength, durability under freeze–thaw cycles, and drying shrinkage of cement mortars. Although the results have proven the viability of the approach, the study had many constraints. Firstly, the stability of the Al_2O_3 nanolayer in high concentrations of the superplasticizer, which is an acidic agent, is unknown so far. Secondly, the effect of the alumina-coated carbon nanofibers on drying shrinkage with different water-to-cement ratios and various temperatures is not included in this study. Finally, the degree of pores' refinements due to Al_2O_3/CNFs incorporation has not been measured by a direct technique such as water absorption or mercury intrusion porosimeter.

5. Conclusions

Based on the results presented in this paper, the following conclusions were drawn:

1. The TEM images revealed a well-coated layer on the CNF surfaces.
2. The SEM results show the nanomaterials were well-distributed in the mortar microstructure.
3. The SEM images indicate the formation of the hydration products gels such as CH, C-S-H, and ettringite. In the case of the Al_2O_3/CNFs-mortar composite, most of the hydrated products were well-distributed C-S-H and were identified from their amorphous appearance, while CNFs-mortar composites showed mostly ettringite and hexagonal CH.
4. With increasing curing age, all samples experienced an increase in their compressive strength. The compressive strength for samples that contained Al_2O_3/CNFs demonstrated progressive improvements compared to other samples evaluated in this research. Relative to control specimens, the addition of Al_2O_3/CNFs-0.125% increased the strength by 36%, 27%, and 19% at the age of 7, 14, and 28 days, respectively. The alumina porous layer improved the binding between the mortar matrix and the CNFs' surfaces and altered the hydration process.
5. TGA-DTA tests showed all CNFs and Al_2O_3/CNF composites exhibited higher degrees of hydration than the control sample. Decreasing calcium hydroxide was observed at an age of 28 days with the addition of the nanomaterials, with a significant effect for the case of coated CNFs. This could be due to the nucleation effect, which produced a good distribution of the gel hydration of C-S-H inside the mortar matrix, which restricts the growth of the CH. Additionally, this may be due to the

pozzolanic effect of the Al_2O_3 layer. There might be some minor overlapping in the signal from CNFs burning and CH diminishing.

6. The XRD measurement at the age of 28 days focused on the intensity of two cementitious moieties: the C_2S and C_3S to be consumed in the hydration reaction. Al_2O_3/CNFs-0.125% showed the lowest intensity of these two compounds, which indicates that they increased the dissolution of C_3S and C_2S, and thus more C-S-H was produced. In addition, XRD data focused on the intensity of CH, as it showed lower intensity in the case of CNFs-0.25% and Al_2O_3/CNFs 0.125% composites. These results are in good agreement with TGA and EDS analyses findings.
7. The addition of Al_2O_3/CNFs-0.125% and Al_2O_3/CNFs-0.5% showed a significant low mass loss of about 1% after 300 freeze–thaw cycles, while the reference sample and CNFs-0.25 showed a mass loss of 9% and 3%, respectively.
8. The compressive strength degradation was the lowest in the case of the Al_2O_3/CNFs-0.125% sample after 150 freeze–thaw cycles at about 14%, while the control and CNFs-0.25% samples showed a degradation of about 45 and 22%, respectively. Such results were attributed to the highly dense microstructure, which was later confirmed by the BET measurements. The BET and BJH tests exhibited low gas adsorption and more refinement in the capillary pores.
9. Incorporation of the optimum amounts of CNFs (0.25%) and Al_2O_3/CNFs (0.125%) led to a decrease in the drying shrinkage due to increasing the stiffness and density of these composites. In addition, the hydrophilicity of the alumina layer explained the significant drying shrinkage mitigation.

Author Contributions: Conceptualization, H.A.Q.; methodology, H.A.Q., H.S. and A.M.J.; investigation, H.A.Q.; resources, H.S.; writing—original draft preparation, H.A.Q.; writing—review and editing, H.S., A.M.J., Y.X. and D.S.; visualization, D.S. and A.M.J.; supervision, H.S.; project administration, H.S.. All authors have read and agreed to the published version of the manuscript.

Funding: This research was partially funded by the University of Missouri's Electron Microscopy Core for SEM and TEM imaging, under award number FY2020, and the APC was funded by H.S.

Institutional Review Board Statement: Not applicable.

Informed Consent Statement: Not applicable.

Data Availability Statement: The data presented in this study are available on request from the corresponding author. The data are not publicly available due to copyright.

Acknowledgments: We acknowledge Orton for assisting in the preparation of the mortar specimens, Fidalgo for the utilization of her lab for the BET measurements, and Ulery for the utilization of his lab the TGA analysis.

Conflicts of Interest: The authors declare no conflict of interest.

References

1. Ateş, A. Mechanical properties of sandy soils reinforced with cement and randomly distributed glass fibers (GRC). *Compos. Part B Eng.* **2016**, *96*, 295–304. [CrossRef]
2. Balaguru, P.N.; Shah, S.P. *Fiber-Reinforced Cement Composites*; McGraw-Hill: New York, NY, USA, 1992.
3. Bentur, A.; Mindess, S. *Fibre Reinforced Cementitious Composites*; CRC Press: Boca Raton, FL, USA, 2006.
4. Banthia, N.; Sappakittipakorn, M. Toughness enhancement in steel fiber reinforced concrete through fiber hybridization. *Cem. Concr. Res.* **2007**, *37*, 1366–1372. [CrossRef]
5. Pigeon, M. *Durability of Concrete in Cold Climates*; CRC Press: Boca Raton, FL, USA, 2014.
6. del Carmen Camacho, M.; Galao, O.; Baeza, F.J.; Zornoza, E.; Garcés, P. Mechanical properties and durability of CNT cement composites. *Mater. Des.* **2014**, *7*, 1640–1651. [CrossRef] [PubMed]
7. Camacho-Ballesta, C.; Galao, Ó.; Baeza, F.J.; Zornoza, E.; Garcés, P. *Durability and Mechanical Properties of CNT Cement Composites*; Springer International Publishing: Cham, Switzerland, 2019; pp. 31–41.
8. Nivethitha, D.; Srividhya, S.; Dharmar, S. Review on mechanical properties of cement mortar enhanced with nanoparticles. *Int. J. Sci. Res. Dev. News R D News* **2016**, *5*, 913–916.

9. Wang, T.; Xu, J.; Meng, B.; Peng, G. Experimental study on the effect of carbon nanofiber content on the durability of concrete. *Constr. Build. Mater.* **2020**, *250*, 118891. [CrossRef]
10. Chuah, S.; Pan, Z.; Sanjayan, J.G.; Wang, C.M.; Duan, W.H. Nano reinforced cement and concrete composites and new perspective from graphene oxide. *Constr. Build. Mater.* **2014**, *73*, 113–124. [CrossRef]
11. Wang, F.; Kong, X.; Jiang, L.; Wang, D. The acceleration mechanism of nano-C-S-H particles on OPC hydration. *Constr. Build. Mater.* **2020**, *249*, 118734. [CrossRef]
12. Ebrahimi, K.; Daiezadeh, M.J.; Zakertabrizi, M.; Zahmatkesh, F.; Korayem, A.H. A review of the impact of micro- and nanoparticles on freeze-thaw durability of hardened concrete: Mechanism perspective. *Constr. Build. Mater.* **2018**, *186*, 1105–1113. [CrossRef]
13. Kumar, S.; Kolay, P.; Malla, S.; Mishra, S. Effect of Multiwalled Carbon Nanotube in Cement Composite on Mechanical Strength and Freeze-Thaw Susceptibility. *Adv. Civ. Eng. Mater.* **2015**, *4*, 257–274. [CrossRef]
14. Lee, H.; Jeong, S.; Cho, S.; Chung, W. Enhanced bonding behavior of multi-walled carbon nanotube cement composites and reinforcing bars. *Compos. Struct.* **2020**, *243*, 112201. [CrossRef]
15. Al-Rub, R.K.A.; Ashour, A.I.; Tyson, B.M. On the aspect ratio effect of multi-walled carbon nanotube reinforcements on the mechanical properties of cementitious nanocomposites. *Constr. Build. Mater.* **2012**, *35*, 647–655. [CrossRef]
16. Cwirzen, A.; Habermehl-Cwirzen, K. The effect of carbon nano-and microfibers on strength and residual cumulative strain of mortars subjected to freeze-thaw cycles. *J. Adv. Concr. Technol.* **2013**, *11*, 80–88. [CrossRef]
17. Metaxa, Z.S.; Konsta-Gdoutos, M.S.; Shah, S.P. Carbon Nanofiber–Reinforced Cement-Based Materials. *Transp. Res. Rec. J. Transp. Res. Board* **2010**, *2142*, 114–118. [CrossRef]
18. Shi, T.; Li, Z.; Guo, J.; Gong, H.; Gu, C. Research progress on CNTs/CNFs-modified cement-based composites—A review. *Constr. Build. Mater.* **2019**, *202*, 290–307. [CrossRef]
19. Givi, A.N.; Rashid, S.A.; Aziz, F.N.A.; Salleh, M.A.M. Experimental investigation of the size effects of SiO2 nano-particles on the mechanical properties of binary blended concrete. *Compos. Part B Eng.* **2010**, *41*, 673–677. [CrossRef]
20. Reches, Y. Nanoparticles as concrete additives: Review and perspectives. *Constr. Build. Mater.* **2018**, *175*, 483–495. [CrossRef]
21. Behfarnia, K.; Salemi, N. The effects of nano-silica and nano-alumina on frost resistance of normal concrete. *Constr. Build. Mater.* **2013**, *48*, 580–584. [CrossRef]
22. El-Gamal, S.M.A.; Abo-El-Enein, S.A.; El-Hosiny, F.I.; Amin, M.S.; Ramadan, M. Thermal resistance, microstructure and mechanical properties of type I Portland cement pastes containing low-cost nanoparticles. *J. Therm. Anal.* **2018**, *131*, 949–968. [CrossRef]
23. Nazari, A.; Riahi, S.; Riahi, S.; Shamekhi, S.F.; Khademno, A. Influence of Al_2O_3 nanoparticles on the compressive strength and workability of blended concrete. *J. Am. Sci.* **2010**, *6*, 6–9.
24. Muzenski, S.; Flores-Vivian, I.; Sobolev, K. Ultra-high strength cement-based composites designed with aluminum oxide nano-fibers. *Constr. Build. Mater.* **2019**, *220*, 177–186. [CrossRef]
25. Fu, X.; Lu, W.; Chung, D.D.L. Ozone treatment of carbon fiber for reinforcing cement. *Carbon* **1998**, *36*, 1337–1345. [CrossRef]
26. Wang, J.; Cheng, Y.; Yuan, L.; Xu, D.; Du, P.; Hou, P.; Zhou, Z.; Cheng, X.; Liu, S.; Wang, Y. Effect of nano-silica on chemical and volume shrinkage of cement-based composites. *Constr. Build. Mater.* **2020**, *247*, 118529. [CrossRef]
27. Abdel-Gawwad, H.A.; Mohammed, M.S.; Alomayri, T. Single and dual effects of magnesia and alumina nano-particles on strength and drying shrinkage of alkali activated slag. *Constr. Build. Mater.* **2019**, *228*, 116827. [CrossRef]
28. Sanchez, F.; Ince, C. Microstructure and macroscopic properties of hybrid carbon nanofiber/silica fume cement composites. *Compos. Sci. Technol.* **2009**, *69*, 1310–1318. [CrossRef]
29. Yazdanbakhsh, A.; Grasley, Z. Utilization of Silica Fume to Stabilize the Dispersion of Carbon Nanofilaments in Cement Paste. *J. Mater. Civ. Eng.* **2014**, *26*, 06014010. [CrossRef]
30. Stynoski, P.; Mondal, P.; Marsh, C. Effects of silica additives on fracture properties of carbon nanotube and carbon fiber reinforced Portland cement mortar. *Cem. Concr. Compos.* **2015**, *55*, 232–240. [CrossRef]
31. Sikora, P.; Elrahman, M.A.; Chung, S.-Y.; Cendrowski, K.; Mijowska, E.; Stephan, D. Mechanical and microstructural properties of cement pastes containing carbon nanotubes and carbon nanotube-silica core-shell structures, exposed to elevated temperature. *Cem. Concr. Compos.* **2019**, *95*, 193–204. [CrossRef]
32. Garg, M.; Das, C.S.; Gupta, R. Use of silica particles to improve dispersion of -COOH CNTs/carbon fibers to produce HyFRCC. *Constr. Build. Mater.* **2020**, *250*, 118777. [CrossRef]
33. Kim, H.-K.; Nam, I.; Lee, H. Enhanced effect of carbon nanotube on mechanical and electrical properties of cement composites by incorporation of silica fume. *Compos. Struct.* **2014**, *107*, 60–69. [CrossRef]
34. Dong, S.; Wang, D.; Ashour, A.; Han, B.; Ou, J. Nickel plated carbon nanotubes reinforcing concrete composites: From nano/micro structures to macro mechanical properties. *Compos. Part A: Appl. Sci. Manuf.* **2021**, *141*, 106228. [CrossRef]
35. Jasim, A.M.; He, X.; White, T.A.; Xing, Y. Nano-layer deposition of metal oxides via a condensed water film. *Commun. Mater.* **2020**, *1*, 9.
36. ASTM C1437-13. Standard Test Method for Flow of Hydraulic Cement Mortar. ASTM: West Conshohocken, PA, USA, 2013.
37. ASTM C109/C109M. Standard Test Method for Compressive Strength of Hydraulic Cement Mortars (Using 2-in. or (50-mm) Cube Specimens). ASTM: West Conshohocken, PA, USA, 1999; Volume 2.
38. ASTM C666–03. Standard Test Method for Resistance of Concrete to Rapid Freezing and Thawing. American Society for Testing Materials International: West Conshohocken, PA, USA, 2008.

39. ASTM C596-18. Standard Test Method for Drying Shrinkage of Mortar Containing Hydraulic Cement. ASTM: West Conshohocken, PA, USA, 2018.
40. ASTM C490-07. Standard Practice for Use of Apparatus for the Determination of Length Change of Hardened Cement Paste, Mortar, and Concrete. ASTM: West Conshohocken, PA, USA, 2011.
41. da Silva Andrade, D.; da Silva Rêgo, J.H.; Morais, P.C.; de Mendonça Lopes, A.N.; Rojas, M.F. Investigation of C-S-H in ternary cement pastes containing nanosilica and highly-reactive supplementary cementitious materials (SCMs): Microstructure and strength. *Constr. Build. Mater.* **2019**, *198*, 445–455. [CrossRef]
42. Al Qader, H.J.N. Enhancing the Mechanical and Durability Properties of Cement Mortars by using Alumina Nanocoating on Carbon Nanofibers. Master Thesis, University of Missouri, Columbia, MO, USA, 2021.
43. Kumar, S.; Rath, T.; Mahaling, R.N.; Das, C.K. Processing and characterization of carbon nanofiber/syndiotactic polystyrene composites in the absence and presence of liquid crystalline polymer. *Compos. Part A Appl. Sci. Manuf.* **2007**, *38*, 1304–1317. [CrossRef]
44. Deboucha, W.; Leklou, N.; Khelidj, A.; Oudjit, M.N. Hydration development of mineral additives blended cement using thermogravimetric analysis (TGA): Methodology of calculating the degree of hydration. *Constr. Build. Mater.* **2017**, *146*, 687–701. [CrossRef]
45. Makar, J.M.; Chan, G.W. Growth of Cement Hydration Products on Single-Walled Carbon Nanotubes. *J. Am. Ceram. Soc.* **2009**, *92*, 1303–1310. [CrossRef]
46. Tafesse, M.; Kim, H.-K. The role of carbon nanotube on hydration kinetics and shrinkage of cement composite. *Compos. Part B: Eng.* **2019**, *169*, 55–64. [CrossRef]
47. Mahinroosta, M.; Allahverdi, A. A Scoping Review on Integrating Inorganic Nanomaterials into Cement Composites. *Adv. Civ. Eng. Mater.* **2019**, *8*, 526–553. [CrossRef]
48. Fehervari, A.; MacLeod, A.J.N.; Garcez, E.O.; Aldridge, L.; Gates, W.P.; Yang, Y.; Collins, F. On the mechanisms for improved strengths of carbon nanofiber-enriched mortars. *Cem. Concr. Res.* **2020**, *136*, 106178. [CrossRef]
49. Peyvandi, A.; Soroushian, P.; Abdol, N.; Balachandra, A.M. Surface-modified graphite nanomaterials for improved reinforcement efficiency in cementitious paste. *Carbon* **2013**, *63*, 175–186. [CrossRef]
50. Li, G.Y.; Wang, P.M.; Zhao, X. Mechanical behavior and microstructure of cement composites incorporating surface-treated multi-walled carbon nanotubes. *Carbon* **2005**, *43*, 1239–1245. [CrossRef]
51. Bharj, J. Experimental study on compressive strength of cement-CNT composite paste. *Indian J. Pure Appl. Phys.* **2015**, *52*, 35–38.
52. Gao, D.; Sturm, M.; Mo, Y. Electrical resistance of carbon-nanofiber concrete. *Smart Mater. Struct.* **2009**, *18*, 095039. [CrossRef]
53. Li, S.; Zhang, Y.; Lin, S.; Yan, J.; Du, S. Effects of nano-SiO2 coated multi-walled carbon nanotubes on mechanical properties of cement-based composites. *Constr. Build. Mater.* **2021**, *281*, 122577. [CrossRef]
54. Liu, J.; Suh, H.; Jee, H.; Xu, J.; Nezhad, E.Z.; Choi, C.-S.; Bae, S. Synergistic effect of carbon nanotube/TiO_2 nanotube multi-scale reinforcement on the mechanical properties and hydration process of portland cement paste. *Constr. Build. Mater.* **2021**, *293*, 123447. [CrossRef]
55. Shafaei, D.; Yang, S.; Berlouis, L.; Minto, J. Multiscale pore structure analysis of nano titanium dioxide cement mortar composite. *Mater. Today Commun.* **2020**, *22*, 100779. [CrossRef]
56. Vardharajula, S.; Ali, S.Z.; Tiwari, P.M.; Eroğlu, E.; Vig, K.; Dennis, V.A.; Singh, S.R. Functionalized carbon nanotubes: Biomedical applications. *Int. J. Nanomed.* **2012**, *7*, 5361.
57. Meng, W.; Khayat, K.H. Effect of graphite nanoplatelets and carbon nanofibers on rheology, hydration, shrinkage, mechanical properties, and microstructure of UHPC. *Cem. Concr. Res.* **2018**, *105*, 64–71. [CrossRef]
58. Engbert, A.; Plank, J. Impact of sand and filler materials on the hydration behavior of calcium aluminate cement. *J. Am. Ceram. Soc.* **2021**, *104*, 1067–1075. [CrossRef]
59. Reches, Y.; Thomson, K.; Helbing, M.; Kosson, D.S.; Sanchez, F. Agglomeration and reactivity of nanoparticles of SiO_2, TiO_2, Al_2O_3, Fe_2O_3, and clays in cement pastes and effects on compressive strength at ambient and elevated temperatures. *Constr. Build. Mater.* **2018**, *167*, 860–873. [CrossRef]
60. Young, J. Physical mechanisms and their mathematical descriptions. In *Mathematical Modelling of Creep and Shrinkage of Concrete*; Bazant, Z.P., Ed.; Wiley: Chichester, UK, 1988; pp. 63–98.
61. Hawreen, A.; Bogas, J.; Dias, A.P.S. On the mechanical and shrinkage behavior of cement mortars reinforced with carbon nanotubes. *Constr. Build. Mater.* **2018**, *168*, 459–470. [CrossRef]
62. Liu, X.; Fang, T.; Zuo, J. Effect of Nano-Materials on Autogenous Shrinkage Properties of Cement Based Materials. *Symmetry* **2019**, *11*, 1144. [CrossRef]

materials

MDPI

Article

Interpretable Machine Learning Algorithms to Predict the Axial Capacity of FRP-Reinforced Concrete Columns

Celal Cakiroglu [1], Kamrul Islam [2], Gebrail Bekdaş [3,*], Sanghun Kim [4] and Zong Woo Geem [5,*]

1 Department of Civil Engineering, Turkish-German University, Istanbul 34820, Turkey; cakiroglu@tau.edu.tr
2 Department of Civil, Geological and Mining Engineering, Polytechnique Montréal, Montreal, QC H3C 3A7, Canada; kamrul.islam@polymtl.ca
3 Department of Civil Engineering, Istanbul University—Cerrahpasa, Istanbul 34320, Turkey
4 Department of Civil and Environmental Engineering, Temple University, Philadelphia, PA 19122, USA; sanghun.kim@temple.edu
5 Department of Smart City & Energy, Gachon University, Seongnam 13120, Korea
* Correspondence: bekdas@iuc.edu.tr (G.B.); geem@gachon.ac.kr (Z.W.G.)

Abstract: Fiber-reinforced polymer (FRP) rebars are increasingly being used as an alternative to steel rebars in reinforced concrete (RC) members due to their excellent corrosion resistance capability and enhanced mechanical properties. Extensive research works have been performed in the last two decades to develop predictive models, codes, and guidelines to estimate the axial load-carrying capacity of FRP-RC columns. This study utilizes the power of artificial intelligence and develops an alternative approach to predict the axial capacity of FRP-RC columns more accurately using data-driven machine learning (ML) algorithms. A database of 117 tests of axially loaded FRP-RC columns is collected from the literature. The geometric and material properties, column shape and slenderness ratio, reinforcement details, and FRP types are used as the input variables, while the load-carrying capacity is used as the output response to develop the ML models. Furthermore, the input-output relationship of the ML model is explained through feature importance analysis and the SHapely Additive exPlanations (SHAP) approach. Eight ML models, namely, Kernel Ridge Regression, Lasso Regression, Support Vector Machine, Gradient Boosting Machine, Adaptive Boosting, Random Forest, Categorical Gradient Boosting, and Extreme Gradient Boosting, are used in this study for capacity prediction, and their relative performances are compared to identify the best-performing ML model. Finally, predictive equations are proposed using the harmony search optimization and the model interpretations obtained through the SHAP algorithm.

Keywords: fiber-reinforced polymer (FRP) rebar; reinforced concrete columns; axial capacity; machine learning; ensemble learning; harmony search optimization

Citation: Cakiroglu, C.; Islam, K.; Bekdaş, G.; Kim, S.; Geem, Z.W. Interpretable Machine Learning Algorithms to Predict the Axial Capacity of FRP-Reinforced Concrete Columns. *Materials* **2022**, *15*, 2742. https://doi.org/10.3390/ma15082742

Academic Editor: Ayman El-Zohairy

Received: 19 March 2022
Accepted: 6 April 2022
Published: 8 April 2022

1. Introduction

Fiber-reinforced polymers (FRP) have gained popularity in construction as a viable replacement for steel rebars in reinforced concrete (RC) members due to their corrosion resistance, higher strength, lightweight, and ease of fabrication [1]. Extensive research works have been carried out in the last two decades to investigate the performance of the different types of FRP reinforcement [2–10]. Some of the studies in this area were experimental in nature [11–19], whereas others aimed to develop more accurate, novel methods of prediction for structural performance using various techniques of artificial intelligence and optimization [20–25]. Murad et al. [23] used gene expression programming on FRP-RC beams to predict flexural strength. As the variables affecting the flexural strength, the beam cross-sectional dimensions, concrete compressive strength, the area, the elasticity modulus, and the ultimate tensile strength of the FRP reinforcement were selected. An experimental database consisting of 116 samples was used to develop the predictive model. Protchenko et al. [24] used an experimental database composed of 102 samples to further improve the existing predictive models in the ACI 440.1R-15 code pertaining to the flexural capacity of FRP-RC beams. While a relatively large amount of research in the literature deals with the experimental study of FRP-RC columns, the literature that deals with the applications of machine learning techniques for the predictive modeling of these structures is somewhat limited. Raza et al. [22] developed a large database of 279 specimens of GFRP-RC columns to develop an improved, ANN-based predictive model to estimate the axial capacity. Bakouregui et al. [25] applied the eXtreme Gradient Boosting (XGBoost) algorithm for the prediction of the axial load-carrying capacity of FRP-reinforced concrete columns. An experimental database consisting of 283 samples was chosen to train the model. The model output was interpreted using the SHAP algorithm to determine the effects of the various input parameters on the model output. In addition to the load eccentricity, the gross cross-sectional area, concrete compressive strength, column slenderness ratio, and the spacing between the transverse reinforcements were found to have the greatest impact on the axial load-carrying capacity of FRP-RC columns.

The current study focuses on the concentric loading case. In addition to the XGBoost algorithm, seven other ML algorithms are applied to a portion of the database compiled by Bakouregui et al. [25], which consists of samples with concentric loading. One of the objectives of the current study is to perform a systematic comparison between different ensemble ML models to predict the compressive capacity of FRP-RC columns, which is missing in the existing literature. The relative computational efficiency of the different ML algorithms is presented as well. To fulfill the objective, we select 117 FRP-RC column test specimens from an existing database [25]. The existing database consists of 283 FRP-RC column test specimens that include both short and slender column specimens subjected to concentric and eccentric loading. The other objective of the current study is to develop opensource, Python-based ML models to estimate the axial load-carrying capacity of FRP-RC columns to further help researchers utilize the developed models to improve the framework once additional experimental data are available.

The input variables of the database used in this paper are the slenderness ratio, gross cross-sectional area, the type of cross-section (circular or rectangular), the type of concrete (light-weight or normal-weight concrete), compressive strength of concrete, type of composite material used in the longitudinal reinforcement (GFRP or CFRP), longitudinal reinforcement ratio, elasticity modulus of the longitudinal reinforcements, the ultimate strength of the longitudinal reinforcements, type of the transverse reinforcement material (GFRP, CFRP, or steel), the configuration of transverse reinforcement (i.e., spirals or ties), and the spacing of the transverse reinforcement. The output variable is the axial load-carrying capacity of the corresponding specimen. The database is split into a training and test set in 90% to 10% ratio, and the performances of the different machine learning models are compared using four different accuracy metrics. Afterward, the SHAP algorithm is utilized to determine the most significant input variables and their interdependencies. Three different, closed-form equation formats are proposed based on the outcome of the SHAP algorithm. All three of these equations depend on a certain number of parameters that can be optimized to minimize the difference between the equation output and the actual axial load-carrying capacity measurements. A well-established, metaheuristic optimization technique called harmony search optimization is applied in this optimization process. Then, the performances of these two equations are compared using different accuracy metrics. The current study is unique in its attempt to obtain closed-form equations for the prediction of the axial load-carrying capacity of FRP-RC columns based on ensemble machine learning algorithms, the SHAP procedure, and metaheuristic optimization. Relatively high accuracies were observed from the obtained equations. The following sections further elaborate on the process of obtaining predictive models for these structures based on experimental data.

2. Machine Learning and Optimization Methodologies

Eight data-driven machine learning models were applied to the database consisting of 117 samples. Each of these samples contained seven continuous-valued input variables, five discrete-valued input variables, and one continuous-valued output variable, the axial load-carrying capacity. The statistical distribution of these variables is shown in Figure 1.

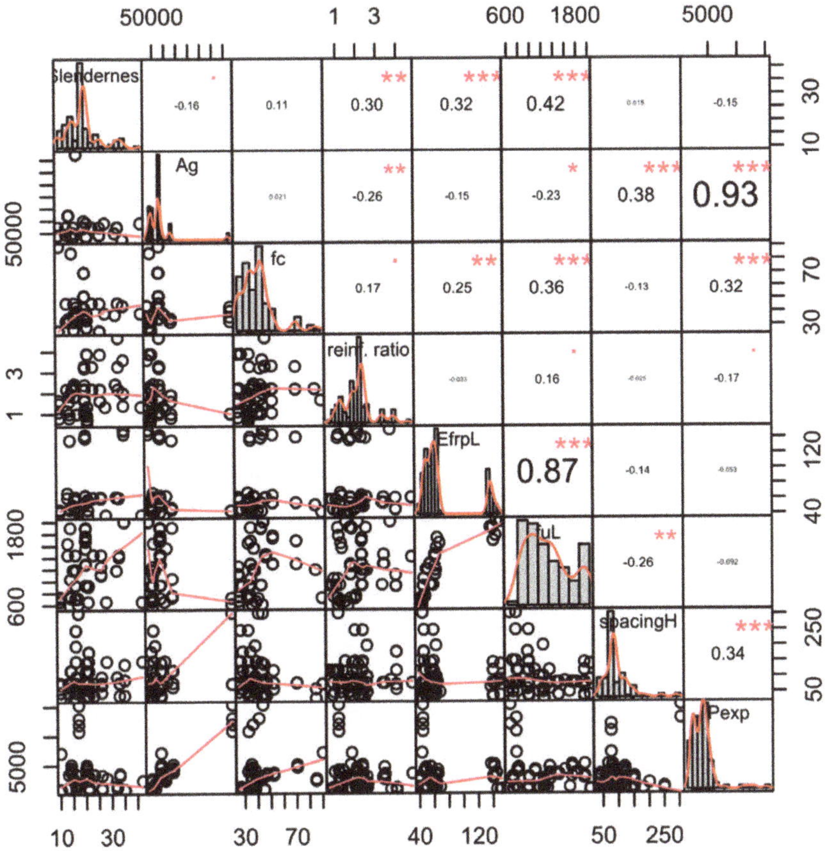

Figure 1. Correlation matrix of the dataset.

Figure 1 shows the correlation plot for the seven non-categorical, continuous-valued input variables together with the output variable Pexp. Ag, f_c', E_{FRP}, fuL, spacingH, and Pexp stand for the gross cross-sectional area, concrete compressive strength at 28 days, elasticity modulus of the longitudinal reinforcements, the ultimate strength of the longitudinal reinforcements, spacing between transverse reinforcements, and the experimental axial load-carrying capacity, respectively. A list of all the variable names and abbreviations used in this paper can be found in Appendix A. In Figure 1, for each variable on a diagonal tile, the scale of this variable is shown either on the left vertical or the right vertical axis. In addition to the vertical axes, for each variable, the same scaling is shown on one of the horizontal axes as well. The lower triangle in this plot contains bivariate scatter plots with regression lines, while the diagonal contains histograms showing the distributions of all variables. The upper triangle contains the Pearson correlation coefficients (r_{xy}) which can be computed using Equation (1). In Equation (1), x and y are two sequences of variable values between which the Pearson correlation is being computed. A Pearson correlation value close to 1 indicates a high correlation between the two variables. The significance of the correlation is indicated with stars in the upper triangle where the number of stars is proportional to the level of significance. According to Figure 1, the gross cross-sectional area Ag has the greatest correlation with Pexp followed by the transverse reinforcement spacing and the concrete compressive strength. The remaining continuous-valued input variables are negatively correlated to Pexp.

$$r_{xy} = \frac{n\sum_{i=1}^{n} x_i y_i - \sum_{i=1}^{n} x_i \sum_{i=1}^{n} y_i}{\sqrt{n\sum_{i=1}^{n} x_i^2 - \left(\sum_{i=1}^{n} x_i\right)^2}\sqrt{n\sum_{i=1}^{n} y_i^2 - \left(\sum_{i=1}^{n} y_i\right)^2}} \quad (1)$$

In what follows, the machine learning models applied in this study are briefly presented.

2.1. Machine Learning Models

All models presented in this section were trained and tested using both the continuous-valued and discrete-valued input variables. The discrete-valued input variables, namely, the type of cross-section, type of concrete, type of material used in the longitudinal and transverse reinforcements, and the configuration of the transverse reinforcements were one-hot encoded before being used in the training and testing of the machine learning models. The machine learning models applied in this study are briefly presented as follows. For further details of the algorithms presented in this section, it is suggested that the reader checks the references [26–37].

2.1.1. Kernel Ridge Regression (KRR)

Let $(P_n, \mathbf{x}_n) \in \mathbb{R} \times \mathbb{R}^{17}$, where $\mathbf{x}_n$ is the n-th training sample containing the information of 17 one-hot encoded and continuous input variables. Let $\mathbf{x}$ be any sample outside of the training set. The kernel ridge prediction $\hat{P}$ for this new sample is calculated using Equation (2) where κ is a kernel function, and N is the total number of samples in the training set [26].

$$\hat{P} = \sum_{n=1}^{N} \hat{\theta}_n \kappa(\mathbf{x}, \mathbf{x}_n) \quad (2)$$

The coefficients $\hat{\theta}_n$ of the kernel ridge regressor are computed using Equation (3), where C is the regularization constant, κ is the kernel matrix, $\mathbf{I}$ is the identity matrix, $\hat{\theta} = \left[\hat{\theta}_1, \ldots, \hat{\theta}_N\right]^T$, and $\mathbf{P} = [P_1, \ldots, P_N]^T$.

$$(\kappa + C\mathbf{I})\hat{\theta} = \mathbf{P} \quad (3)$$

2.1.2. Support Vector Regression (SVR)

The Support Vector Regression technique aims to reduce the computational complexity by limiting the training operations to a subset of the training set. Equation (4) shows the predictive model for a vector $\mathbf{x}$ from outside the training set [27].

$$\hat{P} = \sum_{n=1}^{N} \left(a_n - \hat{a}_n\right) \kappa(\mathbf{x}, \mathbf{x}_n) + b \quad (4)$$

In Equation (4), a_n and $\hat{a}_n$ are the Lagrange multipliers that satisfy the conditions $0 \le a_n \le C$ and $0 \le \hat{a}_n \le C$. The regularization parameter C works towards reducing the noise in the training set as the smaller values of C correspond to greater regularization. In Equation (4), only the training samples for which $a_n - \hat{a}_n \neq 0$ have an impact on the model output, and they are called the support vectors.

2.1.3. Lasso Regression

The Lasso Regression technique is based on the linear predictive model shown in Equation (5) [28].

$$y(x) = \beta_0 + \sum_{n=1}^{17} \beta_n x_n \tag{5}$$

The model coefficients β_n in Equation (5) are estimated through the minimization of the objective function shown in Equation (6). The second term in Equation (6), calculating the L1 norm of the coefficient vector, leads to some of the model coefficients associated with the less significant input variables being set to zero. This yields a predictive model that depends on a smaller number of variables which can also be called a sparse model [28].

$$\sum_{i=1}^{N}\left(P_i - \beta_0 - \sum_{j=1}^{17} \beta_j x_{ij}\right)^2 + \lambda \sum_{j=1}^{17} \left|\beta_j\right| \tag{6}$$

2.1.4. Gradient Boosting Machine (GBM)

GBM is based on the idea of consecutively fitting models to data to increase the model accuracy so that the new base learners $h(x, \theta)$ have maximum correlation with the negative gradient $g_t(x)$ of the loss function $\psi(y, f)$ [29]. In every iteration of the GBM algorithm, the best gradient descent step ρ_t is calculated as in Equation (7), and the estimator function is updated as in Equation (8) [29,30].

$$\rho_t = \operatorname{argmin}_\rho \sum_{i=1}^{N} \psi\left[y_i, \hat{f}_{t-1}(x_i) + \rho h(x_I, \theta_t)\right] \tag{7}$$

$$\hat{f}_t := \hat{f}_{t-1} + \rho_t h(x, \theta_t) \tag{8}$$

2.1.5. Adaptive Boosting (AdaBoost)

AdaBoost is an ensemble learning technique that iteratively runs weak regressors $h_t(x)$ and merges them to produce a strong regressor. The procedure for obtaining the final strong regressor $h_f(x)$ is given in Equations (9) and (10), where ε_t is the loss associated with $h_t(x)$ [31].

$$h_f(x) = \inf\left\{ y \in Y : \sum_{t \cdot h_t(x) \leq y} \log\left(\frac{1}{\beta_t}\right) \geq \frac{1}{2} \sum_t \log(1/\beta_t) \right\} \tag{9}$$

$$\beta_t = \frac{\varepsilon_t}{1 - \varepsilon_t} \tag{10}$$

2.1.6. Random Forest (RF)

Random Forest is another ensemble learning algorithm that merges the output of decision tree models trained on subsets of the training set for a more accurate prediction. In addition to the random subsampling of the training set for the training of the decision trees, in the node splitting phase while generating a tree, a randomly selected subset of the input variables is utilized. The format of the Random Forest model is given in Equation (11), where $\hat{m}_j$ represents one of the decision trees [32,33].

$$\hat{m}(x) = \frac{1}{M} \sum_{i=1}^{N} \hat{m}_j(x) \tag{11}$$

2.1.7. Extreme Gradient Boosting (XGBoost)

XGBoost is a tree boosting algorithm that differentiates itself through high scalability and computational speed. The algorithm is capable of scaling to billions of samples in distributed settings. The output of the XGBoost model can be summarized as in Equation (12), where f_k denotes a regression tree, and K is the total number of regression trees [25,34]. The optimum values of the leaf weights w_j^* that minimize the loss can be computed as in Equation (13), where l denotes the loss function, and I_j is the set that contains the indices of the data samples associated with the j-th leaf [35].

$$\hat{y}_i = \sum_{k=1}^{K} f_k(\mathbf{x}_i) \tag{12}$$

$$w_j^* = -\frac{\sum_{i\in I_j} g_i}{\sum_{i\in I_j} h_i + \lambda}, \quad g_i = \frac{\partial l\left(y_i, \hat{y}_i^{(t-1)}\right)}{\partial \hat{y}_i^{(t-1)}}, \quad g_i = \frac{\partial^2 l\left(y_i, \hat{y}_i^{(t-1)}\right)}{\partial \left(\hat{y}_i^{(t-1)}\right)^2} \tag{13}$$

2.1.8. Categorical Gradient Boosting (CatBoost)

CatBoost is another gradient boosting algorithm that performs particularly well in the presence of categorical data. In addition to the better processing of categorical features, the implementation of ordered boosting contributes to better accuracy in this algorithm. CatBoost addresses a particular issue called prediction shift, which is present in other gradient boosting algorithms such as XGBoost [36,37]. CatBoost overcomes the issue of prediction shift through the implementation of ordered boosting with ordered target statistics, the details of which can be found in [37].

2.2. Harmony Search Optimization

Metaheuristic algorithms have been applied to a large number of engineering problems in recent years. These algorithms are shown to effectively address problems characterized by high nonlinearity or non-differentiability [38]. One of the most widely used and well-established algorithms in this category is the harmony search algorithm. The algorithm was invented by Geem et al. [39] and has been applied to a wide range of engineering problems such as the optimum design of retaining walls [40,41], laminated composite plates [42,43], highway bridge plate girders [44], concrete-filled steel tubular columns [45], reinforced concrete cylindrical walls [46–48], and tuned liquid dampers [49].

The algorithm starts with the generation of an arbitrarily sized population consisting of optimum-solution candidate vectors, also called harmony vectors (HV). The number of harmony vectors in this population is called the harmony memory size (HMS). Each harmony vector has a size equal to the number of parameters being optimized. After the generation of the initial population, all harmony vectors go through the harmony search iterations. These iterations are defined by Equations (14)–(17).

$$k = \text{int}(\text{rand}\cdot\text{HMS}), \ \text{rand} \in (0,1) \tag{14}$$

$$x_{i,new} = x_{i,min} + \text{rand}\cdot(x_{i,max} - x_{i,min}), \quad \text{if HMCR} > \text{rand} \tag{15}$$

$$x_{i,new} = x_{i,k} + \text{rand}\cdot\text{PAR}\cdot(x_{i,max} - x_{i,min}), \quad \text{if HMCR} \le \text{rand} \tag{16}$$

$$\text{HMCR} = 0.5\left(1 - \frac{i}{\max(i)}\right), \ \text{PAR} = 0.05\left(1 - \frac{i}{\max(i)}\right) \tag{17}$$

After each harmony search iteration, the newly generated HVs are compared to the existing ones in terms of their performance and, in cases where they demonstrate better performance, they replace the old HVs. After each iteration, the best-performing HV is identified. The process is repeated until the performances converge to an optimum level. For further details of the harmony search algorithm, it is suggested that the reader see reference [50].

3. Results and Discussions

This section presents the performances of eight different machine learning algorithms in predicting the axial load-carrying capacity of FRP-RC columns based on the experimental database of 117 samples collected from the literature [25]. The performances of these algorithms were measured using four different metrics of accuracy, the details of which have been given in Appendix B. The machine learning algorithms were ranked according to their accuracy in predicting the axial load-carrying capacity and their computational speed. The impacts of different input variables on the machine learning model output were investigated using the SHAP methodology. After the determination of the most significant input variables and their dependencies, three different equation formats were proposed for the prediction of the axial load-carrying capacity. A harmony search algorithm was utilized for the development of these equations. Finally, the performances of the proposed equations are presented.

3.1. Machine Learning Model Performances

The machine learning models were trained after splitting the database of experimental samples into a training set and a test set in a 90% to 10% ratio. The accuracies and computational speed of these models are presented for training and test sets separately.

A list of algorithms and their accuracies with respect to four different metrics of accuracy can be found in Table 1. In Table 1, for each algorithm, the time that elapses from the beginning of the model fitting process to the end of the predictions for the test set is listed. According to Table 1, the Gradient Boosting Machine (GBM) algorithm emerges as the most efficient algorithm considering both prediction accuracy and time elapsed, since this model delivered the most accurate predictions next to XGBoost in 58% of the time. Among the other predictive models, Random Forest (RF) and Lasso Regression can be counted as models that delivered a greater than 95% R^2 score on the test set in less than 3 s. The CatBoost algorithm was observed to take a significantly longer time with a relatively low R^2 score on the test set compared to the top-performing models. The model performances in terms of accuracy metrics and speed of computation are visualized in Figure 2, where the horizontal axes represent the metrics of accuracy and speed. In Figure 3, the experimental axial load-carrying capacity values are plotted against the model predictions. The straight black lines in these plots represent a perfect match between the predicted and actual values, whereas the dotted red lines represent a ±10% deviation from a perfect match.

Table 1. Performance comparison of different machine learning algorithms.

	R^2	RMSE	MAE	MAPE	Time Elapsed
Train (KRR)	0.903	744	492	0.243	2.40 s
Test (KRR)	0.841	768	548	0.299	
Train (SVR)	0.522	1646	604	0.182	2.67 s
Test (SVR)	0.520	1334	652	0.304	
Train (Lasso)	0.971	399	236	0.096	2.36 s
Test (Lasso)	0.955	244	194	0.101	
Train (GBM)	0.999	57.9	44.1	0.022	2.54 s
Test (GBM)	0.975	182.9	146	0.063	
Train (AdaBoost)	0.973	384	284	0.119	2.64 s
Test (AdaBoost)	0.910	345	283	0.117	
Train (RF)	0.993	193	80.7	0.027	2.95 s
Test (RF)	0.969	204	142.4	0.066	
Train (XGBoost)	0.999	22.3	7.5	0.003	4.41 s
Test (XGBoost)	0.982	153.8	112.7	0.054	
Train (CatBoost)	0.999	35.7	28.9	0.013	29.92 s
Test (CatBoost)	0.931	301	197.4	0.119	

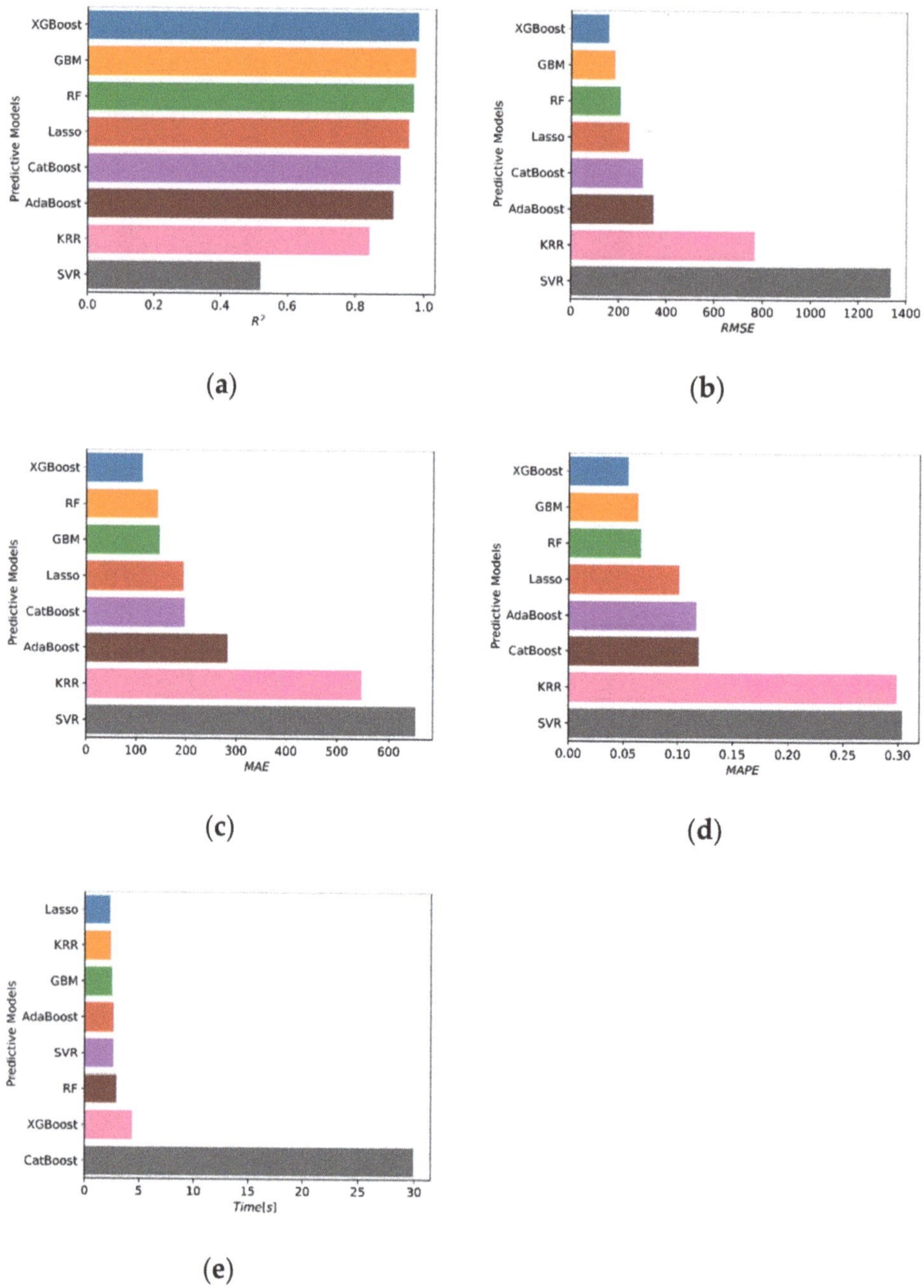

Figure 2. Comparison of predictive models with respect to accuracy and speed; (**a**) R^2; (**b**) RMSE; (**c**) MAE; (**d**) MAPE; (**e**) Time [s].

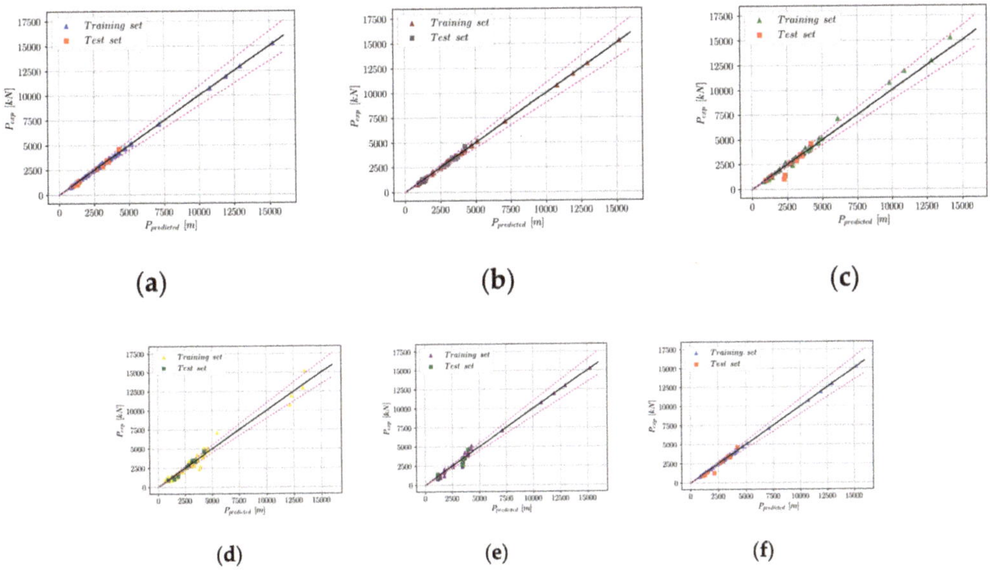

Figure 3. Comparison of experimental and predicted *p* values; (**a**) XGBoost; (**b**) GBM; (**c**) Random Forest; (**d**) Lasso; (**e**) AdaBoost; (**f**) CastBoost.

3.2. Application of the SHAP Algorithm

The SHAP algorithm was applied to the XGBoost model, which demonstrated the best overall performance. The SHAP summary plot is an information-intense representation of how the various design variables affect the axial load-carrying capacity. The SHAP summary plot in Figure 4 shows the SHAP values of the different variables together with a color bar that indicates whether a variable has a high or low value. Figure 4 shows the SHAP values calculated for each variable at each sample point, where every dot corresponds to a sample. Around certain SHAP values, the dots have greater concentration. For the most significant nine variables, the SHAP values were computed separately. For the remaining variables, the SHAP values were summed, and they are represented in Figure 4 as a single variable. For any sample represented in Figure 4, the color of the corresponding dot has a shade of red if a variable has a high value at this sample, and the dot has a shade of blue if the variable has a low value. Positive and negative SHAP values indicate increasing and decreasing effects on the model output, respectively. According to Figure 4, the gross cross-sectional area of the column, the concrete compressive strength, the horizontal reinforcement spacing, and the slenderness ratio (λ) have the greatest impact on the model output. Furthermore, increasing the values of Ag and f_c' also affects the prediction, whereas the opposite can be said for the slenderness ratio.

The feature dependence plots in Figure 5 contain information that supplements the SHAP summary plot. In Figure 5, the feature dependence plots of the four most significant design variables can be seen. For each one of these variables, the variation of another variable that is most dependent on it is presented using color. For each dot in Figure 5, a shade of red indicates a high value of the most dependent variable, whereas a shade of blue indicates a low value of the most dependent variable. The main information contained in Figure 5 is the relationship between the values of a variable and the associated SHAP values. Figure 5 shows that increasing the values of Ag and f_c' also leads to an increase in the corresponding SHAP values, which indicates an increasing effect on the model output. On the other hand, the slenderness ratio and the model output are inversely proportional. Furthermore, increasing the horizontal reinforcement spacing does not have a significant effect on the model output. The developed Python-based ML models to estimate the axial capacity of FRP-RC column can be found at https://github.com/ccakiroglu/FRPRCColumn (accessed on 4 April 2022).

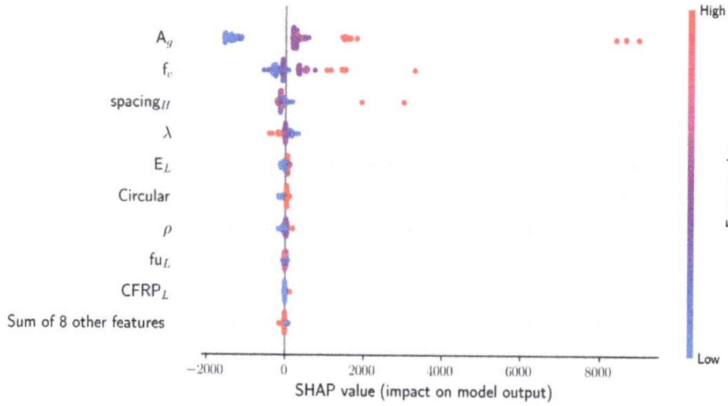

Figure 4. SHAP values of the design variables for the XGBoost model.

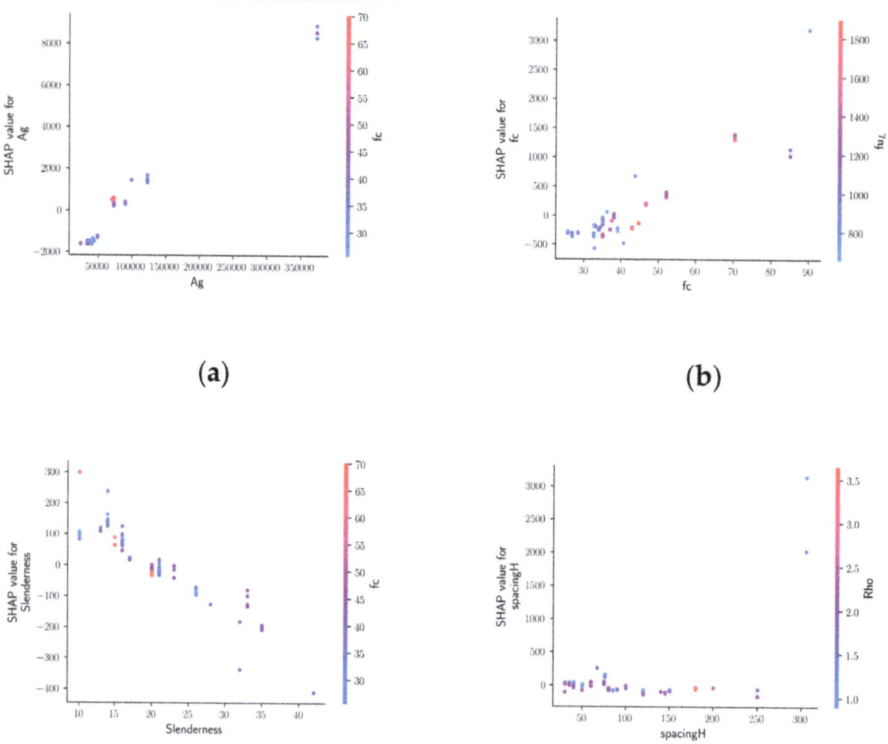

(a) **(b)**

(c) **(d)**

Figure 5. Feature dependence plots for the XGBoost model; (**a**) $A_g - f'_c$; (**b**) $f'_c - f_{uL}$; (**c**) Slenderness $- f'_c$; (**d**) spacing$_H - \rho$.

3.3. *Development of the Predictive Equations*

Based on the SHAP summary plot in Figure 4 and the feature dependence plots in Figure 5, three different equation formats were proposed. The first one of these formats, given in Equation (18), consists of the linear combination of three product terms. The first one of these terms includes, besides the most important input variable A_g, the variables f_c, which is the most dependent variable on A_g, and f_{uL}, which is the most dependent variable on f_c. The second term consists of the multiplication of the slenderness ratio λ and f_c, which is the most dependent variable on λ. Finally, the third term includes the effects of the spacing between the horizontal reinforcements and the longitudinal reinforcement ratio (ρ), which is the most dependent variable on it.

$$P = a_0 + a_1 A_g^{a_2} f_c^{a_3} f_{uL}^{a_4} + a_5 \lambda^{a_6} f_c^{a_7} + a_8 \text{spacing}_H^{a_9} \rho^{a10} \tag{18}$$

An initial population of 20 candidate solution vectors was created. In this population, each solution candidate consists of a randomly generated, unique combination of the coefficients a_0 to a_{10}, where all these coefficients take values in $(-1, +1)$. Afterward, these coefficients went through harmony search iterations. In order to train the coefficients of Equation (18), the entire database of experimental results was split into a training set and a test set in a 90% to 10% ratio. For each member of the population, the difference between the actual experimental axial load-carrying capacities (P_{exp}) and the p values computed by Equation (18) was calculated and stored in a vector that has the length of the entire training set. As the metric that shows the accuracy of the proposed equation, the Euclidean norm of this vector was calculated. Figure 6 shows the decrease in this vector norm throughout the optimization steps.

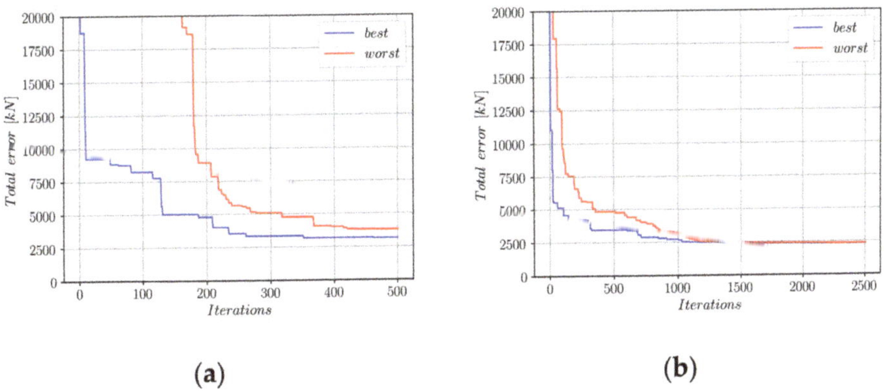

Figure 6. Decrease in the total error throughout the harmony search iterations (Equation (18)). (**a**) First 500 HS iterations; (**b**) First 2500 HS iterations.

In Figure 6, the performances of the best- and worst-performing solution candidates are shown with blue and red colors, respectively. Figure 6b shows that, after 2500 iterations, a convergence to the minimum total error norm could be observed. The development of the coefficients a_0 to a_{10} in the optimization process are shown in Figure 7.

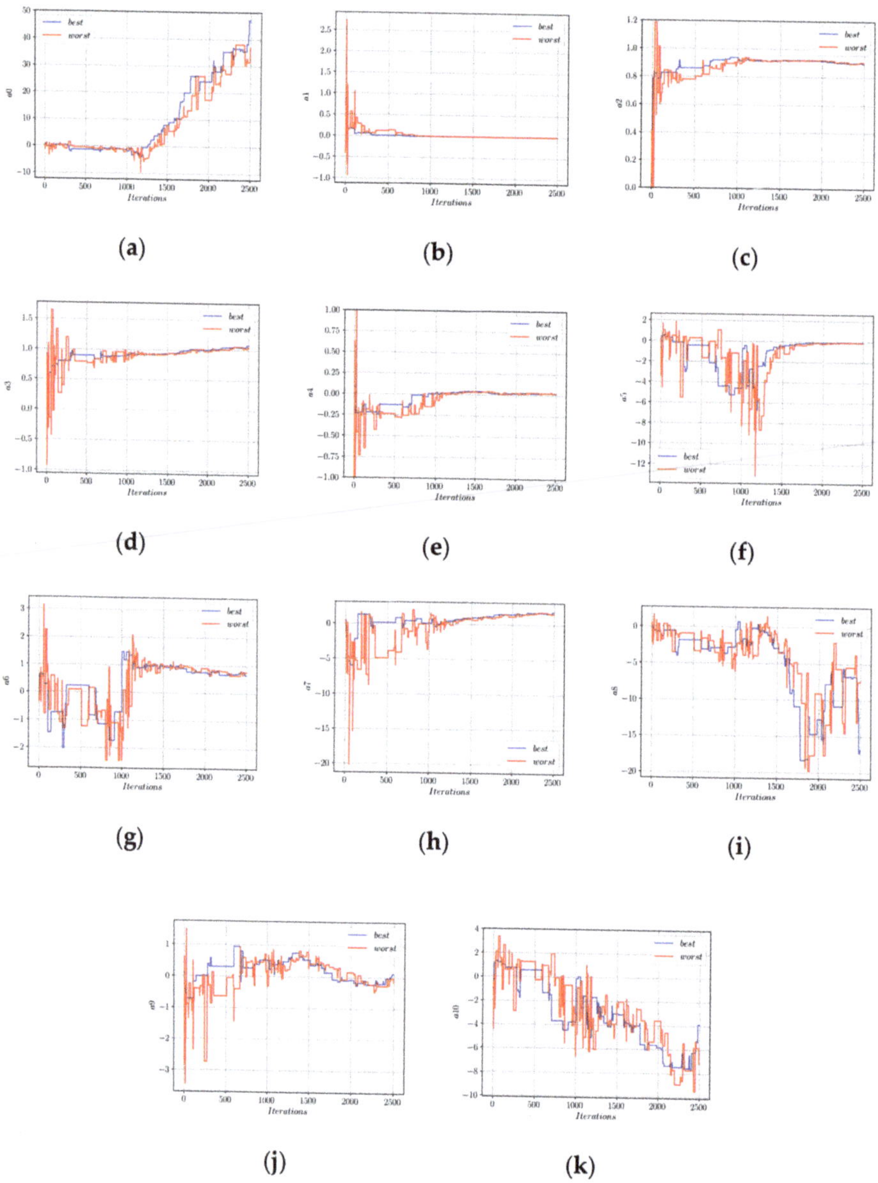

Figure 7. Development of the coefficients in Equation (18); **(a)** a_0; **(b)** a_1; **(c)** a_2; **(d)** a_3; **(e)** a_4; **(f)** a_5; **(g)** a_6; **(h)** a_7; **(i)** a_8; **(j)** a_9; **(k)** a_{10}.

In Figure 7, for each coefficient, the values these coefficients took in the best- and worst-performing members of the population are plotted in blue and red colors, respectively. Using the limit values of the coefficients in Figure 7, Equation (18) can be rewritten as in Equation (19).

$$P = 46.75 + 0.00207A_g^{0.901}f_c^{1.085}f_{uL}^{0.02185} + (-0.0493)\lambda^{0.676}f_c^{1.862} + (-16.84)\mathrm{spacing}_H^{0.1[illegible]}\rho^{0.84} \quad (19)$$

Using Equation (19), the p values were computed for the test set and compared to the actual measured values. This comparison can be seen in Figure 8, where the dotted lines indicate the ±10% deviations from an exact match.

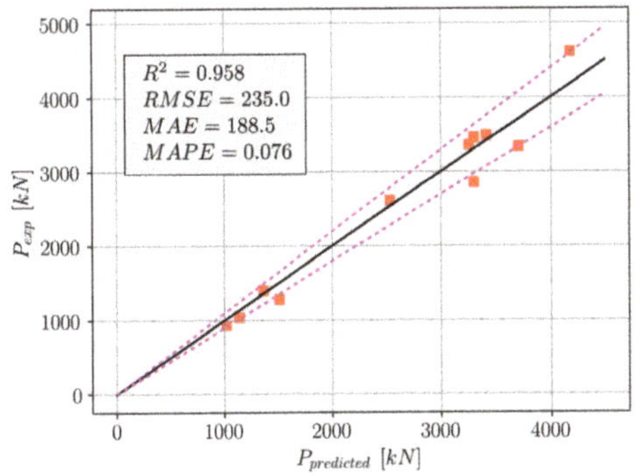

Figure 8. Comparison of experimental and predicted (Equation (19)) p values.

The accuracy metrics in Figure 8 show that Equation (19) outperformed five out of the eight algorithms in Table 1. The machine learning algorithms that performed better than Equation (19) were Random Forest, XGBoost, and GBM. Although good performance could be obtained from Equation (19), a second equation format for practical use was proposed, as in Equation (20).

$$P = a_0 + a_1 A_g^{a_2} f_c^{a_3} + a_4 A_g^{a_5} spacing_H^{a_6} + a_7 A_g^{a_8} \lambda^{a_9} \tag{20}$$

In Equation (20), the number of variables being considered in the prediction and the total number of coefficients were reduced. The variables having the highest impact on the model output according to the SHAP summary plot were selected. Equation (20) consists of three product terms after a_0. In each one of these terms, A_g is multiplied with one of the three most significant variables after itself, according to the SHAP summary plot. The harmony search procedure was repeated for Equation (20).

Figure 9a,b show the development of the error norm during the first 500 and 2500 harmony search iterations, respectively. Using the Euclidean norm of the difference vector between the actual measurements and predictions of Equation (20), a minimum error of 2639 kN could be achieved. This error norm was slightly greater than the 2336 kN that could be achieved using the Equation (18) format. Figure 10 shows the development of the coefficients a_0 to a_9 in Equation (20) during the harmony search process.

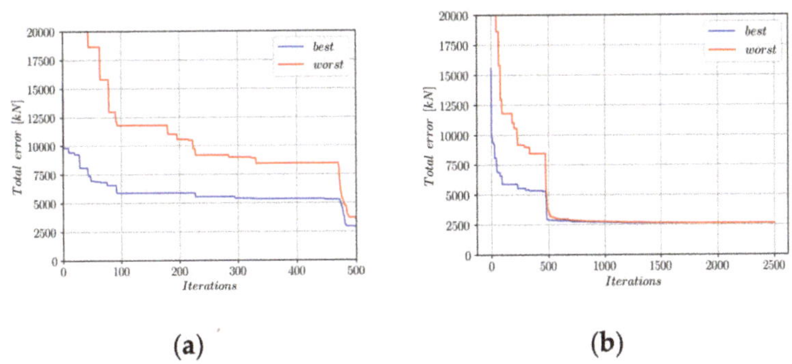

Figure 9. Decrease in the total error throughout the harmony search iterations (Equation (20)). (**a**) First 500 HS iterations; (**b**) First 2500 HS iterations.

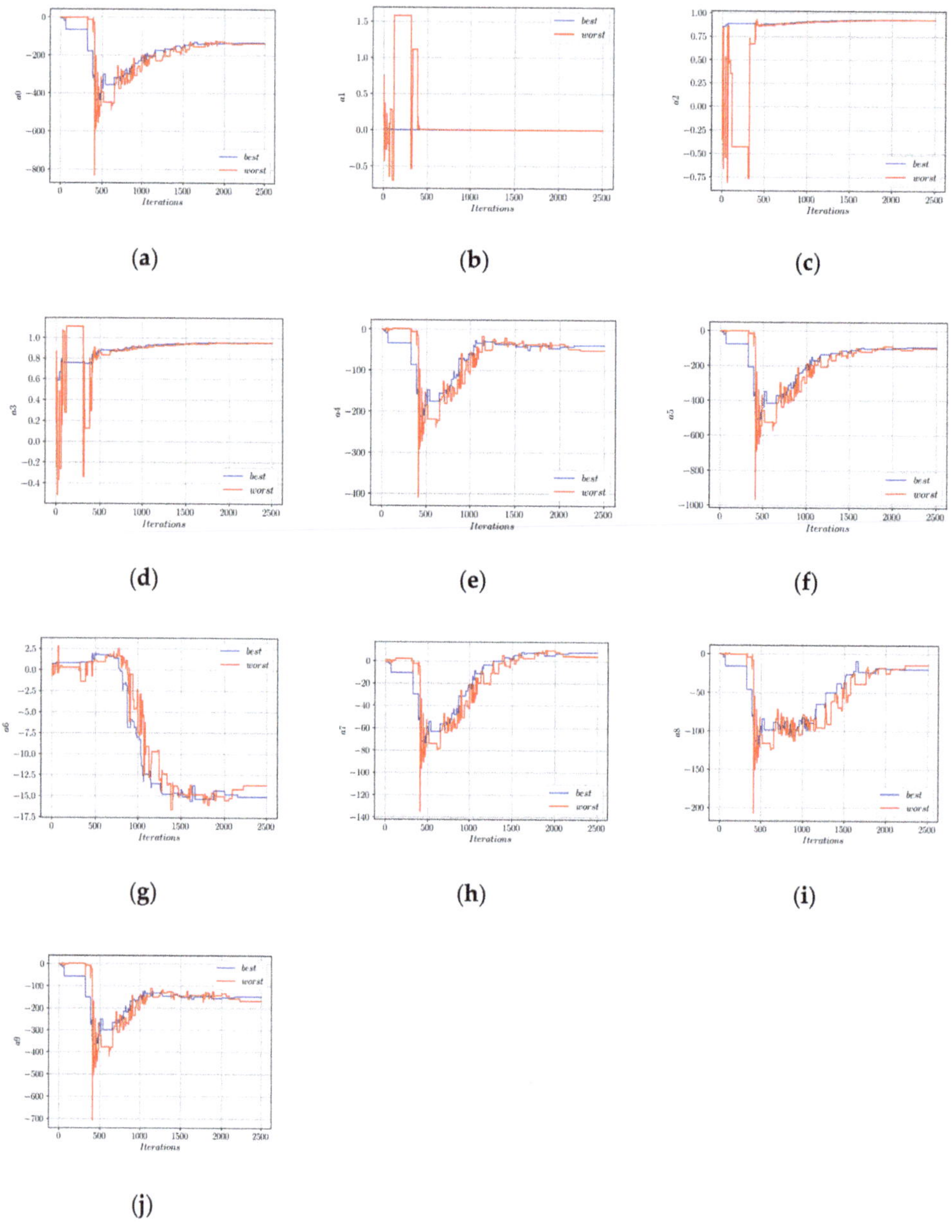

Figure 10. Development of the coefficients in Equation (20); (**a**) a_0; (**b**) a_1; (**c**) a_2; (**d**) a_3; (**e**) a_4; (**f**) a_5; (**g**) a_6; (**h**) a_7; (**i**) a_8; (**j**) a_9.

In both Figures 7 and 10, it can be observed that the coefficients of the worst population member demonstrated much wider fluctuations than the best member. However, after a certain number of iterations, both the best and worst member coefficients converged nearly to the same values. Using the limit values of the best member coefficients, Equation (21) was obtained.

$$P = -131.77 + 0.002608A_g^{0.93058}f_c^{0.955} + (-38.76)A_g^{-93.69}spacing_H^{-15.07} + 7.58A_g^{-19.54}\lambda^{-146.67} \quad (21)$$

The negative powers in the second and third products of Equation (21) make these terms close to zero so that Equation (21) can be used in a simplified format, as in Equation (22).

$$P = -131.77 + 0.002608A_g^{0.93058}f_c^{0.955} \quad (22)$$

Figure 11 shows the comparison of the experimental axial load-carrying capacities to the *p* values predicted by Equation (22). According to the accuracy values in Figure 11, Equation (22) was able to outperform the KRR, SVR, AdaBoost, and CatBoost algorithms. A comparison of Figures 8 and 11 shows that Equation (22) was able to achieve comparable performance to Equation (19) in a much simpler format. In order to eliminate the bias terms in Equations (19) and (22), a third equation format that consists of a single product term was proposed in Equation (23).

$$P = a_0A_g^{a_1}f_c^{a_2}spacing_H^{a_3}\lambda^{a_4}\rho^{a_5}f_{uL}^{a6} \quad (23)$$

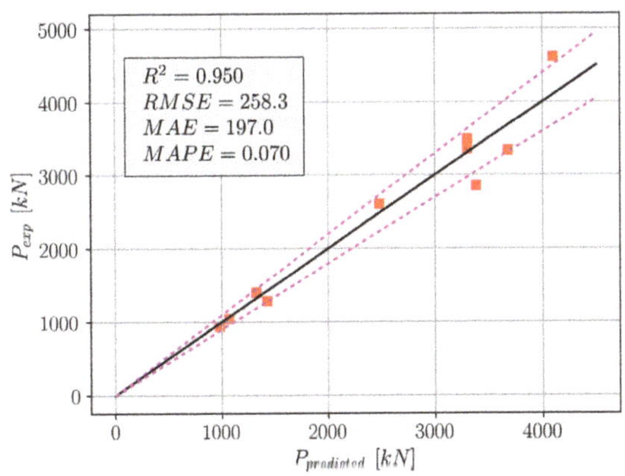

Figure 11. Comparison of experimental and predicted (Equation (22)) *p* values.

Figure 12 shows the development of the error norm for the best- and worst-performing harmony vectors for the first 500 and 2500 iteration steps.

Figure 12 shows that a minimum error norm of 2286 kN could be achieved after 2500 iterations which is less than the error norms achieved through Equations (18) and (20). After inserting the limit values of the coefficients a_0 to a_6 into Equation (23), we obtain Equation (24).

$$P = 0.00123A_g^{0.9946}f_c^{0.9266}spacing_H^{0.01124}\lambda^{-0.1474}\rho^{0.08769}f_{uL}^{0.0589} \quad (24)$$

The performance of Equation (24) in predicting the axial load-carrying capacity has been presented in Figure 13, where the actual *p* values are plotted against the *p* values predicted by Equation (24). Finally, a list of all three equations proposed in this paper can be found in Table 2 with the corresponding R^2 values. A comparison with the R^2 values in Table 1 shows that Equation (24) outperformed all of the predictive models except for the XGBoost model. It should be noted that the proposed equations are data-driven, and the performance of those equations depends on the characteristics of the data used to develop the ML models. The equations are only applicable for the range of maximum and minimum values of the input parameters.

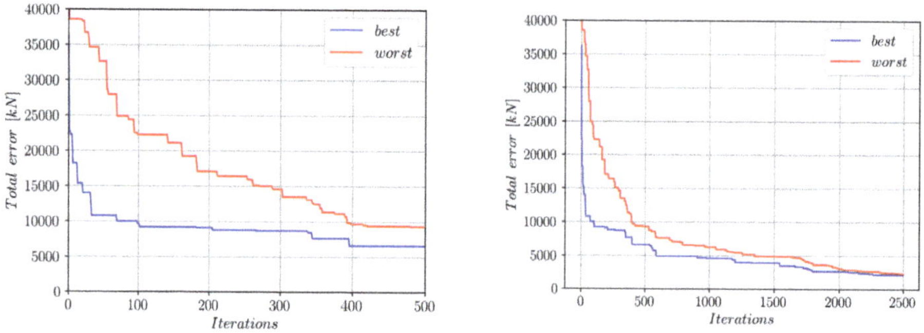

(a) **(b)**

Figure 12. Decrease in the total error throughout the harmony search iterations (Equation (23)); (**a**) First 500 HS iterations; (**b**) First 2500 HS iterations.

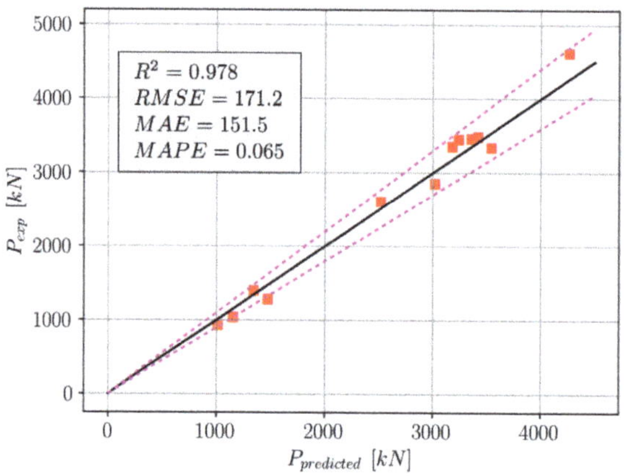

Figure 13. Comparison of experimental and predicted (Equation (24)) p values.

Table 2. Proposed equations.

Equation	R^2
$P = 46.75 + 0.00207A_g^{0.901}f_c^{1.085}f_{uL}^{0.02185} + (-0.0493)\lambda^{0.676}f_c^{1.862} + (-16.84)spacing_H^{0.122}\rho^{-3.84}$	0.958
$P = -131.77 + 0.002608A_g^{0.93058}f_c^{0.955}$	0.950
$P = 0.00123A_g^{0.9946}f_c^{0.9266}spacing_H^{0.01124}\lambda^{-0.1474}\rho^{0.08769}f_{uL}^{0.0589}$	0.978

4. Summary and Conclusions

This work aimed to address an important gap in the research literature dealing with FRP-RC columns under concentric axial loading by applying various machine learning techniques to identify the most significant design variables affecting the axial load-carrying capacity. Particularly, ensemble learning techniques, such as Gradient Boosting Machine,

Random Forest, and XGBoost, were observed to predict the axial load-carrying capacity with high accuracy. The coefficient of determination, root mean square error, mean absolute error, and mean absolute percentage error were used to quantify the accuracy of the predictions made by eight different machine learning models. Based on the predictive model with the best performance, the SHAP algorithm was utilized to identify those variables that have the greatest impact on the structural response. The gross cross-sectional area was found to have the greatest impact on the model output, followed by the concrete compressive strength, spacing of transverse reinforcement, and the slenderness ratio. Once the design variables were ranked according to their impact on the model output, in the next part of the paper, three different equation formats were proposed for the prediction of the axial load-carrying capacity as a function of the most significant design variables. These equations were optimized using the harmony search algorithm, and the equations were observed to have good accuracy. The first one of these equations predicts the axial load-carrying capacity as a function of the gross cross-sectional area, concrete compressive strength, the ultimate strength of the longitudinal reinforcements, slenderness ratio, spacing between transverse reinforcements, and the longitudinal reinforcement ratio. On the other hand, the second equation depends only on the two most significant variables, namely, the gross cross-sectional area and the compressive strength of concrete. Finally, the third equation consists of a single product term that consists of the gross cross-sectional area, the concrete compressive strength, spacing between transverse reinforcements, slenderness ratio, longitudinal reinforcement ratio, and the ultimate strength of the longitudinal reinforcements. With an R^2 score of 0.978, the third equation performed better than the first two equations. Additionally, this equation outperformed all of the predictive machine learning models except for the XGBoost model.

The availability of closed-form equations for an accurate prediction of structural response is beneficial in engineering practice. Due to an R^2 score of 0.95 and its simple format, the second prediction equation developed in this study is convenient for practical application. On the other hand, for more accurate predictions, the third equation can be used.

However, it should be noted that the developed equations are based on an experimental database consisting of 117 samples, and further studies in this area with larger databases are warranted. Furthermore, it should be noted that the results predicted by the developed equations are only valid within the range of the database used. Readers should keep in mind that the proposed equations are data-driven and can violate the mechanics-based capacity prediction model if misrepresented data are used. In addition to experimental studies, the databases could be further enhanced with the help of well-calibrated finite-element models. Besides increasing the size of the database used in model training, future research in this area can focus on the prediction of the axial load-carrying capacity under eccentric axial loading.

Author Contributions: Methodology, G.B. and C.C.; formal analysis (coding), G.B., K.I. and C.C.; writing—original draft preparation, G.B. and C.C.; writing—review and editing, K.I., S.K. and Z.W.G.; visualization, C.C.; supervision, G.B. and Z.W.G.; funding acquisition, Z.W.G. All authors have read and agreed to the published version of the manuscript.

Funding: This work was supported by the National Research Foundation of Korea (NRF) grant funded by the Korea government (MSIT) (2020R1A2C1A01011131).

Data Availability Statement: Data sharing is not applicable for this paper.

Conflicts of Interest: The authors declare no conflict of interest.

Appendix A

Nomenclature			
		XGBoost	Extreme gradient boosting
		KRR	Kernel ridge regression
FRP	Fiber-reinforced polymer	SVR	Support vector regression
GFRP	Glass fiber-reinforced polymer	CatBoost	Categorical gradient boosting
CFRP	Carbon fiber-reinforced polymer	AdaBoost	Adaptive boosting
Ag	Gross cross-sectional area	GBM	Gradient boosting machine
$spacing_H$	Spacing between transverse reinforcements	SHAP	Shapley additive explanations
Pexp	Experimental axial load-carrying capacity	fc'	Concrete compressive strength
MAE	Mean average error	MAPE	Mean average percentage error
RMSE	Root mean square error	R^2	Coefficient of determination
HMS	Harmony memory size	HV	Harmony vector
HMCR	Harmony memory consideration rate	PAR	Pitch adjustment rate
λ	Slenderness ratio	ρ	Longitudinal reinforcement ratio
fuL	Ultimate strength of the longitudinal reinforcements	E_{FRP}	Modulus of elasticity of the longitudinal reinforcements

Appendix B

Definitions of the accuracy metrics used in this paper:

$$R^2 = 1 - \frac{\sum_{i=1}^{N}\left(P_i - \hat{P_i}\right)^2}{\sum_{i=1}^{N}(P_i - P_{mean})^2} \quad \text{(A1)}$$

$$RMSE = \sqrt{\frac{\sum_{i=1}^{N}\left(P_i - \hat{P_i}\right)^2}{N}} \quad \text{(A2)}$$

$$MAPE = \frac{100\sum_{i=1}^{N}\frac{|P_i - P_i|}{|P_i|}}{N} \quad \text{(A3)}$$

$$MAE = \frac{\sum_{i=1}^{N}|P_i - P_i|}{N} \quad \text{(A4)}$$

References

1. Alsayed, S.H.; Alhozaimy, A.M. Ductility of concrete beams reinforced with FRP bars and steel fibers. *J. Compos. Mater.* **1999**, *33*, 1792–1806. [CrossRef]
2. Cousin, P.; Hassan, M.; Vijay, P.V.; Robert, M.; Benmokrane, B. Chemical resistance of carbon, basalt, and glass fibers used in FRP reinforcing bars. *J. Compos. Mater.* **2019**, *53*, 3651–3670. [CrossRef]
3. Hu, X.; Xiao, J.; Zhang, K.; Zhang, Q. The state-of-the-art study on durability of FRP reinforced concrete with seawater and sea sand. *J. Build. Eng.* **2022**, *51*, 104294. [CrossRef]
4. Huang, L.; Chen, J.; Tan, X. BP-ANN based bond strength prediction for FRP reinforced concrete at high temperature. *Eng. Struct.* **2022**, *257*, 114026. [CrossRef]
5. Deifalla, A. Punching shear strength and deformation for FRP-reinforced concrete slabs without shear reinforcements. *Case Stud. Constr. Mater.* **2022**, *16*, e00925. [CrossRef]
6. El Zareef, M.A.; Elbisy, M.S.; Badawi, M. Evaluation of code provisions predicting the concrete shear strength of FRP-reinforced members without shear reinforcement. *Compos. Struct.* **2021**, *275*, 114430. [CrossRef]
7. Truong, G.T.; Choi, K.K.; Kim, C.S. Punching shear strength of interior concrete slab-column connections reinforced with FRP flexural and shear reinforcement. *J. Build. Eng.* **2022**, *46*, 103692. [CrossRef]
8. Arabshahi, A.; Tavakol, M.; Sabzi, J.; Gharaei-Moghaddam, N. Prediction of the effective moment of inertia for concrete beams reinforced with FRP bars using an evolutionary algorithm. *Structures* **2022**, *35*, 684–705. [CrossRef]

9. Ali, A.H.; Mohamed, H.M.; Benmokrane, B. Composite FRP reinforced concrete members with fiber reinforced polymer spirals. *Structures* **2021**, *33*, 1868–1877. [CrossRef]
10. Ali, A.H.; Mohamed, H.M.; Chalioris, C.E.; Deifalla, A. Evaluating the shear design equations of FRP-reinforced concrete beams without shear reinforcement. *Eng. Struct.* **2021**, *235*, 112017. [CrossRef]
11. Almahmood, H.; Ashour, A.; Sheehan, T. Flexural behavior of hybrid steel-GFRP reinforced concrete continuous T-beams. *Compos. Struct.* **2020**, *254*, 112802. [CrossRef]
12. Abdelazim, W.; Mohamed, H.M.; Afifi, M.Z.; Benmokrane, B. Proposed Slenderness Limit for Glass Fiber-Reinforced Polymer-Reinforced Concrete Columns Based on Experiments and Buckling Analysis. *ACI Struct. J.* **2020**, *117*, 241–254. [CrossRef]
13. Adam, M.A.; Said, M.; Mahmoud, A.A.; Shanour, A.S. Analytical and experimental flexural behavior of concrete beams reinforced with glass fiber reinforced polymers bars. *Constr. Build. Mater.* **2015**, *84*, 354–366. [CrossRef]
14. Afifi, M.Z.; Mohamed, H.M.; Benmokrane, B. Strength and axial behavior of circular concrete columns reinforced with CFRP bars and spirals. *J. Compos. Constr.* **2014**, *18*, 04013035. [CrossRef]
15. Abdelazim, W.; Mohamed, H.M.; Benmokrane, B. Inelastic second-order analysis for slender GFRP-reinforced concrete columns: Experimental investigations and theoretical study. *J. Compos. Constr.* **2020**, *24*, 04020016. [CrossRef]
16. Elchalakani, M.; Dong, M.; Karrech, A.; Mohamed Ali, M.S.; Huo, J.S. Circular concrete columns and beams reinforced with GFRP bars and spirals under axial, eccentric, and flexural loading. *J. Compos. Constr.* **2020**, *24*, 04020008. [CrossRef]
17. Zhao, J.; Luo, X.; Wang, Z.; Feng, S.; Gong, X.; Shumuye, E.D. Experimental study on bond performance of carbon-and glass-fiber reinforced polymer (CFRP/GFRP) bars and steel strands to concrete. *Materials* **2021**, *14*, 1268. [CrossRef]
18. AlNajmi, L.; Abed, F. Evaluation of FRP bars under compression and their performance in RC columns. *Materials* **2020**, *13*, 4541. [CrossRef]
19. Abdelkarim, O.I.; Ahmed, E.A.; Mohamed, H.M.; Benmokrane, B. Flexural strength and serviceability evaluation of concrete beams reinforced with deformed GFRP bars. *Eng. Struct.* **2019**, *186*, 282–296. [CrossRef]
20. Zhou, Y.; Zheng, S.; Huang, Z.; Sui, L.; Chen, Y. Explicit neural network model for predicting FRP-concrete interfacial bond strength based on a large database. *Compos. Struct.* **2020**, *240*, 111998. [CrossRef]
21. Golafshani, E.M.; Rahai, A.; Sebt, M.H. Artificial neural network and genetic programming for predicting the bond strength of GFRP bars in concrete. *Mater. Struct.* **2015**, *48*, 1581–1602. [CrossRef]
22. Raza, A.; Shah, S.A.R.; Haq, F.; Arshad, H.; Raza, S.S.; Farhan, M.; Waseem, M. Prediction of axial load-carrying capacity of GFRP-reinforced concrete columns through artificial neural networks. *Structures* **2020**, *28*, 1557–1571. [CrossRef]
23. Murad, Y.; Tarawneh, A.; Arar, F.; Al-Zu'bi, A.; Al-Ghwairi, A.; Al-Jaafreh, A.; Tarawneh, M. Flexural strength prediction for concrete beams reinforced with FRP bars using gene expression programming. *Structures* **2021**, *33*, 3163–3172. [CrossRef]
24. Protchenko, K.; Leśniak, P.; Szmigiera, E.; Urbański, M. New Model for Analytical Predictions on the Bending Capacity of Concrete Elements Reinforced with FRP Bars. *Materials* **2021**, *14*, 693. [CrossRef] [PubMed]
25. Bakouregui, A.S.; Mohamed, H.M.; Yahia, A.; Benmokrane, B. Explainable extreme gradient boosting tree-based prediction of load-carrying capacity of FRP-RC columns. *Eng. Struct.* **2021**, *245*, 112836. [CrossRef]
26. Theodoridis, S. *Machine Learning: A Bayesian and Optimization Perspective*; Academic Press: Amsterdam, The Netherlands, 2015.
27. Bishop, C.M. *Pattern Recognition and Machine Learning*; Springer: New York, NY, USA, 2006.
28. James, G.; Witten, D.; Hastie, T.; Tibshirani, R. *An Introduction to Statistical Learning: With Applications in R*; Spinger: Berlin/Heidelberg, Germany, 2013.
29. Natekin, A.; Knoll, A. Gradient boosting machines, a tutorial. *Front. Neurorobot.* **2013**, *7*, 21. [CrossRef]
30. Friedman, J.H. Greedy function approximation: A gradient boosting machine. *Ann. Statist.* **2001**, *29*, 1189–1232. [CrossRef]
31. Freund, Y.; Schapire, R.E. A decision-theoretic generalization of on-line learning and an application to boosting. *J. Comput. Syst. Sci.* **1997**, *55*, 119–139. [CrossRef]
32. Scikit-Learn Documentation. Available online: https://scikit-learn.org/stable/modules/ensemble.html#forest (accessed on 14 March 2022).
33. Feng, D.C.; Wang, W.J.; Mangalathu, S.; Hu, G.; Wu, T. Implementing ensemble learning methods to predict the shear strength of RC deep beams with/without web reinforcements. *Eng. Struct.* **2021**, *235*, 111979. [CrossRef]
34. Chen, T.; Guestrin, C. XGBoost: A scalable tree boosting system. In Proceedings of the 22nd ACM Sigkdd International Conference on Knowledge Discovery and Data Mining, San Francisco, CA, USA, 13–17 August 2016; pp. 785–794.
35. XGBoost Documentation. Available online: https://xgboost.readthedocs.io/en/stable/tutorials/model.html (accessed on 14 March 2022).
36. Dorogush, A.V.; Ershov, V.; Gulin, A. CatBoost: Gradient boosting with categorical features support. *arXiv* **2018**, arXiv:1810.11363.
37. Prokhorenkova, L.; Gusev, G.; Vorobev, A.; Dorogush, A.V.; Gulin, A. CatBoost: Unbiased boosting with categorical features. *Adv. Neural Inf. Process. Syst.* **2018**, *31*.
38. Ahmadianfar, I.; Heidari, A.A.; Gandomi, A.H.; Chu, X.; Chen, H. RUN beyond the metaphor: An efficient optimization algorithm based on Runge Kutta method. *Expert Syst. Appl.* **2021**, *181*, 115079. [CrossRef]
39. Geem, Z.W.; Kim, J.H.; Loganathan, G.V. A new heuristic optimization algorithm: Harmony search. *Simulation* **2001**, *76*, 60–68. [CrossRef]
40. Kayabekir, A.E.; Arama, Z.A.; Bekdaş, G.; Nigdeli, S.M.; Geem, Z.W. Eco-Friendly Design of Reinforced Concrete Retaining Walls: Multi-objective Optimization with Harmony Search Applications. *Sustainability* **2020**, *12*, 6087. [CrossRef]

41. Yücel, M.; Kayabekir, A.E.; Bekdaş, G.; Nigdeli, S.M.; Kim, S.; Geem, Z.W. Adaptive-Hybrid Harmony Search Algorithm for Multi-Constrained Optimum Eco-Design of Reinforced Concrete Retaining Walls. *Sustainability* **2021**, *13*, 1639. [CrossRef]
42. Cakiroglu, C.; Bekdaş, G.; Geem, Z.W. Harmony Search Optimization of Dispersed Laminated Composite Plates. *Materials* **2020**, *13*, 2862. [CrossRef] [PubMed]
43. Cakiroglu, C.; Islam, K.; Bekdaş, G.; Kim, S.; Geem, Z.W. Metaheuristic Optimization of Laminated Composite Plates with Cut-Outs. *Coatings* **2021**, *11*, 1235. [CrossRef]
44. Cakiroglu, C.; Bekdaş, G.; Kim, S.; Geem, Z.W. Optimization of Shear and Lateral–Torsional Buckling of Steel Plate Girders Using Meta-Heuristic Algorithms. *Appl. Sci.* **2020**, *10*, 3639. [CrossRef]
45. Cakiroglu, C.; Islam, K.; Bekdaş, G.; Billah, M. CO_2 Emission and Cost Optimization of Concrete-Filled Steel Tubular (CFST) Columns Using Metaheuristic Algorithms. *Sustainability* **2021**, *13*, 8092. [CrossRef]
46. Bekdaş, G.; Cakiroglu, C.; Islam, K.; Kim, S.; Geem, Z.W. Optimum Design of Cylindrical Walls Using Ensemble Learning Methods. *Appl. Sci.* **2022**, *12*, 2165. [CrossRef]
47. Bekdas, G. Optimum design of axially symmetric cylindrical reinforced concrete walls. *Struct. Eng. Mech.* **2014**, *51*, 361–375. [CrossRef]
48. Bekdaş, G. Harmony Search Algorithm Approach for Optimum Design of Post-Tensioned Axially Symmetric Cylindrical Reinforced Concrete Walls. *J. Optim. Theory Appl.* **2014**, *164*, 342–358. [CrossRef]
49. Ocak, A.; Nigdeli, S.M.; Bekdaş, G.; Kim, S.; Geem, Z.W. Adaptive Harmony Search for Tuned Liquid Damper Optimization under Seismic Excitation. *Appl. Sci.* **2022**, *12*, 2645. [CrossRef]
50. Kayabekir, A.E.; Bekdaş, G.; Yücel, M.; Nigdeli, S.M.; Geem, Z.W. Harmony Search Algorithm for Structural Engineering Problems. In *Nature-Inspired Metaheuristic Algorithms for Engineering Optimization Applications*; Carbas, S., Toktas, A., Ustun, D., Eds.; Springer Tracts in Nature-Inspired Computing; Springer: Singapore, 2021. [CrossRef]

materials

MDPI

Article

Experimental Assessment and Numerical Modeling of the Bond–Slip Correlation for Steel Rebars in r.c. Members

Pietro Croce *, Paolo Formichi and Filippo Landi

Department of Civil and Industrial Engineering, University of Pisa, Largo Lucio Lazzarino 1, 56122 Pisa, Italy; p.formichi@ing.unipi.it (P.F.); filippo.landi@ing.unipi.it (F.L.)
* Correspondence: p.croce@ing.unipi.it; Tel.: +39-335-5345611

Abstract: Refined non-linear static or dynamic analyses are increasingly used to assess the behavior of new and existing reinforced concrete structures. To perform these analyses, an adequate knowledge of the force–displacement, bending moment–curvature, and bending moment–rotation curves of relevant parts of structural members is needed, and modeling the bond–slip correlation for steel rebars becomes fundamental. The paper presents the results of an experimental campaign on r.c. specimens under tension, aiming, differently from previous studies, to better reproduce the bond–slip relationship accounting for the local confinement and anchorage conditions of real structural members. Resorting to an original numerical procedure allowing us to predict the relative displacement between steel reinforcement and the surrounding concrete in a reinforced concrete element, once assigned the stress in the naked steel bar and the bond–slip law, the experimental results are compared with the numerical outcomes obtained by adopting codified bond–slip laws. The comparison highlights that experimental values of sliding are well below those that are commonly given in existing bond slip laws, such as that adopted by the CEB-FIP Model Code. A new bond–slip model, which is able to satisfactorily predict actual strain fields and slips along the investigated r.c. elements, is thus proposed with the final aim of extending its implementation into non-linear analyses of r.c. structures.

Keywords: bond–slip; reinforced concrete; non-linear behavior; numerical analysis; crack pattern; crack opening; bond stress

check for updates

Citation: Croce, P.; Formichi, P.; Landi, F. Experimental Assessment and Numerical Modeling of the Bond–Slip Correlation for Steel Rebars in r.c. Members. *Materials* **2022**, *15*, 951. https://doi.org/10.3390/ma15030951

Academic Editor: Ayman El-Zohairy

Received: 5 January 2022
Accepted: 25 January 2022
Published: 26 January 2022

Publisher's Note: MDPI stays neutral with regard to jurisdictional claims in published maps and institutional affiliations.

1. Introduction

Non-linear static or dynamic analyses are increasingly used to study the behavior of reinforced concrete structures, especially for seismic designs [1] as well as for the structural assessments of existing buildings [2,3]. Carrying out non-linear analysis requires knowledge of the actual force–displacement, bending moment–curvature, and bending moment–rotation curves of relevant parts of structural members [4–7]. These capacity curves are needed for the evaluation of the available ductility of structural elements as well as for an appropriate estimation of the relevant structural parameters, which are required to perform equivalent elastic analyses, widely used as a simplified approach. Evidently, the curves can be obtained experimentally, through sophisticated and typically expensive full-scale ad hoc tests, or numerically, by resorting to refined theoretical models that have been suitably validated [8].

The non-linear force–displacement or bending moment–rotation relationships of r.c. members depend on the crack widths and on the crack pattern, and on the slip between concrete and reinforcing steel. Advanced models should appropriately consider all these local aspects as key issues for a suitable model [9–14], which are generally ignored in classical simplified approaches. Owing to the increasing interest in the assessment of existing r.c. structures [15,16], refined structural analyses, allowing us to reduce the impact on the structural strengthening interventions of safe sided assumptions often adopted in simplified approaches, are getting extensively used.

Moreover, experimental studies [17,18] have demonstrated that the classical approach to the analysis of reinforced concrete structures, based on the assumption of a perfect bond between steel rebars and concrete, is unsuitable for predicting the actual displacements/rotations of structural members, leading to remarkable overestimates in the evaluation of elements' stiffness. For these reasons, several studies [19–21] have been devoted to improving the approach, to allow a more refined modeling of the behavior of cracked r.c. members, based on specific bond–slip correlations.

In this paper, the results of an experimental campaign carried out on r.c. ties at the University of Pisa are presented and discussed. The experimental campaign and the type of test specimen represent an evolution of previous experimental studies [22–24] and have aimed at simulating the behavior of a real r.c. member under tension.

The test results, in terms of steel strains and the derived slips and bond stresses, are compared with the corresponding theoretical data, which were obtained by implementing a numerical procedure for the solution of the differential equations governing the interaction between steel bars and concrete [25]. The numerical procedure, described in Section 2, allows us to predict the relative displacement between the steel reinforcement and the surrounding concrete in a reinforced concrete tie, once we have assigned the stress in the naked steel bar and the bond–slip law. In fact, the definition of the bond–slip model is a crucial aspect, which has been the subject of numerous studies, leading to the model initially proposed in [9] and then adopted in the CEB-FIP Model Code [26,27].

The bond–slip model provided in the CEB-FIP Model Code, based on the full confinement of concrete members and often adopted in current practice, was adopted in the numerical analyses as a reference model to simulate the results obtained from tests. The comparison showed significant discrepancies from the experimental results, thus indicating the need for a new bond–slip model to predict the behavior of real r.c. members.

In this paper, an alternative effective bond–slip model is proposed with the aim of replicating experimental results and providing a suitable model to be used for the refined structural analyses of real reinforced concrete structures.

2. Bond–Slip Modeling in r.c. Structures

2.1. Analytical Modeling

The classical equations governing the bond–slip problem can be derived considering the equilibrium and compatibility conditions of an infinitesimal r.c. tie (see Figure 1). Considering a tie, the length of which is dx, made up by one centered rebar and by the surrounding portion of concrete in tension, subject to a tensile force F, the equation governing the global equilibrium of the tie is trivially

$$F = F_s + F_c \tag{1}$$

where F_s and F_c are the force acting on steel and on concrete, respectively. Differentiating Equation (1) gives

$$dF = dF_s + dF_c = d\sigma_s \frac{\pi \Phi^2}{4} + A_c(x) d\sigma_c + \sigma_c dA_c(x) \tag{2}$$

and then

$$d\sigma_c = -\rho(x) d\sigma_s - \sigma_c \frac{dA_c(x)}{A_c(x)} \tag{3}$$

where Φ is the diameter of the steel bar, $A_c(x)$ is the concrete area affected by the bond stress diffusion, $\rho(x)$ is the geometric reinforcement ratio, and σ_s and σ_c are the stresses in steel and concrete, respectively.

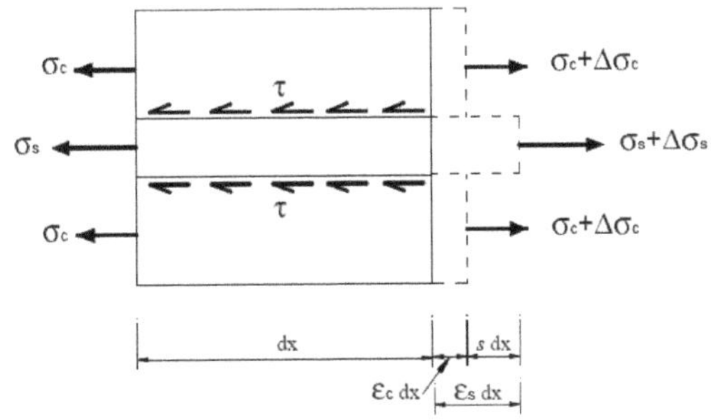

Figure 1. Stresses and elongations on an infinitesimal r.c. tie.

Considering the bond stress $\tau(s)$ and the function of the slip s, the local equilibrium of the rebar over the length dx is given by

$$d\sigma_s \frac{\pi \Phi^2}{4} = \pi \Phi \tau(s) dx \quad (4)$$

and for the compatibility condition,

$$(\varepsilon_s - \varepsilon_c) dx = ds \quad (5)$$

it yields

$$\frac{ds(x)}{dx} = \varepsilon_s - \varepsilon_c, \quad (6)$$

with ε_s and ε_c being the strains in steel and concrete, respectively.

Combining Equations (2), (4) and (6), the bond–slip problem can be described by the following differential equation

$$\frac{d^2}{dx^2} s = \frac{4\tau(s)}{\Phi} \left(\frac{d}{d\sigma_s} \varepsilon_s + \rho(x) \frac{d}{d\sigma_c} \varepsilon_c \right) + \frac{\sigma_c}{A_c(x)} \frac{d}{dx} A_c(x) \quad (7)$$

which can be solved numerically, under appropriate boundary conditions, once the stress–strain relationships for steel and for concrete under tension, the bond–slip law, and the dependence of A_c on x are defined.

2.2. Numerical Modeling

A numerical procedure to solve the non-linear differential equation that governs the bond–slip problem was proposed in [25]. The algorithm is summarized in Figure 2 and is based on an iterative shooting solution of the system of ODEs

$$\begin{cases} \frac{d}{dx} s = \varepsilon_s(\sigma_s) - \varepsilon_c\left(\frac{F - \sigma_s A_s}{A_c(x)}\right) \\ \frac{d}{dx} \sigma_s = \frac{4}{\Phi} \tau(s, x) \end{cases} \quad (8)$$

which can be suitably derived from Equation (7).

The first phase of the algorithm allows us to define the boundary conditions up to the first crack opening in a reinforced concrete tie of length L by means of the following steps:

1. the origin of the x-axis is set at the starting cross-section of the tie, $x = x_0 = 0$, where $\sigma_c = 0$, and half of the tie is subdivided into N intervals of constant length Δx;
2. the constitutive laws $\sigma - \varepsilon$ for steel and concrete under tensile stresses are assigned, as well as the bond–slip law $\tau - s$ and the $A_c(x)$ function;
3. tentative boundary conditions, s_0 and ε_{s0} in $x = 0$, are assigned, and the bond and tensile stress, $\tau_0(s_0)$ and $\sigma_{s0}(\varepsilon_{s0})$, are evaluated using the previously defined bond–slip law and the steel constitutive law;

4. the system of Equation (8) is integrated in the i-th interval using the Runge–Kutta 4th order method [28], determining the slip s_i and the steel stress $\sigma_{s,i}$ at the end of the interval x_i;
5. the bond stress, $\tau_i(s_i)$, in x_i is evaluated using the bond–slip law;
6. the stress in the concrete section, $\sigma_{c,i}$, is calculated through the equilibrium condition

$$\sigma_{c.i} = \frac{(F - \sigma_{s,i} A_s)}{A_c(x_i)} \tag{9}$$

7. the iteration is stopped when $s_i = 0$, and consequently $\tau_i = 0$, or when $i = N$, and convergence is checked (see step 9);
8. the process is iterated from step 4 for the next interval $i = i + 1$;
9. the convergence is checked—if the concrete strain can be considered equal to the steel strain for the assigned tolerance, $\varepsilon_{c.i} = \varepsilon_{s.i}$ in x_i, convergence is achieved, but otherwise, a new value is assigned to s_0, keeping the value of ε_{s0} unchanged, and the process is iterated from step 3 till convergence. The search for s_0 is performed using one of the numerical root-finding methods, such as the secant method or the Newton–Raphson method;
10. once the convergence is achieved, the actual value of $s_0(\varepsilon_{s0})$ is retained together with the actual transfer length $L_a(\varepsilon_{s0})$.

By means of the presented steps, the boundary conditions at $x = 0$, $s_0(\varepsilon_{s0})$ and ε_{s0}, can be found for each stress level in the naked bar, $\sigma_{s0}(\varepsilon_{s0})$. By increasing the value of ε_{s0}, the loading process is simulated until the ultimate tensile strain in the concrete, ε_{ct}, is attained at a distance L_a from the end sections, such as $\varepsilon_c(L_a) = \varepsilon_{ct}$, and the first crack opens at $x = L_f^{(j=1)} \geq L_a$. Actually, since the concrete tensile strain in each section belonging to the interval $L_a \leq x \leq L - L_a$ satisfies the condition $\varepsilon_c = \varepsilon_{ct}$, the location of the first crack is a random section of that interval, but, for the sake of the numerical algorithm, considering that the number of cracks and their distances in the final (stabilized) crack pattern are quite independent of the order in which subsequent cracks appear, the first crack is assumed to occur in the central zone of the tie.

After the first crack has opened, the numerical procedure is suitably modified to evaluate the opening of new cracks. At each step j, the portion of the tie of length $L_f^{(j-1)}$ is considered, and the following operations are carried out:

11. the origin of the x-axis is set at the starting cross-section of the considered portion of the tie, bounded by adjacent cracks spaced $L_f^{(j-1)}$, and the half of the portion is subdivided into M intervals of constant length Δx;
12. the boundary conditions s_0 and ε_{s0} in $x = 0$ are assigned following from the previous step, i.e., after the $(j-1)$-th crack formation phase, leading to $\varepsilon_{s0} = \varepsilon_{s0}^{(j-1)}$, and the bond stress and tensile stress are evaluated accordingly;
13. the system of ODEs in Equation (8) is solved by repeating the same procedure (steps 3 to 9) and checking convergence, which is achieved when slip s vanishes at the midpoint of the tie, i.e., $s\left(L_f^{(j-1)}/2\right) = 0$;
14. if $\varepsilon_c(L_f^{(j-1)}/2)$ has reached the concrete ultimate tensile strain ε_{ct}, a new crack opens at $L_f^{(j-1)}/2$, otherwise the value of ε_{s0} is increased and step 13 is repeated till the condition $\varepsilon_c(L_f^{(j-1)}/2) = \varepsilon_{ct}$ is satisfied. Once new crack opens, the new portion of the tie is set to $L_f^{(j)} = L_f^{(j-1)}/2$ and the process from 11 is iterated until the steel strain ε_{s0} reaches the ultimate value ε_{su}.

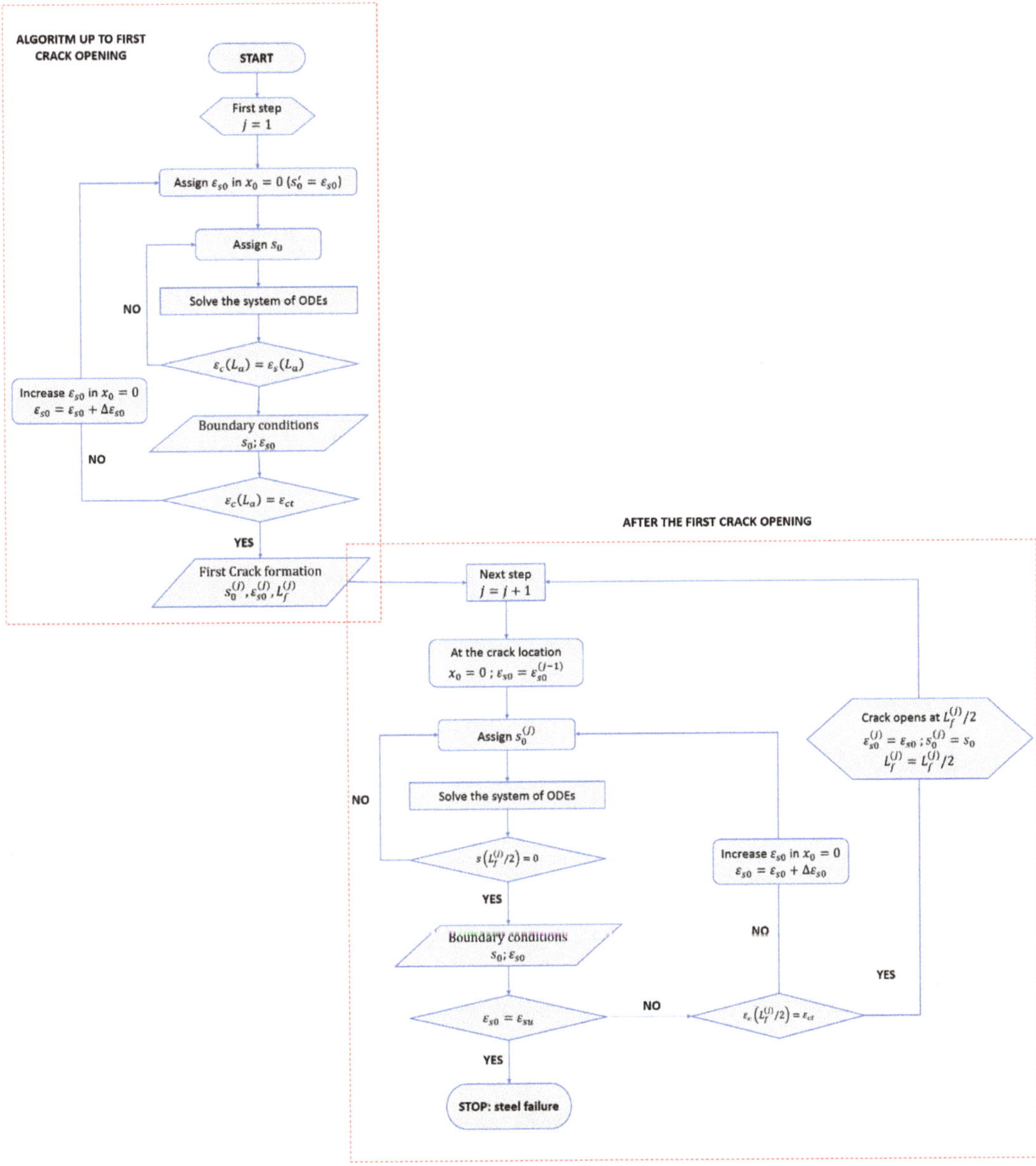

Figure 2. Flowchart of the numerical procedure to solve the bond–slip problem in r.c. elements.

Once one has completed the procedure, the width of each crack is evaluated by summing the quotas of s_0 pertaining to the two parts of the tie delimited by the crack.

An example of an application of the procedure is shown in Figure 3b, considering the reinforced concrete tie illustrated in Figure 3a, where the material mechanical properties (concrete compressive strength f_c, concrete tensile strength f_{ct}, concrete elastic modulus E_{ct}, steel yielding strength f_y, steel ultimate tensile strength f_t, and steel elastic modulus E_s) are also indicated. The r.c. cylindrical tie is characterized by a diameter $\Phi = 112$ mm, length $l = 0.8$ m, and by a 16 mm B450C steel rebar located in the center. In the analysis,

Ramberg–Osgood constitutive laws [29] are assumed for the steel and concrete tensile behavior and the CEB model [26] is adopted for the bond–slip law. The example shows the ability of the algorithm to efficiently predict the cracking of the r.c. tie.

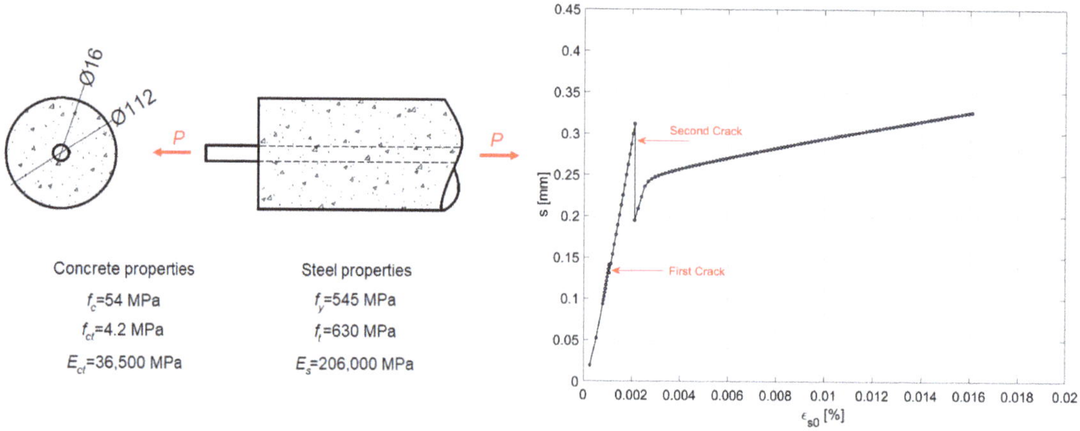

Figure 3. Example of the implementation of the procedure: (**a**) reinforced concrete cylindrical tie under axial load P and experimental materials properties (unit: mm); (**b**) resulting slip s-strain ε_{s0} diagram considering the CEB bond–slip model.

3. Experimental Results

An experimental test campaign was carried out to evaluate the bond–slip correlation for steel rebars in r.c. members. The innovative experimental technique is based on the study of the behavior of long specimens, where rebars are directly instrumented with strain gauges placed on their external surface.

The instrumentation of rebars with strain gauges is a relevant technical issue in the analysis of bond–slip behavior, it being necessary to directly measure the steel stresses at many points along the rebar, closely spaced from each other, allowing the effective placement of the electrical connections and not altering the stress transfer at the rebar surface. Examples of strain gauge arrangements can be found in the literature [22].

The original instrumentation of the steel bars adopted here is obtained by milling the ribbed rebar along two diametrically opposite generators, matching the smooth parts of its external surface (see Figure 4). In this way, the milled section, which is large enough to insert strain gauges with a 3 mm grid length, does not significantly alter the external surface of the bar, and does not modify the ribbed part. Strain gauges are glued into the groove with cyanoacrylate and duly protected and waterproofed so as not to be damaged during the pouring of concrete.

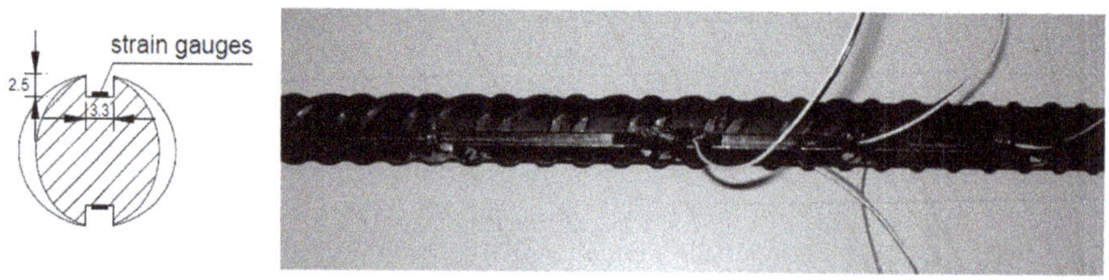

Figure 4. Insertion of electrical strain gauges in the ribbed steel rebar (unit: mm).

The strain gauge connection wires were kept outside the bars and placed radially to the concrete cylinder, which was poured around the bar. The presence of the wires in the concrete mass introduces discontinuities into it, which, however, consist of radial breaks, alternating from the two sides of the bar, along a diametrical plane of the cylinder itself, and do not significantly alter the formation of transverse cracks and of longitudinal ones at a later stage, as was observed during tests.

The arrangement of the strain gauges' wires and the limitation of the milled groove in the bars make it possible to use this technique and this instrumentation even in bars with a lower diameter than those used in the past [22,23]. The cross-section reduction is about 17 mm^2, which represents just over 8% of the resistant section for a 16 mm diameter bar, which is, in any case, considered in the analysis. The reduction in the adherent section on the lateral surface of the bar is equal to $3.3 \times 2 = 6.6$ mm on a nominal perimeter, while that for the same bar φ16 is about 50 mm; the reduction is therefore of the order of 13%. In any case, it does not particularly affect the bond; in fact, the stress transfer is assured by the ribs, which are not affected by the milling of the groove, as clearly results from Figure 4.

The milling is carried out with low-speed milling cutters in a cooling bath to limit the thermally induced stresses generated by the preparation of the specimens.

Inside the two longitudinal millings, the electrical strain gauges were placed at 25 mm intervals, in alternate positions, so that on the same side of the bar there would be a strain gauge every 50 mm, thus reducing the local disturbance effects generated by the presence of the connecting wires.

According to the described technique, bars with lengths of their instrumented parts equal to 100 and 23 cm were prepared. Two diameters were used for the steel rebars: φ16 and φ20. The diameter of the confining concrete cylinder was assumed constant and equal to 132 mm, which corresponds to approximately 6–8 times the diameter of the embedded rebars.

As illustrated in Figure 5, the two ends of the bar are extended beyond the concrete cylinder to allow for clamping into a universal tensile testing machine.

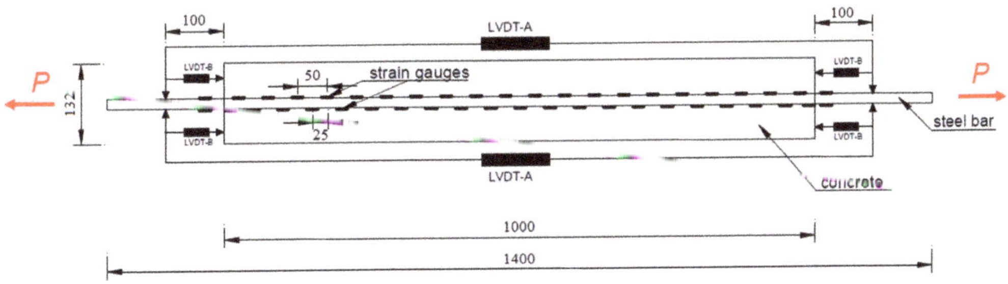

Figure 5. Layout of the test specimen $L = 1000$ mm under axial load P (unit: mm).

In addition to strain gauges, LVDT transducers were used to measure the total elongation of the bar (LVDT A) and the slip at the two end faces of the concrete cylinder (LVDT B).

Although both specimens, the 100 cm- and 23 cm-long ones, can be classified as "long" specimens characterized by a long member behavior [25], the formation of multiple cracks is possible, allowing us to study the evolution of progressive cracking phenomena until their complete stabilization. The shorter specimens, whose length was calibrated to avoid intermediate cracks, were used to verify the evolution of the bond slip law in the absence of transversal cracking.

In Table 1, the main characteristics of the six specimens tested during the experimental campaign are illustrated together with the number of instruments used for each test.

Table 1. Main characteristics of test specimen: length L, steel rebar diameter Φ, r.c. tie diameter Φ_c, number of electrical strain gauges on the naked bar N_1 and on the embedded bar N_2, number of LVDT-A N_{LVDT-A}, and of LVDT-B N_{LVDT-B}.

N.	L (mm)	Φ [1] (mm)	Φ_c (mm)	N_1	N_2	N_{LVDT-A}	N_{LVDT-B}
1	1000	16	132	4	40	4	4
2	235	16	132	1	9	4	2
3	235	16	132	1	9	4	2
4	235	16	132	1	9	4	2
5	1000	20	132	4	40	4	2
6	1000	20	132	4	40	4	-

[1] Notional diameter of the ribbed bar.

The mechanical behaviors of the materials used for each specimen were tested in the laboratory. Steel B450C rebars from the same casting were used for each specimen and mechanical properties were evaluated by means of preliminary tensile tests, conducted on specimens instrumented with a pair of strain gauges, glued in diametrical positions at the mid-section of the specimen. Figure 6 shows one of the experimental stress–strain diagrams obtained during the experimental campaign together with the analytical model obtained by adopting the Ramberg–Osgood constitutive law. Prior to concrete pouring, all the instrumented bars were tested under loads well below their elastic limit to verify the correct functioning of the installed strain gauges.

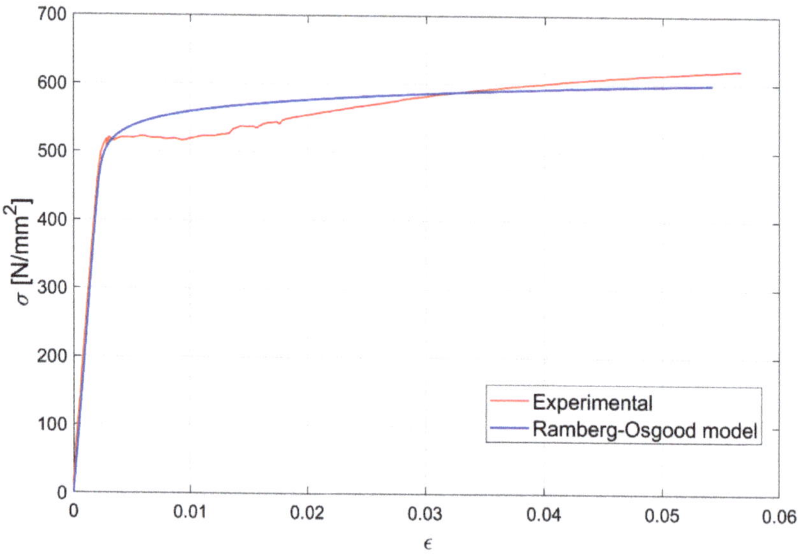

Figure 6. Experimental stress–strain diagram for a φ16 steel rebar used in the test campaign.

Special formworks, consisting of two portions of a tubular steel profile with an internal diameter of 132 mm, were used to prepare the cylindrical specimen. The formworks, shown in Figure 7, were cut along the two generatrixes of a diametrical plane, to allow the arrangement of strain gauge connection wires along the line of conjunction of the two halves. The formworks were completed by a support system, which allowed us at the same time to center the bar along the axis of the cylinder, and to keep it in position during the subsequent phases of pouring and curing. Self-compacting concrete (SCC) was used to avoid any need for the compaction of castings that could potentially damage the instruments. The SCC mix was made from 515 kg/m^3 of hyper fluid expanding cementitious binder, 200 kg/m^3 of gravel, 600 kg/m^3 of fine gravel, 815 kg/m^3 of sand and 240 kg/m^3 of water, so that the water to cement ratio was about 0.466.

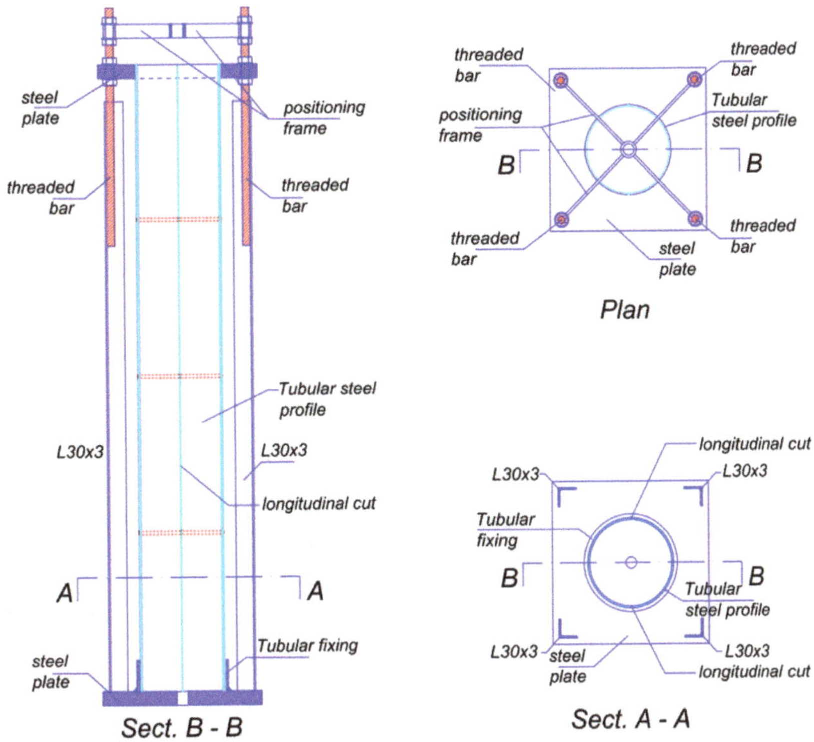

Figure 7. Special formworks used to prepare the cylindrical specimens.

During each casting, two cylindrical concrete samples were taken (except for sample n.1, for which four cylinders were taken) with a height to diameter ratio of 2, to perform compression tests and indirect ("Brazilian") tensile tests. Compression and indirect tensile tests on concrete specimens were performed immediately before the test on each tie, and after the 28th day of concrete curing. The results are summarized in Tables 2 and 3, where R indicates the maximum axial load sustained by the specimen.

In addition to compression and indirect tensile tests on specimens taken during casting, special specimens were prepared for some ties to directly characterize the tensile stress–strain diagram. These specimens are similar to the "long" r.c. specimens used for the tensile tests, except for the steel bar, which is interrupted at about 25 cm ($\approx 2\Phi_c$) across the middle section of the cylinder, and has special ribs welded to the ends of the bar embedded in the concrete. In this way, the concrete section in the middle of the cylinder is subjected to a stress state that can be reasonably assumed to be uniform uniaxial tensile. On the external surface of the cylinder, eight electrical strain gauges for the concrete with an 8 cm-long base were placed in order to record the tensile deformations up to the tensile failure of the concrete. An example of the cylindrical r.c. tie used for the direct tensile test at the beginning of testing is illustrated in Figure 8a, where one can easily distinguish the strain gauges, while Figure 8b shows the tested specimen after the failure.

Table 2. Results of compression tests on concrete specimens.

N.	H (mm)	D (mm)	Mass (kg)	Density (kg/m^3)	R (kN)	f_c (N/mm^2)
1	195	100	3.45	2253	296	37.7
	195	100	3.45	2253	318	40.5
	195	100	3.42	2233	328	41.8
	195	100	3.44	2246	330	42.0
2, 3, 4	193	100	3.36	2217	320	40.7
	196	100	3.40	2209	311	39.6
5	196	100	3.52	2287	254	32.3
	196	100	3.52	2287	299	38.1
6	198	100	3.56	2289	234	29.8
	196	100	3.54	2300	236	30.0

Table 3. Results of indirect tensile tests on concrete specimens.

N.	H (mm)	D (mm)	Mass (kg)	Density (kg/m^3)	R (kN)	$f_{ct,sp}$ (N/mm^2)
1	200	100	3.54	2254	96	3.1
	200	100	3.54	2254	96	3.1
2, 3, 4	200	100	3.44	2190	101	3.2
	200	100	3.50	2228	91	2.9
5	200	100	3.56	2266	106	3.4
	200	100	3.56	2266	100	3.2
6	200	100	3.56	2266	105	3.3
	200	100	3.54	2254	125	4.0

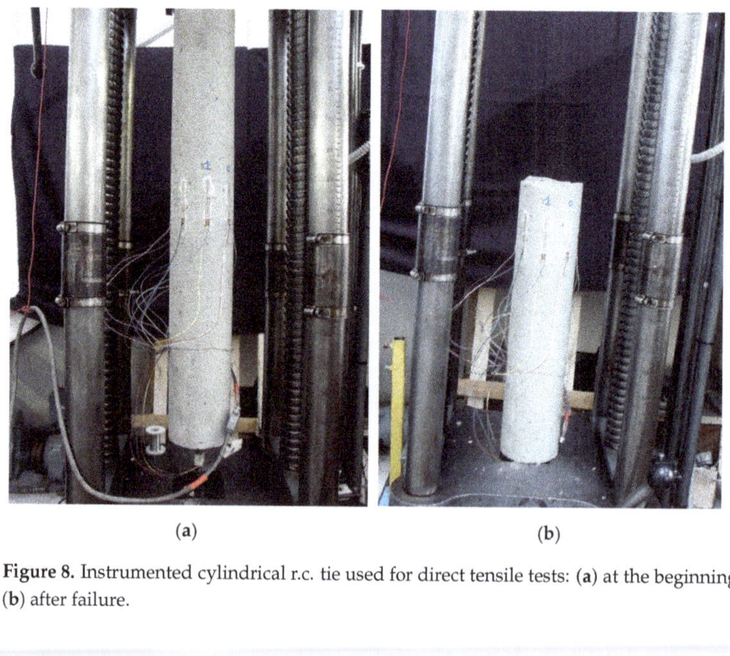

(a) (b)

Figure 8. Instrumented cylindrical r.c. tie used for direct tensile tests: (**a**) at the beginning of the test; (**b**) after failure.

In Figure 9, the experimental stress–strain diagram (in red) obtained for one specimen is compared with the Ramberg–Osgood constitutive law (in blue) after calibration of the parameters.

Table 4. Results of direct uniaxial tensile tests on concrete specimens.

N.	Φ_c (mm)	f_{ct} (N/mm^2)	E_{ct} (N/mm^2)	f_{ctm}(EC2) (N/mm^2)	E_{ct}(EC2) (N/mm^2)
1	132	2.32	29,000	3.5	35,277
2	132	2.16	24,100	2.4	30,835
3	132	2.05	21,200	3.1	33,468

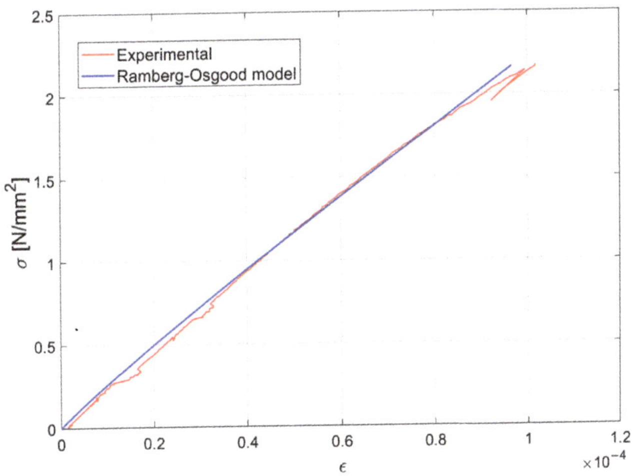

Figure 9. Experimental tensile stress–strain diagram for concrete (specimen n.2 in Table 4).

In Table 4, the results of the tests in terms of tensile strength f_{ct} and elastic modulus E_{ct} are summarized and compared with the values obtained according to the formulations given in Eurocode 2 [30], depending on the compressive strength

$$f_{ctm} = 0.3(f_{cm} - 8)^{\frac{2}{3}} \tag{10}$$

and

$$E_{cm} = 22,000\left(\frac{f_{cm}}{10}\right)^{0.3} \tag{11}$$

The tensile strength determined according to EN formulations is in good agreement with experimental values obtained by the indirect tensile tests, while the experimental values of the elastic modulus are generally lower than those obtained by the EN provisions. A ratio of about 0.75 is obtained between the experimental elastic modulus and that obtained according to Equation (11).

Once we had characterized the mechanical behavior of the concrete and steel rebars used for assembling r.c. specimens, the tensile tests were carried out on the six r.c. ties listed in Table 1. Tests were carried out while controlling the applied load, with an increase rate of about 5 kN/min, in conformity with the loading rate usually adopted in standard laboratory tests. In some cases, to check the slopes of the unloading and reloading cycles and their compliance with the initial slopes, an intermediate unloading cycle was performed, followed by a new reloading, always at constant speed.

3.1. Long Specimen

In Figure 10, an example of a "long" test specimen (n.1 in Table 1) is shown before the test (Figure 10a) and at the end of the test (Figure 10b). In the latter, the transversal cracks, which occurred at increasing values of the load during the test, are clearly visible.

(**a**) (**b**)

Figure 10. Reinforced concrete specimen n.1 tested at the Laboratory of the University of Pisa: (**a**) specimen before the test, (**b**) cracked specimen after the test.

The experimental strains along the steel rebar embedded in the r.c. tie are shown in Figure 11 for different levels of applied load *P*, before (a) and after crack openings (b), while in Figure 12 the experimental load elongation diagram (a) and the location of the cracks (b) are reported. The first three cracks (1, 2 and 3) opened under a load *P* approximately equal to 30 kN, while two subsequent cracks opened at *P* = 58 kN (4–7) and a last crack

(6) opened at P = 74 kN. The load was increased till the yielding of the rebar, but no more cracks opened.

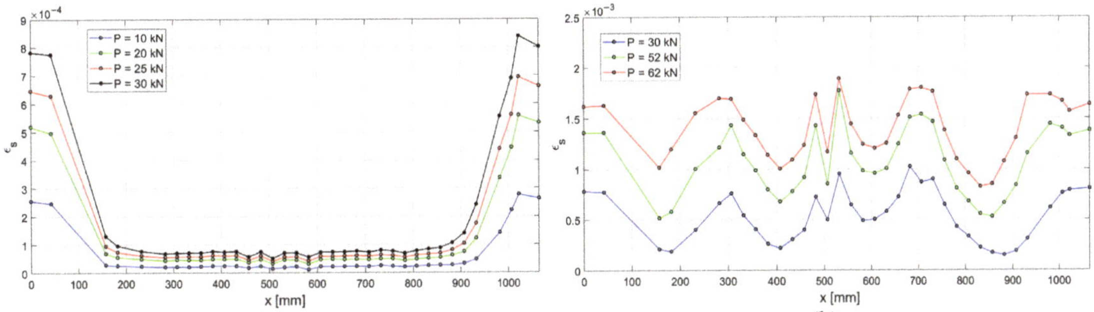

(**a**) (**b**)

Figure 11. Experimental ε-x curves of the n.1 r.c. tie: (**a**) P < 30 kN, (**b**) 30 kN ≤ P ≤ 60 kN.

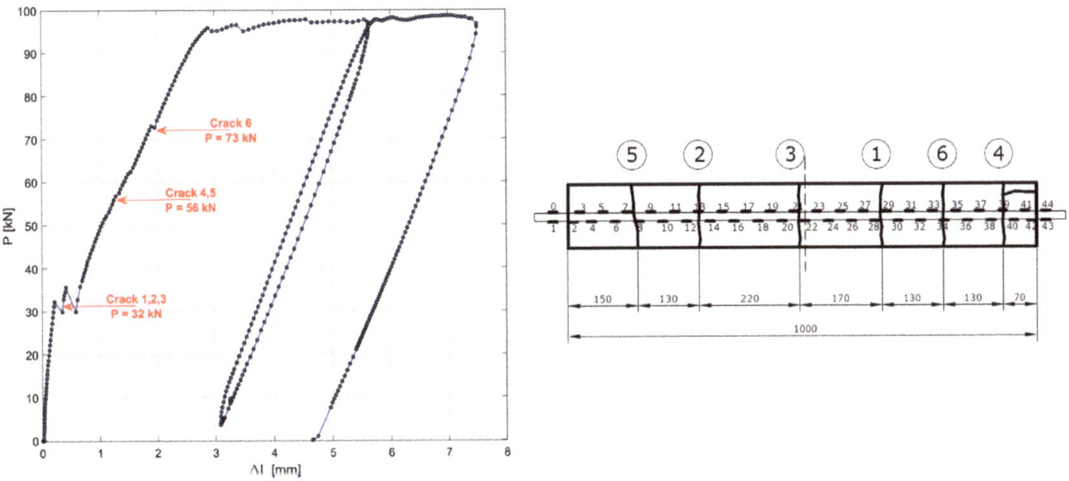

(**a**) (**b**)

Figure 12. Experimental load–displacement diagram specimen n.1 (**a**), and location of the six cracks numbered according to the sequence of occurrence (unit: mm) (**b**).

Looking at Figure 12a, it can be determined that, when a stabilized crack pattern is obtained, the slope of the loading cycle is nearly coincident with the slopes of the unloading and reloading cycles, even when irreversible slips occur.

The experimental results were compared with those obtained by implementing the analytical model and the numerical procedure described in Section 2. In the following numerical analysis, the analytical bond stress–slip relationship defined in the CEB-FIB Model Code 2010 [26] is considered

$$\tau(s) = \begin{cases} \tau_{b,\max}\left(\frac{s}{s_1}\right)^{\alpha} & \text{for } 0 \le s \le s_1 \\ \tau_{b,\max} & \text{for } s_1 \le s \le s_2 \\ \tau_{b,\max} - \left(\tau_{b,\max} - \tau_{bf}\right)\left(\frac{s-s_2}{s_3-s_2}\right)^{\alpha} & \text{for } s_2 \le s \le s_3 \\ \tau_{bf} & \text{for } s > s_4 \end{cases} \quad (12)$$

where the parameters $\tau_{b,\max}$, τ_{bf}, s_1, s_2, s_3, and α are defined in the Model Code as depending mainly on the failure mode—pull-out or splitting—and on the bond conditions within each failure mode. The FIB-MC2010 bond–slip model, which is derived for the investigated specimen, is illustrated in Figure 13.

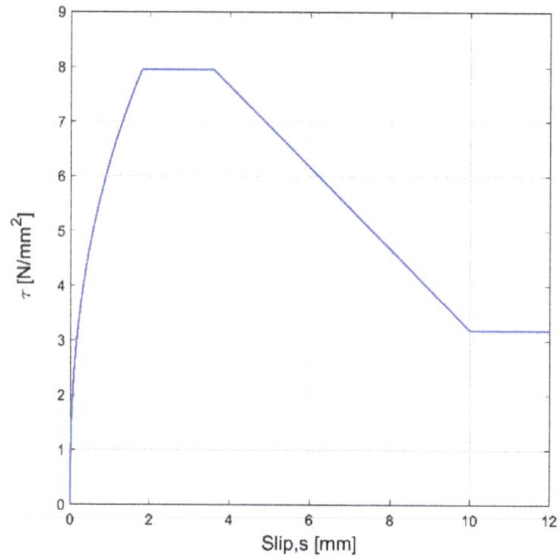

Figure 13. CEB-MC2010 bond stress-slip model, with $\tau_{b,\max} = 1.25\sqrt{f_{cm}} = 7.95\,\text{N/mm}^2$, $\tau_{bf} = 0.4\tau_{b,\max}$, $\alpha = 0.4$, $s_1 = 1.8$ mm, $s_2 = 3.6$ mm, and $s_3 = 10$ mm.

The resulting slip–strain diagram, simulated for specimen n. 1, is shown in Figure 14. A first crack occurs at a distance from the end $L_f = 435$ mm when the applied load P reaches the value of 32.4 kN. When increasing the value of the load, a new crack is recorded in the two lateral portions of the specimen, delimited by the first crack, for $P = 62.9$ kN. It is worth recalling that the loads inducing the cracks in the test were 30 kN, 58 kN and 74 kN respectively. Additionally, the crack spacing is quite different when comparing the numerical simulation and the experimental behavior of the same specimen. On average, the spacing observed during the test is around 300 mm, which is lower than the estimated L_f.

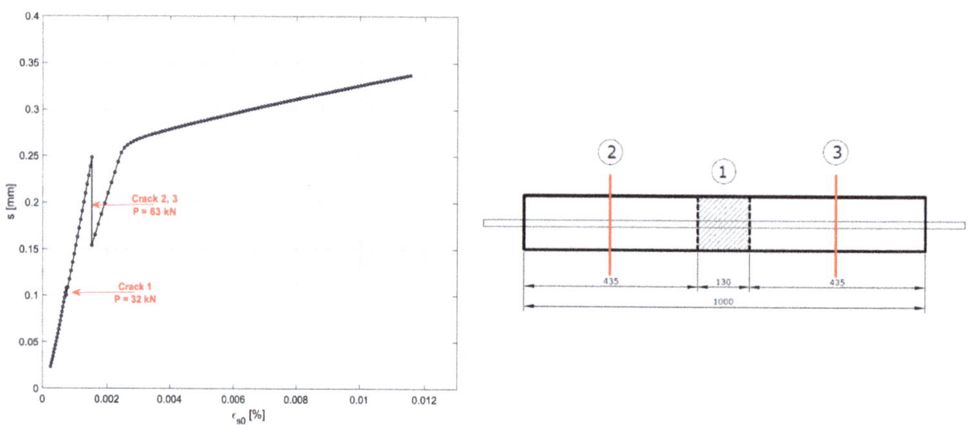

(a) (b)

Figure 14. Analytical slip–strain diagram for specimen n.1 (**a**), and the location of the three cracks numbered according to the sequence of occurrence (unit: mm) (**b**).

Furthermore, the analytical model, adopting the CEB-MC2010 bond–slip relationship, was able to predict only three cracks out of the six observed during the test.

Finally, it is observed that the magnitude of slip values recorded during the experimental campaign is very limited (<1 mm) with respect to those given by the CEB-MC2010 model. In fact, the sliding values that should be reached to activate the bond failure appear to be far too high and are never recorded during the tests.

Similar conclusions can be drawn by checking the results from the remaining long specimens, where the cracks appeared in the same areas, and for similar slip values (see, for example, Figure 15, referring to specimen no. 5, before and after the test). The short specimen conception and results are described in the following section.

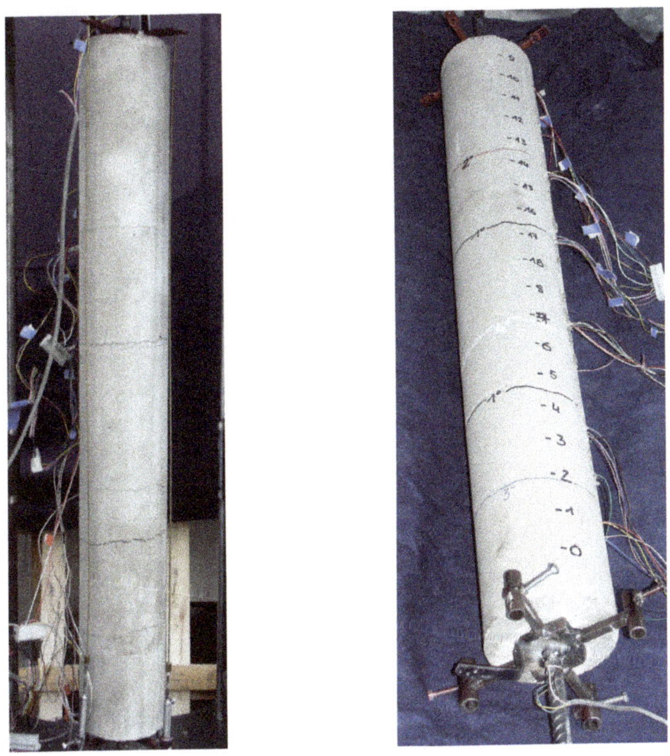

(**a**) (**b**)

Figure 15. Reinforced concrete specimen no.5 tested at the Laboratory of the University of Pisa: (**a**) specimen during the test, (**b**) specimen after the test.

This experimental evidence indicates the need for a new model for the bond–slip behavior calibrated for the simulation of real structural members. Such a new bond–slip model for r.c. members will be presented and discussed in the next section.

3.2. Short Specimen

An example of a "short" specimen (specimen no. 3 in Table 1) is illustrated in Figure 16, where cracks due to splitting were evident at the end of the test. It is important to highlight that cracks due to splitting were obtained for high values of the load close to the end of the test, thus not compromising the experimental test results on the bond–slip behavior of the short specimen.

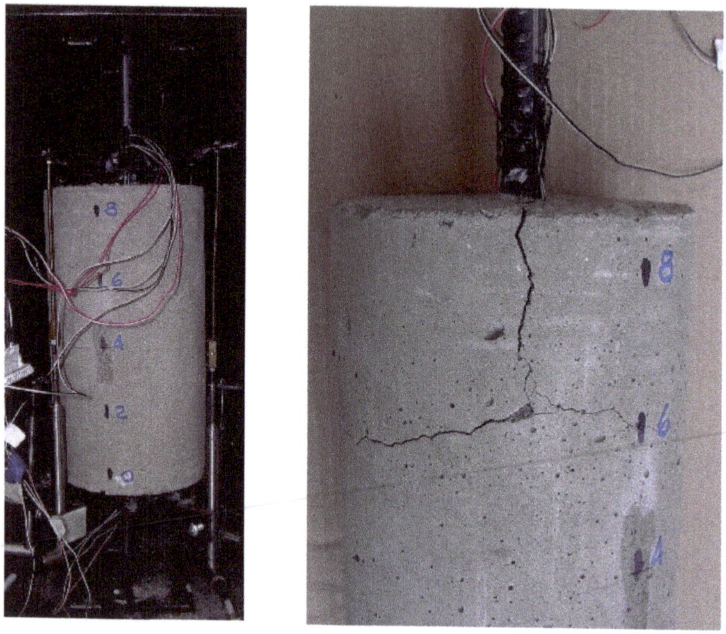

(**a**) (**b**)

Figure 16. Reinforced concrete specimen n.3 tested at the Laboratory of the University of Pisa: (**a**) specimen before the test, (**b**) cracked specimen after the test.

The experimental results measured along the r.c. tie for different levels of the applied load P are shown in Figures 17–19, considering the deformations on the steel rebar ε_s (Figure 17), the slips s (Figure 18), and the variation in shear stress τ (Figure 19). In the same figures, the analytical curves obtained by implementing the numerical procedure described in Section 2 are reported. These curves were derived by adopting, as in the previous case, the CEB-MC2020 bond–slip law.

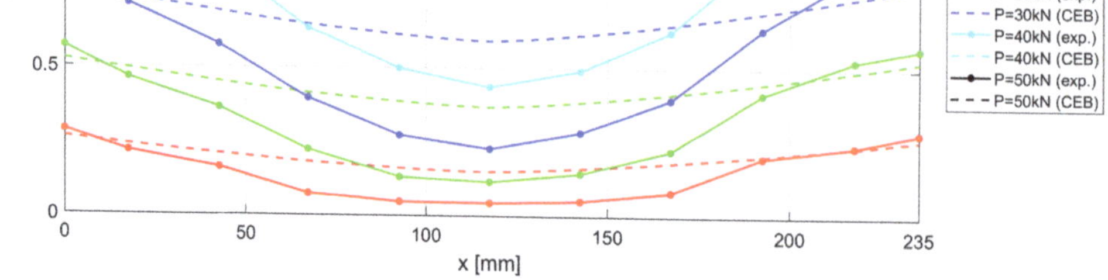

Figure 17. Experimental and numerical $\varepsilon - x$ for increasing load P.

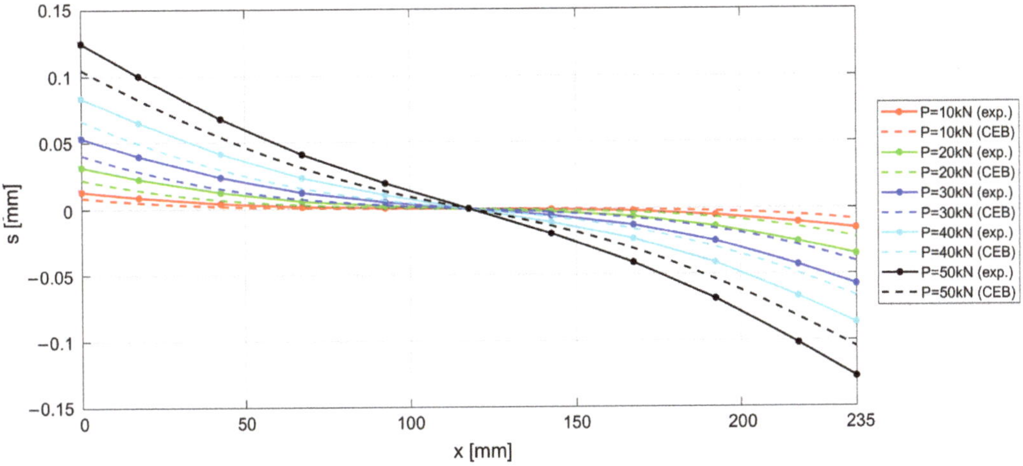

Figure 18. Experimental and numerical $s - x$ for increasing load P.

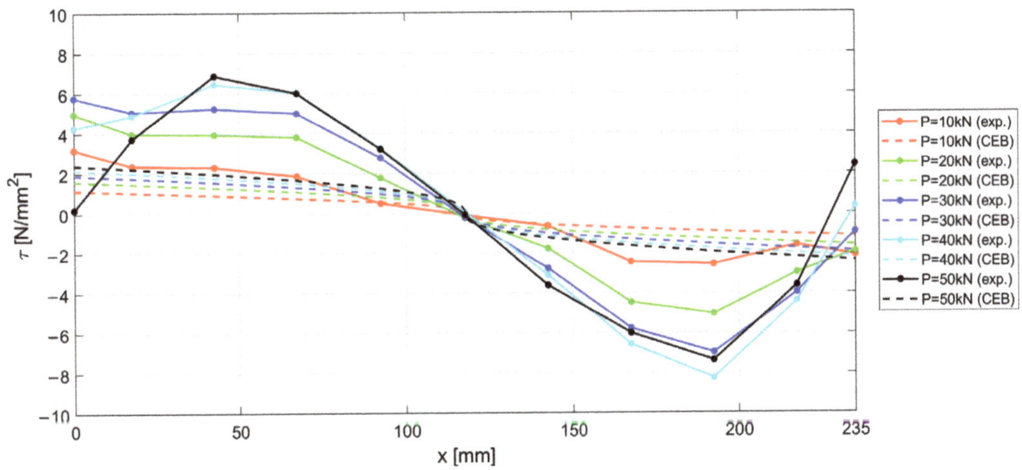

Figure 19. Experimental and numerical $\tau - x$ curves for increasing load P.

The comparison of experimental and numerical curves in Figures 17–19 shows that the use of the CEB bond–slip model has the following effects:

- It leads to an underestimation of the bond stress transfer from steel to concrete. In fact, the steel deformations are characterized by a constant overestimation of the stresses in the bars, especially in the central section of the sample (see Figure 17), thus leading to concrete stresses significantly lower than in the real case;
- Nonetheless, a satisfactory level of approximation is obtained for the slip at the end sections of the r.c. tie, which estimates the crack width (see Figure 18), but the corresponding shear stresses are significantly different from the experimental ones. In fact, maximum slips, which are limited to 0.2 mm, are one order of magnitude lower than the slips associated with the plastic range of the bond–slip CEB model (see Figure 13). This outcome confirms that the actual magnitude of the slip values is well below that corresponding to the horizontal plateau or to the descending branch of the CEB bond–slip model.

From the above considerations, it is concluded that, using the CEB bond–slip model, long sections of confined bar are required to achieve cracking, due to the modest tangential stresses that are activated; consequently, the distance between the cracks increases, in a way that does not appear to be justified on the basis of engineering experience.

4. Discussion

As is highlighted in the Section 3, the results of the experimental campaign suggest the development of a new model for the bond–slip behavior of real r.c. members, especially to capture the local magnitude of slips, which are significantly lower than the theoretical ones provided by the CEB-MC2010 [26].

The whole set of experimental results is represented in terms of bond–slip curves in Figure 20, where the experimental curves are drawn with dashed lines, parametrized according to values of the abscissa x varying between 0 and 60 mm, at 15 mm steps. In the diagram, the color of the curve depends on the value of the parameter x: the blue lines correspond to $x = 0$; red lines to $x = 15$ mm; green lines to $x = 30$ mm; cyan lines to $x = 45$ mm; black lines to $x = 60$ mm.

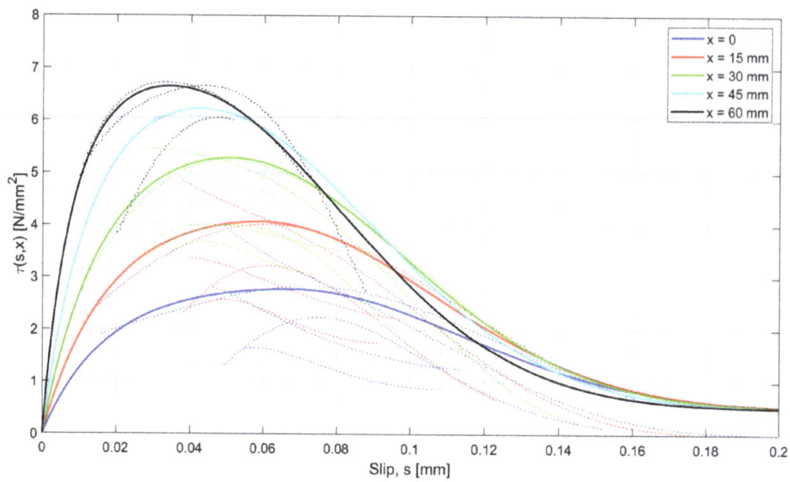

Figure 20. Comparison of the experimental bond–slip $\tau - s$ diagram, dashed lines; the proposed model for different abscissas x of the elements, solid lines.

To fit the experimental curves, a new model for bond–slip behavior has been developed and calibrated. Among the simple analytical models that have been considered and suitably calibrated to reproduce the experimental results, the one providing the most satisfactory predictions results an exponential model expressed by the equations:

$$\tau(s,x) = \begin{cases} \dfrac{\tau_{b,\max}(x)\left(e^{-b_1(x)s_{\max}^2(x)} - e^{-b_1(x)(s-s_{\max}(x))^2}\right)}{e^{-b_1(x)s_{\max}^2(x)} - 1}\ [\text{MPa}] & \text{for } 0 \le s \le s_{\max} \\ \left(\tau_{b,\max}(x) - \tau_{bf}\right)e^{-b_2(x)(s-s_{\max}(x))^2} + \tau_{bf}\ [\text{Mpa}] & \text{for } s > s_{\max} \end{cases} \quad (13)$$

In the present study, the calibration provided the following values of the relevant model parameters:

$$b_1 = -16.06\, e^{0.074\, x} - 350\ \left[\text{mm}^{-2}\right] \quad (14)$$

$$b_2 = 6.33 \cdot 10^{-5} \cdot e^{0.171\, x} + 222\ \left[\text{mm}^{-2}\right] \quad (15)$$

$$\tau_{b,\max} = -1.106 \cdot 10^{-5} \cdot x^3 + 3.542 \cdot 10^{-4} \cdot x^2 + 0.083 \cdot x + 2.778\ [\text{Mpa}] \quad (16)$$

$$s_{max} = 0.067 - 5.462 \cdot 10^{-5} \cdot x \text{ [mm]} \quad (17)$$

where the abscissa x and the slip s are both expressed in mm. In fact, the reference bond stress–slip curves, represented by the solid lines in Figure 20, appear close enough to the experimental ones.

When adopting the proposed bond stress–slip model, the analytical $\varepsilon - x$, $s - x$, and $\tau - x$ curves are derived by means of the numerical procedure introduced in Section 2, for the specimen investigated in Section 3.2, identified as n.3 in Table 1. The results are compared with the experimental curves in Figures 21–23, for steel deformation ε_s, slip s, and shear stresses τ, respectively. In the figures, the analytical curves are shown that were obtained by considering the variation in the bond–slip model for different locations along the tie (point dashed lines), or a fixed bond–slip model ($x = 60$ mm).

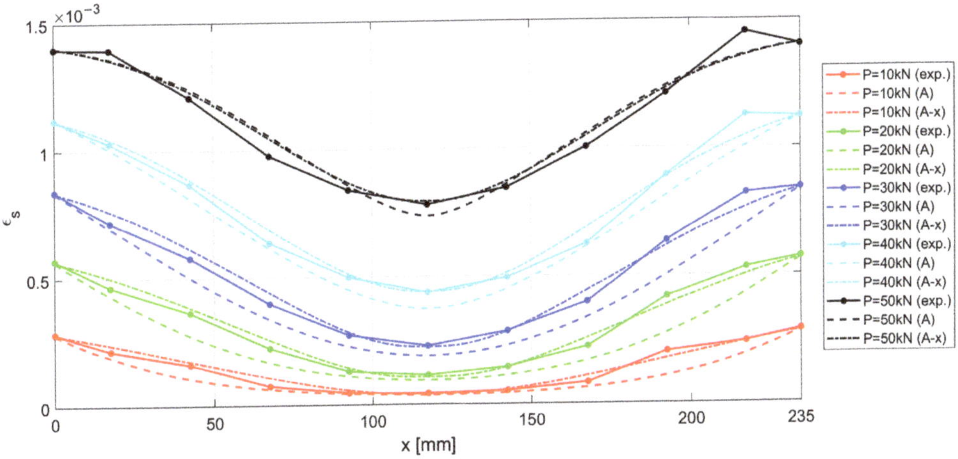

Figure 21. Experimental and numerical $\varepsilon - x$ for increasing load P. Two numerical formulations are presented, obtained by considering a fixed bond–slip model (A) and a model with a varying abscissa (A-x).

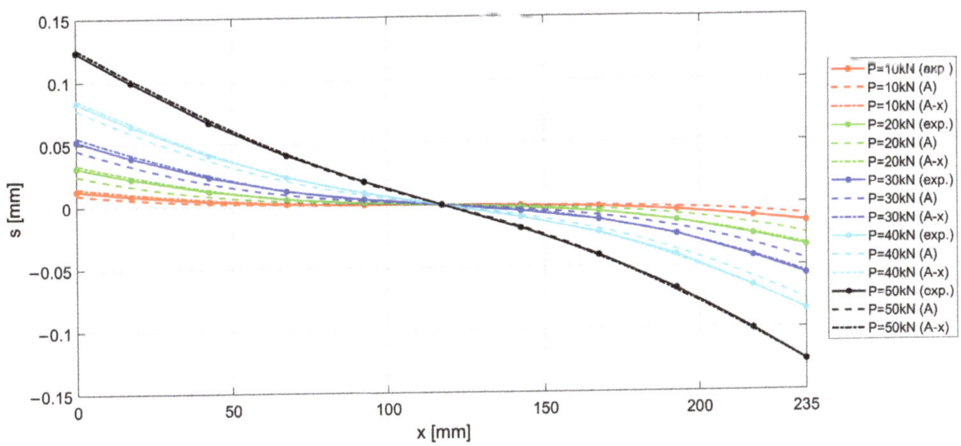

Figure 22. Experimental and numerical $s - x$ for increasing load P. Two numerical formulations are presented, obtained by considering a fixed bond–slip model (A) and a model varying the abscissa (A-x).

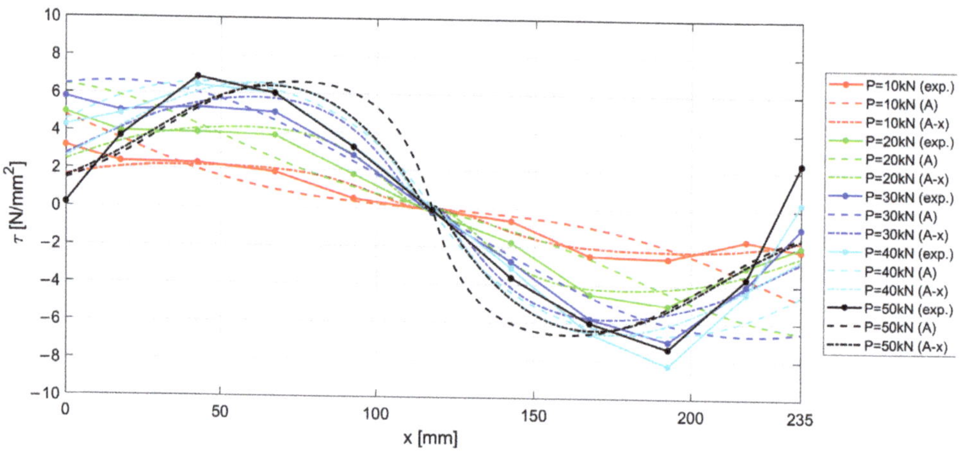

Figure 23. Experimental and numerical $\tau - x$ curves for increasing load P. Two numerical formulations are presented, obtained by considering a fixed bond–slip model (A) and a model varying the abscissa (A-x).

When comparing the curves obtained by numerical integration adopting the proposed exponential bond–slip law (Figures 21–23) with those previously obtained by using the CEB bond slip law (Figures 17–19), it clearly emerges that the proposed model allows not only a better estimate of the actual crack widths, which are the ordinates at the end sections $x = 0$ and $x = 235$ mm, of the $s - x$ diagrams in Figures 18 and 22, respectively, but also a more precise fitting of the experimental local behavior.

5. Conclusions

In this paper, the outcomes of an experimental campaign aiming to evaluate the local bond–slip relationship of rebars in r.c. members are presented and discussed.

The innovative testing arrangement proposed consists of a r.c. tie, in which the steel rebar, suitably prepared, is fully instrumented with strain gauges placed at 25 mm intervals, in alternate positions, allowing one to derive the variation in the bond along the element as a function of the applied load. A refined local bond–slip relationship is thus obtained that can elaborate on the experimental data, thus overcoming the limitation of the standardized bond–slip law.

The implementation of the obtained local bond–slip law in a numerical procedure allows one to reproduce the experimental behavior of the r.c. element under tension, not only at the ends or near the cracks, but also locally, independently of the considered abscissa.

The wide numerical study, presented in Section 3, demonstrates that the bond–slip law proposed by the CEB-FIP Model Code, although able to provide a satisfactory estimate of the crack width, which is the most relevant information from the design point of view, cannot be adopted for refined studies when a precise assessment of the local behavior is required. In this latter case, the original relationship proposed in the paper appears to be more appropriate.

Author Contributions: Conceptualization, P.C., P.F. and F.L.; methodology, P.C., P.F. and F.L.; software, P.C., P.F. and F.L.; validation, P.C., P.F. and F.L.; data curation, P.C., P.F. and F.L.; writing—original draft preparation, P.C., P.F. and F.L.; writing—review and editing, P.C., P.F. and F.L.; visualization, P.C., P.F. and F.L.; resources, P.C. All authors have read and agreed to the published version of the manuscript.

Funding: This research received no external funding.

Institutional Review Board Statement: Not applicable.

Informed Consent Statement: Not applicable.

Data Availability Statement: The data presented in this study are available on request from the corresponding author. The data are not publicly available as they cannot be used for commercial purposes.

Conflicts of Interest: The authors declare no conflict of interest.

References

1. *EN 1998-1. Eurocode 8;* Design of Structures for Earthquake Resistance—Part 1: General Rules, Seismic Actions and Rules for Buildings. European Committee for Standardization (CEN): Brussels, Belgium, 2003.
2. *EN 1998-3. Eurocode 8;* Design of Structures for Earthquake Resistance—Part 3: Assessment and Retrofitting of Buildings. European Committee for Standardization (CEN): Brussels, Belgium, 2003.
3. Marsili, F.; Croce, P.; Friedman, N.; Formichi, P.; Landi, F. Seismic Reliability Assessment of a Concrete Water Tank Based on the Bayesian Updating of the Finite Element Model. *ASCE-ASME J. Risk Uncertain. Eng. Syst. Part B Mech. Eng.* **2017**, *3*, 021004. [CrossRef]
4. Kwak, H.G.; Kim, S.P. Nonlinear analysis of RC beams based on moment–curvature relation. *Comput. Struct.* **2002**, *80*, 615–628. [CrossRef]
5. Carvalho, G.; Bento, R.; Bhatt, C. Nonlinear static and dynamic analyses of reinforced concrete buildings—Comparison of different modelling approaches. *Earthq. Struct.* **2013**, *4*, 451–470. [CrossRef]
6. Croce, P.; Beconcini, M.L.; Formichi, P.; Landi, F.; Puccini, B.; Zotti, V. Bayesian Methodology for Probabilistic Description of Mechanical Parameters of Masonry Walls. *ASCE-ASME J. Risk Uncertain. Eng. Syst. Part A Civ. Eng.* **2021**, *7*, 04021008. [CrossRef]
7. Beconcini, M.L.; Cioni, P.; Croce, P.; Formichi, P.; Landi, F.; Mochi, C. Non-linear static analysis of masonry buildings under seismic actions. In Proceedings of the 22nd World Multi-Conference on Systemics, Cybernetics and Informatics (IMSCI 2018), Orlando, FL, USA, 8–11 July 2018; International Institute of Informatics and Systemics: Winter Garden, FL, USA; pp. 126–131. Available online: http://www.iiisci.org/Journal/CV$/sci/pdfs/EA239AY18.pdf (accessed on 19 November 2021).
8. Murcia-Delso, J.; Benson Shing, P. Bond-Slip Model for Detailed Finite-Element Analysis of Reinforced Concrete Structures. *J. Struct. Eng.* **2015**, *141*, 04014125. [CrossRef]
9. Ciampi, V.; Eligehausen, R.; Bertero, V.V.; Popov, E.P. *Analytical Model for Concrete Anchorages of Reinforced Bars under Generalized Excitations;* Report. No. UCB/EERC83/23; University of California: Berkeley, CA, USA, 1982.
10. Eligehausen, R.; Bigaj-Van Vliet, A. Bond Behaviour and Models. In *Structural Concrete, the Textbook on Behaviour, Design and Performance;* CEB-FIP Bulletins 1, 2, 3; Fédération Internationale du Béton (FIB): Lausanne, Switzerland, 1999.
11. Park, R.; Paulay, T. *Reinforced Concrete Structures;* John Wiley & Sons: New York, NY, USA, 2001.
12. Tastani, S.P.; Pantazopoulou, S.J. Experimental Evaluation of the Direct Tension Pullout Bond Test. In Proceedings of the Symposium on "Bond in Concrete—From Research to Standards, Budapest, Hungary, 20–22 November 2002; pp. 268–276.
13. Borosnyói, A.; Balázs, G.L. Models for Flexural Cracking in Concrete: The State of the Art. *Struct. Concr.* **2005**, *6*, 53–62. [CrossRef]
14. CEB-FIB. *Bond of Reinforcement in Concrete;* Bullettin nr. 10. FIB; Fédération Internationale du Béton (FIB): Lausanne, Switzerland, 2000. [CrossRef]
15. Cairns, J. Local bond–slip model for plain surface reinforcement. *Struct. Concr.* **2021**, *22*, 666–675. [CrossRef]
16. Croce, P.; Formichi, P.; Landi, F. Influence of Reinforcing Steel Corrosion on Life Cycle Reliability Assessment of Existing R.C. Buildings. *Buildings* **2020**, *10*, 99. [CrossRef]
17. Beconcini, M.L.; Croce, P.; Formichi, P. Influence of Bond-Slip on the Behavior of Reinforced Concrete Beam to Column Joints. In Proceedings of the International fib Symposium "Taylor Made Concrete Structures: New Solutions for our Society, Amsterdam, The Netherlands, 19–21 May 2008; pp. 533–539.
18. Formichi, P. *Bond-Slip Laws and Confinement Conditions for Reinforcing Bars in r.c. Elements;* University of Pisa: Pisa, Italy, 2010. (In Italian)
19. Casanova, A.; Jason, L.; Davenne, L. Bond Slip Model for the Simulation of Reinforced Concrete Structures. *Eng. Struct.* **2012**, *39*, 66–78. [CrossRef]
20. Hong, S.; Park, S.K. Uniaxial Bond Stress-Slip Relationship of Reinforcing Bars in Concrete. *Adv. Mater. Sci. Eng.* **2012**, *2012*, 328570. [CrossRef]
21. Tang, C.W.; Cheng, C.K. Modeling Local Bond Stress–Slip Relationships of Reinforcing Bars Embedded in Concrete with Different Strengths. *Materials* **2020**, *13*, 3701. [CrossRef] [PubMed]
22. Mains, R.M. Measurements of the Distribution of Tensile Bond Stresses along Reinforcing Bars. *J. Am. Concr. Inst. Proc.* **1951**, *48*, 225–252.
23. Bresler, B.; Bertero, V. Behavior of Reinforced Concrete under Repeated Load. *J. Struct. Div.* **1968**, *94*, 1567–1590. [CrossRef]
24. Nilson, A.H. Internal measurement of bond slip. *ACI J.* **1972**, *69*, 439–441.
25. Croce, P.; Formichi, P. Numerical Simulation of the Behavior of Cracked Reinforced Concrete Members. *Mater. Sci. Appl.* **2014**, *5*, 883–894. [CrossRef]
26. Fédération Internationale du Béton (FIB). *FIB Model Code for Concrete Structure 2010;* Wilhelm Ernst & Sohn: Berlin, Germany, 2013.
27. Cairns, J. Bond and anchorage of embedded steel reinforcement in fib Model Code 2010. *Struct. Concr.* **2015**, *1*, 45–55. [CrossRef]
28. Butcher, J.C. A history of Runge-Kutta methods. *Appl. Numer. Math.* **1996**, *20*, 247–260. [CrossRef]

29. Ramberg, W.; Osgood, W.R. *Description of Stress-Strain Curves by Three Parameters*; Technical Note No. 902; National Advisory Committee for Aeronautics: Washington, DC, USA, 1943.
30. *EN 1992-1-1. Eurocode 2*; Design of Concrete Structures—Part 1-1: General Rules and Rules for Buildings. European Committee for Standardization (CEN): Brussels, Belgium, 2004.

materials

MDPI

Article

Durability Performance of Self-Compacting Concrete Containing Crumb Rubber, Fly Ash and Calcium Carbide Waste

Sylvia E. Kelechi [1,2], Musa Adamu [2,3,*], Abubakar Mohammed [2], Yasser E. Ibrahim [3,*] and Ifeyinwa I. Obianyo [4]

1 Department of Mechanical and Civil Engineering, Purdue University Northwest, Hammond, IN 46323, USA; skelechi@pnw.edu
2 Department of Civil Engineering, Bayero University, P.M.B 3011, Kano 700006, Nigeria; amohammed.civ@buk.edu.ng
3 Engineering Management Department, College of Engineering, Prince Sultan University, Riyadh 11586, Saudi Arabia
4 Department of Materials Science and Engineering, African University of Science and Technology, Abuja 900109, Nigeria; iobianyo@aust.edu.ng
* Correspondence: madamu.civ@buk.edu.ng or madamu@psu.edu.sa (M.A.); ymansour@psu.edu.sa (Y.E.I.)

Abstract: Waste tire disposal continues to pose a threat to the environment due to its non-biodegradable nature. Therefore, some means of managing waste tires include grinding them to crumb rubber (CR) sizes and using them as a partial replacement to fine aggregate in concrete. However, the use of CR has a series of advantages, but its major disadvantage is strength reduction. This leads to the utilization of calcium carbide waste (CCW) to mitigate the negative effect of CR in self-compacting concrete (SCC). This study investigates the durability properties of SCC containing CR modified using fly ash and CCW. The durability properties considered are water absorption, acid attack, salt resistance, and elevated temperature of the mixes. The experiment was conducted for mixes with no-fly ash content and their replica mixes containing fly ash to replace 40% of the cement. In the mixes, CR was used to partially replace fine aggregate in proportions of 0%, 10%, and 20% by volume, and CCW was used as a partial replacement to cement at 0%, 5%, and 10% by volume. The results indicate that the mixes containing fly ash had higher resistance to acid (H_2SO_4) and salt ($MgSO_4$), with up to 23% resistance observed when compared to the mix containing no fly ash. In addition, resistance to acid attack decreased with the increase in the replacement of fine aggregate with CR. The same principle applied to the salt attack scenario, although the rate was more rapid with the acid than the salt. The results obtained from heating indicate that the weight loss was reduced slightly with the increase in CCW, and was increased with the increase in CR and temperature. Similarly, the compressive strength was observed to slightly increase at room temperature (27 °C) and the greatest loss in compressive strength was observed between the temperature of 300 and 400 °C. However, highest water absorption, of 2.83%, was observed in the mix containing 20% CR, and 0% CCW, while the lowest water absorption, of 1.68%, was found in the mix with 0% CR, 40% fly ash, and 10% CCW. In conclusion, fly ash is recommended for concrete structures immersed in water, acid, or salt in sulphate- and magnesium-prone areas; conversely, fly ash and CR reduce the resistance of SCC to heat beyond 200 °C.

Keywords: calcium carbide residue; crumb rubber; fly ash; self-compacting concrete; acid attack; salt attack

check for updates

Citation: Kelechi, S.E.; Adamu, M.; Mohammed, A.; Ibrahim, Y.E.; Obianyo, I.I. Durability Performance of Self-Compacting Concrete Containing Crumb Rubber, Fly Ash and Calcium Carbide Waste. *Materials* **2022**, *15*, 488. https://doi.org/10.3390/ma15020488

Academic Editor: Karim Benzarti

Received: 14 December 2021
Accepted: 7 January 2022
Published: 9 January 2022

1. Introduction

Environmental pollution causes a great danger to human health as well as contributing to the risk of the climate change. Efficient waste management has become increasingly essential since the industrial revolution. In the recent past, researchers have focused on the valorization of waste materials by turning them into wealth [1]. These waste materials could be recycled and reused for the development of other useful materials [2]. Cement is one of the most utilized construction materials due to its availability, but its production seriously affects environmental sustainability through the emission of high amounts of greenhouse gases such as CO_2. The use of waste materials and by-products that have cementitious and Pozzolanic properties will significantly reduce the demand for cement in the construction industry and hence lower the CO_2 emissions resulting from cement production [3].

Calcium carbide is a dangerous and corrosive waste material obtained as a by-product during the production of acetylene (C_2H_2) gas [4–6], whereas crumb rubber (CR) is produced from used tires (waste tires) that have been abandoned at mechanic shops and dumpsites [7]. On the other hand, fly ash is a waste material generated from coal-fired electric- and steam-generating plants during coal combustion. In a bid to add value to these waste materials as well as to curb the environmental issues associated with the dumping of these wastes, researchers have been exploring ways to recycle and reuse them efficiently. The use of CR as an aggregate replacement in concrete has been necessitated by the continual increase in the demand for concrete aggregates, with its associated environmental impacts. The environmental impact of using conventional aggregates could lead to erosion, landslides, and natural resource depletion (global warming). The utilization of CR as an aggregate in concrete is eco-friendly and would prevent the burning of used tires, which causes air pollution. Thus, the use of these eco-friendly aggregates in concrete has the potential to mitigate global warming through the prevention of the burning of these waste tires, which releases greenhouse gases into the atmosphere [8].

Research works on the utilization of CCW as well as CR in the production of self-compacting concrete (SCC) have been published [9–11].

The durability of SCC is a vital parameter to consider in order to ensure a longer lifespan of the structures constructed with it. This parameter relates to the ability of SCC to retain its original shape, dimension, and quality during its estimated service life when subjected to the toughest weathering conditions, abrasion, and diverse chemical attacks. Tests such as chloride attack/acid attack, sulphate resistance, water absorption, and porosity tests are conducted when assessing the durability of SCC. Several studies have been conducted on the durability of conventional concrete, with a few focused on conventional SCC and very little research conducted on SCC ternary blends. Rajamony Laila et al. [12] investigated the effect of super absorbent polymer (SAP) on the durability and mechanical performance of SCC. They replaced cement using granite Pulver (GP) at 5 to 20% by weight. They used the SAP at 0.1 to 1% by volume fraction of the GP-SCC mixtures. Their results showed that SAP decreased the filling ability and flowability of the SCC. Furthermore, GP increased the early strength development of the SCC at 7 days. The addition of 0.3 and 0.4% SAP significantly improved the flexural, compressive, and splitting tensile strengths of the SCC. In terms of durability, GP improved the resistance of the SCC to acid and sulphate attack. They concluded that up to 15% cement can be replaced by using GP to improve the durability and mechanical properties of the SCC. In another study, Rajamony Laila et al. [13] utilized GP and SAP in conventional concrete and reported similar findings. A study by Surya et al. [14] on the compressive strength of SCC was carried out under elevated temperatures. It was found that as the temperature increases, the compressive strength is decreased. Furthermore, the decrease in the compressive strength was found to be greater at 600 °C and 800 °C, conversely to the effects observed at 200 °C and 400 °C. The study concluded that higher elevated temperature has a negative effect on the compressive strength of SCC. According to Santos et al. [15], an increase in the percentage of recycled coarse aggregate increased chlorine penetration depths. SCC mixes containing 40% recycled coarse aggregate were reported to show better resistance to chloride penetration and acid attack in this study. However, research by Yehia et al. [16] presented low chloride permeability for all mixes, which is an indication of the satisfactory durability of SCC and fiber-reinforced SCC. The low chlorine penetration obtained from their studies was attributed to the high density of the mix. In addition, Babu and Nazeer [17] reported optimum resistance to chloride-ion penetration for SCC ternary blend specimens with 10% silica fume replacement. The results of their studies indicated that there was a high sulphate attack resistance for ternary blended concrete with 5% replacement of silica fumes.

Research carried out by Danish and Ganesh [18] established that the addition of metakaolin and fly ash in SCC mixes reduces water absorption. This could be attributed to the ability of the metakaolin and fly ash to act as fillers by improving the microstructure of the SCC. It was observed that the control mix gave the highest water absorption of 4.84% and 4.04% at 28 and 90 days, respectively, while the incorporation of 15% metakaolin resulted in a lower water absorption of 2.02% and 1.60% at 28 and 90 days, respectively. Hence, the incorporation of metakaolin in the SCC mix resulted in a positive effect of decreasing the water absorption of the mix. Another investigation by Santos et al. [15] showed that the incorporation of a high amount of recycled coarse aggregate in SCC mixes displayed a high initial water absorption level that led to an increase in the permeability values of the mix. The results of the review conducted on the use of waste tire rubber in concrete showed that the water absorption of both SCC and traditional concrete was increased on the addition of waste tire rubber due to the rubber particles' rough surfaces. It was also speculated that increasing the water/cement ratio will increase water absorption [19].

The addition of admixtures such as metakaolin and fly ash has a positive influence on the porosity in SCC mixes [18]. A concrete mix with an increased porosity network tends to result in a decreased strength of the mix as well as a reduced durability [8,20]. Bayuaji et al. [21] investigated the properties of geopolymer paste produced with CCW and fly ash. The results of their research indicated that samples with the highest level of CCW and fly ash showed the lowest porosity at an age of 56 days. Abdurra'uf et al. [22] also reported an increased porosity in mortar with the increase in the CCW content when used as a cement replacement in mortar. A review conducted by Bušić et al. [19] indicated that an increase in rubber aggregate content in SCC mixtures resulted in an increase in porosity and air content of the mix due to the entrapment of air in the interfacial transition zone between the cement paste and rubber particles.

Furthermore, CCW has been used for soil modification and stabilization [23,24]. Fly ash has also been utilized and recommended by researchers for the partial replacement of cement in concrete [18,25,26]. For the purpose of environmental sustainability, specifically to achieve a reduction in the pollution of the environment and a reduction in the demand for natural materials such as aggregates in concrete, CR has been used to partially replace fine or coarse aggregate in concrete. In SCC, CR has been used to partially replace fine aggregate. Several researchers reported the positive effect of the CR in SCC. For example, Hilal [9] reported an improvement in the fracture energy of SCC with the replacement of up to 5% of fine aggregate with CR, and improved ductility with the replacement of up to 25% of fine aggregate using CR. Najim and Hall [27] also reported a significant improvement in the strain energy of SCC, which resulted in reduced crack mouth open displacement. Additionally, they reported an improvement in ductile behavior, energy absorption, and vibration damping in SCC with the replacement of fine aggregate with CR. Khalil et al. [28] reported an improvement in the impact resistance of SCC when up to 30% of fine aggregate was replaced with CR. In a similar study, Adamu et al. [29] also reported improved impact resistance with the replacement of up to 30% of fine aggregate in concrete. However, the major disadvantage of using CR in concrete and SCC is the reduction in mechanical properties and durability performance. These reductions in mechanical properties and durability are more severe with the increment in CR content, as reported in several studies [9,27,30].

As reported, both fly ash and CR have significant advantages when used in SCC; these include a reduction in the cost of the concrete, greater environmental and material sustainability, and improved waste management. However, their main disadvantages include a reduction in the mechanical properties and durability of SCC due to their negative effects. Therefore, in this study, CCW was used as an additive by weight of cementitious materials to mitigate the negative effect of CR on the durability of SCC and activate the pozzolanic reaction of fly ash at early ages to improve the durability of SCC. Additionally, the use of a combination of the three materials (CCW, CR, and fly ash) to produce an environmentally sustainable SCC is yet to be fully explored. Therefore, the objective of this study was to investigate the durability properties of rubberized SCC containing CCW and fly ash as supplementary cementitious materials in normal and adverse conditions. In this research, water absorption, acid resistance, sulphate resistance, and heat resistance tests were carried out to determine the durability of the blended rubberized SCC.

2. Materials and Methods

2.1. Materials

Type 1 Ordinary Portland cement with a specific gravity of 3.5, conforming to the guidelines outlined in BS EN 196-6 [31], was used as the main binder material in this study. The cement has chemical properties, which were established using X-ray Fluorescence and are presented in Table 1. Class F fly ash, conforming to the requirements of ASTM C618 [32], was used as a supplementary cementitious material. The properties of the fly ash are also presented in Table 1. The CCW was collected from a commercial welding workshop disposal site in Kano, Nigeria. After that, it was first air dried for about four days and then oven dried for 24 h at a temperature of 110 ± 5 °C to make it completely dry. The CCW was then ground in a grinding machine and sieved through a No. 325 (45 μm) sieve. The particles retained on the sieve were discarded and the ones that passed through were used. The properties of the CCW were then determined using XRF analysis and the results are presented in Table 1. The fine aggregate used was natural river sand, with a specific gravity of 2.63, a bulk density of 1560 kg/m^3, a water absorption rate of 1.96%, and a mud content of 1.1%. The particle size distribution of the fine aggregate is presented in Figure 1, and belongs to the zone II class based on the grading limit of BS 882 [33]. The CR used had a specific gravity of 0.95, and particle size distribution similar to that of the fine aggregate, as shown in Figure 1. Crushed granite with a maximum size of 19 mm was used as the coarse aggregate. The coarse aggregate had a specific gravity of 2.65, a bulk density of 1450 kg/m^3, and a water absorption rate of 0.94%; its particle size distribution is shown in Figure 1. A high-range water-reducing admixture was added by weight of cementitious materials to achieve the self-compaction. The dosage of the superplasticizer was used based on the manufacturer's specification of 2 to 15 fl.oz/cwt of cementitious materials.

Table 1. Chemical properties of materials.

Oxide	Chemical Compositions (%)		
	Cement	CCW	Fly Ash
SiO_2	12.00	1.10	52.06
Al_2O_3	3.01	0.04	30.48
Fe_2O_3	4.11	0.50	4.47
CaO	74.03	96.46	5.62
MgO	1.30	-	1.32
K_2O	1.28	0.45	1.54
Na_2O	0.19	0.01	0.99
SO_3	2.07	0.29	2.1
Loss of Ignition	1.02	2.24	0.95
Specific Gravity	3.15	2.22	2.27
Specific Surface Area (m^2/g)	325	295	290

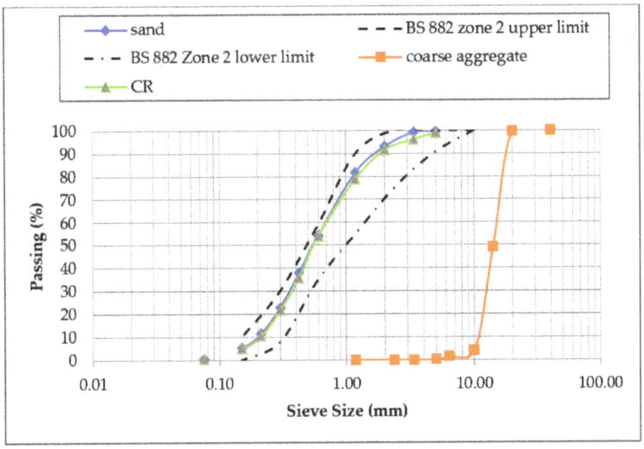

Figure 1. Particle size distribution of coarse aggregate.

2.2. Mix Proportioning

To examine the effect of CR and CCW on the durability performance of SCC under normal and adverse conditions, mixes were prepared by partially replacing fine aggregate with CR at 0%, 10%, and 20% using a volume replacement method, and CCW was used as a partial replacement to cement at 0%, 5%, and 10% by means of a volume replacement method. To further study the effect of fly ash on the durability of SCC blends, fly ash was used to replace 40% cement, with CR replacing fine aggregate at 0%, 10%, and 20% by volume, and CCW replacing cement at 0%, 5%, and 10% by volume. A total of eighteen mixes were prepared, as presented in Table 2. The mixes were tested for acid attack, salt attack, elevated temperature, and water absorption. Each mix was assigned a unique ID; for example, mix M0CR0C is the control mix with 0% CR and 0% CCW, and mix M0CR40F0C is the mix with 0% CR, 40% fly ash, and 0% CCW. Mix M10CR5C is the mix with 10% CR and 5% CCW, and mix M20CR40F10C is the mix with 10% CR, 40% fly ash, and 10% CCW. The superplasticizer dosage used was 1.5% by weight of binder materials.

Table 2. Mix Proportioning.

Mix ID	Cement (kg/m^3)	CCW (kg/m^3)	Fly Ash (kg/m^3)	Fine Aggregate (kg/m^3)	CR (kg/m^3)	Coarse Aggregate (kg/m^3)	Water (kg/m^3)	SP (kg/m^3)	W/B
M0CR0C	520	0	0	880	0	850	192.4	7.80	0.37
M0CR40F0C	312	0	149.89	880	0	850	192.4	6.93	0.42
M0CR5C	494	18.32	0	880	0.00	850	192.4	7.68	0.38
M0CR40F5C	286	18.32	149.89	880	0.00	850	192.4	6.81	0.42
M0CR10C	468	36.65	0	880	0.00	850	192.4	7.59	0.38
M0CR40F10C	260	36.65	149.89	880	0.00	850	192.4	6.70	0.43
M10CR0C	520	0.00	0	792	38.25	850	192.4	7.80	0.37
M10CR40F0C	312	0.00	149.89	792	38.25	850	192.4	6.93	0.42
M10CR5C	494	18.32	0	792	38.25	850	192.4	7.68	0.38
M10CR40F5C	286	18.32	149.89	792	38.25	850	192.4	6.81	0.42
M10CR10C	468	36.65	0	792	38.25	850	192.4	7.59	0.38
M10CR40F10C	260	36.65	149.89	792	38.25	850	192.4	6.70	0.43
M20CR0C	520	0.00	0	704	76.49	850	192.4	7.80	0.37
M20CR40F0C	312	0.00	149.89	704	76.49	850	192.4	6.93	0.42
M20CR5C	494	18.32	0	704	76.49	850	192.4	7.68	0.42
M20CR40F5C	286	18.32	149.89	704	76.49	850	192.4	6.81	0.42
M20CR10C	468	36.65	0	704	76.49	850	192.4	7.59	0.38
M20CR40F10C	260	36.65	149.89	704	76.49	850	192.4	6.70	0.43

2.3. Casting of Specimen

The batching, mixing, and sampling of the fresh concrete was carried out in accordance with the guidelines outlined in BS 1881-125 [34]. A rotating pan mixer was used for the mixing of the fresh concrete. Before mixing, it was ensured that the cement, fly ash, and CCW were completely dried and free from lumps, and that the aggregates were in saturated surface dry condition to avoid the absorption of mixing water. The fine aggregate was first poured into the mixer, followed by the cement, fly ash, and CCW. They were then allowed to mix for about 30 s. After that, the coarse aggregate and half of the mixing water was added, and the mixing continued. The superplasticizer was mixed with the other half of the mixing water and was poured gently to the concrete in the mixer. The mixing was continued for about 2 min until a completely homogenous paste was achieved. The mixing was carried out at a room temperature of 20 $\pm$ 5 °C and controlled relative humidity. Immediately after mixing, the fresh concrete was casted into the molds. Prior to casting, the molds were cleaned and oiled for ease of demolding. The samples were then air dried for 24 h at room temperature before demolding. After demolding, the samples were fully immersed in a clean water for the curing periods prior to testing.

The durability test was carried out in accordance with ASTM C642 [35]. The 100 mm cube specimen, after curing for 28 days, was weighed before being dipped in acid, and was reweighed after 3, 7, and 28 days. H_2SO_4 was used as the acidic medium while $MgSO_4$ was used as the salt medium. The choice of these chemicals was made based on its high reactivity compared to other acids and salt; therefore, it can be considered when designing structures for worst-case scenarios in terms of exposure to high-acid or salt-prone environments. For each mix, three samples were tested, and the mean values were recorded. Cubes were weighed and then subjected to heat for one hour each at a temperature of 100 °C, 200 °C, 300 °C, 400 °C, and 500 °C, respectively, allowed to cool, and then reweighed before being subjected to a compressive strength test. The compressive strength test was carried out on the 100 mm cube samples in accordance with BS EN 12390-3 [36] using a universal testing machine of 2000 kN capacity. Water absorption of the 100 mm cube specimens was also tested after a 28-day curing period in accordance with the method outlined in ASTM C642 [35].

3. Results and Discussions

3.1. Durability of SCC on Immersion in Acid Media

The effect of acid attack on SCC mixes was measured by immersion in H_2SO_4 concentration and the results are presented in Figures 2 and 3. It can clearly be seen that sulphate decreased the weight of all the SCC mixes, as presented in Figures 2 and 3. The decrease in weight (%) can be attributed to the large amount of CaO in the CCW and Portland cement and its hydration product $Ca(OH)_2$, which is primarily responsible for the poor resistance of SCC exposed to acidic attack [37]. The weights of all of the SCC mixes were reduced with the increasing of the immersion period in H_2SO_4. This was a result of continuous attack by the sulphate over time. However, the attack by sulphate was reduced for mixes containing fly ash, as shown in Figure 3. At 28-day immersion age, the M10CR10C mix had the greatest resistance to acid attack for the mixes not containing fly ash, while the M20CR40F10C mix had the greatest resistance to acid attacks for mixes with fly ash, with a 16% reduction in acid penetration upon incorporation of HVFA to the mix. On the other hand, the M20CR0C mix had the least resistance to acid attack for mixes containing no fly ash, while the M0CR40F0C mix had the least resistance to acid attack for mixes with a 40% replacement of cement with fly ash, showing a 23% reduction in acid penetration upon incorporation of fly ash to the mix.

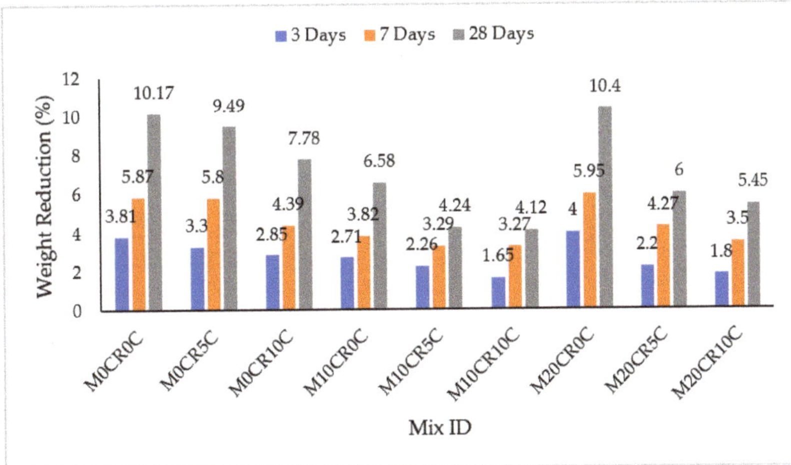

Figure 2. Immersion periods in H_2SO_4 for SCC mixes containing CR and CCW.

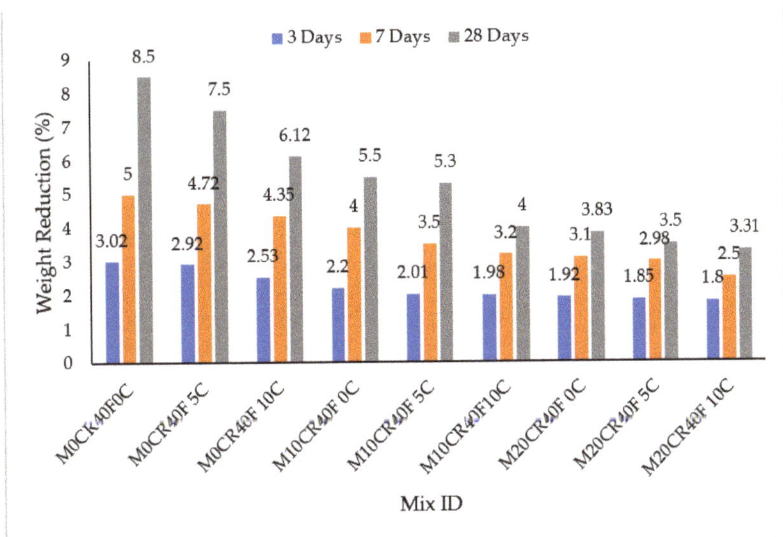

Figure 3. Immersion periods in H_2SO_4 for SCC mixes containing fly ash, CR, and CCW.

The reason for the improvement could be attributed to the pozzolanic reactivity of fly ash in the mix. In addition, the better finishing surface and the minimal voids on the concrete surface of the specimens containing fly ash led to lower penetration of the acid solution into the interior of the concrete and improved its resistance against acid attack [38]. This observation is similar to that of Dhiyaneshwaran et al. [37]. In contrast, attack by sulphate was increased with the increment in the percentage replacement of fine aggregate with CR for mixes without fly ash. This was due to the presence of the micro voids in the samples, which gave room for more penetration of the immersed media. The higher the quantity of the content of the CR, the greater the number of micro voids present, giving rise to more expansions. This is similar to the finding of Xu et al. [39]. It can be seen that the addition of fly ash corrected the sulphate attack associated with the replacement of aggregate with CR for the mix with the least penetration, namely M20CR40F10C.

The addition of CCW to SCC mixes with and without fly ash decreased its resistance to acid attack at all ages of curing, as measured by immersion in H_2SO_4 solution. This can be seen in Figures 2 and 3, where mixes containing higher CCW exhibit lower weight reduction. For instance, compared to the control mix (M0CR0C), the weight losses for mixes M0CR5C and M0CR10C were lower by 6.69% and 23.50% at 28 days, respectively. For mixes with fly ash, the M0CR40F5C and M0C40F10C were lower by 11.76% and 28%, respectively, at 28 days. This weight reduction due to CCW was attributed to the high amount of CaO in the CCW, thereby providing excess $Ca(OH)_2$ in the cement matrix. These calcium compounds are readily attacked by the H_2SO_4, which is a strong acid [40]. Similar findings have been reported by Sata et al. [41] and Hanjitsuwan et al. [40], namely a decreased acid resistance of concrete containing CCW as a binder material. For mixes containing 40% fly ash, the reductions in acid attack (sulphate resistance) of the SCC mixes were lower compared to the mixes without fly ash. This can be observed by comparing the values of weight reductions in Figures 2 and 3. This is attributable to the fact that fly ash decreased the calcium content in the cement matrix via a pozzolanic reaction between the SiO_2 from the fly ash with $Ca(OH)_2$ from the CCW to generate more C-S-H gels.

3.2. Immersion in Salt Media

The effect of salt attack on the SCC mixes was determined by immersing the SCC in a 5% concentration of $MgSO_4$, and the results are presented in Figures 4 and 5. The weight of all the SCC blends was reduced with an increase in immersion age in the $MgSO_4$. This was a result of a continuous attack by the sulphate with time, but the rate of the attack was slower compared to the attack by H_2SO_4. On the other hand, the durability performance of the SCC mixes with or without fly ash in terms of salt resistance decreased with the increment in the partial replacement of fine aggregate using CR. For the mix without fly ash, namely mix M10CR0C, there were reductions in weight by 49.40%, 29.45%, and 26.62% at 3, 7, and 28 days, respectively, compared to the control mix (M0CR0C). For mixes with fly ash, the weight reductions of mix M10CR40F0C were lower by 26.98%, 4.61%, and 17.98% at 3, 7, and 28 days, respectively, compared to mix M0CR40F0C. Similarly for mix M20CR40F0C, the weight losses were lower than that of M0CR40F0C by 47.62%, 9.87%, and 21.49% at 3, 7, and 28 days, respectively. This reduction in salt resistance due to CR was attributed to increased porosity in the hardened cement matrix and the weak interfacial transition zone caused by the CR in the cement matrix. This further instigated the formation of internal pressure by the sulfate-related crystal growth, i.e., the formation of ettringite and gypsum due to the $MgSO_4$; this adversely affected the integrity of the SCC through the further formation of internal cracks, and the escalation of the pores, which was caused by the CR. This consequently resulted in decreased resistance to salt attack ($MgSO_4$) [42].

The addition of CCW to SCC mixes without fly ash further decreased the weight loss due to immersion in $MgSO_4$ at all ages of curing. At 28 days, the weight losses of mixes M0CR5C, M0CR10C, and M20CR10C were lower by 22.3%, 32.01%, and 46.04%, respectively, in comparison to the control mix (M0CR0C). Similarly, for some mixes with fly ash, the weight reduction due to immersion in $MgSO_4$ decreased with the addition of CCW. At 28 days, the weight losses for mixes M0CR40F10C, M10CR40F5C, and M10CR40F10C were lower by 14.47%, 18.86%, 34.21%. The reduction in $MgSO_4$ resistance caused by CCW can be attributed to the high $Ca(OH)_2$ generated by the CCW in the cement matrix. This $Ca(OH)_2$ reacts with the $MgSO_4$ to liberate the hydration of C_3S and C_2S in the cement, thereby instigating gypsum formation, leading to magnesium-gypsum attack on the C-S-H formed during cement hydration, thus causing deterioration [43]. By comparing Figure 4 with Figure 5, it can be observed that the resistance of SCC to salt attack, i.e., immersion in $MgSO_4$, was higher for mixes with fly ash compared to mixes without fly ash. This can be attributed to the pozzolanic reaction of fly ash with CCW, i.e., SiO_2 from fly ash reacting with the excess $Ca(OH)_2$ produced from CCW to generate more C-S-H gels. This consequently reduced the amount of $Ca(OH)_2$ that reacted with $MgSO_4$ to form magnesium based-gypsum, which caused deterioration of the cement matrix [43].

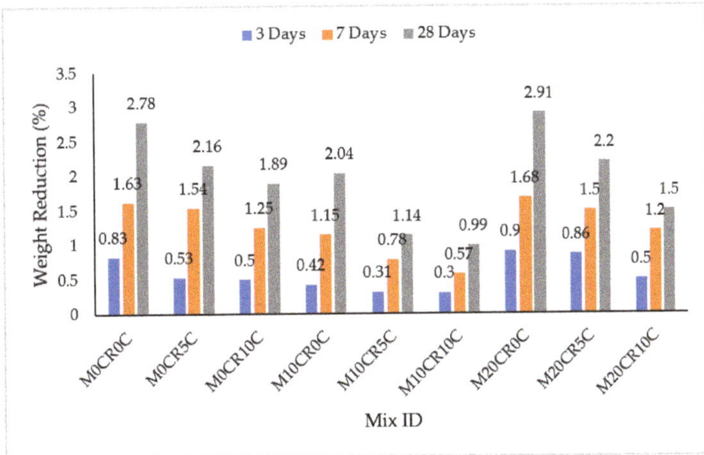

Figure 4. Immersion periods in $MgSO_4$ for SCC mixes containing CR and CCW.

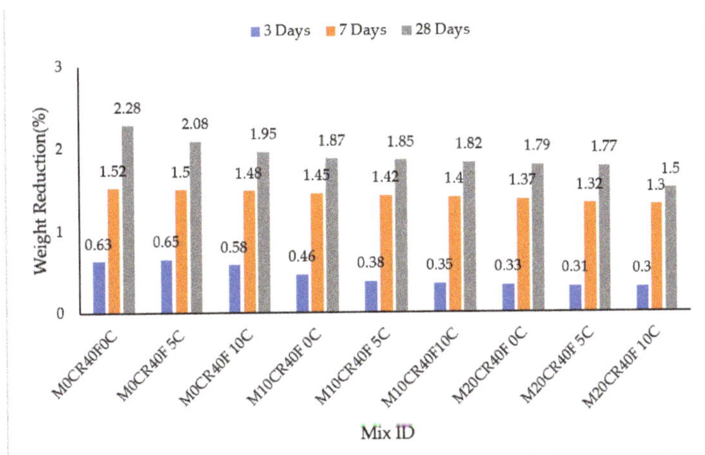

Figure 5. Immersion periods in $MgSO_4$ for SCC mixes containing fly ash, CR, and CCW.

3.3. *Effect of Elevated Temperature on SCC Mixes*

Figures 6 and 7 shows the percentage weight reduction in the SCC mixes at varied temperatures. These figures show that the SCC blends lost more of their weight as the CR content was increased. This observation was noticed as the temperature increased, as shown in the aforementioned figures. The reduction in weight could be a result of spalling caused by internal water pressure at about 300 °C and during the dehydration of hydrated calcium silicate hydrate, which is known to increase internal stresses and induces micro-cracks [44]. The decrease in weight could also have resulted from the expulsion of the excess pore water in the SCC mix [45]. Generally, the addition of 40% optimum fly ash yielded a further decrease in the weight of the mixes by 6%, when compared with the mix without fly ash (M20CR10C) at 500 °C and the M20CR40F10C mix for mixes with fly ash. This confirms a similar result obtained by Fawzy et al. [46] after heating at 200 °C and below.

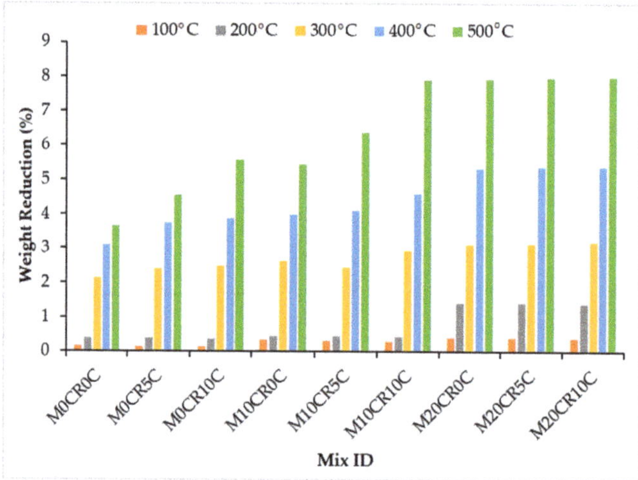

Figure 6. Effects of elevated temperature on weight of SCC mixes containing CR and CCW.

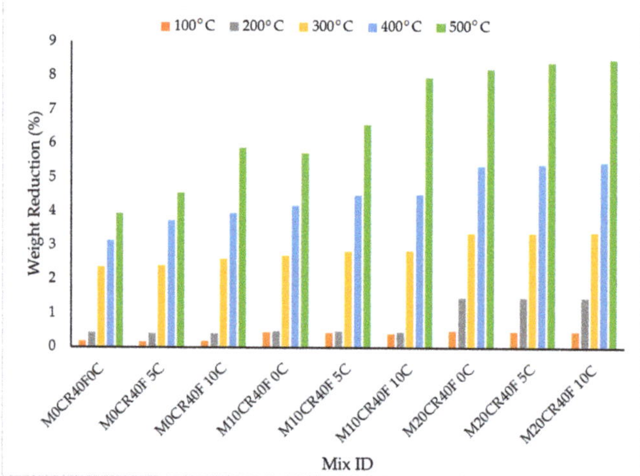

Figure 7. Effects of elevated temperature on weight of SCC mixes containing fly ash, CR, and CCW.

The weight reduction was reduced slightly with the increase in the addition of CCW due to the reaction of CCW with the free lime, which produced more CSH and CAH. The samples exhibited no sign of explosive spalling, but the test samples showed cracks on their surface, which were increased as the temperature increased due to water evaporation, drying shrinkage, and thermal expansion that mounted stress on the concrete, resulting in cracks [22].

In addition, the color of the SCC control mix became brighter because of the water loss at high temperatures, while the mix containing CR was changed from light grey to brownish-black due to the dispersion of carbon black, which is a part of the rubber tires, with the increase in temperature from 400 °C to 500 °C. Certain colors correspond with a specific temperature range, which is an important indicator of the maximum temperature that the structure can be exposed to. At temperatures below 400 °C, the concrete color did not change noticeably, as shown in Figure 8. This observation is similar to the findings of Fawzy et al. [46].

Figure 8. Samples' color changes after heating at elevated temperature. (**a**) M0CR0C (**b**) M0CR40F0C (**c**) M0CR5C (**d**) M0CR40F5C (**e**) M10CR5C (**f**) M10CR40F5C (**g**) M20CR10C (**h**) M20CR40F10C.

The compressive strengths of the various replacements when subjected to elevated temperatures of 100, 200, 300, 400, and 500 °C are graphically represented in Figures 9 and 10. The compressive strength of SCC mixes was decreased with the increase in the temperature. A slight increase in the compressive strength was observed for all mixes and their replicas containing fly ash at control temperature (27 °C) except for the M10CR0C, M10CR0C, M10CR10C, M20CR0C, and M20CR5C mixes, which experienced increases in compressive strength at 100 °C before a further decrease. This was due to the reduction in the calcium hydroxide and un-hydrated area fraction aided by autoclaving (heat healing effect), which enhanced the microstructure [47]. The highest loss in compressive strength was observed at the temperature between 300 and 400 °C due to the calcium silicate hydrate (C-S-H). The incorporation of CR to the mix also reduced the rate of compressive strength loss because the melted rubber particle at 200 °C left space for water vapor to escape and, thus, helped to release the pore pressure. This outcome is similar to the findings of Huang et al. [48]. Generally, it can be observed that CCW slightly increased the compressive strength up to 200 °C and then decreased it after that. This increase was a result of the reaction of CCW with the free lime to produce more CSH and CAH, which deposited in the pore system, and also occurred because of the higher volume of CSH and CAH phases formed and the reduction in $Ca(OH)_2$ content.

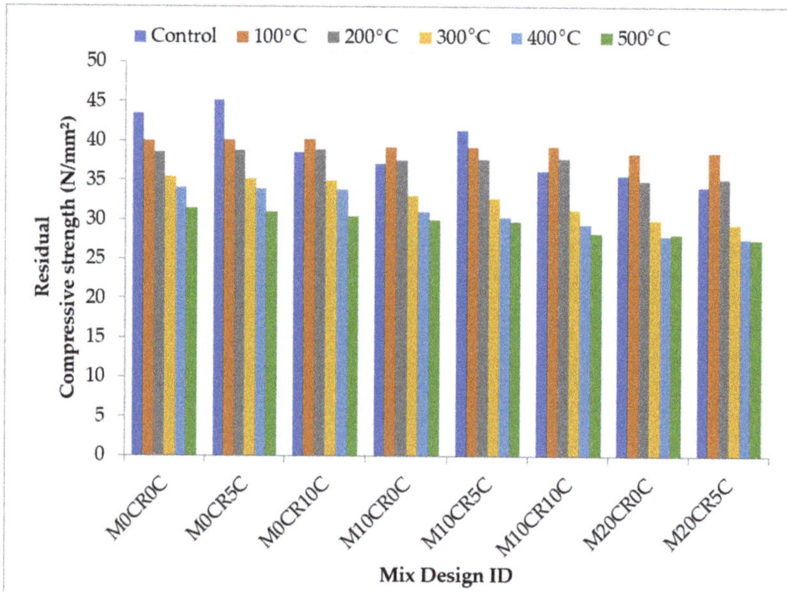

Figure 9. Effect of elevated temperature on the compressive strength of SCC mixes containing CR and CCW.

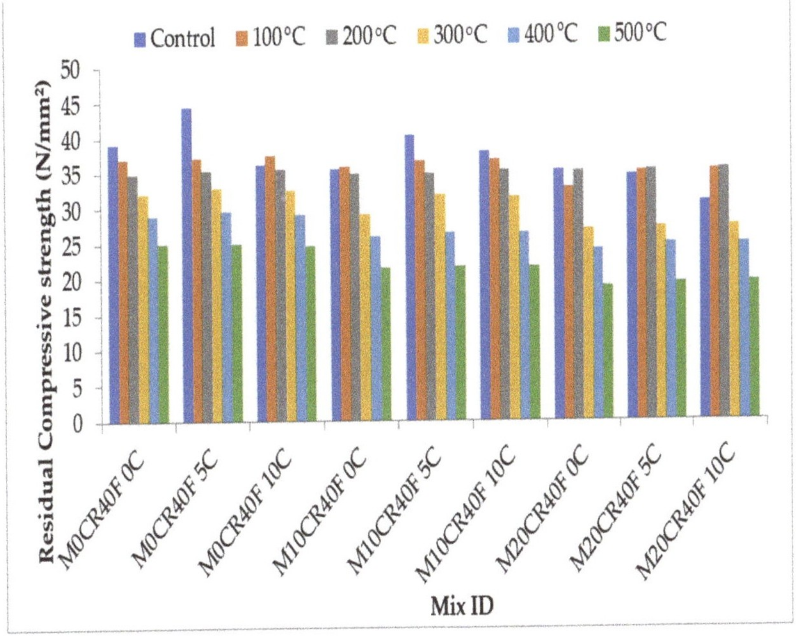

Figure 10. Effect of elevated temperature on the compressive strength of SCC mixes containing fly ash, CR, and CCW.

3.4. Water Absorption

Figures 11 and 12 show the water absorption capacity of the various SCC mixes. The water absorption of the SCC mixes increased with the increment in the partial replacement of fine aggregate with CR for the mixes with and without fly ash. For mixes without fly ash, the water absorption of mixes M10CR0C and M20R0C was higher, by 36.21% and 62.64%, respectively, compared to mix M0CR0C (control mix). Similarly, for mixes with fly ash, the water absorption of mixes M10CR40F0C and M20CR40F0C was lower, by 24.26% and 30.77%, respectively, compared to mix M0CR40F0C. The reduction in water absorption due to CR incorporation can be attributed to the hydrophobic nature of CR, causing it to entrap air on its surface during mixing. When the concrete dried up, the entrapped air left excess voids in the cement matrix, thereby increasing porosity and, hence, leading to higher water absorption [7,8]. The addition of CCW slightly increased the water absorption of mixes without fly ash. For instance, the water absorption of mix M0CR5C was slightly higher than that of the control (by 1.15%), while mix M0CR10C had the same water absorption value as the control mix (M0CR0C). Furthermore, the addition of CCW partially mitigated the negative effect of CR on the water absorption of the SCC mixes with or without fly ash. This can be observed by comparing the water absorption value of M20CR0C (2.83%) with that of M20CR5C (2.19%) and M20CR10C (2.15%), and that of mix M20CR40F0C (2.21%) with M20CR40F5C (2.11%) and M20CR40F10C (2.13%). The reduction in water absorption due to the addition of CCW can be attributed to its finer sizes, thereby acting as a filler material and filling the pores created by the CR and, hence, reducing the water absorption. Another reason could be the higher reactivity of CCW, thereby allowing it to react with cement hydration products to form more C-S-H gels, which filled the nanopores in the cement matrix and, hence, reduced the water absorption [4,6]. By comparing Figure 11 with Figure 12, it can be seen that for the SCC with fly ash, mixes with same amount of CR and CCW had lower water absorption values compared to the SCC mixes without fly ash with the same amount CR and CCW. This was due to the pozzolanic reaction between the SiO_2 from fly ash and $Ca(OH)_2$ from CCW to generate excess C-S-H gels, which filled up the pores in the cement matrix and the decreased water content, similarly to the in the SCC without fly ash. Therefore, it can be concluded that the incorporation of CCW and fly ash reduces the water absorption due to the refined pore system created by fly ash and the CCW microstructure. Overall, the water absorption of the SCC blends containing the replacements in the right proportion resulted in a more durable SCC. This is in line with the research conducted by Mohammed and Adamu [8].

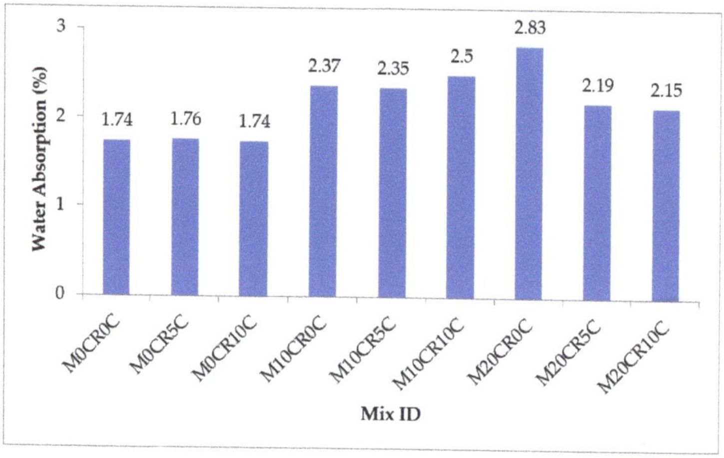

Figure 11. Water absorption capacity of SCC mixes containing CR and CCW.

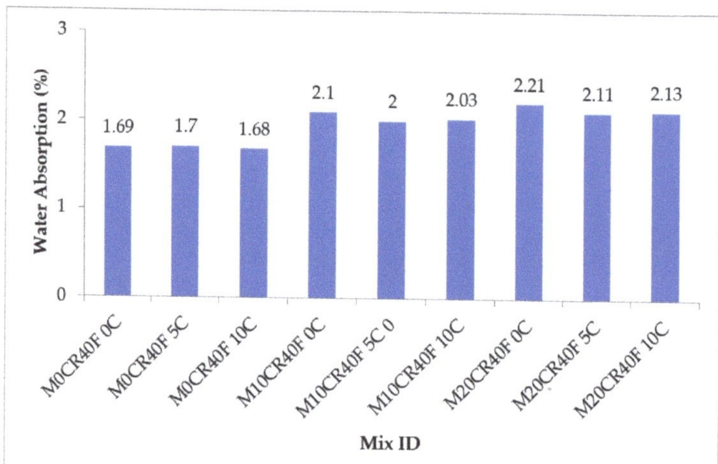

Figure 12. Water absorption of SCC mixes containing fly ash, CR, and CCW.

4. Conclusions

In this study, the effect of the partial replacement of fine aggregate with CR, and of cement with CCW, on the durability performance of SCC in both normal and adverse conditions was investigated. Furthermore, the effect of CR as an aggregate replacement, and of CCW as a cement replacement, on a blend of SCC made produced 40% fly ash as a supplementary cementitious material was evaluated. Based on the experimental results and analysis, the following conclusions can be drawn.

(1) The resistance of the SCC against acid attack, as measured by immersion in H_2SO_4, and the resistance against salt medium, as measured by immersion in $MgSO_4$, all improved with the use of fly ash and CCW as supplementary cementitious materials (SCM).
(2) The partial replacement of fine aggregate with CR in SCC mixes significantly reduced its resistance against acid and salt attacks. This negative effect was more severe in the SCC mixes without fly ash and CCW.
(3) The resistance of the SCC mixes against heat decreased the replacement of 40% cement with fly ash beyond a temperature of 200 °C. Furthermore, the heat resistance of the SCC mixes decreased with the increase in the partial replacement of fine aggregate with CR, and the use of CCW as SCM.
(4) The optimum temperature for the blends was found to be 400 °C; however, the water absorption was decreased with the increasing of the fly ash content and was increased with the increasing of the CR content, but with a rapid increase beyond 10% addition of CR.
(5) Fly ash is recommended for use as a SCM to improve the resistance of SCC mixes containing CR as a partial replacement of fine aggregates for use in acidic and salty environments.

Author Contributions: Conceptualization, S.E.K. and M.A.; methodology, S.E.K., A.M. and I.I.O.; software, S.E.K. and M.A.; validation, M.A. and A.M.; formal analysis, S.E.K., M.A. and I.I.O.; investigation, M.A. and Y.E.I.; resources, M.A. and Y.E.I.; data curation, S.E.K., M.A. and A.M.; writing—original draft preparation, S.E.K. and M.A.; writing—review and editing, Y.E.I.; visualization, M.A.; supervision, M.A. and Y.E.I.; project administration, Y.E.I.; funding acquisition, Y.E.I. All authors have read and agreed to the published version of the manuscript.

Funding: This research was supported by the Structures and Materials (S&M) Research Lab of Prince Sultan University, Saudi Arabia. Furthermore, the authors acknowledge the support of Prince Sultan University in paying the article processing charges (APC) of this publication.

Institutional Review Board Statement: Not applicable.

Informed Consent Statement: Not applicable.

Data Availability Statement: The data presented in this study are available on request from the corresponding author.

Acknowledgments: The authors wish to acknowledge the structures and materials laboratory (S&M Lab) of the College of Engineering, Prince Sultan University, Riyadh, Saudi Arabia, and the Department of Civil Engineering Laboratory, Bayero University Kano, Nigeria, for their vital support.

Conflicts of Interest: The authors declare no conflict of interest.

References

1. Obianyo, I.I.; Mahamat, A.A.; Stanislas, T.T.; Ihekweme, G.O.; Kelechi, S.E.; Onyelowe, K.C.; Onwualu, A.P.; Soboyejo, A.B. Production and utilization of earth-based bricks for sustainable building applications in Nigeria: Status, benefits, challenges and way forward. *J. Build. Pathol. Rehabil.* **2021**, *6*, 37. [CrossRef]
2. Bolden, J.; Abu-Lebdeh, T.; Fini, E. Utilization of recycled and waste materials in various construction applications. *Am. J. Environ. Sci.* **2013**, *9*, 14–24. [CrossRef]
3. Roy, K.; Akhtar, A.; Sachdev, S.; Hsu, M.; Lim, J.; Sarmah, A. Development and characterization of novel biochar-mortar composite utilizing waste derived pyrolysis biochar. *Int. J. Sci. Eng. Res.* **2017**, *8*, 1912–1919.
4. Adamu, M.; Ayeni, K.O.; Haruna, S.I.; Mansour, Y.E.-H.I.; Haruna, S. Durability performance of pervious concrete containing rice husk ash and calcium carbide: A response surface methodology approach. *Case Stud. Constr. Mater.* **2021**, *14*, e00547. [CrossRef]
5. Adamu, M.; Olalekan, S.S.; Aliyu, M.M. Optimizing the Mechanical Properties of Pervious Concrete Containing Calcium Carbide and Rice Husk Ash Using Response Surface Methodology. *J. Soft Comput. Civ. Eng.* **2020**, *4*, 95–118.
6. Adamu, M.; Ibrahim, Y.E.; Al-Atroush, M.E.; Alanazi, H. Mechanical Properties and Durability Performance of Concrete Containing Calcium Carbide Residue and Nano Silica. *Materials* **2021**, *14*, 6960. [CrossRef]
7. Adamu, M.; Mohammed, B.S.; Liew, M.S. Mechanical properties and performance of high volume fly ash roller compacted concrete containing crumb rubber and nano silica. *Constr. Build. Mater.* **2018**, *171*, 521–538. [CrossRef]
8. Mohammed, B.S.; Adamu, M. Mechanical performance of roller compacted concrete pavement containing crumb rubber and nano silica. *Constr. Build. Mater.* **2018**, *159*, 234–251. [CrossRef]

9. Hilal, N.N. Hardened properties of self-compacting concrete with different crumb rubber size and content. *Int. J. Sustain. Built Environ.* **2017**, *6*, 191–206. [CrossRef]
10. Najim, K.B.; Hall, M.R. Workability and mechanical properties of crumb-rubber concrete. *Proc. Inst. Civ. Eng.-Constr. Mater.* **2013**, *166*, 7–17. [CrossRef]
11. Turgut, P.; Yesilata, B. Physico-mechanical and thermal performances of newly developed rubber-added bricks. *Energy Build.* **2008**, *40*, 679–688. [CrossRef]
12. Rajamony Laila, L.; Gurupatham, B.G.A.; Roy, K.; Lim, J.B. Influence of super absorbent polymer on mechanical, rheological, durability, and microstructural properties of self-compacting concrete using non-biodegradable granite pulver. *Struct. Concr.* **2021**, *22*, E1093–E1116. [CrossRef]
13. Rajamony Laila, L.; Gurupatham, B.G.A.; Roy, K.; Lim, J.B.P. Effect of super absorbent polymer on microstructural and mechanical properties of concrete blends using granite pulver. *Struct. Concr.* **2021**, *22*, E898–E915. [CrossRef]
14. Surya, T.R.; Prakash, M.; Satyanarayanan, K.; Celestine, A.K.; Parthasarathi, N. Compressive strength of self compacting concrete under elevated temperature. *Mater. Today Proc.* **2021**, *40*, S83–S87. [CrossRef]
15. Santos, S.; Da Silva, P.; De Brito, J. Self-compacting concrete with recycled aggregates—A literature review. *J. Build. Eng.* **2019**, *22*, 349–371. [CrossRef]
16. Yehia, S.; Douba, A.; Abdullahi, O.; Farrag, S. Mechanical and durability evaluation of fiber-reinforced self-compacting concrete. *Constr. Build. Mater.* **2016**, *121*, 120–133. [CrossRef]
17. Babu, A.; Nazeer, M. Study on the Strength and Durability Properties of Ternary Blended Concrete. *Int. J. Eng. Res. Technol. (IJERT)* **2016**, *3*, 155–160.
18. Danish, P.; Ganesh, M.G. Behaviour of self-compacting concrete using different mineral powders additions in ternary blends. *Rev. Romana Mater.* **2020**, *50*, 232–239.
19. Bušić, R.; Miličević, I.; Šipoš, T.K.; Strukar, K. Recycled rubber as an aggregate replacement in self-compacting concrete—literature overview. *Materials* **2018**, *11*, 1729. [CrossRef]
20. Adamu, M.; Mohammed, B.S.; Shafiq, N.; Liew, M.S. Durability performance of high volume fly ash roller compacted concrete pavement containing crumb rubber and nano silica. *Int. J. Pavement Eng.* **2020**, *21*, 1437–1444. [CrossRef]
21. Bayuaji, R.; Putri, N.; Ming, L.Y.; Mortar, N.A.M. Study Of The Utilization Of Carbide Waste And Fly Ash On Geopolymer Paste. In *IOP Conference Series: Materials Science and Engineering*; IOP Publishing: Bristol, UK, 2020; p. 012042.
22. Abdurra'uf, M.G.; Ogork, E.; Haruna, S.I. Effect of calcium carbide wastes as admixture in mortar. *Sch. J. Eng. Technol.* **2017**, *5*, 655–660.
23. Joel, M.; Edeh, J.E. Soil modification and stabilization potential of calcium carbide waste. In *Advanced Materials Research*; Trans Tech Publications Ltd.: Freienbach, Switzerland, 2013.
24. Liu, Y.; Chang, C.-W.; Namdar, A.; She, Y.; Lin, C.-H.; Yuan, X.; Yang, Q. Stabilization of expansive soil using cementing material from rice husk ash and calcium carbide residue. *Constr. Build. Mater.* **2019**, *221*, 1–11. [CrossRef]
25. Yahyaei, B.; Asadollahfardi, G.; Salehi, A.M. Workability, mechanical, and durability properties of self-compacting concrete using the treated wastewater. *Struct. Concr.* **2021**, *22*, E997–E1008. [CrossRef]
26. Mohammed, B.S.; Adamu, M.; Liew, M.S. Evaluating the effect of crumb rubber and nano silica on the properties of high volume fly ash roller compacted concrete pavement using non-destructive techniques. *Case Stud. Constr. Mater.* **2018**, *8*, 380–391. [CrossRef]
27. Najim, K.B.; Hall, M.R. Mechanical and dynamic properties of self-compacting crumb rubber modified concrete. *Constr. Build. Mater.* **2012**, *27*, 521–530. [CrossRef]
28. Khalil, E.; Abd-Elmohsen, M.; Anwar, A.M. Impact resistance of rubberized self-compacting concrete. *Water Sci.* **2015**, *29*, 45–53. [CrossRef]
29. Adamu, M.; Mohammed, B.S.; Liew, M.S.; Alaloul, W.S. Evaluating the impact resistance of roller compacted concrete containing crumb rubber and nanosilica using response surface methodology and Weibull distribution. *World J. Eng.* **2019**, *16*, 33–43. [CrossRef]
30. Yung, W.H.; Yung, L.C.; Hua, L.H. A study of the durability properties of waste tire rubber applied to self-compacting concrete. *Constr. Build. Mater.* **2013**, *41*, 665–672. [CrossRef]
31. *BS EN 196-6*; Cement Density Test. British Standard Institution: London, UK, 2005.
32. *ASTM C618*; Standard Specification for Coal Fly Ash and Raw or Calcined Natural Pozzolan for Use in Concrete. ASTM International: West Conshohocken, PA, USA, 2019.
33. *BS 882*; Specification for Aggregates from Natural Sources for Concrete. British Standards Institution: London, UK, 1992.
34. *BS 1881-125*; Methods for Mixing and Sampling Concrete in the Laboratory, in Testing Concrete. British Standard Institution: London, UK, 2013.
35. *ASTM C642*; Standard Test Method for Density, Absorption, and Voids in Hardened Concrete. ASTM International: West Conshohocken, PA, USA, 2013.
36. *BS EN 12390-3*; Compressive Strength of Test Specimens, in Testing Hardened Concrete. British Standards Institution: London, UK, 2009.
37. Dhiyaneshwaran, S.; Ramanathan, P.; Baskar, I.; Venkatasubramani, R. Study on durability characteristics of self-compacting concrete with fly ash. *Jordan J. Civ. Eng.* **2013**, *7*, 342–352.

38. Panda, S.; Panigrahi, R.; Narshimam, M. A review on utilization of alkali activated flyash and ggbfs as green concrete. *Adalya J.* **2019**, *8*, 91–96.
39. Xu, J.; Chen, S.; Yu, H.; Wang, Y. Crumb Rubber Concrete Deterioration Caused by Sulphate Attack. In Proceedings of the 3rd International Conference on Material, Mechanical, and Manufacturing Engineering (IC3ME 2015), Guangzhou, China, 27–28 June 2015.
40. Hanjitsuwan, S.; Phoo-ngernkham, T.; Damrongwiriyanupap, N. Comparative study using Portland cement and calcium carbide residue as a promoter in bottom ash geopolymer mortar. *Constr. Build. Mater.* **2017**, *133*, 128–134. [CrossRef]
41. Sata, V.; Sathonsaowaphak, A.; Chindaprasirt, P. Resistance of lignite bottom ash geopolymer mortar to sulfate and sulfuric acid attack. *Cem. Concr. Compos.* **2012**, *34*, 700–708. [CrossRef]
42. Diab, A.M.; Elyamany, H.E.; Abd Elmoaty, M.; Sreh, M.M. Effect of nanomaterials additives on performance of concrete resistance against magnesium sulfate and acids. *Constr. Build. Mater.* **2019**, *210*, 210–231. [CrossRef]
43. Al-Dulaijan, S.U. Sulfate resistance of plain and blended cements exposed to magnesium sulfate solutions. *Constr. Build. Mater.* **2007**, *21*, 1792–1802. [CrossRef]
44. Mohammed, B.S.; Yen, L.Y.; Haruna, S.; Huat, S.; Lim, M.; Abdulkadir, I.; Al-Fakih, A.; Liew, M.; Zawawi, A.; Wan, N.A. Effect of Elevated Temperature on the Compressive Strength and Durability Properties of Crumb Rubber Engineered Cementitious Composite. *Materials* **2020**, *13*, 3516. [CrossRef] [PubMed]
45. Sonebi, M.; Ibrahim, R. Assessment of the durability of medium strength SCC from its permeation properties. In Proceedings of the 5th International RILEM Symposium on Self-Compacting Concrete, Ghent, Belgium, 3–5 September 2007; pp. 677–682.
46. Fawzy, H.; Mustafa, S.; Abd El Badie, A. Effect of Elevated Temperature on Concrete Containing Waste Tires Rubber. *Egypt. J. Eng. Sci. Technol.* **2020**, *29*, 1–13. [CrossRef]
47. Adefemi, A.; Muhammad, U.; Kebbi, U.M.B.; Olugbenga, S. Effect of Admixture on Fire Resistance of Ordinary Portland Cement Concrete. *Civ. Environ. Res.* **2013**, *3*, 302–308.
48. Huang, B.; Li, G.; Pang, S.-S.; Eggers, J. Investigation into waste tire rubber-filled concrete. *J. Mater. Civ. Eng.* **2004**, *16*, 187–194. [CrossRef]

materials

MDPI

Article

Modeling and Testing of a Composite Steel–Concrete Joint for Hybrid Girder Bridges

Bing Shangguan [1,2], Qingtian Su [1,3], Joan R. Casas [4,*], Hang Su [1,4,5], Shengyun Wang [1] and Rongxin Zhao [5]

1 Department of Bridge Engineering, Tongji University, 1239 Siping Road, Shanghai 200092, China; shangguanbing@tongji.edu.cn (B.S.); sqt@tongji.edu.cn (Q.S.); 2010034@tongji.edu.cn (H.S.); 2232463@tongji.edu.cn (S.W.)
2 Guangdong Yejian Construction Drawing Review Center Co., Ltd., Guangzhou 510060, China
3 Shanghai Engineering Research Center of High Performance Composite Bridge, Shanghai 201306, China
4 School of Civil Engineering, Department of Civil and Environmental Engineering, Technical University of Catalunya, 08034 Barcelona, Spain
5 Shanghai Research Institute of Building Sciences Co., Ltd., Shanghai 201108, China
* Correspondence: joan.ramon.casas@upc.edu

Abstract: A hybrid girder bridge adopts a steel segment at the mid-span of the main span of a continuous concrete girder bridge. The critical point of the hybrid solution is the transition zone, connecting the steel and concrete segments of the beam. Although many girder tests revealing the structural behavior of hybrid girders have been conducted by previous studies, few specimens took the full section of a steel–concrete joint due to the large size of prototype hybrid bridges. In this study, a static load test on a composite segment to bridge the joint between the concrete and steel parts of a hybrid bridge with full section was conducted. A finite element model replicating the tested specimen results was established through Abaqus, while parametric studies were also conducted. The test and numerical results revealed that the concrete filling in the composite solution prevented the steel flange from extensive buckling, which significantly improved the load-carrying capacity of the steel–concrete joint. Meanwhile, strengthening the interaction between the steel and concrete helps to prevent the interlayer slip and simultaneously contributes to a higher flexural stiffness. These results are an important basis for establishing a rational design scheme for the steel–concrete joint of hybrid girder bridges.

Keywords: hybrid bridge; steel–concrete joint; static test; load-carrying capacity; numerical study; parametric analysis

Citation: Shangguan, B.; Su, Q.; Casas, J.R.; Su, H.; Wang, S.; Zhao, R. Modeling and Testing of a Composite Steel–Concrete Joint for Hybrid Girder Bridges. *Materials* **2023**, *16*, 3265. https://doi.org/10.3390/ma16083265

Academic Editor: Ayman El-Zohairy

Received: 19 March 2023
Revised: 17 April 2023
Accepted: 19 April 2023
Published: 21 April 2023

1. Introduction

Under the effect of a multifaceted load, a prestressed concrete (PSC) girder bridge [1–4] with a long span has common disadvantages of long-term downward deflection, web cracking and prestress loss, which have an adverse impact on the ultimate strength and serviceability limit states. Hybrid continuous bridges with PSC around the supports and steel in the mid-span area solve this problem by allocating the materials along the bridge where they can perform better (as presented in Figure 1). Steel–concrete hybrid girder requires a composite steel–concrete transition joint to connect the steel and concrete parts along the longitudinal direction. At present, hybrid girders are used in beam-like bridges [5–8], cable-stayed bridges [9–11] and suspension bridges [12,13]. Hybrid girder bridge adopts a steel–concrete composite segment at mid-span of a continuous span, which not only inherits the advantages of easy construction and good economy of the prestressed concrete, but also reduces the self-weight due to the use of steel around the mid-span. In addition, the high weight of the concrete in the side spans contributes to smaller deflection and internal forces in the main span, leading to lower cost and feasibility of the construction of longer spans.

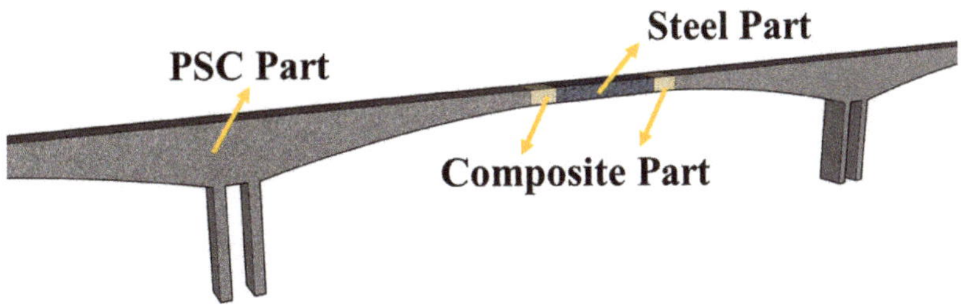

Figure 1. Illustration of a hybrid continuous bridge.

Since the structural form of hybrid girders was proposed, a large number of studies have been carried out to study a better connection type between steel and concrete parts to facilitate the transition of stresses from one to the other. Kim tested three connection types on small-scale steel–PSC hybrid girders, and a connection consisting of parallel perfobond ribs was recommended [14,15]. Chen conducted numerical and analytical analysis on the joint section of hybrid girders and verified the smooth stress transition within the joint section [16]. Xing carried out a numerical study on the effectiveness of shear connectors in the joint section of hybrid girders [17]. The numerical results showed how the shear connectors contributed to a reliable connection between steel and concrete, while the prestressed system is also conducive to strengthen the connection. Jiang conducted model tests on a pressure-resistant shear-transferred steel–concrete joint of hybrid girders and showed a higher bearing capacity of steel–concrete joints [18]. Qin carried out research on the optimal position of the steel–concrete joint section of the hybrid girder bridge through a parametric sensitivity analysis [19]. Zhang proposed a new type of connection form to facilitate the constructability of the steel-to-concrete connection and confirmed its good seismic behavior in precast hybrid steel–concrete girders [20]. He deduced a relationship between the optimal length of the steel–concrete connection and the length ratio of side span to main span [21], which was proved to be an efficient method for the preliminary design of hybrid girder bridges.

The analysis of steel-to-concrete connections refers not only to hybrid beams but also to classical and non-standard composite beams. Su revealed the influence of the steel–concrete interaction on the overall mechanical performance of composite girders [22–24]. Tesser presented 24 lab tests on composite steel trusses and concrete beams with an inferior precast concrete base [25], while the main features of the composite steel trusses and concrete beams are quantitatively and qualitatively discussed. Colajanni investigated the failure modes and the stress transfer mechanism between concrete and embedded steel elements of semi-precast hybrid-steel-trussed concrete beams [26,27]. The influence of steel–concrete interactions on the failure modes and seismic damping capacity were also discussed.

Although many girder tests revealing the structural behavior of hybrid bridges have been conducted, few specimens took the full section of the steel–concrete joint due to the large dimensions of a prototype hybrid bridge. Tests results from a partial section of a girder may reveal the local stress conditions of the steel–concrete joint in the hybrid girder, but its overall structural behavior cannot be obtained directly. When laboratory tests cannot be performed, the size effect on the structural performance were typically evaluated through numerical models. The present research aims to experimentally and numerically investigate the overall mechanical behavior of the steel–concrete joint in hybrid girders. A simply supported composite girder taking the full section of a prototype bridge was tested for a detailed structural performance investigation. A finite element model of the specimen was established through Abaqus and calibrated with the test results. Then, parametric studies were also conducted to derive rational design criteria of the joint section. The objective of the present study is to take the first step in investigating the overall flexural behavior of a steel–concrete joint in hybrid girders and help establish appropriate design criteria for the transition zone and the joint section. The results obtained here need to be complemented with further tests mainly focused on the local mechanisms of response of the joint.

2. Tests Description

2.1. Specimen Design

The steel–concrete joint section in the longitudinal direction proposed in this study is based on a composite steel–concrete solution in the transverse direction. Therefore, a simply supported composite girder was manufactured and tested. The specimen was inspired on the transition segment of the Anhai Bay bridge, which is a key link in the construction of China's first cross-sea high-speed railway. The Anhai bridge, located in Quanzhou City, is a three-span hybrid-beam rigid-frame bridge with a span arrangement of 135 + 300 + 135 m. In the main span, the 103 m long steel box girder is connected monolithically to the concrete box girder by a transition segment (5 m long) at each end with a cross-section depth of approximately 5.9 m (as shown in Figure 2). The bridge has become the world's second largest span hybrid-beam rigid-frame bridge.

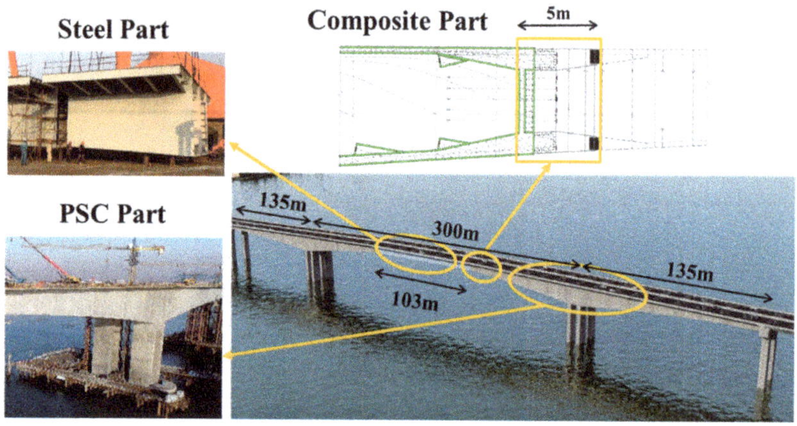

Figure 2. Configuration of Anhai Bay Bridge.

Figure 3 shows the configuration details of the specimen, which was a 1:8 scale model of the Anhai Bay Bridge. The girder test aimed to obtain the sectional bending capacity of the specimen. Thus, the geometric parameters of the section of the specimen were determined as one-eighth of the prototype bridge. As the test model is a small-scale model, the two webs of the box section will be too thin for a proper manufacturing. Thus, the specimen combined the two webs together, while the section was regarded as a steel girder filled with concrete in such a way that the specimen section had the same flexural rigidity as the original section in the bridge. The thickness of all steel plates was 3 mm. The composite girder height was 750 mm, and the width of the top and bottom flanges was 860 mm and 496 mm, respectively (as presented in Figure 3a). Five layers of longitudinal reinforcing bars and two layers of transversal reinforcing bars were included in the web and flanges, respectively. The spacing of the longitudinal and transversal reinforcing bars was 150 mm and 50 mm, respectively, while their diameter was 12 mm. The full girder length was 8 m, including a 250 mm free extension at each end (as presented in Figure 3b). Holes were reserved in advance in the horizontal and vertical stiffeners of the steel beam to facilitate the posterior placement of the rebars. Photos during the specimen construction are shown in Figure 4. The structural form of a perfobond rib was adopted for the steel–concrete connection (as is presented in Figure 4b,c). Figure 4a–c show the fabrication of the steel girder and positioning of the reinforcing bars. Figure 4d,e show the concrete casting.

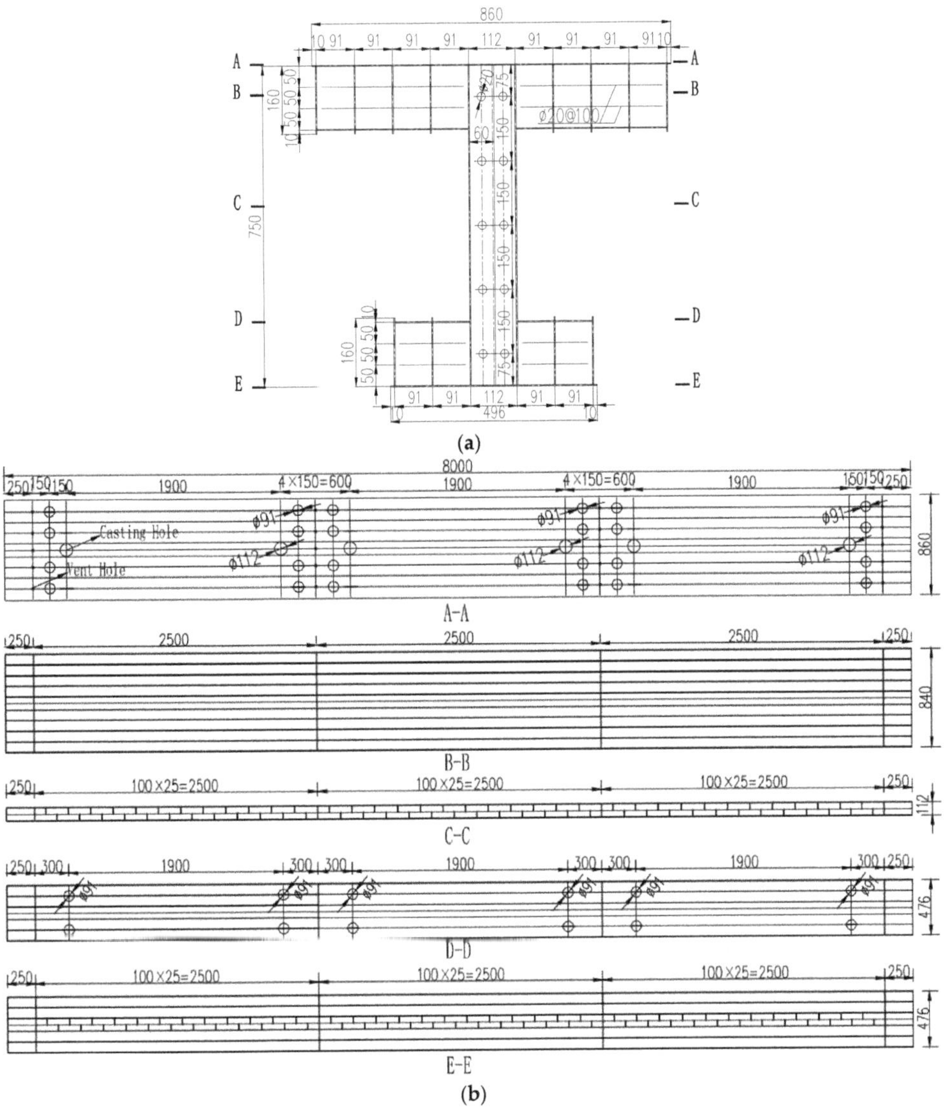

Figure 3. Configuration of the test specimen (mm). (**a**) Composite section and reinforcing bar layout (mm); (**b**) specimen definition.

2.2. Material Properties

The compressive strength of the concrete was 50.4 MPa as obtained from three 150 mm cubic probes tested on the 28th day after casting [28]. Six steel plate specimens and reinforcing bars were also tested. Results showed that the tensile yielding and ultimate strengths of the steel plate was 393 MPa and 573 MPa, respectively. The reinforcing bars exhibited no evident yielding point, while the ultimate strength was 613 MPa.

Figure 4. Photos during specimen construction. (**a**) Steel girder; (**b**) transversal reinforcing bars; (**c**) longitudinal reinforcing bars; (**d**) casting concrete; (**e**) vibrating concrete; and (**f**) fabricated specimen.

2.3. *Loading and Monitoring Setup*

The specimen was loaded at mid-span using a 2000 kN capacity hydraulic jack (as presented in Figure 5). The loading protocol included a 50 kN preload and monotonic loading up to the ultimate state. Figure 6 shows the positions of the sensors, including deflection sensors and strain gauges. The deflections of the specimen were measured using 10 linear variable differential transducers (LVDTs) at mid-span, quarter-span, and side spans. Two transducers were symmetrically mounted in the transverse direction to check load eccentricity. Strain gauges were arranged at mid span and quarter spans as presented in Figure 6b,c. As the concrete was casted into the steel girder, the slip in the interface between concrete and steel could not be monitored directly using the steel–concrete interlayer slip sensor. For this reason, in this paper, the interlayer slip was revealed indirectly by the strain gauges deployed in the specimen.

Figure 5. Loading setup.

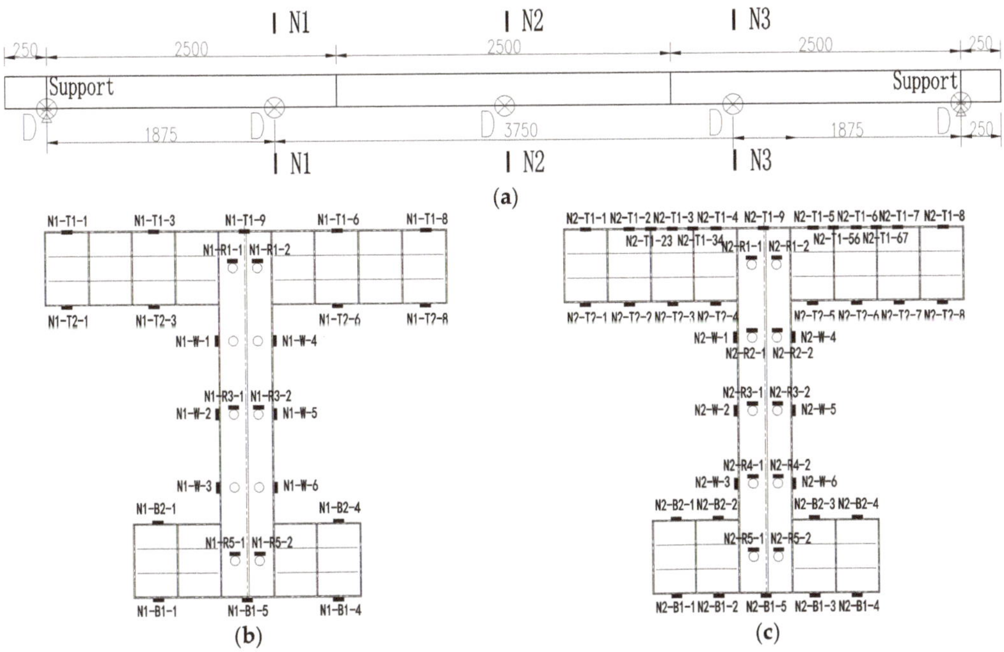

Figure 6. Sensor positions (mm). (**a**) Identification of sections of interest (mm), D: linear variable differential transducers (LVDT); (**b**) strain gauge arrangement at quarter-span; (**c**) strain gauge arrangement at mid-span.

3. Experimental Results

3.1. Load–Deflection Relationship and Failure Mode

The load–deflection curves at mid-span and quarter- span are shown in Figure 7. The deflection value is the average of the two transversally symmetric LVDTs. It can be inferred from Figure 7 that the deflection at the quarter spans were symmetric with each other. The curves generally include an elastic stage, a plastic hardening stage and a final unloading phase. The initial bending stiffness (at 100 kN) of the specimen at mid span and quarter span was 61.1 kN/mm and 83.0 kN/mm, respectively. The sound of the interlayer slip between the steel and concrete appeared at 140 kN. As load kept increasing, steel yielding happened at mid-span and led to nonlinear deflection development. The steel top flange began to buckle at the load level of 780 kN, while the specimen deflection grew drastically. The specimen completely lost carrying capacity at 975.4 kN due to the fracture of the steel bottom flange.

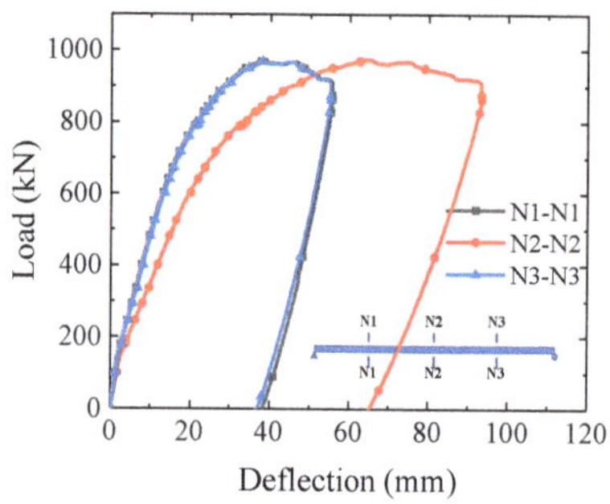

Figure 7. Load–deflection curves at mid-span and quarter-span.

The main cause of the specimen's destruction was the rupture of the steel bottom flange along with steel local buckling in the upper flange, as presented in Figure 8. The specimen was cut through the mid-span after loading to observe the failure mode of the concrete girder. As presented in Figure 9, the steel top and bottom flange was detached from the concrete, and the concrete web was crushed. In addition, diagonal cracks appeared in the concrete web and flanges. In general, the concrete girder prevented the steel top flange from extensive buckling, which significantly improved the load-carrying capacity of the specimen. In addition, strengthening the interaction between the steel and concrete helped to prevent the interlayer slip, which simultaneously contributed to a higher flexural stiffness.

Figure 8. Failure mode of the steel part of the girder.

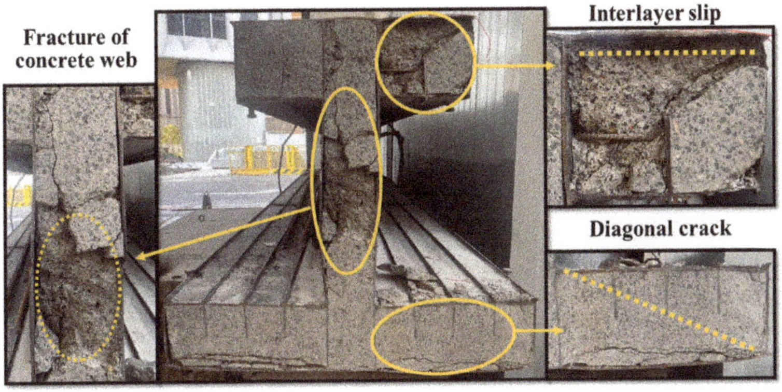

Figure 9. Failure mode of the concrete part of the girder.

3.2. Strains in Steel Flange and Reinforcing Bars

Figure 10 shows the load–strain curves of the steel flanges at quarter-span and mid-span. It can be inferred that both steel top and bottom flanges reached yielding at mid-span, while only steel bottom flange yielded at the quarter-span. In addition, there was an obvious phenomenon of steel–concrete interlayer slip both at mid-span (at 140 kN) and quarter-spans (at 240 kN) in the steel bottom flange, confirming what was visually observed after the test.

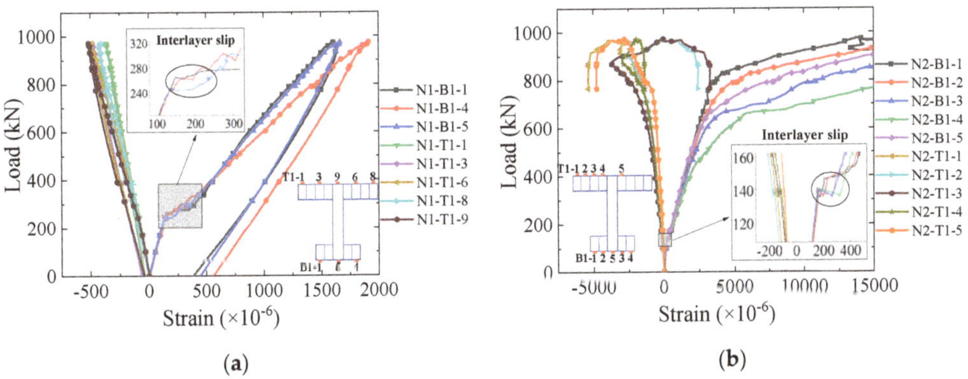

Figure 10. Load–strain curves of the steel flange. (**a**) Quarter-span; (**b**) mid-span.

The load–strain curves of the steel web and the longitudinal reinforcing bars at mid-span are shown in Figure 11 (some strain gauges on reinforcing bars were damaged during construction and are not shown). Strain gauges N2-W-1, N2-R2-1/2, N2-W-2 and N2-R3-1/2 share the same height as the section. Regarding the strain gauge N2-W-1, N2-R2-1/2, N2-W-2 and N2-R3-2, it is evident that the strain of the steel web and reinforcing bars at the same section height is identical during the elastic stage; however, at 200 kN, there is an interlayer slip observed in the nethermost reinforcing bar. The last two rows of reinforcing bars reached yielding at the later loading stage, while the rest remained elastic.

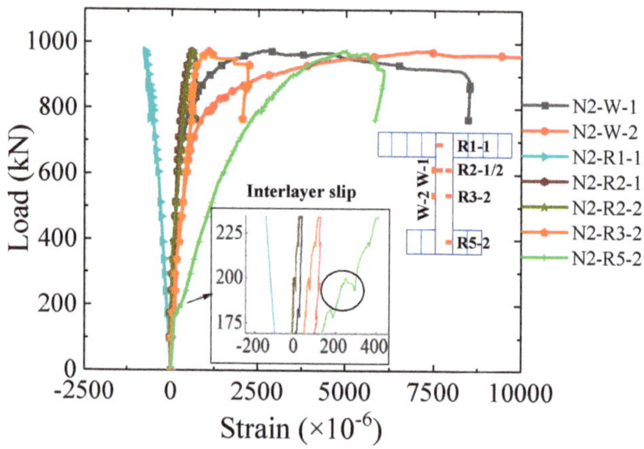

Figure 11. Load–strain curves of the reinforcing bars at mid-span.

3.3. Sectional Strain Distribution

The sectional strain distribution under different load stages at mid-span is shown in Figure 12. The sectional strain distribution up to 100 kN matched well with the plane section assumption from the elastic beam theory. The strain at the bottom of the girder had an uprush at the load stage of 200 kN due to the steel–concrete interlayer slip. The phenomenon of interlayer slip matched with the test observation and the strain monitoring. With the increase of the load, the interlayer slip gradually stretched upward.

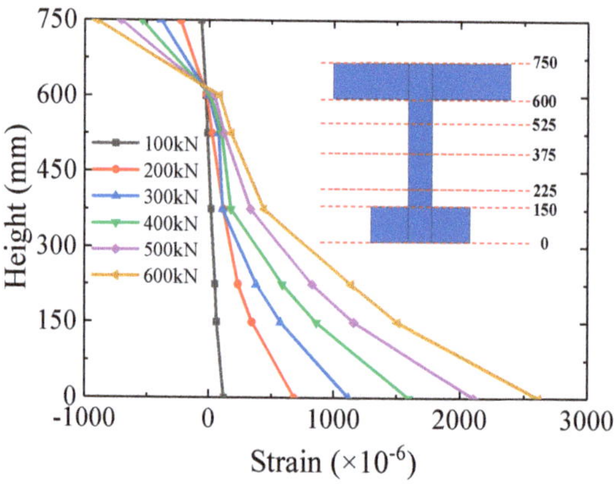

Figure 12. Distribution of sectional strain at mid-span.

As shown in Figure 6c, the steel top flange was divided into four cells by longitudinal stiffeners, and in the mid-span section, strain gauges were deployed both in the support and in the center part of these cells. Figure 13 shows the strain versus loading at the cell supports and cell central part. It can be inferred that the strain profile across the top steel flange is uniform up to 400 kN. As the load kept increasing, the strain at the different locations tends to diverge due to minor local buckling. Generally, the stiffening rib determined the position of the local buckling but did not have a significant impact on the strain variation during the loading.

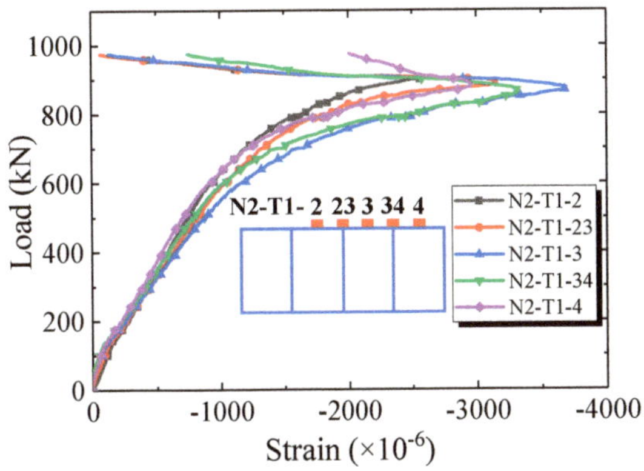

Figure 13. Load–strain curves at support and center of the cells at mid-span.

4. Numerical Analysis

4.1. Numerical Simulation of Composite Girder

A finite element model was developed using Abaqus [29] to further study the structural performance of the composite steel–concrete joint of the hybrid beam bridge. Figure 14 shows the element types and meshes of each component in the model. The concrete girder, steel girder and reinforcing bars were discretized by solid element C3D8R, shell element S4R and truss element T3D2, respectively. The contact between steel and concrete was simulated by penalty frictional formulation along a tangential direction and 'Hard' contact along a normal direction. The friction coefficient was set as 0.1 [30]. The reinforcing bars were constrained with the concrete girder and the stiffening rib in the corresponding position. The loading scheme adopted "uniform" loading in Abaqus, which evenly distributed the load on the loading surface. The displacement boundary conditions took linear constraint support, where all the elements' nodes along the constraint line were restricted. The model comprised a total of 48,050 elements and 56,355 nodes. Dynamic explicit analysis method was employed in the analysis process, while the calculation can be completed within 50 min using a 64-core workstation.

In order to ensure the model's accuracy, sensitivity analyses were conducted on the finite element modelization and presented in Table 1. All eight models listed in the table demonstrated good convergence, as indicated by the values of $A_{\varepsilon}/I_{\varepsilon}$ and $K_{\varepsilon}/I_{\varepsilon}$ being lower than 5%. The results showed that the dilation angle, layers of concrete and constitutive model of concrete had minimal impact on the accuracy of the model. To simplify the meshing work and divide the concrete slab into four layers, a mesh size of 50 mm was used for both concrete and steel. Based on the results, the parameters of the second model in Table 1 were determined for the simulation.

Table 1. Results of sensitivity analysis.

No.	Dilation Angle (°)	Concrete Layers	Constitutive Model	Mesh Size (mm)		A_e/I_e (%)	K_e/I_e (%)
				Concrete	Steel		
1	25	4	fib	50	50	4.1%	2.7%
2	37.5	4	fib	50	50	1.9%	1.3%
3	50	4	fib	50	50	1.5%	1.4%
4	37.5	4	fib	50	50	3.3%	1.6%
5	37.5	4	fib	50	50	3.9%	1.9%
6	37.5	2	fib	100	100	1.6%	0.8%
7	37.5	4	fib	50	100	1.5%	1.0%
8	37.5	4	Secant Modulus	50	50	3.9%	1.9%

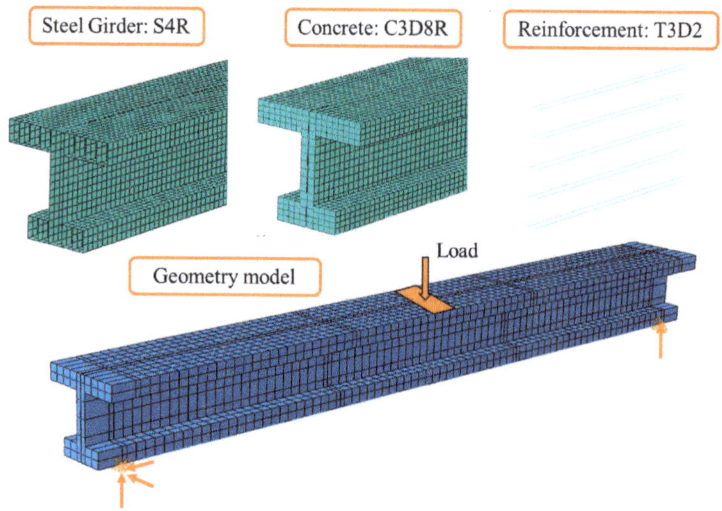

Figure 14. Discretization, element types and loading and boundary conditions in the FEM model.

4.2. Material Properties

The concrete damage plasticity model [31–33] was used for concrete. Figure 15 presents the uniaxial stress–strain curve as determined according to GB50010-2010 [34] using Equation (1) to Equation (6). Equations (2) and (5) describe the coefficients of d_t and d_c. Parameters α_t and α_c were the regulation coefficients of the descending part, which are both empirical parameters as provided in GB50010 [34]. The plastic strain input into Abaqus equals the total strain minus the elastic strain, where $\varepsilon_{t,r}$ and $0.4f_{c,r}$ were taken as the tensile and compressive elastic strains, respectively [35]. Equation (7) defines the damage factor d (input into Abaqus) proposed by Sidoroff [36].

$$\sigma = (1 - d_t)E_c\varepsilon, \tag{1}$$

$$d_t = \begin{cases} 1 - \rho_t[1.2 - 0.2x^5] & x \le 1 \\ 1 - \frac{\rho_t}{\alpha_t(x-1)^{1.7}+x} & x > 1' \end{cases} \tag{2}$$

$$\text{where } x = \frac{\varepsilon}{\varepsilon_{t,r}},\ \rho_t = \frac{f_{t,r}}{E_c\varepsilon_{t,r}}. \tag{3}$$

$$\sigma = (1 - d_c)E_c\varepsilon, \tag{4}$$

$$d_c = \begin{cases} 1 - \frac{\rho_c n}{n-1+x^n} & x \le 1 \\ 1 - \frac{\rho_c}{\alpha_c(x-1)^2+x} & x > 1' \end{cases} \tag{5}$$

$$\text{where } x = \frac{\varepsilon}{\varepsilon_{c,r}},\ \rho_c = \frac{f_{c,r}}{E_c\varepsilon_{c,r}},\ n = \frac{E_c\varepsilon_{c,r}}{E_c\varepsilon_{c,r} - f_{c,r}}. \tag{6}$$

$$d = 1 - \sqrt{\frac{\sigma}{E_0\varepsilon}} \tag{7}$$

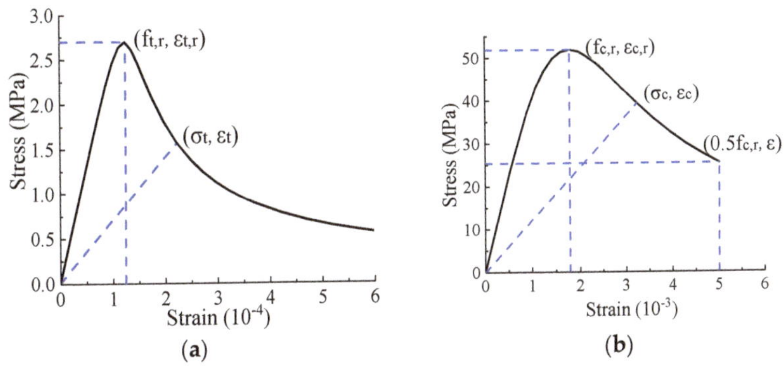

Figure 15. Stress–strain curve for concrete. (**a**) Tensile stress; (**b**) compressive stress.

E_c: Elasticity modulus of concrete.

α_t and α_c: Regulation coefficients of the descending part of concrete.

d_c and d_t: Compressive and tensile damage factor of the concrete.

$f_{t,r}$ and $f_{c,r}$: Ultimate tensile and compressive strengths of concrete, respectively.

$\varepsilon_{t,r}$ and $\varepsilon_{c,r}$: Strains corresponding to the ultimate tensile and compressive strengths of concrete, respectively.

Figure 16a presents a trilinear stress–strain curve for steel plates. The yielding and ultimate stresses σ_{ys} and σ_{us} were obtained from the material property tests. The ultimate strain ε_{us} was 0.6%, while the yield strain ε_{ys} was derived using σ_{ys}/E_s. A bilinear stress–strain model was adopted for the reinforcing bar, as shown in Figure 16b. As the reinforcing bar exhibited no evident yielding point, the yield stress took the ultimate strength from the material property test.

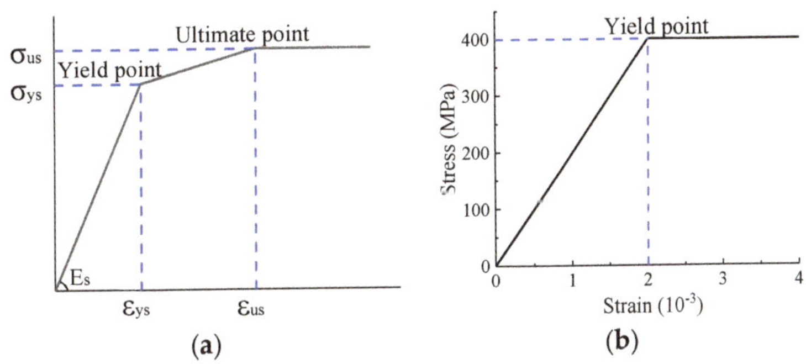

Figure 16. Stress–strain model for steels. (**a**) Steel plate; (**b**) reinforcing bar.

4.3. Validation of FEM Analysis

The load–deflection curve from the FEM model is compared with the test result in Figure 17. The two curves matched well with each other. The phenomenon of interlayer slip around 150 kN appeared in the theoretical curve, which is consistent with the experimental observation and the strain analysis. The initial bending stiffness (at 100 kN) and the load-carrying capacity of the FEM model were 72.4 kN/mm and 1009.6 kN, respectively, which were slightly larger compared to the test specimen. The deviation could be attributed to the accuracy of the material constitutive laws as well as to the simulation of the steel–concrete interlayer interaction.

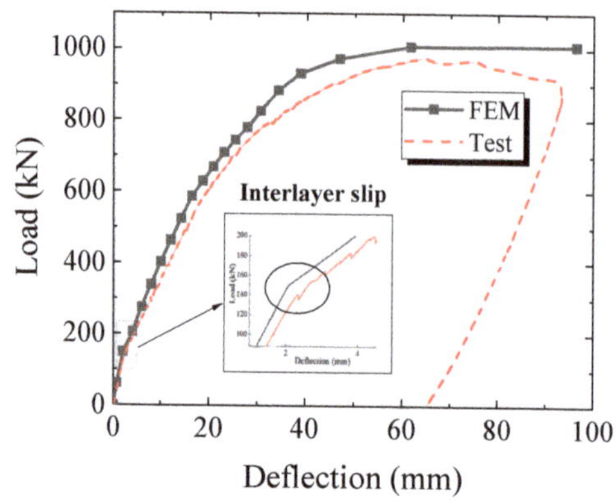

Figure 17. Comparison of numerical and experimental load–deflection curves at mid-span.

4.4. Influence of Composite Action on Flexural Behaviour

Models with solo concrete sections and solo steel sections were established to study the influence of composite action on the flexural behavior. Parametric analysis varying the thickness of the steel plates of the specimen with a solo steel section was also conducted. The rest of parameters in the numerical models remained unchanged from the test specimen.

Figure 18a shows the influence of the composite action on the load-carrying capacity. The load-carrying capacity of the specimen with composite action is much larger than that of specimens with solo concrete or solo steel, and even much larger than the sum of both (see OC + OS3 value in the figure). The huge difference is due to the failure mode of the steel girder. As shown in Figure 18c, the failure mode of the specimen with a solo steel section was local buckling, which did not appear in the specimen with a composite section. As the steel girder provided greater load-carrying capacity, the concrete in the composite section prevented local buckling, which significantly increased the load-carrying capacity of the joint section. It can also be deduced from Figure 18a that adopting a composite section saves around 67% steel flange thickness compared to adopting a solo steel section.

The influence of composite action on initial stiffness (during loading up to 100 kN) is shown in Figure 18b. The increase of steel flange thickness in the specimen with solo steel section had limited influence on the initial stiffness, while the composite section could increase the initial stiffness by nearly three times (compared with OS3).

OC: Specimen with a concrete section only.
OSN: Specimen with a steel section only and a steel plate thickness equal to N mm.
COM: Specimen with a composite section and a steel plate thickness equal to 3 mm.

4.5. Parametric Analysis on Steel Flange Thickness

Models with different steel flange thickness (the other section parameter kept the same as in the test specimen) were established to study its influence on the flexural behavior. Figure 19a shows that the load-carrying capacity presented a linear relation with the steel flange thickness, indicating that the flexural strength depended mainly on the steel girder. As the main failure mode of the test specimen was steel flange fracture, this conclusion is consistent with the experimental results. In other words, the primary function of the concrete filling was to prevent local buckling of the steel, while it had limited influence on the load-carrying capacity of the joint section. In addition, the increase of the steel flange thickness also contributed to an improvement of the initial stiffness (as shown in Figure 19b).

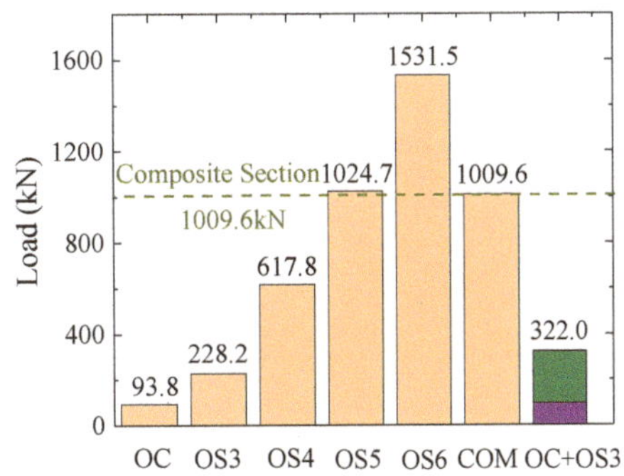

(**a**)

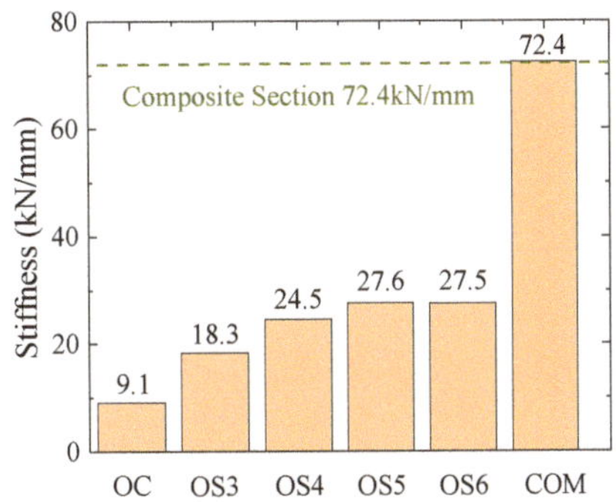

(**b**)

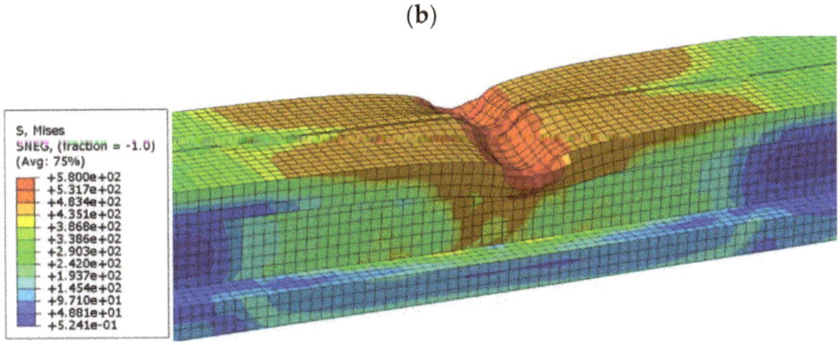

(**c**)

Figure 18. Influence of composite action on flexural behavior. (**a**) Influence of composite action on load-carrying capacity; (**b**) influence of composite action on initial stiffness; (**c**) failure mode of specimen OS6.

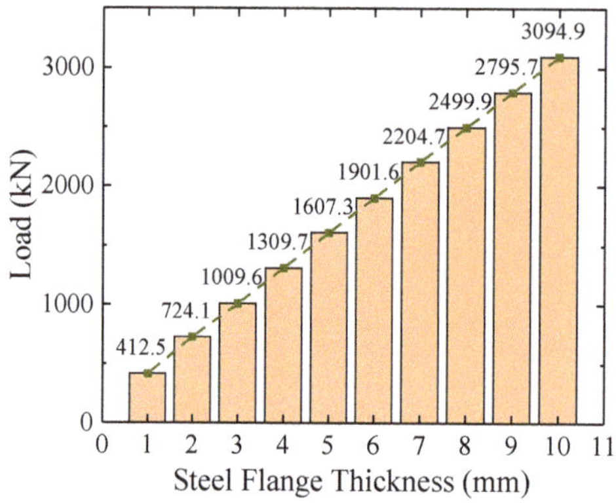

(a)

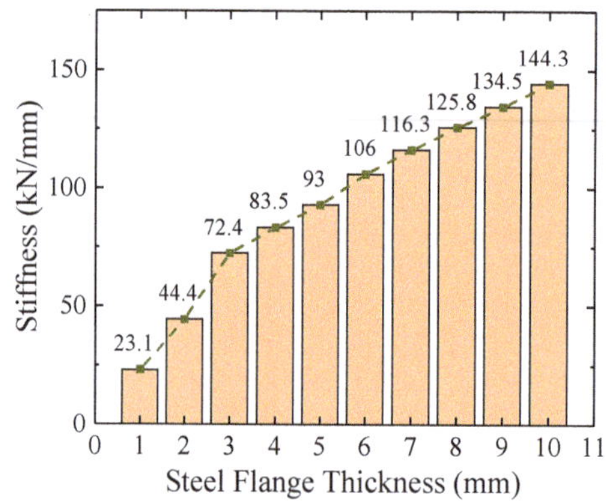

(b)

Figure 19. Influence of steel flange thickness on flexural behavior. (**a**) Influence of steel flange thickness on load-carrying capacity; (**b**) influence of steel flange thickness on initial stiffness.

5. Conclusions

A simply supported composite girder modelling the full section of a prototype hybrid bridge was fabricated and tested in this study. Finite element models of the specimen were built and calibrated for the posterior simulation and parametric analysis. Based on the experimental and numerical results, the following conclusions were drawn:

1. According to the test results, the failure mode of the specimen was fracture of the steel bottom flange along with steel local buckling of the upper flange. The stiffening ribs on the upper flange determined the position of the local buckling but did not have a significant impact on the strain variation during the loading. The concrete girder prevented the steel flange from extensive buckling, which significantly improved the load-carrying capacity of the specimen.
2. According to the test, there was an obvious phenomenon of steel–concrete interlayer slip both at mid-span (at 140 kN) and quarter-span (at 240 kN) in the steel bottom flange. The steel bottom flange was observed to be completely detached from the concrete. Thus, strengthening the interaction between the steel and concrete helps to prevent the interlayer slip and simultaneously contributes to a higher flexural stiffness.

3. Based on the numerical analysis, the FEM model was proved to be reliable with sufficient accuracy. The deviation from the experimental results could be attributed to the lack of accuracy of the material constitutive laws as well as to the not fully correct simulation of the steel–concrete interlayer interaction.
4. Based on the numerical parametric analysis, it is observed how the load-carrying capacity of the specimen with a composite section is much larger than that of specimens with a solo concrete sections and solo steel sections. While the steel girder provided greater load-carrying capacity, the concrete in the composite section prevented local buckling and increased the load-carrying capacity of the joint section. In addition, an increase of steel flange thickness in the specimen with a solo steel section had limited influence on the initial stiffness, while the composite section drastically increases it.
5. Based on the parametric analysis, the load-carrying capacity showed a linear relation with the steel flange thickness, indicating that the flexural strength depended mainly on the steel girder. As the main failure mode of the test specimen was steel flange fracture, this conclusion is consistent with the experimental results. Thus, the primary function of the concrete filling was to prevent the steel local buckling, while it had limited influence on the load-carrying capacity of the joint section. In addition, the increase of the steel flange thickness also contributed to an improvement of the initial stiffness.

Although some relevant conclusions have been already obtained from the study shown in this paper, still some limitations of the presented research have to be pointed out. For instance, for financial reasons, only one specimen was tested in this study. More test specimens with different design parameters are considered in the subsequent studies to obtain additional test data. In addition, as the test was designed to model the global performance and strength of the proposed connection, the small-scale model is hard to analyze the phenomenon of local buckling on the top flange that is due to local stress of the composite section of the specimen. Other local failure mechanisms, for instance the detachment of the steel and concrete in the top and bottom flanges, are due to the constraints imposed by the design of a small-scale specimen. Whether these local mechanisms can occur in the hybrid girder should be investigated by a larger scale test in the future study.

Author Contributions: Conceptualization, B.S. and Q.S.; methodology, Q.S. and J.R.C.; software, H.S. and R.Z.; validation, J.R.C., B.S. and Q.S.; formal analysis, H.S.; investigation, H.S. and S.W.; resources, Q.S.; data curation, J.R.C. and Q.S.; writing—original draft preparation, H.S.; writing—review and editing, J.R.C. and Q.S.; supervision, J.R.C., B.S. and Q.S.; project administration, Q.S.; funding acquisition, B.S. All authors have read and agreed to the published version of the manuscript.

Funding: This research was funded by Spanish Ministry of Science and Innovation (grant number PID2021-126405OB-C31), Shanghai Science and Technology Commission (grant number 20XD1432400) and Guangdong Yejian Construction Drawing Review Center Co., Ltd. (grant number GDYJ-RD-17).

Institutional Review Board Statement: Not applicable.

Informed Consent Statement: Not applicable.

Data Availability Statement: Some or all data that support the findings of this study are available from the corresponding author upon reasonable request.

Conflicts of Interest: The authors declare no conflict of interest.

References

1. Alghazali, H.H.; Myers, J.J. Time-dependent prestress loss behavior of girders in Missouri bridge A7957 compared with a US data set of high-performance concrete bridge girders. *PCI J.* **2017**, *62*, 76–87. [CrossRef]
2. Xiang, Z.; Chan, T.H.T.; Thambiratnam, D.P.; Nguyen, A. Prestress and excitation force identification in a prestressed concrete box-girder bridge. *Comput. Concr.* **2017**, *20*, 617–625.
3. Yu, X.; Ye, J.; Wu, W. Spatial Stress Analysis of Long Span Prestressed Concrete Box Girder Bridges. In *Advances in Civil Engineering II, PTS 1–4*; Trans Tech Publications Ltd.: Guilin, China, 2013; pp. 1693–1696.

4. Yuan, M.; Yan, D.-h. Optimization on Closure Scheme of Multi-Span Prestressed Concrete Box-Girder Bridge. In *Advances in Structures, PTS 1–5*; Trans Tech Publications Ltd.: Guilin, China, 2011; pp. 2369–2375.
5. Azizinarnini, A.; Hash, J.B.; Yake, A.J.; Farimani, R. Shear capacity of hybrid plate girders. *J. Bridge Eng.* **2007**, *12*, 535–543. [CrossRef]
6. Dai, G.; Su, M.; Liu, W.; Chen, Y.F. New Songhua River Bridge: A Continuous Girder, Tied Arch, Hybrid Bridge with Four Rail Tracks in Harbin, China. *Struct. Eng. Int.* **2016**, *26*, 254–259. [CrossRef]
7. Hejll, A.; Taljsten, B.; Motavalli, M. Large scale hybrid FRP composite girders for use in bridge structures—Theory, test and field application. *Compos. Part B Eng.* **2005**, *36*, 573–585. [CrossRef]
8. Mo, Y.L.; Fan, Y.-L. Torsional design of hybrid concrete box girders. *J. Bridge Eng.* **2006**, *11*, 329–339. [CrossRef]
9. Wagner, P. Design developments for roadway bridges based on new hybrid structures—Part 2: Cable stayed bridges. *Stahlbau* **2013**, *82*, 851–867. [CrossRef]
10. Zhang, G.Y. Long-Term Monitoring of a Hybrid Cable-Stayed Bridge. In *Sensing Issues in Civil Structural Health Monitoring*; Springer: Berlin/Heidelberg, Germany, 2005; pp. 485–494.
11. Zhang, X.-J. Investigation on mechanics performance of cable-stayed-suspension hybrid bridges. *Wind. Struct.* **2007**, *10*, 533–542. [CrossRef]
12. Li, X.; Zhong, J.; Jiang, L.; Yuan, W. Retracted: Analysis of the Seismic Vulnerability of Simply-supported Girder Bridge under Two Level Seismic Hazard. In Proceedings of 2018 3rd International Conference on Smart City and Systems Engineering (ICSCSE), Xiamen, China, 29–30 December 2018; pp. 173–177.
13. Paultre, P.; Proulx, J.; Begin, T. Dynamic investigation of a hybrid suspension and cable-stayed bridge. *Earthq. Eng. Struct. Dyn.* **2000**, *29*, 731–739. [CrossRef]
14. Kim, S.-H.; Lee, C.-G.; Ahn, J.-H.; Won, J.-H. Experimental study on joint of spliced steel-PSC hybrid girder, Part I: Proposed parallel-perfobond-rib-type joint. *Eng. Struct.* **2011**, *33*, 2382–2397. [CrossRef]
15. Kim, S.-H.; Lee, C.-G.; Kim, S.-J.; Won, J.-H. Experimental study on joint of spliced steel-PSC hybrid girder, part II: Full-scale test of spliced hybrid I-girder. *Eng. Struct.* **2011**, *33*, 2668–2682. [CrossRef]
16. Chen, W.; Liang, B.; Wang, W. Research on Structure of Steel-concrete Joint Section in Hybrid Girder of a Grand Bridge. In Proceedings of the 5th International Conference on Civil Engineering and Transportation 2015, Guangzhou, China, 28–29 November 2015; pp. 370–374.
17. Xing, W.E.I.; Xiaozhen, L.I.; Jun, L.I.; Shizhong, Q. Using Steel-Concrete Hybrid Structure in Long-Span Continuous Rigid Frame Bridge. *China Railw. Sci.* **2007**, *28*, 43–46.
18. Jiang, X.; Liu, Y.; Sun, X. Model Test and Mechanical Behavior Analysis of Steel-concrete Joint of Hybrid Girder. *J. Civ. Archit. Environ. Eng.* **2014**, *36*, 48–53.
19. Qin, F.J.; Di, J.; Dai, J.; Lu, W.R.; Zhao, M. Study on Rational Position of Joint Section of Steel-Concrete Hybrid Girder Bridge. In *Construction and Urban Planning, PTS 1–4*; Trans Tech Publications Ltd.: Guizhou, China, 2013; pp. 1007–1011.
20. Zhang, X.; Zhang, S.; Niu, S. Experimental studies on seismic behavior of precast hybrid steel-concrete beam. *Adv. Struct. Eng.* **2019**, *22*, 670–686. [CrossRef]
21. He, Z.-Q.; Chen, J.; Liu, Z.; Ma, Z.J. An Explicit Approach for Determining the Rational Length of Steel Portion in Steel-Concrete Hybrid Girder Bridges. *J. Bridge Eng.* **2023**, *28*, 05022011. [CrossRef]
22. Su, H.; Su, Q.; Xu, C.; Zhang, X.; Lei, D. Shear performance and dimension rationalization study on the rubber sleeved stud connector in continuous composite girder. *Eng. Struct.* **2021**, *240*, 112371. [CrossRef]
23. Su, H.; Su, Q.; Casas, J.R.; Xu, C.; Tang, J.; Jiang, C. Influence mechanism of Steel-concrete interlayer partial shear connections on mechanical properties of composite girders. *Structures* **2022**, *46*, 503–520.
24. Su, H.; Su, Q.; Xu, C.; Casas, J.R.; Zhou, G. Mechanical performance based rationalization research on steel-concrete double composite action. *J. Constr. Steel Res.* **2022**, *197*, 107492. [CrossRef]
25. Tesser, L.; Scotta, R. Flexural and shear capacity of composite steel truss and concrete beams with inferior precast concrete base. *Eng. Struct.* **2013**, *49*, 135–145. [CrossRef]
26. Colajanni, P.; La Mendola, L.; Monaco, A. Stress transfer and failure mechanisms in steel-concrete trussed beams: Experimental investigation on slab-thick and full-thick beams. *Constr. Build. Mater.* **2018**, *161*, 267–281. [CrossRef]
27. Colajanni, P.; La Mendola, L.; Monaco, A.; Pagnotta, S. Dissipative connections of RC frames with prefabricated steel-trussed-concrete beams. *Ing. Sismica* **2020**, *37*, 51–63.
28. *GB/T 50081-2002*; Standard for Test Method of Mechanical Properties on Ordinary Concrete. Ministry of Construction People's Republic of China: Beijing, China, 2002.
29. *ABAQUS 6.13 User's Manual*; Dassault Systemes Simulia Corp: Providence, RI, USA, 2014.
30. Rabbat, B.G.; Russell, H.G. Friction coefficient of steel on concrete or grout. *J. Struct. Eng. Asce* **1985**, *111*, 505–515. [CrossRef]
31. Kaminska, I.; Szwed, A. Modification of Concrete Damaged Plasticity Model. Part II: Formulation and Numerical Tests. In *RSP 2017—XXVI R-S-P Seminar 2017 Theoretical Foundation of Civil Engineering*; EDP Sciences: Warsaw, Poland, 2017.
32. Nguyen, G.D.; Korsunsky, A.M. Damage-plasticity modelling of concrete: Calibration of parameters using separation of fracture energy. *Int. J. Fract.* **2006**, *139*, 325–332. [CrossRef]
33. Sarikaya, A.; Erkmen, R.E. A plastic-damage model for concrete under compression. *Int. J. Mech. Sci.* **2019**, *150*, 584–593. [CrossRef]

34. *GB50010-2010*; Code for Design of Concrete Structures. Ministry of Construction People's Republic of China: Beijing, China, 2015.
35. *ENV1992-1-1*; Eurocode-2: Design of Concrete Structures, Part1: General Rules and Rules for Buildings. CEN: Brussels, Belgium, 1992.
36. Mander, J.B.; Priestley, M.J.N.; Park, R. Theoretical stress-strain model for confined concrete. *J. Struct. Eng. Asce* **1988**, *114*, 1804–1826. [CrossRef]

MDPI
St. Alban-Anlage 66
4052 Basel
Switzerland
www.mdpi.com

Materials Editorial Office
E-mail: materials@mdpi.com
www.mdpi.com/journal/materials

www.ingramcontent.com/pod-product-compliance
Lightning Source LLC
Chambersburg PA
CBHW061544220926
51850CB00005B/241